Micaela Morelli • Nicola Simola • Jadwiga Wardas
Editors

The Adenosinergic System

A Non-Dopaminergic Target in Parkinson's Disease

 Springer

Editors
Micaela Morelli
Dept. of Biomedical Sciences Section
of Neuropsychopharmacology
University of Cagliari
Cagliari
Italy

Jadwiga Wardas
Dept. of Neuropsychopharmacology
Institute of Pharmacology
Polish Academy of Sciences
Krakow
Poland

Nicola Simola
Dept. of Biomedical Sciences Section
of Neuropsychopharmacology
University of Cagliari
Cagliari
Italy

ISSN 2363-9563
ISSN 2363-9571 (electronic)
Current Topics in Neurotoxicity
ISBN 978-3-319-20272-3
ISBN 978-3-319-20273-0 (eBook)
DOI 10.1007/978-3-319-20273-0

Library of Congress Control Number: 2015947498

Springer Cham Heidelberg New York Dordrecht London

Printed on acid-free paper

Springer International Publishing AG Switzerland is part of Springer Science+Business Media (www.springer.com)

Preface

Parkinson's disease is the second most common neurodegenerative disorder after Alzheimer's disease, and affects more than 5 million people worldwide. Today, the clinical management of Parkinson's disease chiefly relies on the use of the so called "dopamine replacement therapy" in order to re-establish the function of the dopaminergic system, which is affected by the neurodegeneration underlying the disease. While this approach effectively counteracts the motor deficits featuring Parkinson's disease, the chronic use of dopamine replacement therapy eventually leads to the emergence of motor complications (e.g., dyskinesia and motor fluctuations) that greatly limit its therapeutic potential. Moreover, dopamine replacement therapy has no apparent beneficial effects on the progression of dopaminergic degeneration featuring Parkinson's disease. Based on these considerations, there is a need for the development of alternative therapies that could help to overcome these limitations.

In these years, drugs acting as antagonists of the adenosine A_{2A} receptors have emerged as new promising candidates for the therapy of Parkinson's disease. When evaluated in experimental animal models of the disease, these drugs counteract motor deficits and amplify the beneficial effects of dopaminergic drugs without worsening their dyskinetic effect. Moreover, experimental evidence also indicates that adenosine A_{2A} receptor antagonists might slow down or arrest the dopaminergic degeneration that underlies Parkinson's disease. Building on this evidence, the research in this field has recently made significant progress, leading to the approval of the first A_{2A} receptor antagonist for clinical use as adjunct to L-DOPA (istradefylline, marketed under the name of NOURIAST®), and the ongoing clinical evaluation of other promising drugs (e.g., tozadenant).

This book covers basic biological aspects of the adenosine system relevant to Parkinson's disease, and also discusses recent experimental findings at both the preclinical and clinical level. Attention is dedicated to the localization and function of adenosine A_{2A} receptors, to their interaction with dopaminergic and non-dopaminergic receptors in the brain, and to the development of novel molecules that may target A_{2A} receptors. The critical role of the adenosine system in the regulation of neurotrophic factors, neuroinflammation, and neurotoxicity is also covered, and the relevance of these phenomena to the etiology of Parkinson's disease discussed. Moreover, the book thoroughly describes the effects of adenosine A_{2A} receptor antagonists observed in experimental models of Parkinson's disease on both motor (akinesia, dyskinesia, tremor) and non-motor (cognition, peripheral functions, sleep) symptoms. Finally, attention is dedicated to the clinical relevance of the adenosinergic system, by describing the development of the first ever approved adenosine A_{2A} receptor antagonist (istradefylline), the most advanced clinical trials with these drugs, the use of A_{2A} receptor antagonist in neuroimaging, and the epidemiological evidence that links the adenosine system with the onset and progression of Parkinson's disease.

By gathering updated and high-quality chapters written by world-leading experts in the field, this book provides essential information to preclinical and clinical researchers interested in the development of new therapies against Parkinson's disease and related neurodegenerative disorders.

<table>
<tr><td>Cagliari, Italy</td><td>Micaela Morelli</td></tr>
<tr><td>Cagliari, Italy</td><td>Nicola Simola</td></tr>
<tr><td>Krakow, Poland</td><td>Jadwiga Wardas</td></tr>
</table>

Contents

Contributors

Aderbal Silva Aguiar Laboratório Experimental de Doenças Neurodegenerativas (LEXDON), Centro de Ciências Biológicas (CCB), Universidade Federal de Santa Catarina (UFSC), Florianópolis, SC, Brazil

Laboratório de Neurobiologia do Exercício Físico, Centro de Ciências Biológicas (CCB), Universidade Federal de Santa Catarina (UFSC), Florianópolis, SC, Brazil

Rachit Bakshi Molecular Neurobiology Laboratory, MassGeneral Institute for Neurodegenerative Disease, Massachusetts General Hospital, Boston, MA, USA

Olivier Barret Molecular NeuroImaging, LLC, New Haven, CT, USA

Jordi Bonaventura Integrative Neurobiology Section, National Institute on Drug Abuse, IRP, Intramural Research Program, National Institutes of Health, Baltimore, MD, USA

Daniel Caldeira Laboratory of Clinical Pharmacology and Therapeutics, Faculty of Medicine, University of Lisbon, Lisbon, Portugal

Vicent Casadó Centro de Investigacion Biomedica en Red sobre Enfermedades Neurodegenerativas, Department of Biochemisry and Molecular Biology, Faculty of Biology, University of Barcelona, Barcelona, Spain

Jiang-Fan Chen Department of Neurology, Boston University School of Medicine, Boston, MA, USA

Mercè Correa Behavioral Neuroscience Division, Department of Psychology, University of Connecticut, Storrs, CT, USA

Area de Psicobiologia, Universitat Jaume I, Castello, Spain

Giulia Costa Department of Biomedical Sciences, Section of Neuropsychopharmacology, University of Cagliari, Cagliari, Italy

João Costa Laboratory of Clinical Pharmacology and Therapeutics, Faculty of Medicine, University of Lisbon, Lisbon, Portugal

Clinical Pharmacology Unit, Instituto de Medicina Molecular, Lisbon, Portugal

Evidence Based Medicine Centre, Faculty of Medicine, University of Lisbon, Lisbon, Portugal

Portuguese Collaborating Centre of the Cochrane Iberoamerican Network, Faculty of Medicine, University of Lisbon, Lisbon, Portugal

Rodrigo A. Cunha CNC-Center for Neuroscience and Cell Biology, University of Coimbra, Coimbra, Portugal

FMUC-Faculty of Medicine, University of Coimbra, Coimbra, Portugal

Maria J. Diógenes Institute of Pharmacology and Neurosciences, Faculty of Medicine and Unit of Neurosciences, Instituto de Medicina Molecular, Universidade de Lisboa

Sergi Ferré Integrative Neurobiology Section, National Institute on Drug Abuse, IRP, Intramural Research Program, National Institutes of Health, Baltimore, MD, USA

Joaquim J. Ferreira Laboratory of Clinical Pharmacology and Therapeutics, Faculty of Medicine, University of Lisbon, Lisbon, Portugal

Jimmy George CNC-Center for Neuroscience and Cell Biology, University of Coimbra, Coimbra, Portugal

Catarina Gomes CNC-Center for Neuroscience and Cell Biology, University of Coimbra, Coimbra, Portugal

FMUC-Faculty of Medicine, University of Coimbra, Coimbra, Portugal

Robert A. Hauser USF Byrd Parkinson's Disease and Movement Disorders Center, Byrd Institute, University of South Florida, Tampa, USA

Peter Jenner Neurodegenerative Diseases Research Group, Institute of Pharmaceutical Sciences, Faculty of Life Sciences and Medicine, King's College, London, UK

Michael Lazarus International Institute for Integrative Sleep Medicine (WPI-IIIS), University of Tsukuba, Tsukuba, Ibaraki, Japan

Peter LeWitt Departments of Neurology, Henry Ford Hospital and Wayne State University School of Medicine, West Bloomfield, Michigan, USA

Carme Lluís Centro de Investigacion Biomedica en Red sobre Enfermedades Neurodegenerativas, Department of Biochemisry and Molecular Biology, Faculty of Biology, University of Barcelona, Barcelona, Spain

Robert Logan Molecular Neurobiology Laboratory, MassGeneral Institute for Neurodegenerative Disease, Massachusetts General Hospital, Harvard Medical School, Boston, MA, USA

Lauren L. Long Behavioral Neuroscience Division, Department of Psychology, University of Connecticut, Storrs, CT, USA

Filipe Carvalho Matheus Laboratório Experimental de Doenças Neurodegenerativas (LEXDON), Centro de Ciências Biológicas (CCB), Universidade Federal de Santa Catarina (UFSC), Florianópolis, SC, Brazil

Micaela Morelli Department of Biomedical Sciences, Section of Neuropsychopharmacology, University of Cagliari, Cagliari, Italy

Estefanía Moreno Centro de Investigacion Biomedica en Red sobre Enfermedades Neurodegenerativas, Department of Biochemisry and Molecular Biology, Faculty of Biology, University of Barcelona, Barcelona, Spain

Morgana Moretti Laboratório Experimental de Doenças Neurodegenerativas (LEXDON), Centro de Ciências Biológicas (CCB), Universidade Federal de Santa Catarina (UFSC), Florianópolis, SC, Brazil

Akihisa Mori Medical Affairs Department, Kyowa Hakko Kirin Co., Ltd., Tokyo, Japan

Christa E. Müller PharmaCenter Bonn, Pharmaceutical Chemistry I, Pharmaceutical Institute, Universität Bonn, Bonn, Germany

Gemma Navarro Centro de Investigacion Biomedica en Red sobre Enfermedades Neurodegenerativas, Department of Biochemisry and Molecular Biology, Faculty of Biology, University of Barcelona, Barcelona, Spain

Eric J. Nunes Behavioral Neuroscience Division, Department of Psychology, University of Connecticut, Storrs, CT, USA

Paulo Alexandre de Oliveira Laboratório Experimental de Doenças Neurodegenerativas (LEXDON), Centro de Ciências Biológicas (CCB), Universidade Federal de Santa Catarina (UFSC), Florianópolis, SC, Brazil

Annalisa Pinna Institute of Neuroscience, National Research Council of Italy (CNR), Cagliari, Italy

Samantha J. Podurgiel Behavioral Neuroscience Division, Department of Psychology, University of Connecticut, Storrs, CT, USA

Emmanuelle Pourcher Faculty of Medicine, Clinique Sainte Anne Mémoire et Mouvement, Laval University, Quebec, QC, Canada

Rui Daniel Prediger Laboratório Experimental de Doenças Neurodegenerativas (LEXDON), Centro de Ciências Biológicas (CCB), Universidade Federal de Santa Catarina (UFSC), Florianópolis, SC, Brazil

Daniel Rial Laboratório Experimental de Doenças Neurodegenerativas (LEXDON), Centro de Ciências Biológicas (CCB), Universidade Federal de Santa Catarina (UFSC), Florianópolis, SC, Brazil
CNC-Center for Neuroscience and Cell Biology, University of Coimbra, Coimbra, Portugal

Joaquim A. Ribeiro Institute of Pharmacology and Neurosciences, Faculty of Medicine and Unit of Neurosciences, Instituto de Medicina Molecular, Universidade de Lisboa

Filipe B. Rodrigues Laboratory of Clinical Pharmacology and Therapeutics, Faculty of Medicine, University of Lisbon, Lisbon, Portugal

John D. Salamone Behavioral Neuroscience Division, Department of Psychology, University of Connecticut, Storrs, CT, USA

Michael A. Schwarzschild Molecular Neurobiology Laboratory, MassGeneral Institute for Neurodegenerative Disease, Massachusetts General Hospital, Harvard Medical School, Boston, MA, USA

Ana M. Sebastião Institute of Pharmacology and Neurosciences, Faculty of Medicine and Unit of Neurosciences, Instituto de Medicina Molecular, Universidade de Lisboa

John P. Seibyl Molecular NeuroImaging, LLC, New Haven, CT, USA

Nicola Simola Department of Biomedical Sciences, Section of Neuropsychopharmacology, University of Cagliari, Cagliari, Italy

Ana Cristina Guerra de Souza Laboratório Experimental de Doenças Neurodegenerativas (LEXDON), Centro de Ciências Biológicas (CCB), Universidade Federal de Santa Catarina (UFSC), Florianópolis, SC, Brazil

Gilles D. Tamagnan Molecular NeuroImaging, LLC, New Haven, CT, USA

Adriana Alexandre S. Tavares Molecular NeuroImaging, LLC, New Haven, CT, USA

Yoshihiro Urade International Institute for Integrative Sleep Medicine (WPI-IIIS), University of Tsukuba, Tsukuba, Ibaraki, Japan

Nora D. Volkow National Institute on Alcohol Abuse and Alcoholism, National Institutes of Health, Bethesda, MD, USA

Jadwiga Wardas Department of Neuropsychopharmacology, Institute of Pharmacology, Polish Academy of Sciences, Kraków, Poland

Chapter 1
Adenosine A_{2A} Receptors: Localization and Function

Nicola Simola and Jadwiga Wardas

Abstract Adenosine is an endogenous purine nucleoside present in all mammalian tissues, that originates from the breakdown of ATP. By binding to its four receptor subtypes (A_1, A_{2A}, A_{2B}, and A_3), adenosine regulates several important physiological functions at both the central and peripheral levels. Therefore, ligands for the different adenosine receptors are attracting increasing attention as new potential drugs to be used in the treatment of several diseases.

This chapter is aimed at providing an overview of adenosine metabolism, adenosine receptors localization and their signal transduction pathways. Particular attention will be paid to the biochemistry and pharmacology of A_{2A} receptors, since antagonists of these receptors have emerged as promising new drugs for the treatment of Parkinson's disease. The interactions of A_{2A} receptors with other non-adenosinergic receptors, and the effects of the pharmacological manipulation of A_{2A} receptors on different body organs will be discussed, together with the usefulness of A_{2A} receptor antagonists for the treatment of Parkinson's disease and the potential adverse effects of these drugs.

Keywords Adenylate cyclase · Basal ganglia · Dopamine · G protein-coupled receptors · Heteromeric complexes · Nucleoside · Purine · Striatonigral · Striatopallidal

N. Simola (✉)
Department of Biomedical Sciences, Section of Neuropsychopharmacology,
University of Cagliari, Cagliari, Italy
e-mail: nicola.simola@unica.it

J. Wardas
Department of Neuropsychopharmacology, Institute of Pharmacology, Polish Academy of Sciences, 12 Smętna St., 31-343 Kraków, Poland
e-mail: wardas@if-pan.krakow.pl

© Springer International Publishing Switzerland 2015
M. Morelli et al. (eds.), *The Adenosinergic System,* Current Topics in Neurotoxicity 10,
DOI 10.1007/978-3-319-20273-0_1

Introduction

The concept of purinergic neurotransmission was first introduced by Burnstock in 1972 and subsequently adenosine 5' triphosphate (ATP) was shown to act either as a transmitter or a co-transmitter in most nerves in both the peripheral and central nervous system (CNS) (Abbracchio and Burnstock 1998; Abbracchio et al. 2008; Burnstock 1972, 2013). At present, it is known that ATP acts as a fast excitatory neurotransmitter or neuromodulator, and has a potent long-term trophic role in cell proliferation, growth and development as well as in disease and cytotoxicity (Abbracchio and Burnstock 1998; Abbracchio et al. 2008; Burnstock 2013).

ATP and other nucleotides are stored in secretory and synaptic vesicles, and exocytotic vesicular release of ATP from neurons and astrocytes is well established (Abbracchio et al. 2008; Bowser and Khakh 2007; Burnstock 2013; Pankratov et al. 2006, 2007). There are also evidences indicating additional mechanisms of the release of this nucleotide, including ATP-cassette transporters, connexin or pannexin hemichannels, plasmalemmal voltage-dependent anion channels and the ATP-sensitive P2X7 receptors (Abbracchio et al. 2008; Burnstock 2013). After release, ATP and other nucleotides undergo rapid enzymatic degradation to adenosine by ectonucleotidases (Bonan 2012; Kovacs et al. 2013; Yegutkin 2008; Zimmermann 2006).

Adenosine Metabolism

Adenosine, an endogenous purine ribonucleoside present in all mammalian tissues, modulates a variety of important synaptic processes and signaling pathways, and regulates the functions of several neurotransmitters in the CNS. Adenosine is considered to be a neuromodulator rather than a neurotransmitter, since it is not stored in synaptic vesicles, and is not released from nerve terminals by exocytosis. Adenosine affects neural activity through multiple mechanisms; presynaptically by controlling neurotransmitter release, postsynaptically by hyperpolaryzing or depolarizing neurons, and non-synaptically mainly via regulatory effects on glial cells (Boison et al. 2010; Dare et al. 2007; Fredholm et al. 2005). Although adenosine is generally known to be produced by the ectoenzymatic breakdown of ATP, there might be a subpopulation of neurons and/or astrocytes that release adenosine directly in an activity-dependent manner (Wall and Dale 2007).

It is well established that adenosine may be formed in the CNS either intracelullarly, after degradation of ATP to cyclic-adenosine monophosphate (cAMP) and 5'-AMP, and then transported by nucleotide transporters to the synapse, or extracellularly from nucleotides released into the synaptic cleft (Fig. 1.1). Thus, the formation of adenosine is dependent on the availability of oxygen and energetic compounds as well as on the rate of synthesis and degradation of ATP, released from both neuronal and glial cells. However, it is the release of ATP from astrocytes,

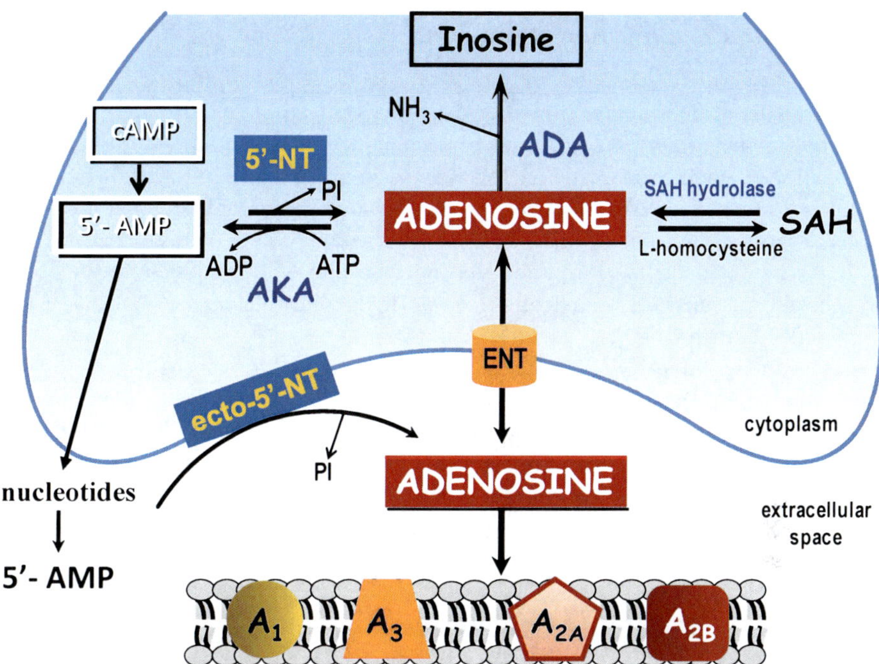

Fig. 1.1 Adenosine synthesis and metabolic pathways. Adenosine is formed both *intracelullarly* from 5'-AMP by the cytosolic 5'-NT, and *extracelullarly* in the metabolism of nucleotides (*ATP*, *ADP*, *AMP*) released from the cell, through the action of ecto-5'-nucleotidase. Another intracellular source of adenosine may be the hydrolysis of SAH by SAH hydrolase. Hence, adenosine formation depends on ATP breakdown and synthesis. Extracellular adenosine is primarily inactivated by uptake through the transporters (*ENT*), which are mainly bidirectional, followed by either phosphorylaton to AMP by AKA (under physiological conditions), or, to a lesser degree, deamination to inosine by ADA. Another possible catabolic pathway of adenosine, though of minor significance, is a reversible reaction catalysed by SAH hydrolase, leading to formation of SAH from adenosine and L-homocysteine (*for more details see the text, and* Abbracchio et al. 2008; Burnstock 2013; Latini and Pedata 2001; Sperlagh and Vizi 2011). *ADA* adenosine deaminase, *AKA* adenosine kinase, *ADP* adenosine diphosphate, *AMP* adenosine monophosphate, *ATP* adenosine 5'-triphosphate, *A$_1$, A$_{2A}$, A$_{2B}$ and A$_3$*—adenosine receptors, *ecto-5'-NT* ecto-5'-nucleotidase, *ENT* nucleoside transporter, *5'-NT* 5'-nucleotidase, *SAH* S-adenosylhomocysteine

either vesicular (Pascual et al. 2005) or via secretion through hemichannels, that is the major source of synaptic adenosine (Kang et al. 2008; Kawamura et al. 2010). Moreover, adenosine can be directly released by nucleoside transporters from astrocytes when its intracellular level is augmented in response to a variety of physiological and pathological stimuli (e.g. increased cellular activity, hypoxia/hypoglycemia, ischemia). Then, adenosine may function as a nonsynaptic signalling molecule that diffuses far away from the site of origin and tonically influences neurotransmission, inflammation, and immune responses, as described below (Bours et al. 2006; Dare et al. 2007; Geiger and Fyda 1991; Sperlagh and Vizi 2011).

Intracellular Formation of Adenosine

In the cell, adenosine may be formed in the process of adenosine monophosphate (AMP) hydrolysis catalyzed by 5'-nucleotidase, which belongs to the family of enzymes called ectonucleotidases (Fig. 1.1, Kovacs et al. 2013; Yegutkin 2008). Seven types of 5'-nucleotidases have been cloned, characterized and demonstrated in various tissues, including brain tissue (Hunsucker et al. 2005; Kovacs et al. 2013). This pathway of adenosine formation via the cytosolic ATP catabolism seems to represent a very sensitive signal of increased metabolic rate or metabolic stress (Latini and Pedata 2001).

Another intracellular source of adenosine may be the hydrolysis of S-adenosyl-homocysteine (SAH) by SAH hydrolase (Fig. 1.1), an enzyme present in brain areas, such as the neocortex, hippocampus, and cerebellum (Latini and Pedata 2001). However, this pathway is not strictly dependent upon the energetic state of the cells, and it does not significantly contribute to adenosine production in the brain under either physiological or ischemic conditions (Latini and Pedata 2001).

Extracellular Formation of Adenosine

The extracellular nucleotide and nucleoside levels in the synaptic cleft are controlled by a cascade of enzymes, belonging to the family of ectonucleotidases. There are four major families of ectonucleotidases, namely ectonucleoside triphosphate diphosphohydrolases (E-NTPDases), ectonucleotide pyrophosphatase/phosphodiesterases (E-NPPs), alkaline phosphatases, and ecto-5'-nucleotidase (ecto-5'-NT) (Bonan 2012; Kovacs et al. 2013; Yegutkin 2008; Zimmerman 2006).

The first step of ATP inactivation is mediated by the family of E-NTPDases, which are able to hydrolyse ATP and adenosine diphosphate (ADP) to AMP (Zimmermann 2006). Moreover, ATP can be dephosphorylated by E-NPPs and alkaline phosphatases which, like E-NTPDases, have widespread distribution in the CNS (Wang and Guidotti 1998; Zimmermann 2006). The next step of extracellular ATP inactivation involves the hydrolysis of AMP to adenosine and phosphate by the ecto-5'-NT, also known as CD73 (Fig. 1.1), which is attached via a GPI anchor to the outer surface of the plasma membrane. Ecto-5'-NT, which is the rate-limiting step in the formation of adenosine (Sperlagh 1996; Sperlagh and Vizi 2007), is also widely expressed in the brain (e.g. in hippocampal and striatal nerve terminals), and it is predominantly associated with glial cells (Cunha et al. 1992; Hunsucker et al. 2005; James and Richardson 1993; Kovacs et al. 2013; Schoen et al. 1987).

Another pathway of extracellular adenosine formation may originate from the cAMP or 5'-AMP released into the synapse. Both these nucleotides are responsible for the slow change in the adenosine concentration. The cAMP can be released through non-specific energy-dependent transporters and then, when in the synapse, it can first be converted to 5'-AMP by ecto-phosphodiesterases and then to adenosine by ecto-5'-NT. Another possibility also exists that the cAMP can be converted to 5'-AMP inside the cell and then 5'-AMP can be released into the synaptic cleft, becoming a source of adenosine (e.g. after the NMDA (N-methyl-D-aspartate) stimulation in cortical sections) (Latini and Pedata 2001; Sperlagh and Vizi 2011).

The process of extracellular adenosine formation is very fast, and occurs within seconds (Dunwiddie et al. 1997). Adenosine is normally present in a concentration between 30–300 nM, but under hypoxic or ischemic conditions adenosine concentrations in the hippocampus can reach 20–30 µM (Dunwiddie et al. 1997; Latini et al. 1999). It seems that *in vivo* a large part of adenosine present in the synapse under basal conditions comes from the extracellular metabolism of nucleotides (Latini and Pedata 2001; Sperlagh and Vizi 2011). In contrast, numerous studies have suggested that in conditions of hypoxia or ischemia adenosine is mainly formed intracellularly and released to the synaptic space by transportes (Latini and Pedata 2001; Sperlagh and Vizi 2011).

Nucleoside Transporters

The level of extracellular adenosine is regulated by the process of bidirectional transport of nucleosides, which allows for rapid exchange between extra and intracellular adenosine. In contrast to conventional neurotransmitters, the reuptake of adenosine does not depend on energy-driven transporter-mediated systems. This transport is driven by chemical gradients and by unidirectional concentrative processes, regulated by sodium electrochemical gradient (Dos Santos-Rodrigues et al. 2014; Parkinson et al. 2011). There are two functionally distinct types of nucleoside transporters:

1. equilibrative nucleoside transporters (ENT), which predominate in the CNS, and carry both purine and pyrimidine nucleosides in both directions across cell membranes, depending on their concentration gradient. Four types of ENT transporters have been characterized: ENT1-2-3-4; type 1 and 2 appear to be present in all cell types, including neurons and glia (Baldwin et al. 2004; Dos Santos-Rodrigues et al. 2014; King et al. 2006; Parkinson et al. 2011).
2. concentrative nucleoside transporters (CNT, sodium-dependent) which mediate the influx of nucleosides under the force of transmembrane sodium gradient (Dos Santos-Rodrigues et al. 2014; Latini and Pedata 2001; Parkinson et al. 2011). Five subtypes of these transporters have been identified, and two types of CNT were cloned and detected in the rat brain, mainly in the posterior hypothalamus, superior colliculus, brainstem, striatum, hippocampus, cerebellum and cortex (Anderson et al. 1996; Dos Santos-Rodrigues et al. 2014; Latini and Pedata 2001; Parkinson et al. 2011).

Since the ENT transporters, which seem to dominate in the CNS, are bi-directional, they can not only increase the flow of adenosine into the cell when its extracellular level exceeds its intracellular one, but they may mediate the efflux of adenosine from the cell, when its intracellular level increases. On the other hand, when the Na$^+$ gradient is reversed, also the concentrative nucleoside transporters can release adenosine from the cell (Dos Santos-Rodrigues et al. 2014; Latini and Pedata 2001; Parkinson et al. 2011).

Adenosine Inactivation

Extracellular adenosine is primarily inactivated by uptake across the neuronal cell membrane, followed by either intracellular phosphorylaton to AMP by adenosine kinase (AKA) or to a lesser degree, deamination to inosine by adenosine deaminase (ADA) (Fig. 1.1).

ADA is a cytosolic enzyme present in many neurons in the brain, but its highest activity is seen in neurons of the basal hypothalamus; ADA can also be expressed extracellularly in various tissues (Desrosiers et al. 2007; Yegutkin 2008). In addition to the enzymatic function, ADA catalyses the irreversible deamination of adenosine to inosine. ADA can also exist in a form associated with the adenosine A_1 receptor, so called ektoADA, which can act as a positive modulator of the adenosine binding and signalling function (Ciruela et al. 1996; Ruiz et al. 2000). Moreover, inosine can be then metabolized to hypoxanthine and finally to urate by xanthine oxidase (Morelli et al. 2010).

AKA is a part of the cycle between adenosine and AMP, which enables the cell to rapidly respond to changes in the concentration of adenosine. AKA can be expressed in both the cytoplasm (short isoform) and the nucleus (long isoform) of astrocytes or neurons, and phosphorylates adenosine to AMP (Boison 2013). In the adult brain, the expression of AKA is largely restricted to astrocytes, with the exception of neurons in the olfactory bulb, which maintain high levels of AKA expression (Boison 2013).

Several lines of evidence indicate that under basal conditions astrocytic AKA is the main regulator of extracellular adenosine, by driving adenosine influx into astrocytes via bi-directional nucleoside transporters (Boison et al. 2010). In contrast, deamination by ADA prevails under conditions in which adenosine levels become excessive (e.g. due to pathologic activity, such as ischemia or hypoxia) (Latini and Pedata 2001).

Another possible metabolic pathway of adenosine involves a reversible reaction catalysed by SAH hydrolase, leading to the formation of SAH and L-homocysteine; however, it represents only a minor pathway of adenosine degradation in physiological conditions, as the level of L-homocysteine and SAH in the brain is very low (Fig. 1.1) (Gharib et al. 1982; Reddington and Pusch 1983).

Once present in the extracellular space, adenosine may diffuse far away and influence its receptors (Abbracchio and Burnstock 1998; Abbracchio et al. 2008; Burnstock 1976; Fredholm et al. 2001, 2011; Ribeiro et al. 2002).

Adenosine Receptors

Currently, four subtypes of adenosine receptors (A_1, A_{2A}, A_{2B}, and A_3), which belong to the family of G protein-coupled receptors (GPCR), have been cloned and characterized (Table 1.1) (for recent review see Chen et al. 2014; Fredholm et al. 2000, 2001, 2011). It has been estimated that under physiological conditions, extracellular

Table 1.1 Adenosine receptors—classification, signal transduction and localization in the CNS

	Adenosine receptors			
	A$_1$	A$_{2A}$	A$_{2B}$	A$_3$
G-protein coupling	Gi/Go	Gs, Golf	Gs	Gi/Go
Transduction mechanisms	$\downarrow$AC; $\downarrow$Ca^{2+}	$\uparrow$AC	$\uparrow$AC	$\downarrow$AC
	$\uparrow$PLC; $\uparrow$K$^+$	$\uparrow$Ca^{2+}	$\uparrow$PLC	$\uparrow$Ca^{2+}; $\uparrow$PLC
Distribution: High	Widespread in the brain; parietal, temporal and occipital cortex, striatum, thalamus	Restricted in the brain: striatum, nucleus accumbens, globus pallidus, olfactory tubercle	Widespread, uniform distribution	Cerebellum, hippocampus
Intermediate— Low	Frontal and cingulate cortex, nucleus accumbens, hippocampus, amygdala, thalamic reticular nuclei, medial geniculate body, globus pallidus, superior colliculus, substantia innominata, substantia nigra, pons, medulla oblongata, spinal cord, cerebellum	Frontal, temporal, parietal and occipital cortex, thalamus, hippocampus, pons, cerebellum, medulla oblongata		Other brain areas

AC adenylate cyclase, *Gi, Golf, Gs* G-proteins, *PLC* phospholipase C, $\uparrow$ stimulation, $\downarrow$ inhibition

levels of adenosine in the rodent CNS (nM range) are sufficient to stimulate both the higher affinity A$_1$ and A$_{2A}$ receptors. Under pathological conditions, like hypoxia/ischemia and seizures, adenosine level rises markedly to concentrations that can stimulate both the lower affinity A$_3$ and A$_{2B}$ receptors.

Signal Transduction

The main adenosine-mediated intracellular signalling pathways involve the formation of cAMP, with A$_1$ and A$_3$ receptor stimulation causing (through Gi and Go proteins) the inhibition of adenylate cyclase (AC) and the decrease of cAMP production, which leads to reduced protein kinase A (PKA) activity and cyclic AMP response element binding protein (CREB) phosphorylation. On the other hand, stimulation of A$_{2A}$ and A$_{2B}$ receptors activates AC through Gs/olf proteins, resulting in activation of PKA and CREB phosphorylation (Table 1.1, Fig. 1.2) (Cunha 2001; Fredholm et al. 2001, 2011).

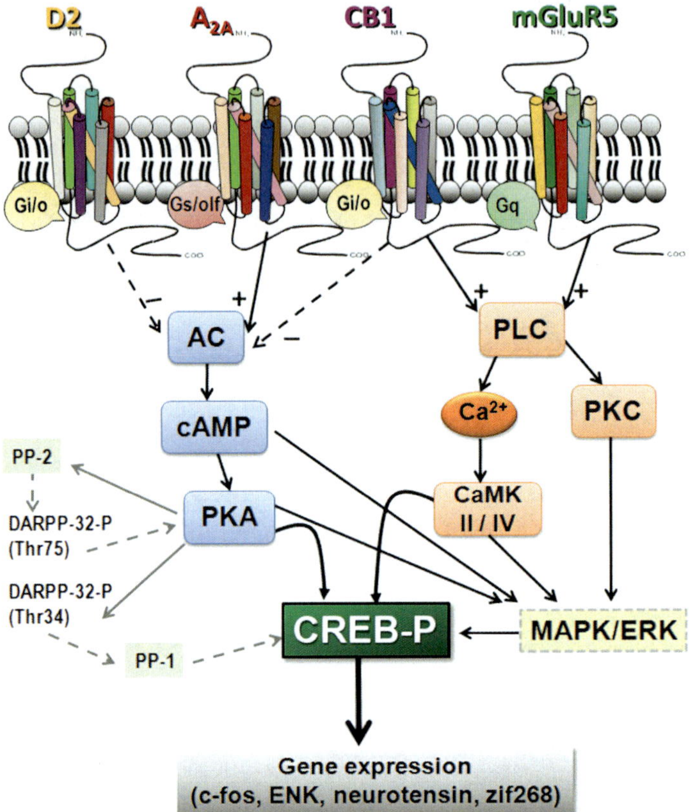

Fig. 1.2 Functional interactions among dopamine D_2, adenosine A_{2A}, cannabinoid CB_1 and metabotropic glutamate mGlu5 receptors in striato-pallidal neurons. At the intramembrane level, adenosine A_{2A} receptors interact antagonistically with D_2 and CB_1 receptors. These receptors also exert an opposing effect on the AC level and AC-regulated downstream molecules, such as PKA, DARPP-32, CREB-P, and early genes. MGlu5 and A_{2A} receptors act synergistically to counteract the D_2 dopamine receptor signalling in striato-pallidal neurons. Synergistic interactions exist between A_{2A} and mGlu5 receptors at the level of early gene expression (e.g. c-fos), MAP kinases and phosphorylation of DARPP-32 protein. *dashed lines*, inhibitory effect; '+', stimulation; '−', inhibition. *AC* adenylyl cyclase, *Ca²⁺* calcium ions, *CaMK* II/IV calcium/calmodulin-dependent protein kinase type II/IV, *cAMP* cyclic AMP, *CREB* cAMP response element-binding protein, *DARPP-32* dopamine- and cAMP-regulated phosphoprotein, *DARPP-32-P (Thr75) and DARPP-32-P (Thr34)* DARPP32-phopshorylated at threonine residues 75 and 34, respectively, *Gi Go*, inhibitory G proteins, *Gq, Gs, Golf* stimulatory G proteins, *MAPK* mitogen-activated protein kinase, *PKA* protein kinase A, *PKC* protein kinase C, *PLC* phospholipase C, *PP-1* protein phosphatase-1, *PP-2* protein phosphatase-2

Besides CREB, the dopamine- and cAMP-regulated phosphoprotein of 32 kDa (DARPP-32), abundantly expressed in striatal projection neurons, is another downstream target of PKA activation induced by A_{2A} receptor stimulation. Activation of A_{2A} receptors increases the phosphorylation of DARPP-32 protein at the threonine residue 34 (Thr34), which converts this protein into a potent inhibitor of protein phosphatase-1 (PP-1) (Fig. 1.2) (Fredholm et al. 2007; Svenningsson et al. 2000, 2004). In turn, blockade of A$_{2A}$ receptors reduces the effect of D$_2$ receptor blockade on DARPP-32 phosphorylation at Thr34 and, at the same time, increases the phosphorylation of this protein at the threonine residue 75, which converts DARPP-32 into an inhibitor of PKA (Fredholm et al. 2007; Svenningsson et al. 2000, 2004). Thus, DARPP-32 has the unique property of being a dual-function protein, acting as an inhibitor of either PP-1 or of PKA.

Other mechanisms, such as voltage-sensitive Ca^{2+} channels (types Q, N, and P), K$^+$ channels and phospholipase C, are also involved in signal transduction by each of the adenosine receptors (Table 1.1; Fig. 1.2) (Dunwiddie and Masino 2001; Fredholm et al. 2001, 2011; Ralevic and Burnstock 1998). Additionally, the involvement of mitogen-activated protein kinase (MAPK) pathway was also shown in Chinese hamster ovary cells (CHO) and COS-7 fibroblast-like cells (Dickenson et al. 1998; Schulte and Fredholm 2000, 2003).

Adenosine A$_1$, A$_{2B}$ and A$_3$ Receptors Localization

The inhibitory A$_1$ receptors, which are expressed on both neurons and glial cells, are the most abundant adenosine receptors in many regions of the brain. These receptors are localized both pre- and postsynaptically. The highest expression of A$_1$ receptors has been found in the cortex, striatum, thalamus, cerebellum and hippocampus (Table 1.1) (Fastbom et al. 1987; Fredholm et al. 2005; Ochiishi et al. 1999; Schindler et al. 2001; Sebastiao and Ribeiro 2009b). Moreover, the A$_1$ receptor mRNA is also present in basal ganglia (BG) structures, including the striatum, globus pallidus, and subthalamic nucleus (Dixon et al. 1996). These receptors are also present on astrocytes, oligodendrocytes, and microglia (Biber et al. 1997; Dare et al. 2007; Gebicke-Haerter et al. 1996; Othman et al. 2003). In the striatum, adenosine A$_1$ receptors are present on both direct and indirect GABAergic efferent neurons as well as on cholinergic interneurons (Alexander and Reddington 1989; Ferré et al. 1996; Rivkees et al. 1995). Moreover, presynaptic A$_1$ receptors are present on glutamatergic cortico-striatal and dopaminergic nigro-striatal afferents but also on nerve terminals in the globus pallidus, substantia nigra and hippocampus, where they modulate the release of neurotransmitters, such as glutamate, acetylcholine, serotonin and GABA (Cunha 2001; Fastbom et al. 1987; Rebola et al. 2003).

Adenosine A$_{2B}$ receptors are mainly present in peripheral organs, like the bowel, bladder, lung, and vas deferens, but can also be found in the spinal cord and brain (Feoktistov and Biaggioni 1997; Pierce et al. 1992; Ralevic and Burnstock 1998). In the brain, A$_{2B}$ receptors are present in hippocampal CA1 and CA3 neurons, in the hypothalamic, thalamic, and striatal neurons; low levels of these receptors are also expressed on glial cells (Table 1.1) (Dare et al. 2007; Feoktistov and Biaggioni 1997; Fredholm et al. 2001; Pierce et al. 1992; Ralevic and Burnstock 1998).

The distribution and physiological functions of A$_3$ receptors in the brain are still unclear, although these receptors are widely distributed in peripheral organs (mainly in the testis and lung) (Dixon et al. 1996; Rivkees et al. 2000; Shearman and Weaver

1997). A relatively low level of A_3 receptors and their mRNA was detected in the hippocampus, cortex, cerebellum and striatum with cellular localization on neurons, astrocytes, and microglia (Table 1.1) (Brand et al. 2001; Daré et al. 2007; Dixon et al. 1996; Fredholm et al. 2011; Hammarberg et al. 2003; Wittendorp et al. 2004).

Adenosine A_{2A} Receptors and their Localization in the Brain

In contrast to the widespread distribution of A_1 receptors in the CNS, the A_{2A} receptors are highly abundant in the striatum and nucleus accumbens. Moreover, positron emission tomography (PET) studies in humans showed that, like in rodents, A_{2A} receptors were concentrated in the caudate-putamen and nucleus accumbens (Brooks et al. 2008). However, studies performed with more sensitive techniques have demonstrated the presence of A_{2A} receptors and corresponding mRNAs, albeit at lower level of expression, in several other brain areas, such as the hippocampus, cerebral cortex, extended amygdala, thalamic nuclei, and substantia nigra (Cunha et al. 1994; Dixon et al. 1996; Jarvis and Williams 1989; Rebola et al. 2005; Rosin et al. 1998, 2003; Svenningsson et al. 1998, 1999). It is noteworthy that A_{2A} receptors are also present on glial cells, and that about 3 % of their total number are located on striatal astrocytes (Dare et al. 2007; Hettinger et al. 2001; Matos et al. 2012, 2013; Rosin et al. 2003).

In the striatum, A_{2A} receptors are homogeneously distributed throughout the lateral and medial parts and display dense labelling of the neuropil (Rosin et al. 1998, 2003). These receptors are mainly localized postsynaptically on the GABAergic medium-sized spiny neurons of the indirect pathway projecting to the globus pallidus external segment (GPe). These latter neurons also express a high density of dopamine D_2 receptors and enkephalin (Augood and Emson 1994; Fink et al. 1992; Rebola et al. 2005; Rosin et al. 2003; Schiffmann et al. 1991, 2007; Svenningsson et al. 1998). Conversely, neurons of the direct striato-nigral pathway, which selectively express dopamine D_1 receptors and the peptide dynorphin, do not contain a significant level of A_{2A} receptors (Schiffmann et al. 1991). Morphologically, A_{2A} receptors in the striatum predominate on dendrites and dendritic spines and are expressed to a lesser extent on axons and axon terminals of recurrent collaterals projecting back to the striatum or from the cortical areas (Rebola et al. 2005).

The A_{2A} receptors in the striatum are also localized presynaptically on glutamatergic terminals that contact medium-sized spiny neurons of the GABAergic direct striato-nigral pathway (Quiroz et al. 2009; Rodrigues et al. 2005; Rosin et al. 2003), where they heteromerize with A_1 receptors and regulate the release of glutamate (Ciruela et al. 2006; Quiroz et al. 2009). Such a co-expression of adenosine A_{2A} and A_1 receptor mRNAs was also found on the glutamatergic nerve terminals in the hippocampus (Rebola et al. 2005), where these receptors may control glutamate release. Moreover, A_{2A} receptors located on GABAergic collateral axons may modulate in an inhibitory way the GABA release from medium-sized spiny projection neurons, likely relieving a GABA-mediated inhibition of these neurons (Mori et al. 1996). In turn, A_{2A} receptors located on striatal cholinergic nerve terminals modulate acetylcholine (Ach) release (Brown et al. 1990; Kurokawa et al. 1994, 1996). A$_{2A}$ receptor agonists enhance, and A$_{2A}$ receptor antagonists reduce the Ach release *in vivo* (Kurokawa et al. 1996), which is modulated by the dopaminergic transmission (Kurokawa et al. 1996).

Regarding the nucleus accumbens (which is part of the so-called ventral striatum), A$_{2A}$ receptors follow the same pattern of distribution as the dopamine D$_2$ receptors, and the shell of the nucleus accumbens displays a density of adenosine A$_{2A}$ receptors by about 40 % lower than that in the dorsal striatum (Rosin et al. 2003). A distinction between the dorsal and ventral striatum has already been suggested by others. The dorsal part seems to be the most important for the control of dopamine-mediated motor behaviour (Groenewegen 2007; Joel and Weiner 2000; Voorn et al. 2004). On the other hand, the so-called ventral striatum, which comprises the nucleus accumbens, the ventromedial part of the striatum, and the olfactory tubercle, is a region connected with limbic structures, and seems to be strongly associated with emotional and motivational aspects of behaviour (Groenewegen 2007; Joel and Weiner 2000; Voorn et al. 2004).

Homo- and Heteromeric Complexes Formed by Adenosine A$_{2A}$ Receptors

A growing body of evidence indicates that A$_{2A}$ receptors, like many other GPCR not only form homodimers, and heterodimers with A$_1$ receptors, but also interact with other non-adenosinergic receptors (Ferré et al. 2011; Fredholm et al. 2007; Sebastiao and Ribeiro 2009a, b). Such heteromers are presently regarded as a molecular basis for the known direct and indirect (via adapter proteins) intramembrane receptor/receptor interactions. The best-known heterodimeric interactions involve A$_{2A}$ and dopamine D$_2$ receptors (see Chap. 2).

A direct evidence for A$_{2A}$/D$_2$ heteromers, in addition to A$_{2A}$ homomeric complexes, within the plasma membrane came from fluorescent and bioluminescent resonance energy transfer (FRET and BRET) analyses (Canals et al. 2003). Such a heteromer represents one of the possible molecular mechanisms for the functional antagonism between A$_{2A}$/D$_2$ receptors, demonstrated earlier at different levels, including the receptor and second messenger systems (Fig. 1.2) (Ferré et al. 1997, 2011; Fuxe et al. 2003; Morelli et al. 1995; Sebastiao and Ribeiro 2009a,b; Svenningsson et al. 2000).

Moreover, heterodimerization between A$_{2A}$ and metabotropic glutamate mGlu5 receptors has been detected in glutamatergic striatal terminals *in vivo*, and in striatal neurons by *in vitro* studies, and has been suggested to play a role in striatal plasticity and in modulation of the activity of striato-pallidal neurons (Ferré et al. 2002; Rodrigues et al. 2005). Unlike in heteromers composed of A$_{2A}$ and dopamine D$_2$ receptors, which interact in an opposing functional way, the A$_{2A}$/mGlu5 receptor interaction may account for the synergism found after combined treatments with agonists or antagonists, demonstrated at both the biochemical and behavioural levels (Fig. 1.2) (Ferré et al. 2002; Nishi et al. 2003; Popoli et al. 2001). A molecular mechanism underlying this functional interaction may be based on the fact that co-activation of mGlu5 and A_{2A} receptors by agonists synergistically increases DARPP-32 phosphorylation (Nishi et al. 2003). This potentiation of A_{2A}/DARPP-32 signalling by mGlu5 receptors seems to results from the ability of mGlu5 to enhance the A_{2A}-mediated cAMP formation in an extracellular signal-regulated kinase (ERK1/2)-dependent manner. Since A_{2A}, D_2 and mGlu5 receptors co-localize on the dendritic spines of the indirect striato-pallidal GABA pathway, the interactions between them may have a major role in the control of these striatal output neurons. In addition, presynaptic interactions between A_{2A} and mGlu5 receptors on striatal glutamatergic nerve terminals may also contribute to the described interaction by synergistic regulation of glutamate release (Rodrigues et al. 2005).

Further interaction was reported between A_{2A} and cannabinoid CB_1 receptors, which may also form heteromeric complexes and in this way A_{2A} receptor activation facilitates CB_1 receptor signalling in the striatum (Fig. 1.2) (Carriba et al. 2007; Ferré et al. 2010; Sebastiao and Ribeiro 2009a). Accordingly, A_{2A} receptor blockade was found to counteract the motor depressant effects produced by intrastriatal administration of CB_1 receptor agonists (Carriba et al. 2007; Ferré et al. 2010).

Recently, the existence of receptor heteromultimers has been proposed. Thus, using a sequential resonance energy transfer (SRET) and bimolecular fluorescence complementation plus BRET, evidence for A_{2A}-CB_1-D_2 and A_{2A}-D_2-mGlu5 receptor heteromers in transfected cells has been obtained (Cabello et al. 2009; Carriba et al. 2008). Such interactions at both pre- and postsynaptic levels play an important role in the control of neurotransmission and signalling in different brain structures, and provide selective targets for drug development in many disorders of the CNS. However, recently Pinna et al. (2014) showed that the interactions between A_{2A}, CB_1, and D_2 receptors may be disrupted by L-DOPA administration in hemiparkinsonian rats, which could question the relevance of receptor heteromultimers to the therapy of motor dysfunctions in Parkinson's disease (PD).

Physiological Functions of Adenosine and Adenosine A_{2A} Receptors

Adenosine receptors regulate several important physiological functions at both the central and peripheral levels. However, the specific influence of each receptor subtype on these functions may vary, due to differences in both receptor distribution in the various body organs and affinity for endogenous adenosine, as described above. Remarkably, adenosine A_{2A} receptors have recently attracted a great deal of attention as potential targets of drugs for different pathological conditions. Further in this chapter, we will summarize the best-characterized biological functions of adenosine A_{2A} receptors. The effects mediated by adenosine A_{2A} receptors that are more relevant to the pathological features of PD will be extensively discussed in other chapters of this book.

Central Effects of Adenosine A$_{2A}$ Receptors

A major branch of the research on adenosine A$_{2A}$ receptors focuses on the modulation of motor behaviour, based on the fact that these receptors are highly expressed in the striatum, a key nucleus of the BG circuitry (Fig. 1.3), where they are almost exclusively located on the GABAergic neurons of the striato-pallidal (or indirect)

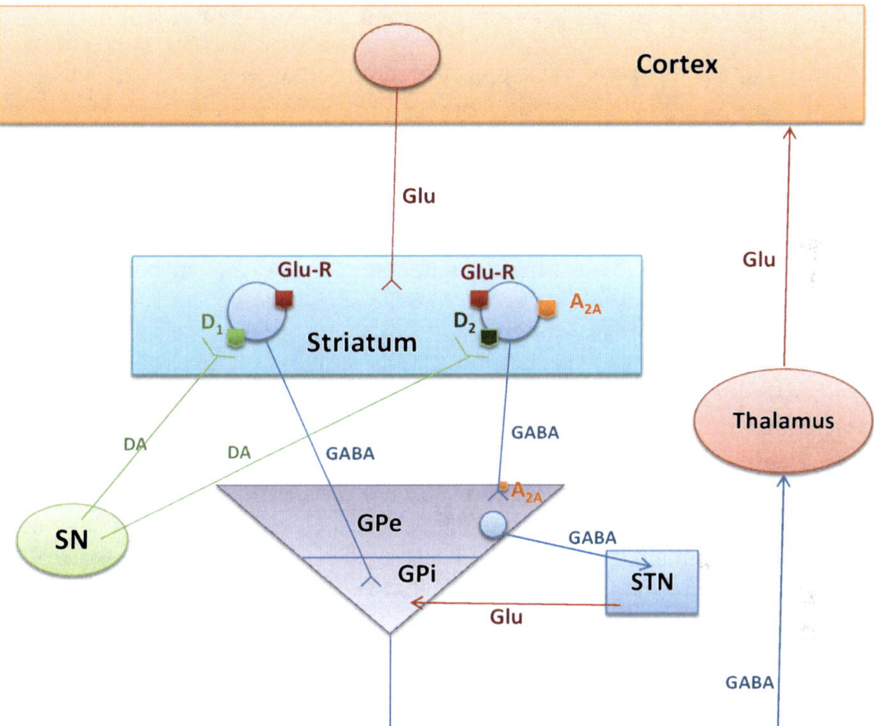

Fig. 1.3 Schematic representation of the basal ganglia circuitry and cellular localization of A$_{2A}$ receptors in the striatum. The picture shows the two major striatal GABAergic output pathways. A$_{2A}$ receptors are almost selectively localized to GABAergic neurons that express dopamine D$_2$ receptors and project to the GPe (striato-pallidal neurons). By contrast, GABAergic neurons that project directly to the GPi and express D$_1$ receptors (striato-nigral neurons), display scarce levels of A$_{2A}$ receptors. Dopamine depletion in the striatum that is characteristic of PD, results in a reduced stimulation of both dopamine D$_1$ and D$_2$ receptors, leading to a disinhibition of GABAergic striato-pallidal neurons, a reduced stimulation of GABAergic striato-nigral neurons, and to a reduction of the inhibitory control on the GPi. The disinhibition of GPe neurons amplifies the excitatory glutamate transmission of the STN. The resulting imbalance between the activity of the two main striatal efferent pathways, leads to a marked increase in the inhibitory output from the GPi, and to an excessive inhibition of Th-Cortex neurons, resulting in reduced movement performance. Blockade of A$_{2A}$ receptors in PD mitigates the overactivity of striato-pallidal and STN-GPi neurons, restoring some balance between the activity of the indirect and direct pathways. *DA* dopamine, *GABA* γ-aminobutyric acid, *Glu* glutamate, *GPe* globus pallidus *pars externa*; *GPi* globus pallidus *pars interna*, *PD* Parkinson's disease, *STN* subthalamic nucleus, *Th* thalamus

pathway that project to the GPe (Hettinger et al. 2001). At this level, adenosine A_{2A} receptors can interact in an opposing way with dopamine D_2 receptors (Svenningsson et al. 1999), in such a way that the stimulation of A_{2A} receptors depresses the D_2 receptors-dependent signalling (Díaz-Cabiale et al. 2001; Ferré et al. 1997). In line with these observations, and considering that dopamine D_2 receptors crucially regulate movement execution, the stimulation of adenosine A_{2A} receptors results in motor depressant effects, while the blockade of these receptors stimulates movement (Ferré et al. 1997; Hauber and Münkle 1997). Importantly, and notwithstanding their almost exclusive expression on the striato-pallidal neurons, A_{2A} receptors, by acting on BG loops, can also influence the effects mediated by dopamine D_1 receptors, located on GABAergic neurons belonging to the striato-nigral (or direct) pathway, which play a crucial role in motor control, as well (Ferré et al. 1997; Le Moine et al. 1997). Taken together, these findings justify the intensive studies of adenosine A_{2A} receptor antagonists as new drugs for the treatment of the motor deficits occurring in PD (see Chaps. 7, 9, 14).

Besides motor control, adenosine A_{2A} receptors regulate important non-motor central functions. Studies with caffeine, a non-selective A_1/A_{2A} adenosine receptor antagonist, have clearly demonstrated that the adenosine system is involved in the regulation of attention and motivation. Data obtained from both experimental animals and humans indicate that caffeine augments alertness and wakefulness, reduces the perception of fatigue, and delays the need for sleep (Fredholm et al. 1999; Snel and Lorist 2011). Interestingly, additional studies in experimental animals have demonstrated that adenosine A_{2A} receptors play a critical role in caffeine-induced arousal and increased alertness (Higgins et al. 2007; Lazarus et al. 2011). Moreover, caffeine improves the performance in memory tasks in both experimental animals and humans, and similar effects have been described for selective adenosine A_{2A} receptor antagonists in experimental animals (Kadowaki Horita et al. 2013; Prediger et al. 2005, see also Chap. 10), although others failed to observe beneficial effects of A_{2A} receptor antagonists on memory (O'Neill and Brown 2007). Furthermore, A_{2A} receptors play a crucial role in reward, motivation, and perception of stimuli, and both caffeine and selective A_{2A} receptor antagonists facilitate these phenomena (Fredholm et al. 1999; Higgins et al. 2007; Mott et al. 2009). In line with these data, other studies have demonstrated that adenosine A_{2A} receptors may influence the effects of psychostimulant drugs of abuse, such as cocaine, methamphetamine and nicotine (Cauli et al. 2003; Justinova et al. 2009; Kobayashi et al. 2010; Simola et al. 2006; Wells et al. 2012).

Adenosine A_{2A} receptors have also been implicated in depression, as suggested by the beneficial effects of either genetic deletion or pharmacological blockade of these receptors in animal models of this pathology (El Yacoubi et al. 2001; Yamada et al. 2013). Another crucial function which appears to be regulated by adenosine A_{2A} receptors is epileptogenesis, as indicated by the experimental and clinical evidences showing that caffeine and theophylline, another non-selective adenosine receptor antagonist, may induce and/or aggravate seizures (Boison 2011). However, the precise role of adenosine A_{2A} receptors in epileptogenesis is still debated, as studies in experimental animals have demonstrated that these receptors can have either facilitatory or inhibitory effects on seizures, depending on the experimental model utilized (Ates et al. 2004; Souza et al. 2013; Tchekalarova et al. 2010). In addition, adenosine A$_{2A}$ receptors can modulate nociception, and either blockade or genetic deletion of these receptors has been shown to elevate the pain threshold in experimental models (Hussey et al. 2007; Ledent et al. 1997), likely by an action on central nociceptive pathways. It has to be mentioned that adenosine A$_{2A}$ receptors can also be found in peripheral nerves, where their stimulation decreases the pain threshold, likely by facilitating the transmission at the level of the primary afferent pathways (Khasar et al. 1995).

The regulation of neuron homeostasis and survival is another major function of adenosine A$_{2A}$ receptors in the CNS. A number of studies in experimental models of neurodegenerative diseases, such as Alzheimer's disease, Huntington's disease (HD), and PD (Espinosa et al. 2013; Popoli et al. 2008; Schwarzschild et al. 2003), cerebral ischemia (Chen and Pedata 2008), and spinal cord trauma (Cassada et al. 2002) have consistently demonstrated that genetic and/or pharmacological manipulation of A$_{2A}$ receptors may counteract the neurodegeneration and neuroinflammation associated with these conditions. However, it has to be mentioned that A$_{2A}$ receptors may differently influence these processes depending on the specific experimental model used. Thus, A$_{2A}$ receptor blockade has consistently been shown to attenuate neuronal death and inflammatory damage in models of cerebral ischemia and neurodegenerative diseases. Conversely, the stimulation, rather than blockade, of A$_{2A}$ receptors affords neuroprotection in experimental models of spinal trauma. Furthermore, evidences also exist suggesting that stimulation of A$_{2A}$ receptors may protect neurons in models of HD (Popoli et al. 2008). It has been hypothesized that adenosine A$_{2A}$ receptors may modulate neuronal homeostasis by attenuating either glutamate-induced excitotoxicity or glial activation (or both), two mechanisms that are known to play a crucial role in neurodegenerative and neuroinflammatory phenomena (Halliday and Stevens 2011; Milanese et al. 2009).

Peripheral Effects of Adenosine A$_{2A}$ Receptors

In addition to the protective effects elicited in the CNS, studies in experimental animals have indicated that adenosine A$_{2A}$ receptors can modulate inflammation and tissue damage in different peripheral organs, including the heart, kidney, lung, and intestine, as observed in several *in vitro* and *in vivo* models of inflammatory diseases. The modulation of inflammatory responses by A$_{2A}$ receptors can be explained considering that many cells of the immune system, such as basophils, lymphocytes, mast cells, monocytes, and neutrophils express A$_{2A}$ receptors, and that these receptors profoundly influence the function of immune cells (Haskó et al. 2008; Hershfield 2005; Revan et al. 1996). Among the functions regulated by A$_{2A}$ receptors are the induction of pro-inflammatory mediators (Pouliot et al. 2002; Sullivan et al. 2001), activation of T cells (Sevigny et al. 2007), mast cell migration (Duffy et al. 2007), and monocyte secretion (Link et al. 2000). Anti-inflammatory effects are usually observed following the stimulation of A_{2A} receptors, although these receptors have complex effects on inflammation (Antonioli et al. 2008; Trevethick et al. 2008), and data also exist showing that blockade of A_{2A} receptors may attenuate inflammation in peripheral organs (Katebi et al. 2008).

Besides their effects on inflammation, A_{2A} receptors can modulate other important functions of peripheral organs. Adenosine A_{2A} receptors regulate several aspects of cardiovascular physiology, although some of these effects are ascribable to either the cross-talk between A_{2A} and other adenosine receptor subtypes, or to extracardiac A_{2A} receptors (Headrick et al. 2013). Stimulation of A_{2A} receptors has been reported to enhance the contractility of cardiomyocytes, to elicit a positive inotropic action (Dobson and Fenton 1997), and to promote dilation of different vessels, including the coronary arteries (Belardinelli et al. 1998; Rump et al. 1999; Sato et al. 2005). Remarkably, the A_{2A} receptor agonist regadenoson is currently the most commonly used vasodilator in the U.S.A. (Ghimire et al. 2013). Adenosine A_{2A} receptors have also been suggested to participate in angiogenesis by promoting the generation of vascular endothelial growth factor (VEGF) (Adair et al. 2005), in atherosclerosis, by inhibiting the formation of foam-cells (Bingham et al. 2010), and in cardioprotection during ischemia, owing to their ability to modulate cell infiltration and inflammatory responses (Glover et al. 2005). Adenosine A_{2A} receptors are also expressed at the level of the intestine, where they may influence some aspects of enteric function, such as contractility and secretion, although inconsistent results have been reported (Fornai et al. 2009; Storr et al. 2002; Tomaru et al. 1995). However, the best-characterized effect of A_{2A} receptors at this level is related to the modulation of intestinal inflammation, and a marked up-regulation of high-affinity A_{2A} receptors has been observed in experimental colitis (Antonioli et al. 2006, 2008). Importantly, independent studies have shown that the stimulation of A_{2A} receptors attenuates inflammatory responses in the colon (Antonioli et al. 2010; Odashima et al. 2005; Rahimian et al. 2010), although it has to be acknowledged that others failed to observe this effect (Selmeczy et al. 2007). Adenosine A_{2A} receptors have also been shown to modulate inflammation and tissue damage in the lung, as demonstrated by several preclinical studies (Eckle et al. 2009; Trevethick et al. 2008; Wilson et al. 2009). This effect may be particularly relevant to some diseases, like chronic obstructive pulmonary disease (COPD) and asthma, both of which involve major inflammatory mechanisms, as well as to acute lung trauma. The efficacy of A_{2A} receptor agonists has indeed been demonstrated in preclinical models of these diseases (Bonneau et al. 2006; Fozard et al. 2002; LaPar et al. 2011), and these drugs are currently under clinical evaluation, though with inconsistent results (Salgado Garcia et al. 2014; Trevethick et al. 2008) Adenosine A_{2A} receptors also regulate kidney physiology, by modulating the dilation of efferent arterioles, renal blood flow, and glomerular filtration rate (Al Mashhadi et al. 2009; Carlström et al. 2011; Levens et al. 1991), as well as by influencing renal inflammation (Awad et al. 2006; Garcia et al. 2011; Okusa et al. 1999). Finally, adenosine A_{2A} receptors have been suggested to participate in other physiopathological functions, such as ocular hemodynamics and protection from ischemic retinal damage (Zhong et al. 2013), wound healing (Katebi et al. 2008; Squadrito et al. 2014), inflammation in experimental models of arthritis (Mazzon et al. 2011), and tumour growth (Kalhan et al. 2012; Montinaro et al. 2013).

As described above, adenosine A$_{2A}$ receptors regulate several physiological functions at both the central and peripheral level. Therefore, considering clinical prospects for chronic use of drugs binding to A$_{2A}$ receptors, these effects should receive a great deal of scrutiny. However, it should be mentioned that instances of side effects, in particular from the cardiovascular system, have been relatively frequently observed with A$_{2A}$ receptor agonists. On the other hand, A$_{2A}$ receptor antagonists, that show the most promising antiparkinsonian potential among adenosinergic ligands, appear generally well-tolerated, as confirmed by clinical trials (LeWitt et al. 2008; Mizuno et al. 2013).

References

Abbracchio MP, Burnstock G (1998) Purinergic signalling: pathophysiological roles. Jpn J Pharmacol 78:113–145

Abbracchio MP, Burnstock G, Verkhratsky A, et al (2008) Purinergic signaling in the nervous system: an overview. Trends Neurosci 32:19–29

Adair TH, Cotten R, Gu JW et al (2005) Adenosine infusion increases plasma levels of VEGF in humans. BMC Physiol 5:10

Alexander SP, Reddington M (1989) The cellular localization of adenosine receptors in rat neostriatum. Neuroscience 28:645–651

Al-Mashhadi RH, Skøtt O, Vanhoutte PM et al (2009) Activation of A(2) adenosine receptors dilates cortical efferent arterioles in mouse. Kidney Int 75:793–799

Anderson CM, Xiong W, Young JD et al (1996) Demonstration of the existance of mRNAs encoding N1/cif and N2/cit sodium/nucleoside cotransporters in rat brain. Mol Brain Res 42:358–361

Antonioli L, Fornai M, Colucci R et al (2006) A2a receptors mediate inhibitory effects of adenosine on colonic motility in the presence of experimental colitis. Inflamm Bowel Dis 12:117–122

Antonioli L, Fornai M, Colucci R et al (2008) Regulation of enteric functions by adenosine: pathophysiological and pharmacological implications. Pharmacol Ther 120:233–253

Antonioli L, Fornai M, Colucci R et al (2010) The blockade of adenosine deaminase ameliorates chronic experimental colitis through the recruitment of adenosine A2A and A3 receptors. J Pharmacol Exp Ther 335:434–442

Ates N, Sahin D, Ilbay G (2004) Theophylline, a methylxanthine derivative, suppresses absence epileptic seizures in WAG/Rij rats. Epilepsy Behav 5:645–648

Augood SJ, Emson PC (1994) Adenosine A2a receptor mRNA is expressed by enkephalin cells but not somatostatin cells in rat striatum: a co-expression study. Mol Brain Res 22:204–210

Awad AS, Huang L, Ye H et al (2006) Adenosine A2A receptor activation attenuates inflammation and injury in diabetic nephropathy. Am J Physiol Renal Physiol 290(4):F828–837

Baldwin SA, Beal PR, Yao SY et al (2004) The equilibrative nucleoside transporter family, SLC29. Pflugers Arch 447:735–743

Belardinelli L, Shryock JC, Snowdy S et al (1998) The A2A adenosine receptor mediates coronary vasodilation. J Pharmacol Exp Ther 284:1066–1073

Biber K, Klotz KN, Berger M et al (1997) Adenosine A1 receptor-mediated activation of phospholipase C in cultured astrocytes depends on the level of receptor expression. J Neurosci 17:4956–4964

Bingham TC, Fisher EA, Parathath S et al (2010) A2A adenosine receptor stimulation decreases foam cell formation by enhancing ABCA1-dependent cholesterol efflux. J Leukoc Biol 87:683–690

Boison D (2011) Methylxanthines, seizures, and excitotoxicity. Handb Exp Pharmacol 200:251–266

Boison D (2013) Adenosine kinase: exploitation for therapeutic gain. Pharmacol Rev 65:906–943

Boison D, Chen JF, Fredholm BB (2010) Adenosine signalling and function in glial cells. Cell Death Differ 17:1071–1082

Bonan CD (2012) Ectonucleotidases and nucleotide/nucleoside transporters as pharmacological targets for neurological disorders. CNS Neurol Dis–Drug Target 11:739–750

Bonneau O, Wyss D, Ferretti S et al (2006) Effect of adenosine A2A receptor activation in murine models of respiratory disorders. Am J Physiol Lung Cell Mol Physiol 290:L1036–L1043

Bours MJ, Swennen EL, Di Virgilio F et al (2006) Adenosine 5'-triphosphate and adenosine as endogenous signaling molecules in immunity and inflammation. Pharmacol Ther 112:358–404

Bowser DN, Khakh BS (2007) Two forms of single-vesicle astrocyte exocytosis imaged with total internal reflection fluorescence microscopy. Proc Natl Acad Sci U S A 104:4212–4217

Brand A, Vissiennon Z, Eschke D et al (2001) Adenosine A(1) and A(3) receptors mediate inhibition of synaptic transmission in rat cortical neurons. Neuropharmacology 40:85–95

Brooks DJ, Doder M, Osman S et al (2008) Positron emission tomography analysis of [11C]KW-6002 binding to human and rat adenosine A2A receptors in the brain. Synapse 62:671–681

Brown SJ, James S, Reddington M et al (1990) Both A1 and A2a purine receptors regulate striatal acetylcholine release. J Neurochem 55:31–38

Burnstock G (1972) Purinergic nerves. Pharmacol Rev 24:509–581

Burnstock G (1976) Purinergic receptors. J Theor Biol 62:491–503

Burnstock G (2013) Introduction to purinergic signalling in the brain. Adv Exp Med Biol 986:1–12

Cabello N, Gandia J, Bertarelli DC et al (2009) Metabotropic glutamate type 5, dopamine D2 and adenosine A2A receptors form higher- order oligomers in living cells. J Neurochem 109:1497–1507

Canals M, Marcellino D, Fanelli F et al (2003) Adenosine A2A-dopamine D2 receptor-receptor heteromerization: qualitative and quantitative assessment by fluorescence and bioluminescence energy transfer. J Biol Chem 278:46741–46749

Carlström M, Wilcox CS, Welch WJ (2011) Adenosine A2A receptor activation attenuates tubuloglomerular feedback responses by stimulation of endothelial nitric oxide synthase. Am J Physiol Renal Physiol 300:F457–464

Carriba P, Ortiz O, Patkar K et al (2007) Striatal adenosine A2A and cannabinoid CB1 receptors form functional heteromeric complexes that mediate the motor effects of cannabinoids. Neuropsychopharmacology 32:2249–2259

Carriba P, Navarro G, Ciruela F et al (2008) Detection of heteromerization of more than two proteins by sequential BRET-FRET. Nat Methods 5:727–733

Cassada DC, Tribble CG, Young JS et al (2002) Adenosine A2A analogue improves neurologic outcome after spinal cord trauma in the rabbit. J Trauma 53:225–229

Cauli O, Pinna A, Valentini V et al (2003) Subchronic caffeine exposure induces sensitization to caffeine and cross-sensitization to amphetamine ipsilateral turning behavior independent from dopamine release. Neuropsychopharmacology 28:1752–1759

Chen JF, Pedata F (2008) Modulation of ischemic brain injury and neuroinflammation by adenosine A2A receptors. Curr Pharm Des 14:1490–1499

Chen JF, Lee CF, Chern Y (2014) Adenosine receptor neurobiology: overview. Int Rev Neurobiol 119:1–49

Ciruela F, Saura C, Canela EI et al (1996) Adenosine deaminase affects ligand-induced signalling by interacting with cell surface adenosine receptors. FEBS Lett 380:219–223

Ciruela F, Casado V, Rodrigues RJ et al (2006) Presynaptic control of striatal glutamatergic neurotransmission by adenosine A1-A2A receptor heteromers. J Neurosci 26:2080–2087

Cunha RA (2001) Adenosine as a neuromodulator and as a homeostatic regulator in the nervous system: Different roles, different sources and different receptors. Neurochem Int 38:107–125

Cunha RA, Sebastiao AM, Ribeiro JA (1992) Ecto-5'-nucleotidase is associated with cholinergic nerve terminals in the hippocampus but not in the cerebral cortex of the rat. J Neurochem 59:657–666

Cunha RA, Johansson B, van der Ploeg I et al (1994) Evidence for functionally important adenosine A2a receptors in the rat hippocampus. Brain Res 649:208–216

Daré E, Schulte G, Karovic O et al (2007) Modulation of glial cell functions by adenosine receptors. Physiol Behav 92:15–20

Desrosiers MD, Cembrola KM, Fakir MJ et al (2007) Adenosine deamination sustains dendritic cell activation in inflammation. J Immunol 179:1884–1892

Díaz-Cabiale Z, Hurd Y, Guidolin D et al (2001) Adenosine A2A agonist CGS 21680 decreases the affinity of dopamine D2 receptors for dopamine in human striatum. Neuroreport 12:1831–1834

Dickenson JM, Blank JL, Hill SJ (1998) Human adenosine A1 receptor and P2Y2-purinoceptor-mediated activation of the mitogen-activated protein kinase cascade in transfected CHO cells. Br J Pharmacol 124:1491–1499

Dixon AK, Gubitz AK, Sirinathsinghji DJ et al. (1996) Tissue distribution of adenosine receptor mRNAs in the rat. Br J Pharmacol 118:1461–1468

Dobson JG Jr, Fenton RA (1997) Adenosine A2 receptor function in rat ventricular myocytes. Cardiovasc Res 34:337–347

Dos Santos-Rodrigues A, Grañé-Boladeras N, Bicket A et al (2014) Nucleoside transporters in the purinome. Neurochem Int 73:229–237

Duffy SM, Cruse G, Brightling CE et al (2007) Adenosine closes the K+ channel KCa3.1 in human lung mast cells and inhibits their migration via the adenosine A2A receptor. Eur J Immunol 37:1653–1662

Dunwiddie TV, Masino SA (2001) The role and regulation of adenosine in the central nervous system. Ann Rev Neurosci 24:31–55

Dunwiddie TV, Diao L, Proctor WR (1997) Adenine nucleotides undergo rapid, quantitative conversion to adenosine in the extracellular space in rat hippocampus. J Neurosci 17:7673–7682

El Yacoubi ME, Ledent C, Parmentier M et al (2001) Adenosine A2A receptor antagonists are potential antidepressants: evidence based on pharmacology and A2A receptor knockout mice. Br J Pharmacol 134:68–77

Eckle T, Koeppen M, Eltzschig HK (2009) Role of extracellular adenosine in acute lung injury. Physiology (Bethesda) 24:298–306

Espinosa J, Rocha A, Nunes F et al (2013) Caffeine consumption prevents memory impairment, neuronal damage, and adenosine A2A receptors upregulation in the hippocampus of a rat model of sporadic dementia. J Alzheimers Dis 34:509–518

Fastbom J, Pazos A, Palacios JM (1987) The distribution of adenosine A1 receptors and nucleotidase in the brain of some commonly used experimental animals. Neuroscience 22:813–826

Feoktistov I, Biaggioni I (1997) Adenosine A2B receptors. Pharmacol Rev 49:381–402

Ferré S, O'Connor WT, Svenningsson P et al (1996) Dopamine D1 receptor-mediated facilitation of GABAergic neurotransmission in the rat strioentopenduncular pathway and its modulation by adenosine A1 receptor-mediated mechanisms. Eur J Neurosci 8:1545–1553

Ferré S, Fredholm BB, Morelli M et al (1997) Adenosine-dopamine receptor-receptor interactions as an integrative mechanism in the basal ganglia. Trends Neurosci 20:482–487

Ferré S, Karcz-Kubicha M, Hope BT et al (2002) Synergistic interaction between adenosine A2A and glutamate mGlu5 receptors: implications for striatal neuronal function. Proc Natl Acad Sci U S A 99:11940–11945

Ferré S, Lluis C, Justinova Z et al (2010) Adenosine-cannabinoid receptor interactions. Implications for striatal function. Br J Pharmacol 160:443–453

Ferré S, Quiroz C, Orru M et al (2011) Adenosine A(2A) receptors and A(2A) receptor heteromers as key players in striatal function. Front Neuroanat 5:1–8

Fink JS, Weaver DR, Rivkees SA et al (1992) Molecular cloning of the rat A2 adenosine receptor: selective co-expression with D2 dopamine receptors in rat striatum. Mol Brain Res 14:186–195

Fornai M, Antonioli L, Colucci R et al (2009) A1 and A2a receptors mediate inhibitory effects of adenosine on the motor activity of human colon. Neurogastroenterol Motil 21:451–466

Fozard JR, Ellis KM, Villela Dantas MF et al (2002) Effects of CGS 21680, a selective adenosine A2A receptor agonist, on allergic airways inflammation in the rat. Eur J Pharmacol 438:183–188

Fredholm BB, Bättig K, Holmén J et al (1999) Actions of caffeine in the brain with special reference to factors that contribute to its widespread use. Pharmacol Rev 51:83–133

Fredholm BB, Arslan G, Halldner L et al (2000) Structure and function of adenosine receptors and their genes. Naunyn Schmiedebergs Arch Pharmacol 362:364–374

Fredholm BB, Jacobson KA, Klotz KN et al (2001) International Union of Pharmacology. XXV. Nomenclature and classification of adenosine receptors. Pharmacol Rev 53:527–552

Fredholm BB, Chen JF, Cunha RA et al (2005) Adenosine and brain function. Int Rev Neurobiol 63:191–270

Fredholm BB, Chern Y, Franco R et al (2007) Aspects of the general biology of adenosine A2A signaling. Prog Neurobiol 83:263–276

Fredholm BB, IJzerman AP, Jacobson KA et al (2011) International union of basic and clinical pharmacology. LXXXI. Nomenclature and classification of adenosine receptors—an update. Pharmacol Rev 63:1–34

Fuxe K, Agnati LF, Jacobsen K et al (2003) Receptor heteromerization in adenosine A2A receptor signaling: relevance for striatal function and Parkinson's disease. Neurology 61:S19–S23

Garcia GE, Truong LD, Chen JF et al (2011) Adenosine A(2A) receptor activation prevents progressive kidney fibrosis in a model of immune-associated chronic inflammation. Kidney Int 80:378–388

Gebicke-Haerter PJ, Christoffel F, Timmer J et al (1996) Both adenosine A1- and A2-receptors are required to stimulate microglial proliferation. Neurochem Int 29:37–42

Geiger JD, Fyda DM (1991) Adenosine transport in nervous system tissues. In: Stone TW (ed) Adenosine in the nervous system. Academic, San Diego, pp 1–23

Gharib A, Sarda N, Chabannes B et al (1982) The regional concentrations of S-adenosyl-L-methionine, S-adenosyl-L-homocysteine, and adenosine in rat brain. J Neurochem 38:810–815

Ghimire G, Hage FG, Heo J et al (2013) Regadenoson: a focused update. J Nucl Cardiol 20:284–288

Glover DK, Riou LM, Ruiz M et al (2005) Reduction of infarct size and postischemic inflammation from ATL-146e, a highly selective adenosine A2A receptor agonist, in reperfused canine myocardium. Am J Physiol Heart Circ Physiol 288:H1851–H1858

Groenewegen HJ (2007) The ventral striatum as an interface between the limbic and motor systems. CNS Spectr 12:887–892

Halliday GM, Stevens CH (2011) Glia: initiators and progressors of pathology in Parkinson's disease. Mov Disord 26:6–17

Hammarberg C, Schulte G, Fredholm BB (2003) Evidence for functional adenosine A3 receptors in microglia cells. J Neurochem 86:1051–1054

Haskó G, Linden J, Cronstein B et al (2008) Adenosine receptors: therapeutic aspects for inflammatory and immune diseases. Nat Rev Drug Discov 7:759–770

Hauber W, Münkle M (1997) Motor depressant effects mediated by dopamine D2 and adenosine A2A receptors in the nucleus accumbens and the caudate-putamen. Eur J Pharmacol 323:127–131

Headrick JP, Ashton KJ, Rose'meyer RB et al (2013) Cardiovascular adenosine receptors: expression, actions and interactions. Pharmacol Ther 140:92–111

Hershfield MS (2005) New insights into adenosine-receptor-mediated immunosuppression and the role of adenosine in causing the immunodeficiency associated with adenosine deaminase deficiency. Eur J Immunol 35:25–30

Hettinger BD, Lee A, Linden J et al (2001) Ultrastructural localization of adenosine A2A receptors suggests multiple cellular sites for modulation of GABAergic neurons in rat striatum. J Comp Neurol 431:331–346

Higgins GA, Grzelak ME, Pond AJ et al (2007) The effect of caffeine to increase reaction time in the rat during a test of attention is mediated through antagonism of adenosine A2A receptors. Behav Brain Res 185:32–42

Hunsucker SA, Mitchell BS, Spychala J (2005) The 5'-nucleotidases as regulators of nucleotide and drug metabolism. Pharmacol Ther 107:1–30

Hussey MJ, Clarke GD, Ledent C et al (2007) Reduced response to the formalin test and lowered spinal NMDA glutamate receptor binding in adenosine A2A receptor knockout mice. Pain 129:287–294

James S, Richardson PJ (1993) Production of adenosine from extracellular ATP at the striatal cholinergic synapse. J Neurochem 60:219–227

Jarvis MF, Williams M (1989) Direct autoradiographic localization of adenosine A2 receptors in the rat brain using the A2-selective agonist, (3H)CGS 21680. Eur J Pharmacol 8:243–246

Joel D, Weiner I (2000) The connections of the dopaminergic system with the striatum in rats and primates: an analysis with respect to the functional and compartmental organization of the striatum. Neuroscience 96:451–474

Justinova Z, Ferré S, Barnes C et al (2009) Effects of chronic caffeine exposure on adenosinergic modulation of the discriminative-stimulus effects of nicotine, methamphetamine, and cocaine in rats. Psychopharmacology 203:355–367

Kadowaki Horita T, Kobayashi M, Mori A et al (2013) Effects of the adenosine A2A antagonist istradefylline on cognitive performance in rats with a 6-OHDA lesion in prefrontal cortex. Psychopharmacology 230:345–352

Kalhan A, Gharibi B, Vazquez M et al (2012) Adenosine A2A and A2B receptor expression in neuroendocrine tumours: potential targets for therapy. Purinergic Signal 8:265–274

Kang J, Kang N, Lovatt D et al (2008) Connexin 43 hemichannels are permeable to ATP. J Neurosci 28:4702–4711

Katebi M, Fernandez P, Chan ES et al (2008) Adenosine A2A receptor blockade or deletion diminishes fibrocyte accumulation in the skin in a murine model of scleroderma, bleomycin-induced fibrosis. Inflammation 31:299–303

Kawamura M Jr, Ruskin DN, Masino SA (2010) Metabolic autocrine regulation of neurons involves cooperation among pannexin hemichannels, adenosine receptors, and KATP channels. J Neurosci 30:3886–3895

Khasar SG, Wang JF, Taiwo YO et al (1995) Mu-opioid agonist enhancement of prostaglandin-inducedhyperalgesia in the rat: a G-protein beta gamma subunit-mediated effect? Neuroscience 67:189–195

King AE, Ackley MA, Cass CE et al (2006) Nucleoside transporters: from scavengers to novel therapeutic targets. Trends Pharmacol Sci 27:416–425

Kobayashi H, Ujike H, Iwata N et al (2010) The adenosine A2A receptor is associated with methamphetamine dependence/psychosis in the Japanese population. Behav Brain Funct 6:50

Kovacs Z, Dobolyi A, Kekesi A et al (2013) 5-nucleotidases, nucleosides and their distribution in the brain: pathological and therapeutic implications. Curr Med Chem 20:4217–4240

Kurokawa M, Kirk IP, Kirkpatrick KA et al (1994) Inhibition by KF17837 of adenosine A2A receptor mediated modulation of striatal GABA and Ach release. Br J Pharmacol 113:43–48

Kurokawa M, Koga K, Kase H (1996) Adenosine A2a receptor-mediated modulation of striatal acetylcholine release in vivo. J Neurochem 66:1882–1888

LaPar DJ, Laubach VE, Emaminia A et al (2011) Pretreatment strategy with adenosine A2A receptor agonist attenuates reperfusion injury in a preclinical porcine lung transplantation model. J Thorac Cardiovasc Surg 142:887–894

Latini S, Pedata F (2001) Adenosine in the central nervous system: release mechanisms and extracellular concentrations. J Neurochem 79:463–484

Latini S, Bordoni F, Pedata F et al (1999) Extracellular adenosine concentrations during *in vitro* ischaemia in rat hippocampal slices. Br J Pharmacol 127:729–739

Lazarus M, Shen HY, Cherasse Y et al (2011) Arousal effect of caffeine depends on adenosine A2A receptors in the shell of the nucleus accumbens. J Neurosci 31:10067–10075

Le Moine C L, Svenningsson P, Fredholm BB et al (1997) Dopamine-adenosine interactions in the striatum and the globus pallidus: inhibition of striatopallidal neurons through either D2 or A2A receptors enhances D1 receptor-mediated effects on c-fos expression. J Neurosci 17:8038–8048

Ledent C, Vaugeois JM, Schiffmann SN et al (1997) Aggressiveness, hypoalgesia and high blood pressure in mice lacking the adenosine A2a receptor. Nature 388:674–678

Levens N, Beil M, Jarvis M (1991) Renal actions of a new adenosine agonist, CGS 21680A selective for the A2 receptor. J Pharmacol Exp Ther 257:1005–1012

LeWitt PA, Guttman M, Tetrud JW et al (2008) Adenosine A2A receptor antagonist istradefylline (KW-6002) reduces "off" time in Parkinson's disease: a double-blind, randomized, multicenter clinical trial (6002-US-005). Ann Neurol 63:295–302

Link AA, Kino T, Worth JA et al (2000) Ligand-activation of the adenosine A2a receptors inhibits IL-12 production by human monocytes. J Immunol 164:436–442

Matos M, Augusto E, Santos-Rodrigues AD et al (2012) Adenosine A2A receptors modulate glutamate uptake in cultured astrocytes and gliosomes. Glia 60:702–716

Matos M, Augusto E, Agostinho P et al (2013) Antagonistic interaction between adenosine A2A receptors and Na+/K+-ATPase-a2 controlling glutamate uptake in astrocytes. J Neurosci 33:18492–18502

Mazzon E, Esposito E, Impellizzeri D et al (2011) CGS 21680, an agonist of the adenosine (A2A) receptor, reduces progression of murine type II collagen-induced arthritis. J Rheumatol 38:2119–2129

Milanese M, Bonifacino T, Zappettini S et al (2009) Glutamate release from astrocytic gliosomes under physiological and pathological conditions. Int Rev Neurobiol 85:295–318

Mizuno Y, Kondo T, Japanese Istradefylline Study Group (2013) Adenosine A2A receptor antagonist istradefylline reduces daily OFF time in Parkinson's disease. Mov Disord 28:1138–1141

Montinaro A, Iannone R, Pinto A et al (2013) Adenosine receptors as potential targets in melanoma. Pharmacol Res 76:34–40

Morelli M, Pinna A, Wardas J et al (1995) Adenosine A2 receptors stimulate c-fos expression in striatal neurons of 6-hydroxydopamine-lesioned rats. Neuroscience 67:49–55

Morelli M, Carta AR, Kachroo A et al (2010) Pathophysiological roles for purines: adenosine, caffeine and urate. Prog Brain Res 183:183–208

Mori A, Shindou T, Ichimura M et al (1996) The role of adenosine A2a receptors in regulating GABAergic synaptic transmission in striatal medium spiny neurons. J Neurosci 16:605–611

Mott AM, Nunes EJ, Collins LE et al (2009) The adenosine A2A antagonist MSX-3 reverses the effects of the dopamine antagonist haloperidol on effort-related decision making in a T-maze cost/benefit procedure. Psychopharmacology 204:103–112

Nishi A, Liu F, Matsuyama S et al (2003) Metabotropic mGlu5 receptors regulate adenosine A2A receptor signaling. Proc Natl Acad Sci U S A 100:1322–1327

Ochiishi T, Chen L, Yukawa A et al (1999) Cellular localization of adenosine A1 receptors in rat forebrain: immunohistochemical analysis using adenosine A1 receptor-specific monoclonal antibody. J Comp Neurol 411:301–316

Odashima M, Bamias G, Rivera-Nieves J et al (2005) Activation of A2A adenosine receptor attenuates intestinal inflammation in animal models of inflammatory bowel disease. Gastroenterology 129:26–33

Okusa MD, Linden J, Macdonald T et al (1999) Selective A2A adenosine receptor activation reduces ischemia-reperfusion injury in rat kidney. Am J Physiol 277:F404–412

O'Neill M, Brown VJ (2007) The effect of striatal dopamine depletion and the adenosine A2A antagonist KW-6002 on reversal learning in rats. Neurobiol Learn Mem 88:75–81

Othman T, Yan H, Rivkees SA (2003) Oligodendrocytes express functional A1 adenosine receptors that stimulate cellular migration. Glia 44:166–172

Pankratov Y, Lalo U, Verkhratsky A et al (2006) Vesicular release of ATP at central synapses. Pflugers Arch 452:589–597

Pankratov Y, Lalo U, Verkhratsky A et al (2007) Quantal release of ATP in mouse cortex. J Gen Physiol 129:257–265

Parkinson FE, Damaraju VL, Graham K et al (2011) Molecular biology of nucleoside transporters and their distributions and functions in the brain. Curr Top Med Chem 11:948–972

Pascual O, Casper KB, Kubera C et al (2005) Astrocytic purinergic signaling coordinates synaptic networks. Science 310:113–116

Pierce KD, Furlong TJ, Selbie LA et al (1992) Molecular cloning and expression of an adenosine A2b receptor from human brain. Biochem Biophys Res Commun 187:86–93

Pinna A, Bonaventura J, Farre D et al (2014) L-DOPA disrupts adenosine A(2A)-cannabinoid CB(1)-dopamine D(2) receptor heteromer cross-talk in the striatum of hemiparkinsonian rats: biochemical and behavioral studies. Exp Neurol 253:180–191

Popoli P, Pezzola A, Torvinen M et al (2001) The selective mGlu(5) receptor agonist CHPG inhibits quinpirole-induced turning in 6-hydroxydopamine-lesioned rats and modulates the binding characteristics of dopamine D(2) receptors in the rat striatum: interactions with adenosine A(2a) receptors. Neuropsychopharmacology 25:505–513

Popoli P, Blum D, Domenici MR et al (2008) A critical evaluation of adenosine A2A receptors as potentially "druggable" targets in Huntington's disease. Curr Pharm Des 14:1500–1511

Pouliot M, Fiset ME, Massé M et al (2002) Adenosine up-regulates cyclooxygenase-2 in human granulocytes: impact on the balance of eicosanoid generation. J Immunol 169:5279–5286

Prediger RD, Fernandes D, Takahashi RN (2005) Blockade of adenosine A2A receptors reverses short-term social memory impairments in spontaneously hypertensive rats. Behav Brain Res 159:197–205

Quiroz C, Lujan R, Uchigashima M et al (2009) Key modulatory role of presynaptic adenosine A2A receptors in cortical neurotransmission to the striatal direct pathway. Sci World J 9:1321–1344

Rahimian R, Fakhfouri G, Daneshmand A et al (2010) Adenosine A2A receptors and uric acid mediate protective effects of inosine against TNBS-induced colitis in rats. Eur J Pharmacol 649:376–381

Ralevic V, Burnstock G (1998) Receptors for purines and pyrimidines. Pharmacol Rev 50:413–492

Rebola N, Pinheiro PC, Oliveira CR et al (2003) Subcellular localization of adenosine A(1) receptors in nerve terminals and synapses of the rat hippocampus. Brain Res 987:49–58

Rebola N, Canas PM, Oliveira CR et al (2005) Different synaptic and subsynaptic localization of adenosine A2A receptors in the hippocampus and striatum of the rat. Neuroscience 132:893–903

Reddington M, Pusch R (1983) Adenosine metabolism in a rat hippocampal slice preparation: incorporation into S-adenosylhomocysteine. J Neurochem 40:285–290

Revan S, Montesinos MC, Naime D et al (1996) Adenosine A2 receptor occupancy regulates stimulated neutrophil function via activation of a serine/threonine protein phosphatase. J Biol Chem 271:17114–17118

Ribeiro JA, Sebastiao AM, de Mendonça A (2002) Adenosine receptors in the nervous system: pathophysiological implications. Prog Neurobiol 68:377–392

Rivkees SA, Price SL, Zhou FC (1995) Immunohistochemical detection of A1 adenosine receptors in rat brain with emphasis on localization in the hippocampal formation, cerebral cortex, cerebellum, and basal ganglia. Brain Res 677:193–203

Rivkees SA, Thevananther S, Hao H (2000) Are A3 adenosine receptors expressed in the brain? Neuroreport 11:1025–1030

Rodrigues RJ, Alfaro TM, Rebola N et al (2005) Colocalization and functional interaction between adenosine A2A and metabotropic group 5 receptors in glutamatergic nerve terminals of the striatum. J Neurochem 92:433–441

Rosin DL, Robeva A, Woodard RL et al (1998) Immunohistochemical localization of adenosine A2A receptors in the rat central nervous system. J Comp Neurol 401:163–186

Rosin DL, Hettinger BD, Lee A et al (2003) Anatomy of adenosine A2A receptors in brain: morphological substrates for integration of striatal function. Neurology 61:S12–S18

Ruiz MA, Escriche M, Lluis C et al (2000) Adenosine A(1) receptor in cultured neurons from rat cerebral cortex: colocalization with adenosine deaminase. J Neurochem 75:656–664

Rump LC, Jabbari-T J, von Kügelgen I et al (1999) Adenosine mediates nitric-oxide-independent renal vasodilation by activation of A2A receptors. J Hypertens 17:1987–1993

Salgado Garcia C, Jimenez Heffernan A, Sanchez de Mora et al (2014) Comparative study of the safety of regadenoson between patients with mild/moderate chronic obstructive pulmonary disease and asthma. Eur J Nucl Med Mol Imaging 41:119–125

Sato A, Terata K, Miura H et al (2005) Mechanism of vasodilation to adenosine in coronary arterioles from patients with heart disease. Am J Physiol Heart Circ Physiol 288:H1633–H1640

Schiffmann SN, Jacobs O, Vanderhaeghen JJ (1991) Striatal restricted adenosine A2 receptor (RDC8) is expressed by enkephalin but not by substance P neurons: an *in situ* hybridization histochemistry study. J Neurochem 57:1062–1067

Schiffmann SN, Fisone G, Moresco R et al (2007) Adenosine A2A receptors and basal ganglia physiology. Prog Neurobiol 83:277–292

Schindler M, Harris CA, Hayes B et al (2001) Immunohistochemical localization of adenosine A1 receptors in human brain regions. Neurosci Lett 297:211–215

Schoen SW, Graeber MB, Reddington M et al (1987) Light and electron microscopical immunocytochemistry of 5'-nucleotidase in rat cerebellum. Histochemistry 87:107–113

Schulte G, Fredholm BB (2000) Human adenosine A(1), A(2A), A(2B), and A(3) receptors expressed in Chinese hamster ovary cells all mediate the phosphorylation of extracellular-regulated kinase 1/2. Mol Pharmacol 58:477–482

Schulte G, Fredholm BB (2003) Signalling from adenosine receptors to mitogen-activated protein kinases. Cell Signal 15:813–827

Schwarzschild MA, Xu K, Oztas E et al (2003) Neuroprotection by caffeine and more specific A2A receptor antagonists in animal models of Parkinson's disease. Neurology 61:S55–S61

Sebastiao AM, Ribeiro JA (2009a) Tuning and fine-tuning of synapses with adenosine. Curr Neuropharmacol 7:180–194

Sebastiao AM, Ribeiro JA (2009b) Adenosine receptors and the central nervous system. Handb Exp Pharmacol 193:471–534

Selmeczy Z, Csóka B, Pacher P et al (2007) The adenosine A2A receptor agonist CGS 21680 fails to ameliorate the course of dextran sulphate-induced colitis in mice. Inflamm Res 56:204–209

Sevigny CP, Li L, Awad AS et al (2007) Activation of adenosine 2A receptors attenuates allograft rejection and alloantigen recognition. J Immunol 178:4240–4249

Shearman LP, Weaver DR (1997) [125I]4-aminobenzyl- 5'-N-methylcarboxamidoadenosine (125I) AB-MECA) labels multiple adenosine receptor subtypes in rat brain. Brain Res 745:10–20

Simola N, Cauli O, Morelli M (2006) Sensitization to caffeine and cross-sensitization to amphetamine: influence of individual response to caffeine. Behav Brain Res 172:72–79

Snel J, Lorist MM (2011) Effects of caffeine on sleep and cognition. Prog Brain Res 190:105–117

Souza MA, Mota BC, Gerbatin RR et al (2013) Antioxidant activity elicited by low dose of caffeine attenuates pentylenetetrazol-induced seizures and oxidative damage in rats. Neurochem Int 62:821–830

Sperlagh B (1996) Neuronal synthesis, storage and release of ATP. Semin Neurosci 8:175–186

Sperlagh B, Vizi ES (2007) Extracellular interconversion of nucleotides reveals an ecto-adenylate kinase activity in the rat hippocampus. Neurochem Res 32:1978–1989

Sperlagh B, Vizi ES (2011) The role of extracellular adenosine in chemical neurotransmission in the hippocampus and basal ganglia: pharmacological and clinical aspects. Curr Top Med Chem 11:1034–1046

Squadrito F, Bitto A, Altavilla D et al (2014) The effect of PDRN, an adenosine receptor A2A agonist, on the healing of chronic diabetic foot ulcers: results of a clinical trial. J Clin Endocrinol Metab 99:E746–E753

Storr M, Thammer J, Dunkel R et al (2002) Modulatory effect of adenosine receptors on the ascending and descending neural reflex responses of rat ileum. BMC Neurosci 3:21

Sullivan GW, Rieger JM, Scheld WM et al (2001) Cyclic AMP-dependent inhibition of human neutrophil oxidative activity by substituted 2-propynylcyclohexyl adenosine A(2A) receptor agonists. Br J Pharmacol 132:1017–1026

Svenningsson P, Le Moine C, Aubert I et al (1998) Cellular distribution of adenosine A2A receptor mRNA in the primate striatum. J Comp Neurol 399:229–240

Svenningsson P, LeMoine C, Fisone G et al (1999) Distribution, biochemistry and function of striatal adenosine A2A receptors. Prog Neurobiol 59:355–396

Svenningsson P, Lindskog M, Ledent C et al (2000) Regulation of the phosphorylation of the dopamine- and cAMP-regulated phosphoprotein of 32 kDa *in vivo* by dopamine D1, dopamine D2, and adenosine A2A receptors. Proc Natl Acad Sci U S A 97:1856–1860

Svenningsson P, Nishi A, Fisone G et al (2004) DARPP-32: an integrator of neurotransmission. Annu Rev Pharmacol Toxicol 44:269–296

Tchekalarova J, Kubová H, Mares P (2010) Effects of early postnatal caffeine exposure on seizure susceptibility of rats are age- and model-dependent. Epilepsy Res 88:231–238

Tomaru A, Ina Y, Kishibayashi N, Karasawa A (1995) Excitation and inhibition via adenosine receptors of the twitch response to electrical stimulation in isolated guinea pig ileum. Jpn J Pharmacol 69:429–433

Trevethick MA, Mantell SJ, Stuart EF et al (2008) Treating lung inflammation with agonists of the adenosine A2A receptor: promises, problems and potential solutions. Br J Pharmacol 155:463–474

Voorn P, Vanderschuren LJ, Groenewegen HJ et al (2004) Putting a spin on the dorsal-ventral divide of the striatum. Trends Neurosci 27:468–474

Wall MJ, Dale N (2007) Auto-inhibition of rat parallel fibre-Purkinje cell synapses by activity-dependent adenosine release. J Physiol 581:553–565

Wang TF, Guidotti G (1998) Widespread expression of ecto-apyrase (CD39) in the central nervous system. Brain Res 790:318–322

Wells L, Opacka-Juffry J, Fisher D et al (2012) In vivo dopaminergic and behavioral responses to acute cocaine are altered in adenosine A(2A) receptor knockout mice. Synapse 66:383–390

Wilson CN, Nadeem A, Spina D et al (2009) Adenosine receptors and asthma. Handb Exp Pharmacol 193:329–362

Wittendorp MC, Boddeke HW, Biber K (2004) Adenosine A3 receptor-induced CCL2 synthesis in cultured mouse astrocytes. Glia 46:410–418

Yamada K, Kobayashi M, Mori A et al (2013) Antidepressant-like activity of the adenosine A(2A) receptor antagonist, istradefylline (KW-6002), in the forced swim test and the tail suspension test in rodents. Pharmacol Biochem Behav 114–115:23–30

Yegutkin GG (2008) Nucleotide- and nucleoside-converting ectoenzymes: important modulators of purinergic signaling cascade. Biochim Biophys Acta 1783:673–694

Zhong Y, Yang Z, Huang WC et al (2013) Adenosine, adenosine receptors and glaucoma: an updated overview. Biochim Biophys Acta 1830:2882–2890

Zimmermann H (2006) Ectonucleotidases in the nervous system. Novartis Found Symp 276:113–128

Chapter 2
Allosteric Mechanisms in the Adenosine A_{2A}-Dopamine D_2 Receptor Heteromer

Sergi Ferré, Gemma Navarro, Jordi Bonaventura, Estefanía Moreno, Nora D. Volkow, Carme Lluís and Vicent Casadó

Abstract The pentameric structure constituted by one G protein coupled receptor (GPCR) homodimer and one heterotrimeric G protein provides a main functional unit and oligomeric entities can be viewed as multiples of dimers. For GPCR heteromers, experimental evidence supports a tetrameric structure, comprised of two different homodimers, each able to signal with their preferred G protein. GPCR homomers and heteromers can act as the conduit of allosteric interactions of orthosteric ligands. One ligand binding to one of the receptor units (protomer) modulates the properties of the same or another orthosteric ligand binding to another protomer. The agonist/agonist interaction in the adenosine A_{2A} receptor (A_{2A}R)-dopamine D_2 receptor (D_2R) heteromer, by which A_{2A}R agonists decrease the affinity of D_2R agonists, constitutes a well-known example and gave the first rationale for the use of A_{2A}R antagonists in Parkinson's disease. We review most recent studies that extend those findings to, first, ligand-independent allosteric modulations of the D_2R protomer that result in changes of the binding properties of A_{2A}R ligands in the A_{2A}R-D_2R heteromer; second, the differential modulation of the intrinsic efficacy of D_2R ligands for G protein-dependent and independent signaling; and third, the existence of the canonical antagonistic Gs-Gi interaction within the frame of the A_{2A}R-D_2R heteromer. These studies support the heterotetrameric structure of GPCR heteromers.

S. Ferré (✉) · J. Bonaventura
Integrative Neurobiology Section, National Institute on Drug Abuse, IRP, Intramural Research Program, National Institutes of Health, Triad Technology Building, 333 Cassell Drive, Baltimore, MD 21224, USA
e-mail: sferre@intra.nida.nih.gov

G. Navarro · E. Moreno · C. Lluís · V. Casadó
Centro de Investigacion Biomedica en Red sobre Enfermedades Neurodegenerativas, Department of Biochemisry and Molecular Biology, Faculty of Biology, University of Barcelona, Barcelona, Spain

N. D. Volkow
National Institute on Alcohol Abuse and Alcoholism, National Institutes of Health, Bethesda, MD, USA

© Springer International Publishing Switzerland 2015
M. Morelli et al. (eds.), *The Adenosinergic System,* Current Topics in Neurotoxicity 10, DOI 10.1007/978-3-319-20273-0_2

Keywords Adenosine A_{2A} receptor · Dopamine D_2 receptor · Heterotetramer · Allosterism

The A_{2A}R- D_2R Heteromer as a Model to Understand Allosterism within GPCR Oligomers

John Newport Langley and Paul Ehrlich independently introduced the "receptor" concept in 1878. Since then receptors have mostly been considered as single functional units. But we know now that receptors form multimolecular aggregates that include other receptors with the formation of receptor oligomers (Ferré et al. 2009). Most evidence indicates that, as for family C G protein-coupled receptors (GPCRs), family A GPCRs form homo- and heteromers (Ferré et al. 2009, 2014; Milligan and Bouvier 2005; Pin et al. 2007). Receptor oligomer is defined as a macromolecular complex composed of at least two (functional) receptor units (protomers) with biochemical properties that are demonstrably different from those of its individual components (Ferré et al. 2009).

To understand the unique biochemical properties of receptor oligomers we need to understand the basis of allosterism, which is defined as the process by which the interaction of a chemical or protein at one location on a protein or macromolecular complex (the allosteric site) influences the binding or function of the same or another chemical or protein at a topographically distinct site (Smith and Milligan 2010). In this respect, it is useful to consider ligands as modulators and modulated entities, and the receptors or receptor oligomers as the conduits of the allosteric modulation (Kenakin and Miller 2010). An orthosteric agonist (which binds to the same receptor site as the endogenous transmitter) has two main properties: affinity (the avidity to bind to the receptor) and intrinsic efficacy (the power with which the agonist, once bound to the receptor, produces the functional response). In classical allosterism, the allosteric ligand, by binding to a non-orthosteric site, can modify either of these properties. In this frame, the GPCR is the conduit of the allosteric modulation and is usually considered as a monomeric entity.

A first important concept that arises from the new field of GPCR oligomerization is that the pentameric structure constituted by one GPCR homodimer and one heterotrimeric G protein provides a main functional unit, and oligomeric entities can be viewed as multiples of dimers (Ferré et al. 2014). Then, in the frame of GPCR homodimers, allosterism implies that the dimer can act as the conduit of the allosteric modulation by an orthosteric ligand, which binds to one of the protomers, to the same or another orthosteric ligand, which binds to the second protomer. The realization of these interactions is leading to a profound modification of classical pharmacology. For instance, application of new models of analysis of radioligand binding experiments that consider the homodimer as a fundamental functional unit is allowing a better understanding of complex binding saturation or competition curves. Particularly, the two-state dimer model (Casadó et al. 2007; Ferré et al. 2014) is a practical model to analyze allosteric modulations of one ligand molecule binding on the affinity of a second ligand molecule binding to a GPCR homodimer. From saturation experiments, the two-state dimer model provides an index of cooperativity of the radioligand (degree of modulation exerted by the first ligand molecule binding to the first protomer on the affinity of the second ligand molecule binding to the second protomer in the homodimer). From competition experiments, the two-state dimer model provides three more indexes: an index of cooperativity of the competing ligand; and index of the modulation of the affinity of the competing ligand binding to the second protomer by the radioligand binding to the first protomer; and an index of the modulation of the affinity of the radioligand binding to the second protomer by the competing ligand binding to the first protomer (Casadó et al. 2007; Ferré et al. 2014).

When considering receptor heteromers as conduits of allosteric interactions, two possible scenarios should be considered (Kenakin and Miller 2010). In the first scenario, a ligand binding to one of the receptors in the heteromer leads to changes in the properties (affinity or intrinsic efficacy) of a ligand binding to the second molecularly different receptor. The best example is the allosteric antagonistic interaction between adenosine A$_{2A}$ receptor (A$_{2A}$R) agonists on dopamine D$_2$ receptor (D$_2$R) agonists in the A$_{2A}$R-D$_2$R heteromer, by which A$_{2A}$R agonists decrease the affinity of D$_2$R agonists. This is probably the most quoted and reproduced allosteric modulation in a GPCR heteromer (Dixon et al. 1997; Ferré et al. 1991; Kudlacek et al. 2003). The A$_{2A}$R-D$_2$R heteromer is selectively localized in the GABAergic striato-pallidal neuron (also called indirect medium spiny neuron or iMSNs) (Azdad et al. 2009; Ferré et al. 2007; Trifilieff et al. 2011). It has been hypothesized that allosteric interactions between A$_{2A}$R and D$_2$R agonists within the A$_{2A}$R-D$_2$R heteromer provide a mechanism responsible for the behavioral depressant effects of adenosine analogues and for the psychostimulant effects of selective adenosine A$_{2A}$R antagonists and the non-selective adenosine receptor antagonist caffeine, with implications for several neuropsychiatric disorders (Ferré 2008; Ferré et al. 2004, 2008). In fact, the same mechanism provided the first rationale for the use of A$_{2A}$R antagonists in Parkinson's disease (Armentero et al. 2011; Ferré et al. 1992; Müller and Ferré 2007). In the second scenario of allosteric modulation within GPCR heteromers, the modulator is not a ligand, but a protein (see the above-mentioned definition of allosterism): one of the receptors acts as modulator of a ligand binding to the other molecularly different receptor (Kenakin and Miller 2010). It is this allosteric modulation that can theoretically allow the selective targeting of different subpopulations of a particular receptor, like pre- *versus* postsynaptic receptors (see below). Again, the A$_{2A}$R-D$_2$R provides a valuable example. Screening with various *in vitro* and *in vivo* techniques led to the finding of very different qualitative properties of several selective A$_{2A}$R antagonists. The most striking finding was a decrease in the affinity of SCH 442416 for A$_{2A}$R when forming heteromers with D$_2$R, compared to when not forming heteromers or forming heteromers with adenosine A$_1$ receptor (A$_1$R) (Orru et al. 2011a). Application of the two-state dimer model indicated that SCH 442416 binds with low affinity due a strong negative cooperativity that appears when the D$_2$R binds to the A$_{2A}$R in the heteromer (Ferré et al. 2014; Orru et al. 2011a), strongly suggesting that the A$_{2A}$R-D$_2$R comprises at least two A$_{2A}$R protomers.

Being a weak ligand for the $A_{2A}R$-D_2R heteromer, SCH 442416 would not be useful in Parkinson's disease. Nevertheless, SCH 442416 acts preferentially on presynaptic striatal $A_{2A}R$ localized in cortico-striatal glutamatergic terminals that forming heteromers with A_1R. By blocking presynaptic $A_{2A}R$, SCH 442416 potently blocks cortico-striatal glutamatergic neurotransmission at doses that do not produce locomotor activation, that do not block postsynaptic $A_{2A}R$ (Orru et al. 2011a). The opposite pharmacological profile was obtained with KW 6002, which produced strong locomotor activity at doses that would be ineffective at blocking cortico-striatal glutamatergic neurotransmission (Orru et al. 2011a). KW 6002 would therefore be a promising antiparkinsonian agent. In fact, KW 6002 is already being successfully used in the treatment of Parkinson's disease (Jenner 2014; Pinna 2014).

The possibility of selectively targeting A_1R-$A_{2A}R$ heteromers with SCH 442416 was used to identify an important contributor to the reinforcing effects of cannabinoids: cortico-striatal glutamatergic neurotransmission. Initially, a paradoxical result had ben reported, by which the $A_{2A}R$ antagonist MSX-3 decreases THC and anandamide self-administration in squirrel monkeys at a relatively low dose, while a three-fold higher dose produced the opposite effect (Justinová et al. 2011). Based on results obtained in rats (Orru et al. 2011a), it was hypothesized that the different dose-dependent effects of MSX-3 could be related to a slightly selective presynaptic effect at lower doses with an overriding postsynaptic effect at larger doses. This hypothesis was confirmed by testing the effects of SCH-442416 and KW-6002 (Justinová et al. 2014). SCH-442416 produced a significant shift to the right of the THC self-administration dose-response curves, consistent with antagonism of the reinforcing effects of THC. On the other hand, KW-6002 produced a significant shift to the left, consistent with potentiation of the reinforcing effects of THC. These results show that selectively blocking presynaptic $A_{2A}R$ could provide a pharmacological approach to the treatment of marijuana dependence, and underscore cortico-striatal glutamatergic neurotransmission as a possible main mechanism involved in the rewarding effects of THC. At a more general level, these results also show that while the concept of using GPCR heteromers to target specific cell types is relatively new, it is a promising approach for targeting specific cell types to modulate specific symptoms of SUD.

Functional Significance and Regulation of Allosteric Interactions in the $A_{2A}R$-D_2R Heteromer

Demonstration of the functional significance of receptors heteromers is becoming an important goal in GPCR research. One main reason is their possible use as targets for drug development. The allosteric interactions in GPCR heteromers determine the specific biochemical properties of these heteromers, conferring their functional and pharmacoldgcical significance. In order to ascertain a biochemical property of the GPCR heteromer, which can then be used as a "biochemical fingerprint" for its identification in native tissues, the putative biochemical property should be disrupted with molecular or chemical tools that destabilize the quaternary structure of the heteromer (Ferré et al. 2009, 2014). This can be achieved by introducing mutations that modify key determinant residues at the oligomerization interfaces or using competing peptides with the sequence of specific receptor domains putatively involved in receptor oligomerization (Azdad et al. 2009; Banères and Parello 2003; Guitart et al. 2014; He et al. 2011; Hebert et al. 1996; Pei et al. 2010). Studies of peptide-peptide interactions using biophysical methods (such as Bioluminescence Resonance Energy Transfer or BRET) and mass spectrometry, led to the identification of intracellular epitopes of the D_2R (an arginine-rich epitope of the third intracellular loop or 3IL) and the $A_{2A}R$ (a distal C-terminal epitope containing a phosphorylated serine, serine-374) that establish a strong electrostatic interaction and are important determinants of the quaternary structure of the $A_{2A}R$-D_2R heteromer (Borroto-Escuela et al. 2010; Ciruela et al. 2004; Navaro et al. 2010; Woods and Ferré 2005). In BRET, a bioluminescence donor molecule, *Renilla* luciferase (Rluc), emits light upon addition of its substrate coelenterazine H. If in very close proximity (less than 10 nm), this emission transfers energy to a fluorescence acceptor molecule, such as yellow fluorescence protein (YFP). When studying GPCR heteromerization, Rluc is fused to one of the receptors and YFP is fused to the other receptor unit. Heteromerization of $A_{2A}R$-Rluc and D_2R-YFP was then demonstrated in transfected cells (Canals et al. 2003). Subsequent studies showed that transfection with a mutant $A_{2A}R$ with substitution of serine-374 by alanine ($A_{2A}R^{A374}$-Rluc, instead of $A_{2A}R$-Rluc) and D_2R-YFP, significantly reduced BRET values (Borroto-Escuela et al. 2010; Navarro et al. 2010), and the potency of the $A_{2A}R$ agonist CGS 21680 to decrease the affinity of D_2R for dopamine agonists (Bonaventura et al. 2014; Borroto-Escuela et al. 2010). These results demonstrated that the agonist-agonist allosteric interaction constitutes a biochemical property of the $A_{2A}R$-D_2R heteromer. Therefore, its demonstration in striatal tissue indicates the presence of the $A_{2A}R$-D_2R heteromer in the brain (Ferré et al. 1991).

A peptide approach was then used to evaluate the neuronal localization and functional significance of the $A_{2A}R$-D_2R heteromer. A very effective antagonistic interaction between $A_{2A}R$ and D_2R agonists was demonstrated with patch-clamp experiments (using knock-in mice expressing GFP) in D_2R-containing neurons in striatal slices (Azdad et al. 2009). CGS 21680 completely counteracted the ability of the D_2R agonist R(-)-propylnorapomorphine hydrochloride (NPA) to block NMDA-induced neuronal firing. This effect was selectively counteracted by the application of a small peptide with an amino acid sequence corresponding the epitope of the $A_{2A}R$ that includes serine-374 (Azdad et al. 2009). These results would suggest that this pharmacological interaction is determined by the agonist-agonist allosteric interaction in the $A_{2A}R$-D_2R heteromer, since both depend on the electrostatic interaction between intracellular domains of the $A_{2A}R$ and D_2R involved in the establishment of the quaternary structure of the $A_{2A}R$-D_2R heteromer. However, just a decrease in the affinity of NPA could not explain by itself the ability of CGS 21680 to abolish the decrease in excitability of D_2R-containing neurons induced by the high concentration of the D_2R agonist used, which should overcome the decrease in affinity. A decrease in the intrinsic efficacy of the D_2R agonist was therefore also involved

(Azdad et al. 2009). Importantly, we should not conclude from the peptide experiments that the electrostatic interactions between intracellular domains are the only ones determining the quaternary structure of GPCR heteromers, including $A_{2A}R$-D_2R heteromers. Also from experiments with peptides, it is becoming clear that interactions between specific transmembrane domains are also involved (as recently shown for the dopamine D_1R-D_3R heteromer; Guitart et al. 2014).

An enigma to be resolved about the function of $A_{2A}R$-D_2R heteromers is the possibility of simultaneous antagonistic reciprocal interactions between the two different receptor units. As mentioned above, in the striatum, stimulation of $A_{2A}R$ counteracts a D_2R agonist-induced inhibitory modulation of NMDA receptor-mediated effects (Azdad et al. 2009, see also Higley and Sabatini 2010). But other studies have reported the ability of D_2R activation to potently inhibit $A_{2A}R$ adenylyl-cyclase signaling in transfected cells (Hillion et al. 2002; Kull et al. 1999) and it is not entirely clear if this canonical interaction between G_s- and G_i-mediated signaling pathways takes place in the frame of the $A_{2A}R$-D_2R heteromer, as recently suggested for other receptor heteromers (Cristóvão-Ferreira et al. 2013; Guitart et al. 2014). In the striatum, under normal conditions, the ability of $A_{2A}R$ to activate adenylyl-cyclase (and consequent expression of genes such as *c-fos* or *preproenkephalin* by the striato-pallidal neuron) seems to be restrained by a strong tonic inhibitory effect of endogenous dopamine on striatal D_2R, which efficiently inhibits $A_{2A}R$-mediated adenylyl-cyclase activation (Karcz-Kubicha et al. 2003; Svenningsson et al. 1999). Pharmacological or genetic blockade of D_2R produces a significant activation of the adenylyl-cyclase-cAMP-PKA cascade, and the consequent depressant motor effects and biochemical effects (such as increase in striatal *c-fos* or *preproenkephalin* expression) can be counteracted by genetic or pharmacologic blockade of $A_{2A}R$ (Bertran-Gonzalez et al. 2009; Chen et al. 2001; Håkansson et al. 2006). To explain the co-existence of these simultaneous reciprocal antagonistic interactions between striatal $A_{2A}R$ and D_2R, we previously postulated that they were mediated by two different subpopulations of $A_{2A}R$, forming and not forming heteromers with D_2R (Ferré et al. 2008; Orru et al. 2011b).

However, from recent experiments we could provide a heuristic model that allows understanding the possibility of different and simultaneous reciprocal interactions between $A_{2A}R$ and D_2R withion the $A_{2A}R$-D_2R heteromer. Depending on the intracellular Ca^{2+} levels, the neuronal Ca^{2+}-binding proteins NCS-1 and calneuron-1 exert a differential modulation of two different signaling pathways in the $A_{2A}R$-D_2R heteromer. Both Ca^{2+}-binding proteins were found to compete for the same binding sites in the $A_{2A}R$-D_2R heteromer. We first found that, in the absence of Ca^{2+}-binding proteins, an $A_{2A}R$ agonist decreases the intrinsic efficacy of a D_2R agonist-mediated G protein-dependent inhibition of adenylyl-cyclase and G protein-independent MAPK activation (Navarro et al. 2014). Thus, in transfected HEK-293 cells, the D_2R agonist quinpirole could not counteract the ability of the $A_{2A}R$ agonist CGS 21680 to induce cAMP accumulation, due to the allosteric modulation by which $A_{2A}R$ activation counteracts D_2R-mediated G protein-dependent signaling. However, this allosteric modulation was absent when cells were co-transfected with NCS-1 or calneuron-1 in the presence of low or high intracellular Ca^{2+} levels, respectively.

The same biochemical interactions were also found in striatal cells, where low or high intracellular Ca^{2+} levels determined if either NCS-1 or calneuron-1 bind to the A$_{2A}$R-D$_2$R heteromer. Knocking down the expression of NCS-1 or calneuron-1 led to the reappearance of the allosteric interaction under conditions of low or high intracellular Ca^{2+} levels, respectively, and quinpirole could not counteract the ability of CGS 21680 to stimulate adenylyl-cyclase (Navarro et al. 2014) (Fig. 2.1).

A different scenario was observed in relation to MAPK signaling. In transfected HEK-293 cells, MAPK activation (ERK1/2 phosphorylation) was similar under conditions of activation of either A$_{2A}$R or D$_2$R or co-activation of both receptors.

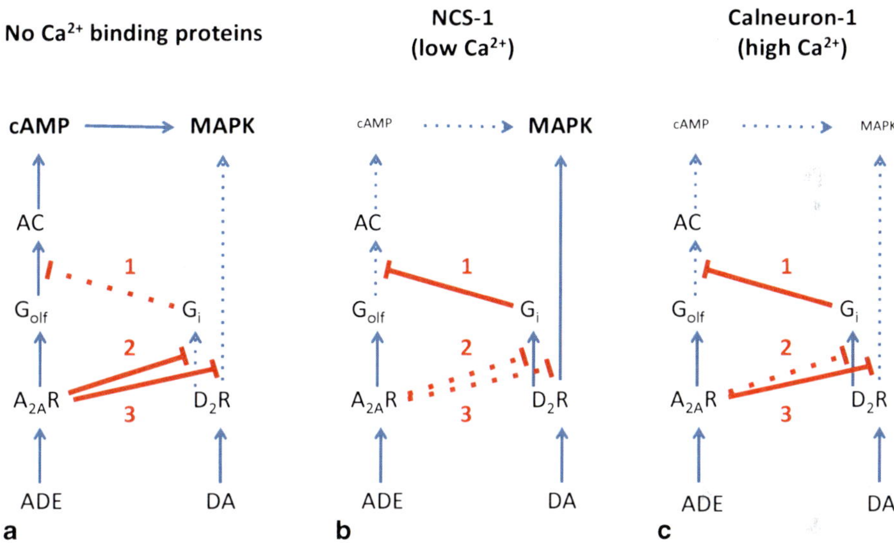

Fig. 2.1 Model representing the differential role of NCS-1 and calneuron-1 in A$_{2A}$R-D$_2$R heteromer signaling. Depending on the intracellular levels of Ca^{+2}, the neuronal Ca^{+2}-binding proteins NCS-1 and calneuron-1 exert a differential modulation of the A$_{2A}$R-D$_2$R heteromer signaling. In the absence of neuronal Ca^{+2}-binding proteins (a; non-transfected HEK-293 cells or knocking down protein expression in striatal cells in culture), the D$_2$R agonist cannot counteract the ability of the A$_{2A}$R agonist to induce cAMP accumulation (*1*), due to an allosteric modulation by which A$_{2A}$R activation inhibits D$_2$R-mediated G protein-dependent signaling (*2*). Under these conditions, A$_{2A}$R activation also inhibits the D$_2$R agonist-mediated G protein-independent MAPK activation (*3*). These two allosteric modulations (*2* and *3*) are absent when NCS-1 binds to the receptor heteromer in the presence of low intracellular Ca^{+2} levels (b; transfected HEK-293 cells or striatal cells, where low intracellular Ca^{+2} levels determine the binding of NCS-1 to the A$_{2A}$R-D$_2$R heteromer). Under these conditions, co-activation of both receptors in the A$_{2A}$R-D$_2$R heteromer does not produce cAMP accumulation but still induces MAPK activation. When calneuron-1 binds to A$_{2A}$R-D$_2$R heteromer (c; transfected HEK-293 cells or striatal cells where high intracellular Ca^{+2} levels determine the binding of calneuron-1 to the A$_{2A}$R-D$_2$R heteromer), the allosteric modulation at the level of G protein-dependent signaling (*2*) is selectively disrupted, since the allosteric modulation at the level of G protein-independent signaling (*1*) is maintained. This results in very low activation of both MAPK signaling and cAMP production upon co-activation of both receptors in the A$_{2A}$R-D$_2$R heteromer, since A$_{2A}$R agonist-mediated MAPK activation (*3*, which is dependent on adenylyl-cyclase signaling) is also inhibited.

The absence of at least an additive effect of $A_{2A}R$ and D_2R agonists would indicate some degree of antagonistic interaction. But, under conditions of high intracellular Ca^{2+} levels and in the presence of calcineuron-1, co-activation of $A_{2A}R$ and D_2R did not produce a noticeable ERK1/2 phosphorylation (Navarro et al. 2014). Since, as described previously (Canals et al. 2005; Klinger et al. 2002), we also found $A_{2A}R$-mediated MAPK activation be mostly dependent on G-protein-adenylyl-cyclase signaling (Fig. 2.1), these results indicated that high intracellular Ca^{2+} levels allows calcineuron-1 to selectively facilitate an allosteric interaction in the $A_{2A}R$-D_2R heteromer by which $A_{2A}R$ agonists also blocks a G-protein-independent D_2R-mediated ERK1/2 phosphorylation. The same mechanisms were also found to operate in striatal cells and no ERK1/2 phosphorylation was observed upon co-activation of $A_{2A}R$ and D_2R under conditions of high intracellular Ca^{2+} levels (which induce binding of calcineuron-1 to the $A_{2A}R$-D_2R heteromer). MAPK activation was nevertheless very significant under the same conditions but knocking down the expression of calcineuron-1 (Navarro et al. 2014). Therefore, as recently found for the dopamine D_1R-D_3R heteromer (Guitart et al. 2014), we found functional selectivity of allosteric interactions within the $A_{2A}R$-D_2R heteromer, and this functional selectivity was found to be dependent on intracellular Ca^{+2} levels (Navarro et al. 2014). The functional inhibition by D_2R agonists of NMDA receptor-mediated Ca^{2+}-dependent effects observed in striatal tissue preparations (Azdad et al. 2009; Higley and Sabatini 2010), which can be counteracted by $A_{2A}R$ activation, should depend largely on G-protein-independent D_2R-mediated signaling.

These results allow understanding the co-existence of reciprocal antagonistic interactions between striatal $A_{2A}R$ and D_2R, considering only one predominant population of $A_{2A}R$, which forms heteromers with D_2R. This could account for different G protein-dependent or independent functional responses, which could be differentially modulated by intracellular Ca^{+2} levels. Apart from adenosine and dopamine, the Ca^{+2}-dependent modulation of $A_{2A}R$-D_2R heteromer function allows further integration of other neurotransmitter systems such as glutamate (through NMDA receptor activation) and acetylcholine (through G_q-coupled muscarinic receptors) (Tozzi et al. 2011).

As mentioned before the existence of negative cooperativity of the $A_{2A}R$ antagonist SCH 442416 (Orrú et al. 2011a) strongly suggested that the $A_{2A}R$-D_2R comprises at least two $A_{2A}R$ protomers. Also, it would be difficult for two GPCR protomers to simultaneously accommodate two trimeric G-protein molecules due to steric hindrance (Maurice et al. 2011). Therefore, the results on allosteric interactions in the $A_{2A}R$-D_2R heteromer at the level of adenylyl cyclase signaling supports a tetrameric structure, comprised of two different homodimers, each able to signal with their preferred G protein. This molecular arrangement would allow the canonical interaction between G_s- and G_i-mediated signaling to take place in the frame of the heteromer (Ferré et al. 2014; Guitart et al. 2014).

Conclusions

GPCR oligomerization is a reality and it is becoming obvious that GPCR homodimers constitute not only functional but also structural building blocks. In this way, receptor heteromers would be comprised of two different homodimers, each able to signal with their preferred G protein. We postulate that the canonical interaction between G$_s$- and G$_i$-mediated signaling is in fact a biochemical property of GPCR heteromer. Experiments are now in progress to validate this hypothesis. But what it is already obvious, and here exemplified from the studies on A$_{2A}$R-D$_2$R heteromers, is that allosteric mechanisms in the frame of GPCR heterotetramers provide them with multiple unique biochemical properties, including ligand and functional selectivity. These properties allow understanding complex experimental results with pharmacological significance, such as: the existence of reciprocal interactions between activated A$_{2A}$R and D$_2$R, which are differentially modulated by intracellular Ca^{2+}, making the A$_{2A}$R-D$_2$R heterotetramer a cellular device that integrates signals from the extracellular and intracellular compartments (dopamine, adenosine and Ca^{2+}) to produce a specific functional response; the selective negative cooperativity of the A$_{2A}$R antagonist SCH 442416, which provides the proof of concept of the possibility that different GPCR heteromers can account for pharmacologically different subpopulations of receptors. In fact, SCH 44416 has been successfully used to target selectively striatal presynaptic A$_{2A}$R in a non-human primate model of addiction to cannabinoids (Justinová et al. 2014). Research is in progress to obtain molecules that selectively target striatal postsynaptic A$_{2A}$R, i.e. the A$_{2A}$R-D$_2$R heterotetramer.

Acknowledgements Work supported by the intramural funds of the National Institute on Drug Abuse and from Spanish "Ministerio de Ciencia y Tecnología" (SAF2011-23813), from the Government of Catalonia (2009-SGR-12) and a grant (CB06/05/0064) from "Centro de Investigación Biomédica en Red sobre Enfermedades Neurodegenerativas" (CIBERNED).

References

Armentero MT, Pinna A, Ferré S et al (2011) Past, present and future of A(2A) adenosine receptor antagonists in the therapy of Parkinson's disease. Pharmacol Ther 132:280–299

Azdad K, Gall D, Woods AS (2009) Dopamine D2 and adenosine A2A receptors regulate NMDA-mediated excitation in accumbens neurons through A2A-D2 receptor heteromerization. Neuropsychopharmacology 34:972–986

Banères JL, Parello J (2003) Structure-based analysis of GPCR function: evidence for a novel pentameric assembly between the dimeric leukotriene B4 receptor BLT1 and the G-protein. J Mol Biol 329:815–829

Bertran-Gonzalez J, Håkansson K, Borgkvist A et al (2009) Histone H3 phosphorylation is under the opposite tonic control of dopamine D2 and adenosine A2A receptors in striatopallidal neurons. Neuropsychopharmacology 34:1710–1720

Borroto-Escuela DO, Romero-Fernandez W, Tarakanov AO et al (2010) Characterization of the A2AR-D2R interface: focus on the role of the C-terminal tail and the transmembrane helices. Biochem Biophys Res Commun 402:801–807

Canals M, Marcellino D, Fanelli F et al (2003) Adenosine A2A-dopamine D2 receptor-receptor heteromerization: qualitative and quantitative assessment by fluorescence and biolumines- cence energy transfer. J Biol Chem 278:46741–46749

Canals M, Angulo E, Casadó V et al (2005) Molecular mechanisms involved in the adenosine A and A receptor-induced neuronal differentiation in neuroblastoma cells and striatal primary cultures. J Neurochem 92:337–348

Casadó V, Cortés A, Ciruela F et al (2007) Old and new ways to calculate the affinity of ago- nists and antagonists interacting with G-protein-coupled monomeric and dimeric receptors: the receptor-dimer cooperativity index. Pharmacol Ther 116:343–354

Chen JF, Moratalla R, Impagnatiello F et al (2001) The role of the D(2) dopamine receptor (D(2)R) in A(2A) adenosine receptor (A(2A)R)-mediated behavioral and cellular responses as revealed by A(2A) and D(2) receptor knockout mice. Proc Natl Acad Sci U S A 98:1970–1975

Ciruela F, Burgueño J, Casadó V et al (2004) Combining mass spectrometry and pull-down tech- niques for the study of receptor heteromerization. Direct epitope-epitope electrostatic interac- tions between adenosine A2A and dopamine D2 receptors. Anal Chem 76:5354–5363

Cristóvão-Ferreira S, Navarro G, Brugarolas M et al (2013) A1R-A2AR heteromers coupled to Gs and G i/0 proteins modulate GABA transport into astrocytes. Purinergic Signal 9:433–449

Dixon AK, Widdowson L, Richardson PJ (1997) Desensitisation of the adenosine A1 receptor by the A2A receptor in the rat striatum. J Neurochem 69:315–321

Ferré S (2008) An update on the mechanisms of the psychostimulant effects of caffeine. J Neuro- chem 105:1067–1079

Ferré S, von Euler G, Johansson B et al (1991) Stimulation of high-affinity adenosine A2 receptors decreases the affinity of dopamine D2 receptors in rat striatal membranes. Proc Natl Acad Sci U S A 88:7238–7241

Ferré S, Fuxe K, von Euler G et al (1992) Adenosine-dopamine interactions in the brain. Neurosci- ence 51:501–512

Ferré S, Ciruela F, Canals M et al (2004) Adenosine A2A-dopamine D2 receptor-receptor het- eromers. Targets for neuro-psychiatric disorders. Parkinsonism Relat Disord 10:265–271

Ferré S, Ciruela F, Woods AS et al (2007) Functional relevance of neurotransmitter receptor het- eromers in the central nervous system. Trends Neurosci 30:440–446

Ferré S, Ciruela F, Borycz J et al (2008) Adenosine A1-A2A receptor heteromers: new targets for caffeine in the brain. Front Biosci 13:2391–2399

Ferré S, Baler R, Bouvier B et al (2009) Building a new conceptual framework for receptor het- eromers. Nat Chem Biol 5:131–134

Ferré S, Casado V, Devi LA et al (2014) G protein-coupled receptor oligomerization revisited: functional and pharmacological perspectives. Pharmacol Rev 66:413–434

Ginés S, Hillion J, Torvinen M et al (2000) Dopamine D1 and adenosine A1 receptors form func- tionally interacting heteromeric complexes. Proc Natl Acad Sci U S A 97:8606–8611

Guitart X, Navarro G, Moreno E et al (2014) Functional selectivity of allosteric interactions within GPCR oligomers: the dopamine D_1-D_3 receptor heterotetramer. Mol Pharmacol 86:417–429

Håkansson K, Galdi S, Hendrick J et al (2006) Regulation of phosphorylation of the GluR1 AMPA receptor by dopamine D2 receptors. J Neurochem 96:482–488

He SQ, Zhang ZN, Guan JS et al (2011) Facilitation of μ-opioid receptor activity by preventing d-opioid receptor-mediated codegradation. Neuron 69:120–131

Hebert TE, Moffett S, Morello JP et al (1996) A peptide derived from a beta2-adrenergic recep- tor transmembrane domain inhibits both receptor dimerization and activation. J Biol Chem 271:16384–16392

Higley MJ, Sabatini BL (2010) Competitive regulation of synaptic Ca2+ influx by D2 dopamine and A2A adenosine receptors. Nat Neurosci 13:958–966

Hillion J, Canals M, Torvinen M et al (2002) Coaggregation, cointernalization, and codesensitiza- tion of adenosine A2A receptors and dopamine D2 receptors. J Biol Chem 277:18091–18097

Jenner P (2014) An overview of adenosine A2A receptor antagonists in Parkinson's disease. Int Rev Neurobiol 119:71–86

Justinová Z, Ferré S, Redhi GH et al (2011) Reinforcing and neurochemical effects of cannabinoid CB1 receptor agonists, but not cocaine, are altered by an adenosine A2A receptor antagonist. Addict Biol 16:405–415

Justinová Z, Redhi GH, Goldberg SR et al (2014) Differential effects ofpresynaptic versus post-synaptic adenosine A2A receptor blockade on Δ9-tetrahydrocannabinol (THC) self-administration in squirrel monkeys. J Neurosci 34:6480–6484

Karcz-Kubicha M, Quarta D, Hope BT et al (2003) Enabling role of adenosine A1 receptors in adenosine A2A receptor-mediated striatal expression of c-fos. Eur J Neurosci 18:296–302

Kenakin T, Miller LJ (2010) Seven transmembrane receptors as shapeshifting proteins: the impact of allosteric modulation and functional selectivity on new drug discovery. Pharmacol Rev 62:265–304

Kerppola TK (2006) Visualization of molecular interactions by fluorescence complementation. Nat Rev Mol Cell Biol 7:449–456

Klinger M, Kudlacek O, Seidel MG et al (2002) MAP kinase stimulation by cAMP does not require RAP1 but SRC family kinases. J Biol Chem 277:32490–32497

Kudlacek O, Just H, Korkhov VM et al (2003) The human D2 dopamine receptor synergizes with the A2A adenosine receptor to stimulate adenylyl cyclase in PC12 cells. Neuropsychopharmacology 28:1317–1327

Kull B, Ferré S, Arslan G et al (1999) Reciprocal interactions between adenosine A2A and dopamine D2 receptors in Chinese hamster ovary cells co-transfected with the two receptors. Biochem Pharmacol 58:1035–1045

Lebon G, Warne T, Edwards PC et al (2011) Agonist-bound adenosine A2A receptor structures reveal common features of GPCR activation. Nature 474:521–525

Maurice P, Kamal M, Jockers R (2011) Asymmetry of GPCR oligomers supports their functional relevance. Trends Pharmacol Sci 32:514–520

Milligan G, Bouvier M (2005) Methods to monitor the quaternary structure of G protein-coupled receptors. FEBS J 272:2914–2925

Müller CE, Ferré S (2007) Blocking striatal adenosine A2A receptors: a new strategy for basal ganglia disorders. Recent Pat CNS Drug Discov 2:1–21

Navarro G, Ferré S, Cordomi A et al (2010) Interactions between intracellular domains as key determinants of the quaternary structure and function of receptor heteromers. J Biol Chem 285:27346–27359

Navarro G, Aguinaga D, Moreno E et al (2014) Intracellular calcium levels determine functional selectivity within adenosine A2A-dopamine D2 receptor heteromers in striatal neurons. Chem Biol 21:1546–1556

Orru M, Bakesova J, Brugarolas M et al (2011a) Striatal pre- and postsynaptic profile of adenosine A(2A) receptor antagonists. PLoS ONE 6:e16088

Orrú M, Quiroz C, Guitart X et al (2011b) Pharmacological evidence for different populations of postsynaptic adenosine A2A receptors in the rat striatum. Neuropharmacology 61:967–974

Pei L, Li S, Wang M et al (2010) Uncoupling the dopamine D1-D2 receptor complex exerts anti-depressant-like effects. Nat Med 16:1393–1395

Pin JP, Neubig R, Bouvier M et al (2007) International union of basic and clinical pharmacology. LXVII. Recommendations for the recognition and nomenclature of G protein-coupled receptor heteromultimers. Pharmacol Rev 59:5–13

Pinna A (2014) Adenosine A2A receptor antagonists in Parkinson's disease: progress in clinical trials from the newly approved istradefylline to drugs in early development and those already discontinued. CNS Drugs 28:455–474

Robitaille M, Héroux I, Baragli A et al (2009) Novel tools for use in bioluminescence resonance energy transfer (BRET) assays. Methods Mol Biol 574:215–234

Smith NJ, Milligan G (2010) Allostery at G protein-coupled receptor homo- and heteromers: uncharted pharmacological landscapes. Pharmacol Rev 62:701–725

Svenningsson P, Fourreau L, Bloch B et al (1999) Opposite tonic modulation of dopamine and adenosine on c-fos gene expression in striatopallidal neurons. Neuroscience 89:827–837

Tozzi A, de Iure A, Di Filippo M et al (2011) The distinct role of medium spiny neurons and cholinergic interneurons in the D_2/A_2A receptor interaction in thestriatum: implications for Parkinson's disease. J Neurosci 31:1850–1862

Trifilieff P, Rives ML, Urizar E et al (2011) Detection of antigen interactions ex vivo by proximity ligation assay: endogenous dopamine D2-adenosine A2A receptor complexes in the striatum. Biotechniques 5:111–118

Woods AS, Ferré S (2005) Amazing stability of the arginine-phosphate electrostatic interaction. J Proteome Res 4:1397–1402

Chapter 3
Adenosine A_{2A} Receptor Antagonists in Drug Development

Christa E. Müller

Abstract The first A_{2A} adenosine receptor antagonist, istradefylline, was approved in 2013 in Japan for the treatment of Parkinson's disease (PD). This will allow long-term studies to elucidate the neuroprotective potential of A_{2A} antagonists in patients. New A_{2A} antagonists are in clinical evaluation for PD. Additional promising indications for A_{2A} antagonists are being explored, including Alzheimer's disease (AD) and other neurodegenerative diseases, depression, and attention deficit hyperactivity disease (ADHD). A_{2A} antagonists may be useful for the treatment of several rare neurodegenerative diseases, and their clinical evaluation for those diseases is warranted. Dual- and multi-target drugs combining A_{2A} antagonism with A_1 antagonism, MAO-B inhibition, dopamine receptor activation and/or NMDA receptor blockade may be advantageous for the treatment of PD and perhaps also for other brain diseases. X-ray structure of the human A_{2A} adenosine receptor in complex with several antagonists and agonists provide a basis for understanding drug-receptor interactions and support the development of new drugs.

Keywords Adenosine · Antagonists, Dual-drug target approach · Multi-drug target approach · Neurodegenerative diseases · Rare diseases · X-ray structure

Introduction

Adenosine receptors (ARs) are G protein-coupled receptors (GPCRs) activated by the nucleoside adenosine. They belong to the largest subgroup of GPCRs, the rhodopsin-like class A receptors. Four different AR subtypes exist designated A_1, A_{2A}, A_{2B} and A_3. All are coupled to adenylate cyclase (AC), A_1 and A_3 via $G_{i/o}$ proteins mediating inhibition of AC, A_{2A} and A_{2B} via G_s or G_{olf} that mediate stimulation of AC resulting in an increase in intracellular cAMP concentration. Further coupling has been described (Fredholm et al. 2011), e.g. A_{2B}ARs can couple to $G_{q/11}$ leading

C. E. Müller (✉)
PharmaCenter Bonn, Pharmaceutical Chemistry I, Pharmaceutical Institute, Universität Bonn, An der Immenburg 4, 53121 Bonn, Germany
e-mail: christa.mueller@uni-bonn.de

© Springer International Publishing Switzerland 2015
M. Morelli et al. (eds.), *The Adenosinergic System,* Current Topics in Neurotoxicity 10, DOI 10.1007/978-3-319-20273-0_3

to an activation of phospholipase C liberating inositol trisphosphate (IP_3) and subsequently to mobilization of intracellular calcium, while A_{2A}ARs can also increase IP_3 by activation of $G_{\alpha 15}$ and $G_{\alpha 16}$. All AR subtypes are involved in cell differentiation, growth, survival and death by activation of mitogen-activated protein kinase (MAPK) (Jacobson and Gao 2006).

Upon activation ARs show desensitization: A_{2A} and A_{2B} ARs have been found to exhibit 50% desensitization in chromaffin cells within 20 min of agonist stimulation (Mundell and Kelly 1998).

The A_{2A}AR is the largest AR subtype consisting of 412 amino acids in humans and 410 amino acids in rat and mouse. In contrast to the other AR subtypes it contains a long intracellular N-terminal tail consisting of >120 amino acid residues, which is not required for G_s coupling, but may interact with modulatory proteins (Navarro et al. 2009; Zezula and Freissmuth 2008). The A_{2A}AR is most closely related to the A_{2B}AR subtype (59% sequence identity for the human A_{2A} and A_{2B} subtypes). The percentage of amino acid sequence identity of A_{2A}AR in the three species is as follows: human vs. rat 82%, human vs. mouse 82%, and rat vs. mouse 96%.

ARs are found throughout the body in every tissue. However, the four AR subtypes show distinct expression patterns. While the A_1 and A_{2A} AR subtypes are expressed in high density in the central nervous system, A_{2B} and A_3 ARs are only weakly expressed in the brain under normal conditions. Whereas the A_1AR is highly expressed in many parts of the brain including cortex, striatum, and hippocampus, the A_{2A}AR displays a more restricted expression pattern and is found only in the caudate-putamen (striatum), olfactory tubercle, and nucleus accumbens in high density, and in much lower density in other brain areas (de Lera Ruiz et al. 2014; Fredholm et al. 2011). This restricted expression of the A_{2A}AR contributes to its attractivity as a drug target for the treatment of Parkinson's disease (PD). However, A_{2A}AR expression may be altered under pathological conditions, e.g. an upsurge of A_{2A}AR expression in the hippocampus has been observed in Alzheimer's disease (AD) patients (Flaten et al. 2014). In the periphery, A_{2A}ARs are expressed for example, on blood platelets (mediating aggregation), in blood vessels (causing reduction in blood pressure), on T-lymphocytes (leading to immunosuppression) (de Lera Ruiz et al. 2014) and on brown adipose tissue (inducing thermogenesis) (Gnad et al. 2014). A_{2A}ARs like A_{2B}ARs can be upregulated under hypoxic conditions (Brown et al. 2011; Ma et al. 2010).

It is increasingly recognized that GPCRs typically assemble to form homo- or heteromeric structures that consist of two or more receptor proteins. The A_{2A}AR was shown to form a heteromeric receptor complex with dopamine D_2 receptors (A_{2A}/D_2 heteromer), which displays an altered pharmacology as compared to the homomeric receptors (Armentero et al. 2011; Ferre et al. 2004). The C-terminal tail of the A_{2A}AR was postulated to bind to the intracellular loop 3 of the D_2 receptor by electrostatic interactions (Borroto-Escuela et al. 2010). The A_{2A}/D_2 heteromers can form a complex with a G_s protein; they also show fast β-arrestin2 recruitment (Borroto-Escuela et al. 2011). The A_{2A}AR appears to heteromerize with several other GPCRs as well, including A_1, D_3, CB_1, and mGluR5 receptors (de Lera Ruiz et al. 2014). A_{2A}AR-containing heteromeric complexes with D_2 and CB_1 receptors in the striatum were found to be disrupted upon treatment with L-DOPA (Pinna et al. 2014). Multivalent drugs have recently been designed based on A$_{2A}$ antagonists bound to a nanoscaffold allowing multivalent interaction with receptor complexes (Dix et al. 2014).

Despite enormous efforts in developing drugs for ARs in the last decades, only few drugs have been approved so far. The short-acting physiological agonist adenosine (Adenoscan®) and the A$_{2A}$-selective agonist regadenoson (Lexiscan®) are used in myocardial stress imaging acting on A$_{2A}$ARs of coronary blood vessels which leads to dilation and a drop in blood pressure. The A$_{2A}$AR represents one of the most important regulators of the innate immune response. A$_{2A}$ARs inhibit the secretion of proinflammatory mediators by immune cells thereby dampening inflammatory reactions and therefore have potential as anti-inflammatory, anti-rheumatic, and immunosuppressive drugs. Separation of their anti-inflammatory and hypotensive effects has recently been achieved by a prodrug approach (El-Tayeb et al. 2009; Flögel et al. 2012).

The non-selective AR antagonist caffeine (and to a minor extent theophylline) is widely consumed as a brain-stimulatory natural product present in coffee and tea. Furthermore, caffeine is applied as a drug, e.g. as a central stimulant to improve mental alertness, and in combination with analgesics for the treatment of pain including migraine, while theophylline is used as an antiasthmatic drug. Recently, caffeine has been approved for the treatment of apnea in preterm infants, and it has been found to have additional positive effects on brain development (Maitre and Stark 2012).

A$_{2A}$AR antagonists have been developed for the treatment of PD. The A$_{2A}$-selective antagonist istradefylline (NOURIAST®) was approved in Japan for the treatment of Parkinson's disease (PD) in combination with L-DOPA or other dopamine agonists, and more A$_{2A}$AR antagonists are currently in clinical development (de Lera Ruiz et al. 2014). For further potential indications of A$_{2A}$AR antagonists see below.

Therapeutic Indications for A$_{2A}$ Adenosine Receptor Antagonists

Neurodegenerative Diseases

Caffeine, a non-selective AR antagonist which blocks all four AR subtypes in humans at micromolar concentrations, has shown potent neuroprotection in retrospective and prospective human studies and a number of animal models (Carman et al. 2014; Chen 2014; Chen and Chern 2011; Flaten et al. 2014). The development of AD and PD was inversely correlated with the consumption of coffee and caffeine, and the blockade of A$_{2A}$ARs appears to be responsible for the cognitive-enhancing and neuroprotective effects.

Parkinson's Disease (PD)

The most advanced indication for $A_{2A}AR$ antagonists is PD (Pinna 2014). Activation of postsynaptic A_{2A} ARs negatively modulates dopamine D_2 receptors in A_{2A}/D_2 heteromers present in the striatum. Blockade of A_{2A}ARs therefore positively modulates D_2 receptor signaling. This is most apparent upon co-treatment of an A_{2A} antagonist with a D_2 agonist. Like D_2 agonists (G_i-coupled), A_{2A} antagonists will lead to reduced intracellular cAMP levels. $A_{2A}AR$ antagonists and caffeine have shown positive effects on motoric symptoms of PD without inducing, or even preventing dyskinesia (Jones et al. 2013; Kanda and Uchida 2014; Wills et al. 2013). Additional non-motor effects, e.g. improvement of cognition, contribute to the beneficial effects of A_{2A} antagonists in PD (Chen 2014; Nomoto et al. 2014; Uchida et al. 2014). Moreover, A_{2A} antagonists displayed neuroprotective properties in animal studies (Cerri et al. 2014) which may be explained by increased glutamate uptake induced by A_{2A}ARs expressed on astrocytes (Matos et al. 2012, 2013).

$A_{2A}AR$ in striatum are colocalized with ecto-5'-nucleotidase (CD73), which is responsible for providing adenosine for $A_{2A}AR$ stimulation by catalyzing the hydrolysis of AMP (Augusto et al. 2013). CD73 inhibitors may therefore be useful for indirect inhibition of ARs.

A_{2A} antagonists may also be valuable for the treatment restless legs syndrome (Decerce et al. 2007).

Alzheimer's Disease and Impaired Memory and Cognition

An upsurge of $A_{2A}AR$ expression in the hippocampus has been observed in aged people and in Alzheimer patients, and this may contribute to the symptoms of this disastrous neurodegenerative disease, in particular to impaired memory and cognition (Flaten et al. 2014). In animal models caffeine and A_{2A}-selective antagonists ameliorated tau-induced as well as beta-amyloid-induced pathology and led to improved memory (Laurent et al. 2014a, b; Li et al. 2015)

Machado–Joseph Disease

Machado–Joseph disease (MJD) or spinocerebellar ataxia type 3 (SCA3), is a rare autosomal, dominantly inherited neurodegenerative disease that causes progressive cerebellar ataxia, which results in a lack of muscle control and coordination. The symptoms are caused by a genetic mutation that results in an abnormal form of the protein ataxin which causes degeneration of cells in the hindbrain. Some symptoms are similar to those in PD. In a mouse model of MJD caffeine as well as A_{2A} knockout decreased the pathology. A_{2A} antagonists may therefore be the first therapeutic approach for treating this fatal rare disease (Goncalves et al. 2013).

Attention Deficit Hyperactivity Disorder (ADHD)

In a rat model of attention deficit hyperactivity disorder (ADHD) caffeine and A_{2A}- as well as A_1-selective AR antagonists improved cognitive and attention deficits (Pandolfo et al. 2013; Pires et al. 2009). A recent study reported on a possible association between A_{2A}AR polymorphisms and ADHD in humans (Molero et al. 2013). Therefore, A_{2A} antagonists are currently in development for ADHD.

Depression

Caffeine and A_{2A}-selective antagonists showed potential for the treatment of depression and possibly also for anxiety in preclinical studies (Yamada et al. 2013, 2014a, b).

Addiction

Adenosine is involved in the signaling induced in the brain by addictive drugs. The effects of A_{2A}AR antagonists on heroin, cocaine, amphetamine, Δ^9-tetrahydrocannabinol and alcohol addiction, and on food seeking behavior, appear to be complex and are not fully understood yet (Brown and Short 2008; Brown et al. 2009; Justinova et al. 2014; Lopez-Cruz et al. 2013; O'Neill et al. 2014; Wydra et al. 2015a, b; Yao et al. 2006).

Peripheral Diseases

A_{2A}AR antagonists may also be helpful for several peripheral diseases, including liver fibrosis (Wang et al. 2014), stimulation of the immune system and cancer (Eltzschig et al. 2012; Hatfield et al. 2015; Linden and Cekic 2012; Sitkovsky et al. 2014), and even scar treatment (Perez-Aso et al. 2012).

Side-Effects of Caffeine and A_{2A} Adenosine Receptor Antagonists

A_{2A}AR antagonists evaluated in clinical trials for PD have shown a remarkable safety profile with little side-effects (Chen 2014). In contrast to L-DOPA and dopamine agonists, they do not induce dyskinesia. Initially observed nausea disappeared with time.

Based on preclinical studies with A_{2A} antagonists and on experience with caffeine use in patients, contraindications may be pregnancy and lactation since caffeine was found to alter fetal brain development in mice (Silva et al. 2013), epilepsy because A_{2A} antagonists may increase the susceptibility to seizures, and severe cardiovascular problems.

Development of Drugs Targeting A_{2A} Adenosine Receptors

The field of $A_{2A}AR$ antagonists is mature looking back to more than two decades of successful activities in the design and development of potent and selective compounds (Armentero et al. 2011; Cristalli et al. 2009; de Lera Ruiz et al. 2014; Fredholm et al. 2011; Jorg et al. 2014; Müller and Ferre 2007; Muller and Jacobson 2011a, b; Müller and Scior 1993).

Caffeine and Theophylline

Caffeine (**1**) and theophylline (**2**) are non-selective AR antagonists (see Fig. 3.1). In humans, they block all four AR subtypes with similar potency in the micromolar concentration range (see Table 3.1). However, in rodents, both methylxanthine derivatives are virtually inactive at A_3ARs and only block A_1, A_{2A} and A_{2B} receptors. Caffeine has recently been found to act as an antagonist with inverse agonistic activity at $A_{2A}ARs$ (Fernandez-Duenas et al. 2014). Caffeine consumption protects from neurodegenerative diseases such as AD and PD (Chen and Chern 2011).

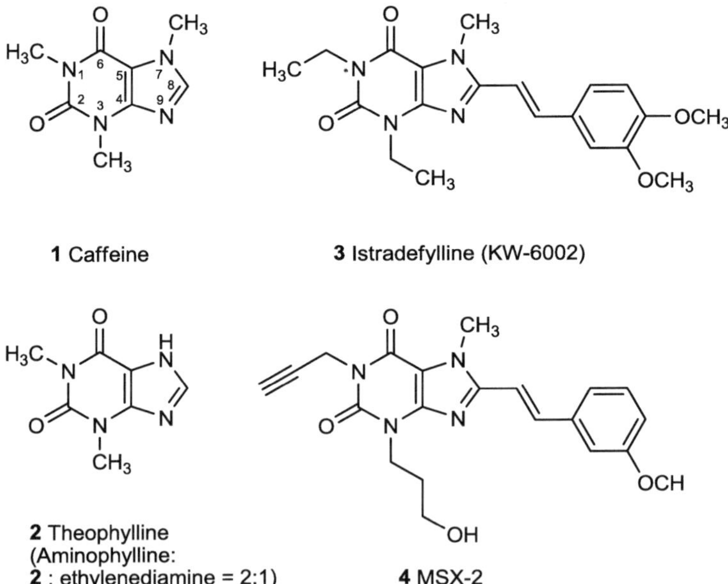

Fig. 3.1 Xanthine derivatives

Table 3.1 Adenosine receptor affinities of antagonists (*h* human, *r* rat, *m* mouse)

		K$_i$ (nM)^a			
		A$_1$	A$_{2A}$	A$_{2B}$	A$_3$
Non-selective antagonists					
1	Caffeine	10,700–44,900 (h)	9560–23,400 (h)	10,400–33,800 (h)	13,300 (h) > 100,000 (r)
		41,000 (r)	45,000 (r)	30,000 (r)	
				13,000 (m)	
2	Theophylline	6770 (h)	6700 (h)	9070 (h)	22,300 (h) > 100,000 (r)
		14,000 (r)	22,000 (r)	15,100 (r)	
				5630 (m)	
A$_{2A}$-selective antagonists					
3	Istradefylline (KW6002)	841 (h)^c	12 (h)	> 10,000 (h)^c	4470 (h)^c
		230 (r)^c	2.2 (r)		
4	MSX-2	2500 (h)	5.38 (h)	> 10,000 (h)	> 10,000 (h)
		900 (r)	8.04 (r)		
5	CGS 15943	3.5 (h)	1.2 (h)	32.4 (h)	35 (h)
		6.4 (r)		9.07 (m)	
6	SCH-58261	725 (h)	5.0 (h)	1110 (h)	1200 (h)
7	SCH-442416	1110 (h)	4.1 (h)	> 10,000 (h)	> 10,000 (h)
8	Preladenant (SCH-420814)	> 1000 (h)	0.9 (h)	> 1000 (h)	> 1000 (h)
9	ZM-241385	774 (h)	1.6 (h)	75 (h)	743 (h)
10	Vipadenant (BIIB014, V2006)	68 (h)	1.3 (h)	63 (h)	1005 (h)
11	ST-1535	71.8 (h)	6.6 (h)	352.3 (h)	> 1000 (h)
12	Tozadenant (SYN-115)	nd	nd	nd	nd
13	TC-G-1004	85 (h)	0.44 (h)		
Dual- and multi-target drugs					
14	Lu AA41063	410 (h)	5.9 (h)	260 (h)	> 10,000 (h)
18	ASP5854	9.03 (h)	1.76 (h)	nd	> 557 (h)
		12.48 (r)	1.24 (r)		
		7.89 (m)	1.62 (m)		
19	JNJ-40255293	48 (h)	6.5 (h)	230 (h)	9200 (h)
20	Pyrimidopurinedione derivative	249 (h)	253 (h)	3520 (h)	> 10,000 (h)
		135 (r)	533 (r)		
21	CSC (K$_i$ MAO-B Ki 80.6 nM) [164]	28,000 (r)	54 (r)	8200	> 10,000 (r)
22	Benzothiazinone derivative (hMAO-B IC$_{50}$ 34.9 nM)	2500 (h)	39.5 (h)	> 1000 (h)	> 1000 (h)
23	Pyrimidopurinedione derivative hMAO-B IC$_{50}$ 1.80 μM	605 (h)	417 (h)	> 1000 (h)	4410 (h)
		1060 (r)	641 (r)		
24	Pyrazinopurinedione derivative (hMAO-B IC$_{50}$ 508 nM)	217 (h)	268 (h)	> 1000 (h)	> 300 (h)
		111 (r)	603 (r)		
25	Pyrazinopurinedione derivative (rMAO-B IC$_{50}$ 260 nM)	791 (h)	1510 (h)	> 300 (h)	> 1000 (h)
		315 (r)	322 (r)		
27	DP-L-A$_{2A}$ANT (D2 > 5000)	< 5000 (h)	7.32 (h)	> 5000 (h)	> 5000 (h)
			2.07 (r)		

Selective A$_{2A}$-Adenosine Receptor Antagonists

Substitution of xanthines at the 8-position with (*E*)-configurated styryl groups has led to selective A$_{2A}$AR antagonists. The 8-styrylxanthine istradefylline (**3**, KW6002) was among the first A$_{2A}$AR antagonists reported (Fig. 3.1). Istradefylline was clinically evaluated and approved in May 2013 in Japan for the adjunctive treatment of PD in combination with L-DOPA (Dungo and Deeks 2013); approval in other countries is still awaited and will require additional clinical studies (Tao and Liang 2015) (Müller 2013). A more potent and selective xanthine A$_{2A}$ antagonist with increased polarity due to the hydroxyl substituent is MSX-2 (**4**), which has also been prepared in tritium-labeled form for radioligand binding studies (Müller et al. 2000).

Various amino-substituted heterocyclic ring systems which bear similarity to adenine were developed with high affinity and selectivity for the A$_{2A}$AR. An early example of an amino-substituted heterotricyclic structure proposed as an A$_{2A}$AR antagonist was the triazoloquinazoline CGS 15943 (**5**), which was later demonstrated to be only slightly A$_{2A}$-selective. Modification of the triazoloquinazoline by addition of a third ring or alteration of the heterocyclic system greatly improved the A$_{2A}$AR selectivity. The pyrazolotriazolopyrimidines SCH-58261 (**6**), SCH442416 (**7**), and preladenant (SCH 420814, **8**) are examples of very potent A$_{2A}$AR antagonists. Preladenant was clinically evaluated for the treatment of PD, but lacked significant effects in Phase III clinical trials and its development was therefore stopped. Preladenant is one of the most potent A$_{2A}$ antagonists with exceptionally high selectivity. Structurally related A$_{2A}$ antagonists include the triazolotriazine ZM241385 (**9**), the triazolopyrimidine vipadenant (**10**, BII014, V2006), and the triazolyl-substituted adenine derivative ST-1535 (**11**). ZM241385 (**9**) also binds to the human A$_{2B}$AR with moderate affinity, and has been used as a radioligand at that receptor [71].

An example for a further non-xanthine A$_{2A}$ antagonists that is, however, structurally unrelated to the xanthine or the adenine derivatives, is the benzothiazole derivative tozadenant (SYN-115, **12**). The scaffold was discovered by a high-throughput screening approach. Tozadenant is clinically evaluated for PD (Phase IIB successfully completed). Another novel structure is represented by TC-G 1004 (**13**), a commercially available acetylamino-substituted pyrimidine derivative which shows high A$_{2A}$ affinity and good selectivity (Zhang et al. 2008). The benzamide **14** (Lu AA4163) was developed by Lundbeck (Mikkelsen et al. 2015) and converted to a water-soluble phosphate prodrug (see below) (Figs. 3.2 and 3.3).

Water-Soluble Prodrugs

The phosphate prodrug MSX-3 (**15**) and the *L*-valine ester prodrug MSX-4 (**16**) have been prepared as water-soluble prodrugs of the potent and selective A$_{2A}$ antagonist MSX-2 (**4**) (Sauer et al. 2000; Vollmann et al. 2008). Both are now broadly used as pharmacological tools in particular for in vivo studies. Another phosphate prodrug of an A$_{2A}$-selective antagonist, Lu AA 47070 (**17**) was developed by Lun-

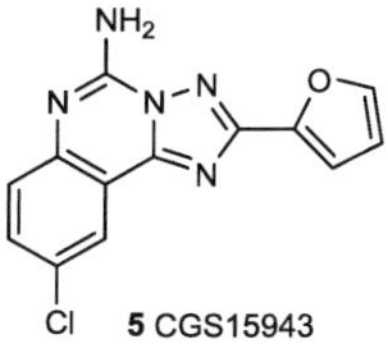
5 CGS15943

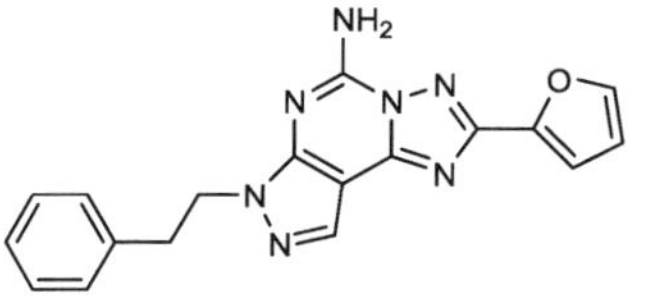
6 SCH-58261

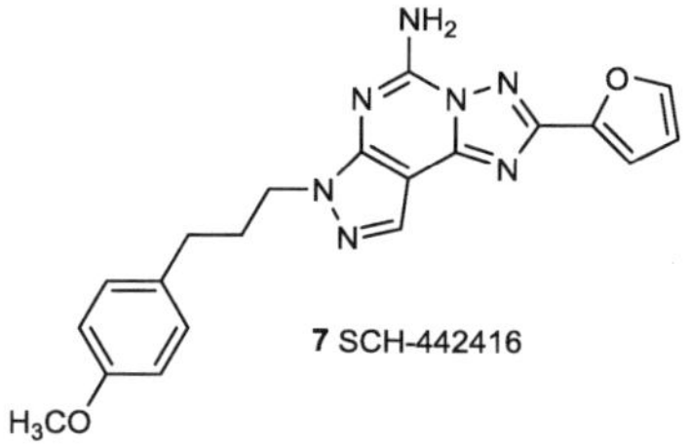
7 SCH-442416

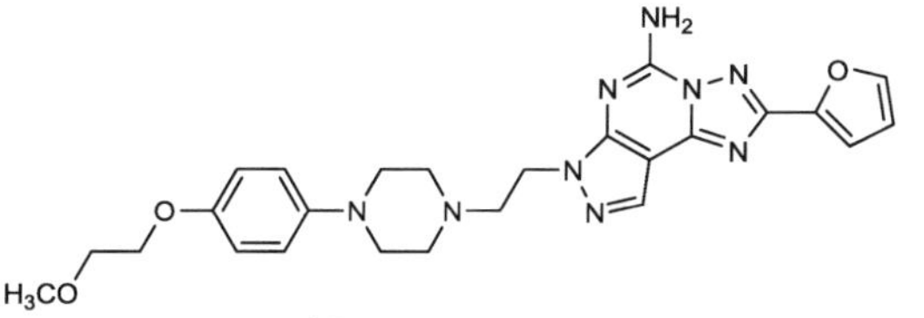
8 Preladenant (SCH-420814)

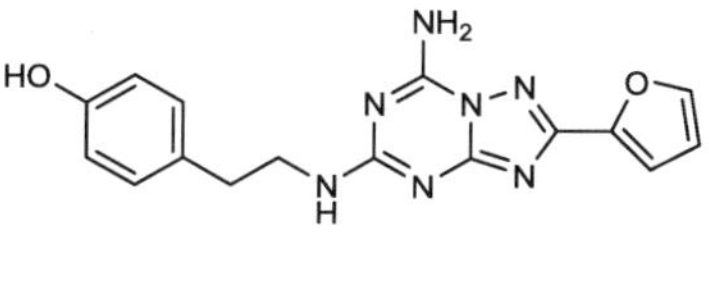
9 ZM-241385

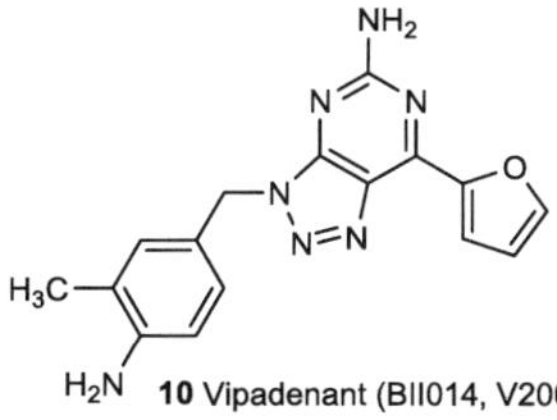
10 Vipadenant (BII014, V2006)

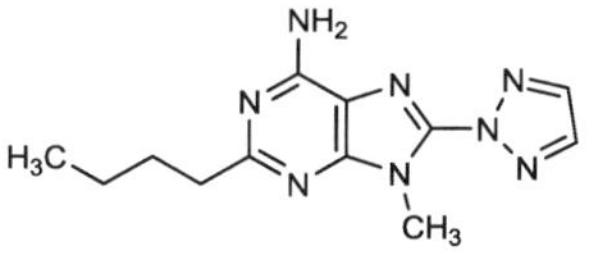
11 ST-1535

Fig. 3.2 A_{2A} adenosine receptor antagonists with adenine-like structure

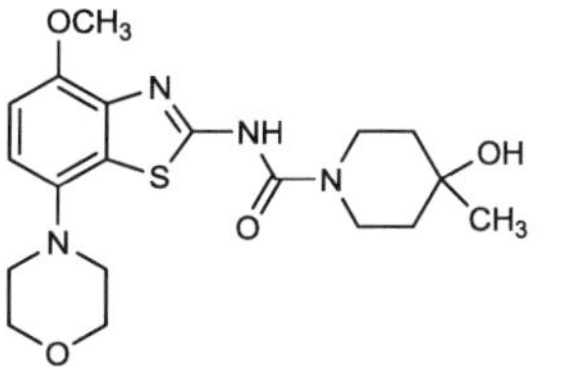
12 Tozadenant (SYN-115)

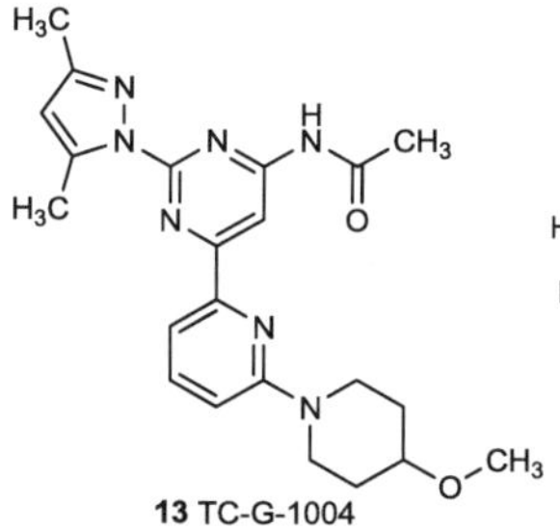
13 TC-G-1004

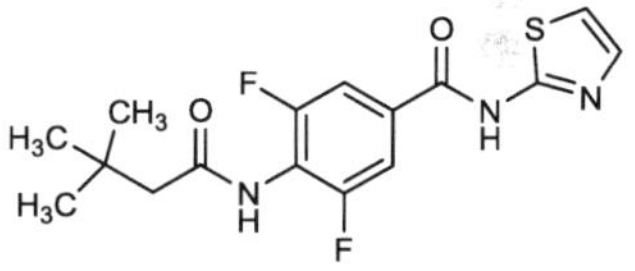
14 Lu AA41063

Fig. 3.3 A_{2A} adenosine receptor antagonists with amide structure

dbeck (Sams et al. 2011). The phosphate prodrugs are highly water-soluble as disodium salts, which can be prepared from the phosphoric acid, e.g., by treatment with the appropriate amount of sodium hydroxide. All of these prodrugs are perorally bioavailable after hydrolysis (Fig. 3.4).

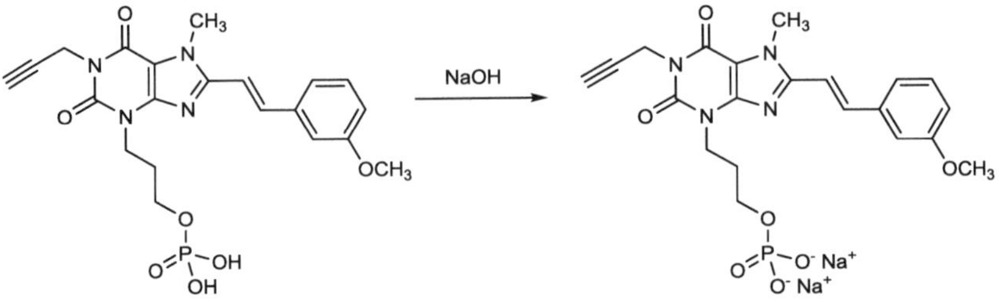

15a MSX-3 (free acid) **15b** MSX-3 (di-sodium salt)

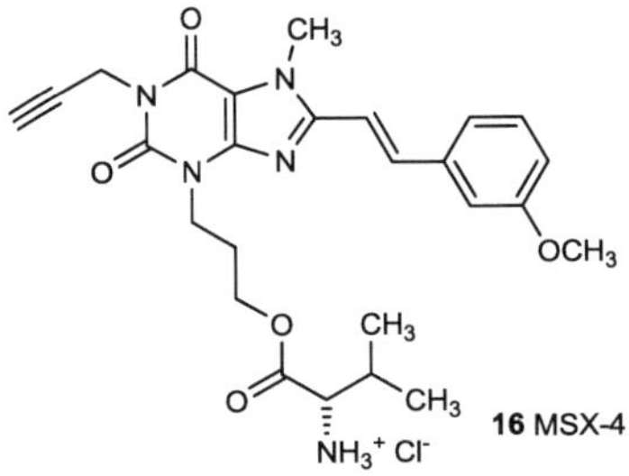

16 MSX-4

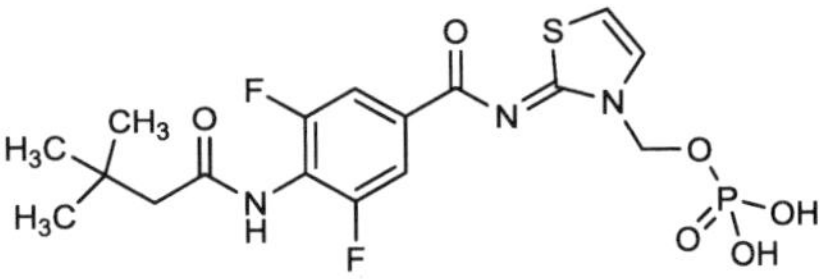

17 Lu AA47070

Fig. 3.4 Water-soluble prodrugs of A_{2A} adenosine receptor antagonists

Negative Allosteric Modulators of A_{2A} Adenosine Receptors

The concept of allosteric modulation of GPCRs is quite new and only recently, allosteric modulators for a number of GPCRs have been developed (De Amici et al. 2010; Jacobson et al. 2011; Müller et al. 2012). For A_{2A}ARs antibodies have been described that act as allosteric inverse agonists locking the receptor in an inactive conformation (Hino et al. 2012).

Dual- and Multi-Target Approaches

Several clinical trials with selective A_{2A} antagonists which were to be developed for the treatment of PD have failed. More than 200 different clinical trials evaluating new drugs for AD addressing different targets have failed in the past 10 years. This has led to the worry that addressing a single target may be inefficient in the treatment of complex diseases. Multi-target approaches which modulate biological networks might be more promising (Geldenhuys and Van der Schyf 2013). These have been successful in cancer therapy (multi-kinase inhibitors), and in many infectious diseases (e.g. HIV and tuberculosis therapy). Many potent central nervous system drugs, such as antidepressants and neuroleptics, interact with multiple targets. Moreover, bioactive natural products, e.g. caffeine, curcumin, resveratrol and many more, often interact with several target proteins at similar concentrations. Activity of a drug at multiple targets may therefore result in additive or even synergistic effects and may be associated with lower side-effects because reduced doses can be applied.

A simple approach to hit several targets is drug combination. However, a better strategy is to combine several properties in a single drug molecule and to develop multi-target drugs. Advantages of a dual- or multi-target drug approach include simplification of drug regimen, improved compliance, less side-effects and reduced toxicity, more predictable pharmacology, reduced drug-drug interactions, less complex pharmacokinetics, and easier manufacturing and formulation.

Dual- or multi-target drugs can be obtained by two principal strategies: (i) a bivalent drug approach connecting pharmacophore structures by a linker, and (ii) a common pharmacophore approach in which the pharmacophoric structures for the targets are merged.

Dual A$_1$/A$_{2A}$ Adenosine Receptor Antagonists

Dual A$_1$/A$_{2A}$ antagonists have been developed to combine the anti-PD activities of A$_{2A}$AR blockade (improvement of motor impairment and neuroprotective effects) with improvement of cognitive function by A$_1$AR antagonism. Examples for dual A$_1$/A$_{2A}$ antagonists include ASP5854 (**18**) (Mihara et al. 2007), JNJ-40255293 (**19**) (Atack et al. 2014), and the tricyclic pyrimidopurinedione **20** (Koch et al. 2013).

Dual A$_{2A}$ Antagonists/MAO-B Inhibitors

Monoamineoxidase B (MAO-B) inhibitors, including selegiline and rasagiline, are clinically used for the treatment of PD, mostly in combination with L-DOPA or dopamine agonists. Both MAO-B inhibitor show an irreversible mode of action. The first reversible MAO-B inhibitor, safinamide, has recently been approved for the treatment of PD in Germany. Since MAO-B inhibitors show only weak effects on PD symptoms, multi-target ligands have been proposed which display additional activities, e.g. A$_{2A}$AR blockade (Pisani et al. 2011). Some 8-styrylxanthine derivatives, such as 8-(3-chlorostyryl)caffeine (CSC, **21**) were accidentally found to inhibit MAO-B in addition to the A$_{2A}$AR (Pretorius et al. 2008). Structure-activity relationships for this class of compounds has been extensively studied and analyzed by computational methods (Azam et al. 2012; Petzer and Petzer 2015).

The first non-xanthine-derived dual A$_{2A}$/MAO-B inhibitors have recently been described (Stössel et al. 2013). The most potent compound of a series of benzothiazinones was **22**.

Triple A$_1$/A$_{2A}$ Antagonists/MAO-B Inhibitors

Several series of tricyclic pyrimido- and pyrazino-purinediones have been developed with triple inhibition of MAO-B, A$_{2A}$- and A$_1$ARs (Brunschweiger et al. 2014; Koch et al. 2013). The best triple-active compounds were **23–25** (see Table 3.1). It should be noted that some of the compounds showed considerable species differences (Fig. 3.5).

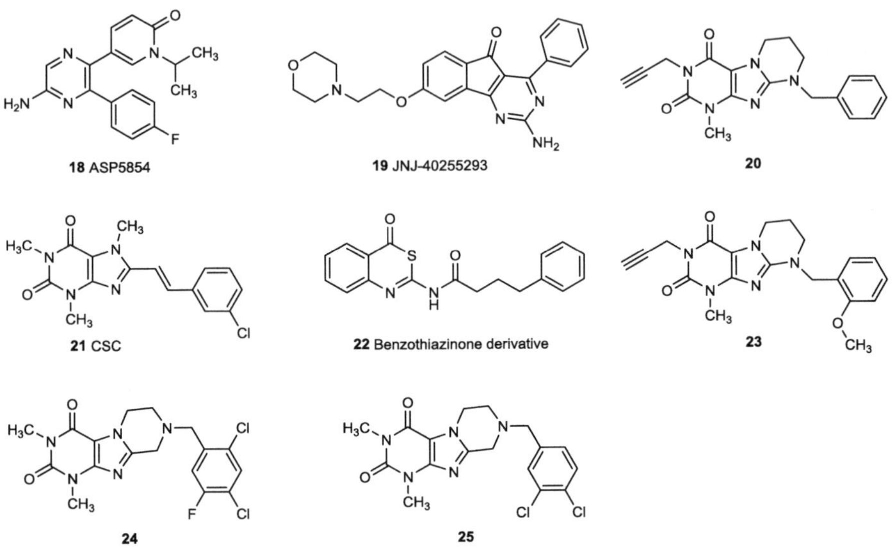

Fig. 3.5 Dual- and multi-target drugs: A_{2A} antagonists with additional A_1-antagonistic and/or MAO-B inhibitory activity

Dual A_{2A} Antagonists and Dopamine Agonists

Activation of dopamine D_2 receptors and blockade of A_{2A}ARs is expected to be beneficial for the treatment of PD. The merging of pharmacophores for an agonist at one receptor (D_2) and an antagonist at another receptor (A_{2A}) is difficult or even impossible. Therefore all published approaches have connected two pharmacophores, one for each receptor, by linkers of different length (Dalpiaz et al. 2012; Jorg et al. 2015; Soriano et al. 2009). Compound **27** is a dopamine prodrug which releases dopamine after amide hydrolysis (Fig. 3.6).

A_{2A} and NMDA Antagonists

A combination of an A_{2A} antagonist and NMDA receptor antagonist with selectivity for the NR2B receptor subtype resulted in synergistic effects in a rat model of PD (Michel et al. 2014).

Structure Elucidation by X-Ray Crystallography

In 2008 the first X-ray structure of an AR was solved in a 2.6 Å resolution by Ray Stevens and coworkers: that of the A_{2A}AR in complex with the antagonist ZM-241385 (**9**) (Jaakola et al. 2008). The surprise of that structure was the orientation of

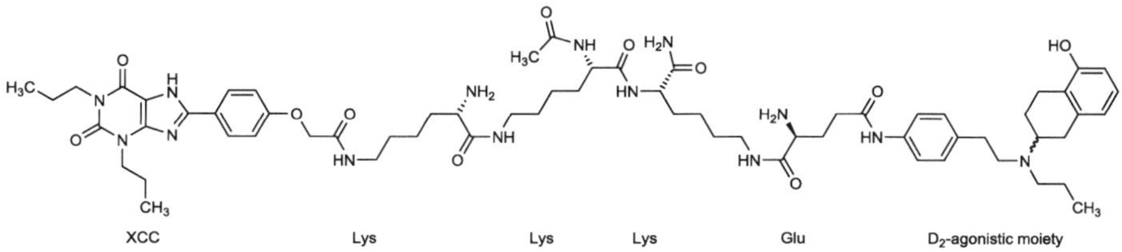

26 A$_{2A}$ antagonist-D$_2$ agonist peptide conjugate

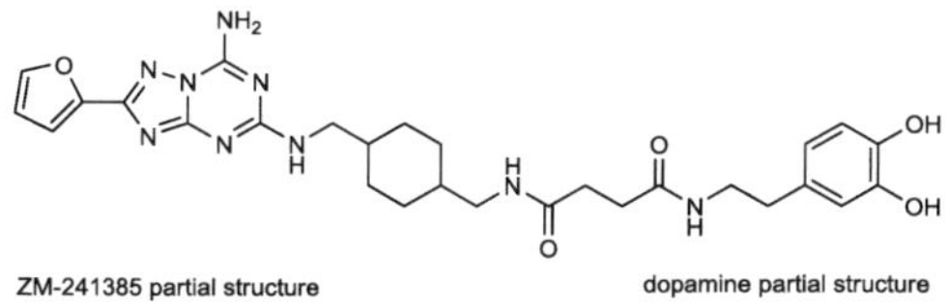

27 DP-L-A$_{2A}$ANT (prodrug)

Fig. 3.6 Conjugates of A$_{2A}$ antagonists D$_2$-agonists

the ligand in the binding pocket, which was very different from that of the biogenic amine receptors. The ligand was not arranged parallel to the (extracellular) surface of the receptor, but perpendicular. Three years later, the agonist-bound structure of the A$_{2A}$AR in its activated conformation was reported with a similarly high resolution by the same group. (Xu et al. 2011) In the same year, Marshall and coworkers from Hepares Therapeutics published several antagonist-bound A$_{2A}$AR X-ray structures (Dore et al. 2011), and in 2012 Stevens and coworkers discovered the sodium binding site in a high resolution structure (1.8 Å) of the A$_{2A}$AR (Liu et al. 2012). Several other X-ray structures of the A$_{2A}$AR with various ligands, including the agonists adenosine and NECA, have been determined (see http://www.ebi.ac.uk/ interpro/protein/P29274/structures;jsessionid=089D40DB436ED1ED1BB9C2C63 A4D41B7). These structures have allowed comformational dynamic, docking and virtual screening studies and have contributed to the identification and development of new A$_{2A}$ receptor ligands, particularly antagonists (Bacilieri et al. 2013; Carlsson et al. 2010; Chen et al. 2013; Pang et al. 2013; Rodriguez et al. 2015; Wei et al. 2010).

Conclusions

The development of drugs targeting A$_{2A}$ARs has made huge progress in the last years. The first A$_{2A}$ antagonist, istradefylline, is marketed in Japan for the treatment of PD. Further drugs are in development for various indications. Despite initial failures it appears that A$_{2A}$ antagonists will have a bright future.

Acknowledgements CEM is grateful for support by BMBF (BioPharma—Neuroallianz), DFG, DAAD, Alzheimer Forschung Initiative (AFI), Alexander von Humboldt foundation (AvH).

References

Armentero MT, Pinna A, Ferré S et al (2011) Past, present and future of A (2A) adenosine receptor antagonists in the therapy of Parkinson's disease. Pharmacol Ther 132:280–299

Atack JR, Shook BC, Rassnick S et al (2014) JNJ-40255293, a novel adenosine A2A/A1 antagonist with efficacy in preclinical models of Parkinson's disease. ACS Chem Neurosci 5:1005–1019

Augusto E, Matos M, Sévigny J et al (2013) Ecto-5'-nucleotidase (CD73)-mediated formation of adenosine is critical for the striatal adenosine A2A receptor functions. J Neurosci 33:11390–11399

Azam F, Madi AM, Ali HI (2012) Molecular docking and prediction of pharmacokinetic properties of dual mechanism drugs that block MAO-B and adenosine A(2A) receptors for the treatment of Parkinson's disease. J Young Pharm 4:184–192

Bacilieri M, Ciancetta A, Paoletta S et al (2013) Revisiting a receptor-based pharmacophore hypothesis for human A(2A) adenosine receptor antagonists. J Chem Inf Model 53:1620–1637

Borroto-Escuela DO, Marcellino D, Narvaez M et al (2010) A serine point mutation in the adenosine A2AR C-terminal tail reduces receptor heteromerization and allosteric modulation of the dopamine D2R. Biochem Biophys Res Commun 394:222–227

Borroto-Escuela DO, Romero-Fernandez W, Tarakanov AO et al (2011) On the existence of a possible A2A-D2-beta-arrestin2 complex: A2A agonist modulation of D2 agonist-induced beta-arrestin2 recruitment. J Mol Biol 406:687–699

Brown RM, Short JL (2008) Adenosine A(2A) receptors and their role in drug addiction. J Pharm Pharmacol 60:1409–1430

Brown RM, Short JL, Cowen MS et al (2009) A differential role for the adenosine A2A receptor in opiate reinforcement vs opiate-seeking behavior. Neuropsychopharmacology 34:844–856

Brown ST, Reyes EP, Nurse CA (2011) Chronic hypoxia upregulates adenosine 2a receptor expression in chromaffin cells via hypoxia inducible factor-2alpha: role in modulating secretion. Biochem Biophys Res Commun 412:466–472

Brunschweiger A, Koch P, Schlenk M et al (2014) 8-Benzyltetrahydropyrazino[2,1-f]purinediones: water-soluble tricyclic xanthine derivatives as multitarget drugs for neurodegenerative diseases. ChemMedChem 9:1704–1724

Carlsson J, Yoo L, Gao ZG et al (2010) Structure-based discovery of A2A adenosine receptor ligands. J Med Chem 53:3748–3755

Carman AJ, Dacks PA, Lane RF et al (2014) Current evidence for the use of coffee and caffeine to prevent age-related cognitive decline and Alzheimer's disease. J Nutr Health Aging 18:383–392

Cerri S, Levandis G, Ambrosi G et al (2014) Neuroprotective potential of adenosine A2A and cannabinoid CB1 receptor antagonists in an animal model of Parkinson disease. J Neuropathol Exp Neurol 73:414–424

Chen JF (2014) Adenosine receptor control of cognition in normal and disease. Int Rev Neurobiol 119:257–307

Chen JF, Chern Y (2011) Impacts of methylxanthines and adenosine receptors on neurodegeneration: human and experimental studies. Handb Exp Pharmacol 200:267–310

Chen D, Ranganathan A, IJzerman AP et al (2013) Complementarity between in silico and biophysical screening approaches in fragment-based lead discovery against the A(2A) adenosine receptor. J Chem Inf Model 53:2701–2714

Cristalli G, Muller CE, Volpini R (2009) Recent developments in adenosine A2A receptor ligands. Handb Exp Pharmacol 193:59–98

Dalpiaz A, Cacciari B, Vicentini CB et al (2012) A novel conjugated agent between dopamine and an A2A adenosine receptor antagonist as a potential anti-Parkinson multitarget approach. Mol Pharm 9:591–604

De Amici M, Dallanoce C, Holzgrabe U et al (2010) Allosteric ligands for G protein-coupled receptors: a novel strategy with attractive therapeutic opportunities. Med Res Rev 30:463–549

De Lera Ruiz M, Lim YH, Zheng J (2014) Adenosine A2A receptor as a drug discovery target. J Med Chem 57:3623–3650

Decerce J, Smith LF, Gonzalez W et al (2007) Effectiveness and tolerability of istradefylline for the treatment of restless legs syndrome: an exploratory study in five female patients. Curr Ther Res Clin Exp 68:349–359

Dix AV, Moss SM, Phan K et al (2014) Programmable nanoscaffolds that control ligand display to a G-protein-coupled receptor in membranes to allow dissection of multivalent effects. J Am Chem Soc 136:12296–12303

Doré AS, Robertson N, Errey JC et al (2011) Structure of the adenosine A(2A) receptor in complex with ZM241385 and the xanthines XAC and caffeine. Structure 19:1283–1293

Dungo R, Deeks ED (2013) Istradefylline: first global approval. Drugs 73:875–882

El-Tayeb A, Iqbal J, Behrenswerth A et al (2009) Nucleoside-5'-monophosphates as prodrugs of adenosine A2A receptor agonists activated by ecto-5'-nucleotidase. J Med Chem 52:7669–7677

Eltzschig HK, Sitkovsky MV, Robson SC (2012) Purinergic signaling during inflammation. N Engl J Med 367:2322–2333

Fernández-Dueñas V, Gómez-Soler M, López-Cano M et al (2014) Uncovering caffeine's adenosine A2A receptor inverse agonism in experimental parkinsonism. ACS Chem Biol 9:2496–2501

Ferré S, Ciruela F, Canals M et al (2004) Adenosine A2A-dopamine D2 receptor-receptor heteromers. Targets for neuro-psychiatric disorders. Parkinsonism Relat Disord 10:265–271

Flaten V, Laurent C, Coelho JE et al (2014) From epidemiology to pathophysiology: what about caffeine in Alzheimer's disease? Biochem Soc Trans 42:587–592

Flögel U, Burghoff S, van Lent PL et al (2012) Selective activation of adenosine A2A receptors on immune cells by a CD73-dependent prodrug suppresses joint inflammation in experimental rheumatoid arthritis. Sci Transl Med 4:146ra108

Fredholm BB, IJzerman AP, Jacobson KA et al (2011) International union of basic and clinical pharmacology. LXXXI. Nomenclature and classification of adenosine receptors–an update. Pharmacol Rev 63:1–34

Geldenhuys WJ, Van der Schyf CJ (2013) Rationally designed multi-targeted agents against neurodegenerative diseases. Curr Med Chem 20:1662–1672

Gnad T, Scheibler S, von Kügelgen I et al (2014) Adenosine activates brown adipose tissue and recruits beige adipocytes via A2A receptors. Nature 516:395–399

Gonçalves N, Simões AT, Cunha RA et al (2013) Caffeine and adenosine A(2A) receptor inactivation decrease striatal neuropathology in a lentiviral-based model of Machado–Joseph disease. Ann Neurol 73:655–666

Hatfield SM, Kjaergaard J, Lukashev D et al (2015) Immunological mechanisms of the antitumor effects of supplemental oxygenation. Sci Transl Med 7:277ra230

Hino T, Arakawa T, Iwanari H et al (2012) G-protein-coupled receptor inactivation by an allosteric inverse-agonist antibody. Nature 482:237–240

Jaakola VP, Griffith MT, Hanson MA et al (2008) The 2.6 angstrom crystal structure of a human A2A adenosine receptor bound to an antagonist. Science 322:1211–1217

Jacobson KA, Gao ZG (2006) Adenosine receptors as therapeutic targets. Nat Rev Drug Discov 5:247–264

Jacobson KA, Gao ZG, Göblyös A et al (2011) Allosteric modulation of purine and pyrimidine receptors. Adv Pharmacol 61:187–220

Jones N, Bleickardt C, Mullins D et al (2013) A2A receptor antagonists do not induce dyskinesias in drug-naive or L-dopa sensitized rats. Brain Res Bull 98:163–169

Jorg M, Scammells PJ, Capuano B (2014) The dopamine D2 and adenosine A2A receptors: past, present and future trends for the treatment of Parkinson's disease. Curr Med Chem 21:3188–3210

Jörg M, May LT, Mak FS et al (2015) Synthesis and pharmacological evaluation of dual acting ligands targeting the adenosine A2A and dopamine D2 receptors for the potential treatment of Parkinson's disease. J Med Chem 58:718–738

Justinová Z, Redhi GH, Goldberg SR et al (2014) Differential effects of presynaptic versus postsynaptic adenosine A2A receptor blockade on Delta9-tetrahydrocannabinol (THC) self-administration in squirrel monkeys. J Neurosci 34:6480–6484

Kanda T, Uchida S (2014) Clinical/pharmacological aspect of adenosine A2A receptor antagonist for dyskinesia. Int Rev Neurobiol 119:127–150

Koch P, Akkari R, Brunschweiger A et al (2013) 1,3-Dialkyl-substituted tetrahydropyrimido[1,2-f]purine-2,4-diones as multiple target drugs for the potential treatment of neurodegenerative diseases. Bioorg Med Chem 21:7435–7452

Laurent C, Eddarkaoui S, Derisbourg M et al (2014a) Beneficial effects of caffeine in a transgenic model of Alzheimer's disease-like tau pathology. Neurobiol Aging 35:2079–2090

Laurent C, Burnouf S, Ferry B et al (2014b) A adenosine receptor deletion is protective in a mouse model of Tauopathy. Mol Psychiatry. doi:10.1038/mp.2014.151 (in press)

Li P, Rial D, Canas PM et al (2015) Optogenetic activation of intracellular adenosine A receptor signaling in the hippocampus is sufficient to trigger CREB phosphorylation and impair memory. Mol Psychiatry. doi:10.1038/mp.2014.182 (in press)

Linden J, Cekic C (2012) Regulation of lymphocyte function by adenosine. Arterioscler Thromb Vasc Biol 32:2097–2103

Liu W, Chun E, Thompson AA et al (2012) Structural basis for allosteric regulation of GPCRs by sodium ions. Science 337:232–236

López-Cruz L, Salamone JD, Correa M (2013) The impact of caffeine on the behavioral effects of ethanol related to abuse and addiction: a review of animal studies. J Caffeine Res 3:9–21

Ma DF, Kondo T, Nakazawa T et al (2010) Hypoxia-inducible adenosine A2B receptor modulates proliferation of colon carcinoma cells. Hum Pathol 41:1550–1557

Maitre NL, Stark AR (2012) Neuroprotection for premature infants?: another perspective on caffeine. JAMA 307:304–305

Matos M, Augusto E, Santos-Rodrigues AD et al (2012) Adenosine A2A receptors modulate glutamate uptake in cultured astrocytes and gliosomes. Glia 60:702–716

Matos M, Augusto E, Agostinho P et al (2013) Antagonistic interaction between adenosine A2A receptors and Na+/K+ -ATPase-alpha2 controlling glutamate uptake in astrocytes. J Neurosci 33:18492–18502

Michel A, Downey P, Nicolas JM et al (2014) Unprecedented therapeutic potential with a combination of A2A/NR2B receptor antagonists as observed in the 6-OHDA lesioned rat model of Parkinson's disease. PLoS One 9:e114086

Mihara T, Mihara K, Yarimizu J et al (2007) Pharmacological characterization of a novel, potent adenosine A1 and A2A receptor dual antagonist, 5-[5-amino-3-(4-fluorophenyl)pyrazin-2-yl]-1-isopropylpyridine-2(1 H)-one (ASP5854), in models of Parkinson's disease and cognition. J Pharmacol Exp Ther 323:708–719

Mikkelsen GK, Langgård M, Schrøder TJ et al (2015) Synthesis and SAR studies of analogues of 4-(3,3-dimethyl-butyrylamino)-3,5-difluoro-N-thiazol-2-yl-benzamide (Lu AA41063) as adenosine A2A receptor ligands with improved aqueous solubility. Bioorg Med Chem Lett 25:1212–1216

Molero Y, Gumpert C, Serlachius E et al (2013) A study of the possible association between adenosine A2A receptor gene polymorphisms and attention-deficit hyperactivity disorder traits. Genes Brain Behav 12:305–310

Muller T (2013) Suitability of the adenosine antagonist istradefylline for the treatment of Parkinson's disease: pharmacokinetic and clinical considerations. Expert Opin Drug Metab Toxicol 9:1015–1024

Muller CE, Ferré S (2007) Blocking striatal adenosine A2A receptors: a new strategy for basal ganglia disorders. Recent Pat CNS Drug Discov 2:1–21

Muller CE, Jacobson KA (2011a) Xanthines as adenosine receptor antagonists. Handb Exp Pharmacol 200:151–199

Muller CE, Jacobson KA (2011b) Recent developments in adenosine receptor ligands and their potential as novel drugs. Biochim Biophys Acta 1808:1290–1308

Muller CE, Scior T (1993) Adenosine receptors and their modulators. Pharm Acta Helv 68:77–111

Müller CE, Maurinsh J, Sauer R (2000) Binding of [3H]MSX-2 (3-(3-hydroxypropyl)-7-methyl-8-(m-methoxystyryl)-1-propargylxanthine) to rat striatal membranes–a new, selective antagonist radioligand for A(2A) adenosine receptors. Eur J Pharm Sci 10:259–265

Müller CE, Schiedel AC, Baqi Y (2012) Allosteric modulators of rhodopsin-like G protein-coupled receptors: opportunities in drug development. Pharmacol Ther 135:292–315

Mundell SJ, Kelly E (1998) Evidence for co-expression and desensitization of A2a and A2b adenosine receptors in NG108-15 cells. Biochem Pharmacol 55:595–603

Navarro G, Aymerich MS, Marcellino D et al (2009) Interactions between calmodulin, adenosine A2A, and dopamine D2 receptors. J Biol Chem 284:28058–28068

Nomoto M, Nagai M, Nishikawa N (2014) Clinical nonmotor aspect of A2A antagonist in PD treatment. Int Rev Neurobiol 119:191–194

O'Neill CE, Hobson BD, Levis SC et al (2014) Persistent reduction of cocaine seeking by pharmacological manipulation of adenosine A1 and A2A receptors during extinction training in rats. 231:3179–3188

Pandolfo P, Machado NJ, Köfalvi A et al (2013) Caffeine regulates frontocorticostriatal dopamine transporter density and improves attention and cognitive deficits in an animal model of attention deficit hyperactivity disorder. Eur Neuropsychopharmacol 23:317–328

Pang X, Yang M, Han K (2013) Antagonist binding and induced conformational dynamics of GPCR A2A adenosine receptor. Proteins 81:1399–1410

Perez-Aso M, Chiriboga L, Cronstein BN (2012) Pharmacological blockade of adenosine A2A receptors diminishes scarring. FASEB J 26:4254–4263

Petzer JP, Petzer A (2015) Caffeine as a lead compound for the design of therapeutic agents for the treatment of Parkinson's disease. Curr Med Chem 22:975–988

Pinna A (2014) Adenosine A2A receptor antagonists in Parkinson's disease: progress in clinical trials from the newly approved istradefylline to drugs in early development and those already discontinued. CNS Drugs 28:455–474

Pinna A, Bonaventura J, Farré D et al (2014) L-DOPA disrupts adenosine A(2A)-cannabinoid CB(1)-dopamine D(2) receptor heteromer cross-talk in the striatum of hemiparkinsonian rats: biochemical and behavioral studies. Exp Neurol 253:180–191

Pires VA, Pamplona FA, Pandolfo P et al (2009) Adenosine receptor antagonists improve short-term object-recognition ability of spontaneously hypertensive rats: a rodent model of attention-deficit hyperactivity disorder. Behav Pharmacol 20:134–145

Pisani L, Catto M, Leonetti F et al (2011) Targeting monoamine oxidases with multipotent ligands: an emerging strategy in the search of new drugs against neurodegenerative diseases. Curr Med Chem 18:4568–4587

Pretorius J, Malan SF, Castagnoli N Jr et al (2008) Dual inhibition of monoamine oxidase B and antagonism of the adenosine A(2A) receptor by (E, E)-8-(4-phenylbutadien-1-yl)caffeine analogues. Bioorg Med Chem 16:8676–8684

Rodríguez D, Gao ZG, Moss SM et al (2015) Molecular docking screening using agonist-bound GPCR structures: probing the A2A adenosine receptor. J Chem Inf Model 55:550–563

Sams AG, Mikkelsen GK, Larsen M et al (2011) Discovery of phosphoric acid mono-{2-[(E/Z)-4-(3,3-dimethyl-butyrylamino)-3,5-difluoro-benzoylimino]-thiazol-3 -ylmethyl} ester (Lu AA47070): a phosphonooxymethylene prodrug of a potent and selective hA(2A) receptor antagonist. J Med Chem 54:751–764

Sauer R, Maurinsh J, Reith U et al (2000) Water-soluble phosphate prodrugs of 1-propargyl-8-styrylxanthine derivatives, A(2A)-selective adenosine receptor antagonists. J Med Chem 43:440–448

Silva CG, Métin C, Fazeli W et al (2013) Adenosine receptor antagonists including caffeine alter fetal brain development in mice. Sci Transl Med 5:197ra104

Sitkovsky MV, Hatfield S, Abbott R et al (2014) Hostile, hypoxia-A2-adenosinergic tumor biology as the next barrier to overcome for tumor immunologists. Cancer Immunol Res 2:598–605

Soriano A, Ventura R, Molero A et al (2009) Adenosine A2A receptor-antagonist/dopamine D2 receptor-agonist bivalent ligands as pharmacological tools to detect A2A-D2 receptor heteromers. J Med Chem 52:5590–5602

Stössel A, Schlenk M, Hinz S et al (2013) Dual targeting of adenosine A(2A) receptors and mono-amine oxidase B by 4H-3,1-benzothiazin-4-ones. J Med Chem 56:4580–4596

Tao Y, Liang G (2015) Efficacy of adenosine A2A receptor antagonist istradefylline as augmentation for Parkinson's disease: a meta-analysis of randomized controlled trials. Cell Biochem Biophys 71:57–62

Uchida S, Kadowaki-Horita T, Kanda T (2014) Effects of the adenosine A2A receptor antagonist on cognitive dysfunction in Parkinson's disease. Int Rev Neurobiol 119:169–189

Vollmann K, Qurishi R, Hockemeyer J et al (2008) Synthesis and properties of a new water-soluble prodrug of the adenosine A2A receptor antagonist MSX-2. Molecules 13:348–359

Wang H, Guan W, Yang W et al (2014) Caffeine inhibits the activation of hepatic stellate cells induced by acetaldehyde via adenosine A2A receptor mediated by the cAMP/PKA/SRC/ERK1/2/P38 MAPK signal pathway. PLoS One 9:e92482

Wei J, Qu W, Ye Y et al (2010) 3D pharmacophore based virtual screening of A2A adenosine receptor antagonists. Protein Pept Lett 17:332–339

Wills AM, Eberly S, Tennis M et al (2013) Caffeine consumption and risk of dyskinesia in CALM-PD. Mov Disord 28:380–383

Wydra K, Suder A, Borroto-Escuela DO et al (2015a). On the role of A2A and D2 receptors in control of cocaine and food-seeking behaviors in rats. Psychopharmacology 232:1767–1778

Wydra K, Gołembiowska K, Suder A et al (2015b) On the role of adenosine (A)(2)A receptors in cocaine-induced reward: a pharmacological and neurochemical analysis in rats. Psychopharmacology 232:421–435

Xu F, Wu H, Katritch V et al (2011) Structure of an agonist-bound human A2A adenosine receptor. Science 332:322–327

Yamada K, Kobayashi M, Mori A et al (2013) Antidepressant-like activity of the adenosine A(2A) receptor antagonist, istradefylline (KW-6002), in the forced swim test and the tail suspension test in rodents. Pharmacol Biochem Behav 114–115:23–30

Yamada K, Kobayashi M, Kanda T (2014a) Involvement of adenosine A2A receptors in depression and anxiety. Int Rev Neurobiol 119:373–393

Yamada K, Kobayashi M, Shiozaki S et al (2014b) Antidepressant activity of the adenosine A2A receptor antagonist, istradefylline (KW-6002) on learned helplessness in rats. Psychopharmacology 231:2839–2849

Yao L, McFarland K, Fan P et al (2006) Adenosine A2a blockade prevents synergy between mu-opiate and cannabinoid CB1 receptors and eliminates heroin-seeking behavior in addicted rats. Proc Natl Acad Sci U S A 103:7877–7882

Zezula J, Freissmuth M (2008) The A(2A)-adenosine receptor: a GPCR with unique features? Br J Pharmacol 153:S184–S190

Zhang X, Tellew JE, Luo Z et al (2008) Lead optimization of 4-acetylamino-2-(3,5-dimethylpyrazol-1-yl)-6-pyridylpyrimidines as A2A adenosine receptor antagonists for the treatment of Parkinson's disease. J Med Chem 51:7099–7110

Chapter 4
Adenosine A_{2A} Receptors and Neurotrophic Factors: Relevance for Parkinson's Disease

Maria J. Diógenes, Joaquim A. Ribeiro and Ana M. Sebastião

Abstract Neurotrophic factors (NTF) or drugs able to boost NTF actions have been frequently considered as promising therapies for neurodegenerative diseases namely for Parkinson's disease (PD).

A considerable number of data was published demonstrating that there is a cross talk between NTF and a particular type of adenosine receptors, the A_{2A} receptors (A_{2A}R). Together, those studies show that relevant actions of NTF are dependent on or facilitated by activation of A_{2A}R, so that most NTF actions on synapses are lost upon blockade of A_{2A}R. These findings suggest caution in the use of A_{2A}R antagonists whenever NTF actions are demanded and place the A_{2A}R agonists in a suitable position as a pharmacologic strategy to potentiate NTF mediated actions in neurodegenerative diseases, including PD. However, the negative interaction between A_{2A}R and dopamine D_2 receptors in the striatum, together with the A_{2A}R-mediated exacerbation of excitotoxicity mechanisms, points towards the therapeutic potential of A_{2A}R antagonists in PD. Indeed, clinical trials with A_{2A}R antagonists were already conducted.

Here we detail the existing, molecular and functional, evidence for the cross-talk between NTF and A_{2A}R and discuss its possible relevance for the treatment of PD. Available data highlights the need for considering appropriate time windows for the different strategies to fight the disease to avoid losing endogenous neurotrophic support in the early phases of the disease where synapses and neurons are to struggling for life.

Keywords Adenosine · A_{2A} Receptors · Neurotrophic factors · Neuromodulation

A. M. Sebastião (✉) · M. J. Diógenes · J. A. Ribeiro
Institute of Pharmacology and Neurosciences, Faculty of Medicine and Unit of Neurosciences,
Instituto de Medicina Molecular, Universidade de Lisboa
e-mail: anaseb@medicina.ulisboa.pt

© Springer International Publishing Switzerland 2015
M. Morelli et al. (eds.), *The Adenosinergic System,* Current Topics in Neurotoxicity 10,
DOI 10.1007/978-3-319-20273-0_4

Cross Talk Between Receptors for Neurotrophic Factors and Adenosine A$_{2A}$ Receptors

Introduction

Neurotrophic factors (NTF) are secreted proteins that actively promote the growth and survival of developing neurons, whilst playing a housekeeping role in the homeostatic maintenance of mature neuronal circuits. The NTF family comprises four distinct major groups: (1) the neurotrophin family [brain-derived neurotrophic factor (BDNF), nerve growth factor (NGF), neurotrophin-3 (NT-3), and neurotrophin-4/5 (NT-4)]; (2) the glial cell line-derived neurotrophic factor (GDNF) family of ligands (GFLs); (3) neurotrophic cytokines (neurokines); and (4) the family of cerebral dopamine neurotrophic factor (CDNF) and mesencephalic astrocyte derived neurotrophic factor (MANF).

In the 90s several papers reported the role of cyclic AMP (cAMP), a second messenger that is increased following adenosine A$_{2A}$ receptor (A$_{2A}$R) activation, on NTF mediated actions (e.g. (Boulanger and Poo 1999; Meyer-Franke et al. 1995) or on NTF expression levels (e.g. (Yamamoto et al. 1993)). The first evidence for a possible cross-talk between adenosine receptors and NTF arose in 1997 with two studies showing that the activation of A$_{2A}$R increased NGF expression and release from primary glial cultures (Heese et al. 1997) and that PC12 cells differentiation mediated by NGF was accompanied by a decrease in A$_{2A}$R-mediated cAMP accumulation (Arslan et al. 1997). In 2001 Lee and Chao, pioneered the field describing that activation of A$_{2A}$R can, in the absence of neurotrophins, induce phosphorylation of TrkA and TrkB receptors, in PC12 cells and in hippocampal neurons, respectively (Lee and Chao 2001). The functional influence of A$_{2A}$R on NTF effects on synaptic transmission was first reported by Diógenes et al. (2014), who showed that the facilitation of BDNF actions on synaptic transmission requires A$_{2A}$R activation, since the effect of BDNF is blocked by A$_{2A}$R blockade and it is exacerbated by enhancement of ambient levels of adenosine. This action of A$_{2A}$R is mediated by the activation of the cAMP/protein kinase A (PKA) signaling cascade (Diogenes et al. 2004). As discussed (Assaife-Lopes et al. 2014; Diógenes et al. 2004; Sebastião et al. 2013), this process may not involve Trk transactivation (Trk phosphorylation in the absence of neurotrophins), since it occurs within a time frame ($\sim$30 min) faster than transactivation. Transactivation of TrkB by A$_{2A}$Rs requires around 3 h of agonist exposure and involves mostly immature, intracellular Trk receptors located in Golgi-associated membranes (Rajagopal et al. 2004). Facilitation of BDNF synaptic actions can also be induced by a presynaptic depolarization, a process also dependent on A$_{2A}$R activation, through cAMP formation and PKA activity (Diógenes et al. 2004). The initial study by Diógenes et al. (2004) was followed by a number of studies evaluating the molecular mechanisms and further detailing functional evidences for the A$_{2A}$R/NTF cross-talk, at the central and peripheral nervous system (see Table 4.1).

Table 4.1 Evidence for the cross-talk between neurotrophic factors (NTF) and adenosine A$_{2A}$ receptors (A$_{2A}$R)

Evidences for the cross talk A$_{2A}$R/NTF	Main observation	References
1-Molecular evidences		
1.1-NTF receptors phosphorylation		
Trk transactivation	A$_{2A}$R activation transactivates TrkA in PC12 cells and TrkB in cultured hippocampal neurons	(Lee and Chao 2001)
	Trk receptors transactivation mediated by A$_{2A}$R occurs in intracellular membranes	(Rajagopal et al. 2004)
	A$_{2A}$R activation transactivates TrkB in motoneurons	(Wiese et al. 2007)
	Spinal A$_{2A}$R activation transactivates TrkB in rat cervical spinal cord near phrenic motoneurons	(Golder et al. 2008)
GDNF receptor phosphorylation	Cortical stimulation induces GDNF receptor phosphorylation in the striatum, an action that requires A$_{2A}$R tonic activation	(Gomes et al. 2009)
1.2-Trk translocation	A$_{2A}$R agonists increase TrkB levels in lipid rafts of cortical membranes	(Assaife-Lopes et al. 2014)
1.3-Levels of NTF and receptors	In primary glial cultures, A$_{2A}$R activation increases NGF expression and release	(Heese et al. 1997)
	Tonic activation of A$_{2A}$R is required for normal BDNF levels in hippocampus	(Tebano et al. 2008)
	A$_{2A}$R activation up-regulates BDNF expression in rat primary cortical neurons	(Jeon et al. 2011)
	BDNF production is prevented by the blockade of A$_{2A}$R activation	(Jeon et al. 2012)
	Chronic In vivo administration of A$_{2A}$R antagonist reduced rat hippocampal content on TrkB-FL receptors mRNA and protein	(Jeronimo-Santos et al. 2014)
2-Functional evidences		
2.1-Synaptic actions		
2.1.1-Synaptic transmission	A$_{2A}$R activation facilitates BDNF excitatory actions in CA1 area of young rat hippocampal slices	(Diógenes et al. 2004)
	BDNF excitatory actions are dependent on A$_{2A}$R activation in CA1 area of rat hippocampal slices	(Diogenes et al. 2007)
	BDNF decreases α7 nicotinic acetylcholine receptor responses in a mechanism dependent on A$_{2A}$R activation in interneurons of hippocampal CA1 stratum radiatum	(Fernandes et al. 2008)
	BDNF excitatory effect is abolished in A$_{2A}$R KO mice in CA1 area of hippocampal slices	(Tebano et al. 2008)
	BDNF excitatory effect is loss in adult mice over-expressing ADK and abolished in mice underexpressing ADK in CA1 area of hippocampal slices	(Diógenes et al. 2014)
2.1.2- Synaptic plasticity		
Ex vivo LTP	Facilitatory action of exogenous BDNF upon CA1 hippocampal LTP is dependent on A$_{2A}$R activation	(Fontinha et al. 2008)

Table 4.1 (continued)

Evidences for the cross talk $A_{2A}R$/NTF	Main observation	References
	Facilitatory action of endogenous BDNF upon CA1 hippocampal LTP is increased in aging and it is dependent on $A_{2A}R$ activation	(Diógenes et al. 2011)
	Chronic in vivo blockade of $A_{2A}R$ inhibits the facilitatory action of BDNF upon CA1 hippocampal LTP	(Jeronimo-Santos et al. 2014)
Ex vivo LTD	LTD attenuation induced by BDNF in CA1 hippocampal area is dependent on $A_{2A}R$ activation	(Rodrigues et al. 2014a)
2.1.3-Neuromuscular transmission	BDNF promotes enhancement of neuromuscular transmission by a mechanism dependent on $A_{2A}R$ activation in innervated rat diaphragm	(Pousinha et al. 2006)
2.2-Long-lasting phrenic motor facilitation	$A_{2A}R$ activation induces long-lasting phrenic motor facilitation via TrkB phosphorylation	(Golder et al. 2008)
2.3-Neurotransmitter dynamics	BDNF inhibits GAT-1-mediated GABA transport by nerve endings an action that can be enhanced by TrkB/$A_{2A}R$ receptor cross talk	(Vaz et al. 2008)
	In cultured astrocytes, BDNF enhances GAT-1-mediated GABA transport in a mechanism that requires active $A_{2A}R$	(Vaz et al. 2011)
	GDNF facilitates dopamine release from rat striatal synaptosomes in a manner dependent on $A_{2A}R$ activation	(Gomes et al. 2006)
	GDNF facilitates glutamate release from rat striatal synaptosomes in an $A_{2A}R$-dependent manner	(Gomes et al. 2009)
2.4-Neuronal differentiation and survival	$A_{2A}R$ activation increases cell survival of PC12 cells or hippocampal neurons after NGF or BDNF withdrawal	(Lee and Chao 2001)
	$A_{2A}R$ activation in PC12 cells rescues the blockade of NGF-induced neurite outgrowth	(Cheng et al. 2002)
	$A_{2A}R$ mediates neurite outgrowth in PC12 cells which depends on NGF mediated signaling	(Charles et al. 2003)
	$A_{2A}R$ agonist reduces ADA-induced and NGF-withdrawal-induced apoptosis of rat superior cervical ganglion cultures	(Ramirez et al. 2004)
	$A_{2A}R$ and Trk antagonism protects motor neurons from toxic insults	(Mojsilovic-Petrovic et al. 2006)
	$A_{2A}R$ contributes to motoneuron survival by TrkB transactivation	(Wiese et al. 2007)
	BDNF promotes cellular survival, synapse formation and neurite in a mechanism dependent on $A_{2A}R$ activation	(Jeon et al. 2012)
2.5-Neuronal inflamation	LPS-mediated increase of BDNF and microglia proliferation is dependent on $A_{2A}R$	(Gomes et al. 2013)
	Exogenous BDNF-induced microglia proliferation is dependent on $A_{2A}R$	
2.6-Behaviour	Altered fear and anxiety-like behaviors in fb-$A_{2A}R$ KO mice associated with a reduction of BDNF levels in hippocampus	(Wei et al. 2014)

Table 4.1 (continued)

Evidences for the cross talk A$_{2A}$R/NTF	Main observation	References
3-Evidences from animal models of diseases		
3.1-HD	In HD animal models the blockade of A$_{2A}$R significantly reduces striatal BDNF levels	(Potenza et al. 2007)
	Blockade of A$_{2A}$R abolishes the neuroprotective actions mediated by BDNF against NMDA toxicity	(Martire et al. 2013)
3.2-ALS	In ALS mice model the selective A$_{2A}$R agonist slows the onset of motor neuron degeneration and muscle weakness similarly to BDNF TrkB.T1 receptor removal	(Yanpallewar et al. 2012)

A$_{2A}$R adenosine A$_{2A}$ receptors, *ADA* adenosine deaminase *ADK* adenosine kinase, *ALS* amyotrophic lateral sclerosis, *BDNF* brain-derived neurotrophic factor, *fb* forebrain, *CDNF* cerebral-dopamine neurotrophic factor, *GDNF* glial-derived neurotrophic factor, *HD* Huntington's disease, *KO* knockout, *LPS* lipopolysaccharide, *LTD* long-term depression, *LTP* long-term potentiation, *NTF* neurotrophic factors, *NGF* nerve growth factor

Molecular Evidences

Neurotrophic Factor Receptors Phosphorylation

As mentioned previously, the first evidence that neurotrophin Trk receptors could be phosphorylated by a mechanism dependent on A$_{2A}$R activation in the absence of neurotrophins appeared in 2001 (Lee and Chao 2001). In this work it was demonstrated that NGF TrkA receptors, in PC12 cells, or BDNF TrkB receptors, in cultured hippocampal neurons could be transactivated by a prolonged exposure (~90 min) of A$_{2A}$R agonists (Lee and Chao 2001). Later, the pool of Trk receptors that undergos the transactivation process was identified as being in intracellular locations particularly associated with Golgi membranes (Rajagopal et al. 2004). Further studies revealed that A$_{2A}$R activation mediates transactivation of TrkB in motoneurons (Wiese et al. 2007) and in the cervical ventral horn (Golder et al. 2008). Nor only Trk receptors were shown to become phosphorylated through A$_{2A}$R activation but also GDNF receptor phosphorylation after cortical stimulation was showed to be dependent on A$_{2A}$R activation (Gomes et al. 2009). Interestingly, in vivo stimulation of corticostriatal afferents leads to activation of a canonical NTF pathway, phosphorylation mitogen-activated protein kinase (ERK1/2), and this is prevented by in vivo A$_{2A}$R blockade (Quiroz et al. 2006).

Trk Translocation

Lipid rafts are cholesterol-rich membrane domains that form an organized portion of the membrane that is thought to concentrate signaling molecules. These specialized domains have been implicated in the regulation of signal transduction in multiple cell types, including neurons, by promoting close proximity or segregation of signaling molecules (Sebastião et al. 2013). Indeed, the translocation of BDNF TrkB receptors to lipid rafts, is known to be required for BDNF effects upon glutamate release, synaptic fatigue (Suzuki et al. 2004) and for the activation of the phospholipase C pathway (Pereira and Chao 2007).

$A_{2A}R$ activation was shown to enhance the levels of TrkB receptors in the lipid raft fraction of cortical membranes (Assaife-Lopes et al. 2014) (Fig. 4.1). This may involve two processes: one, that is BDNF-independent and does not involve phosphorylation of TrkB receptors, and another that results from potentiation of subthreshold actions of BDNF that, in the absence of $A_{2A}R$ activation, induce mild TrkB receptor phosphorylation and poor or no TrkB translocation to lipid rafts. This suggests that the increased concentration of TrkB receptors in the lipid rafts, as a consequence of $A_{2A}R$ activation, leads to enhanced proximity of TrkB receptors promoting auto-phosphorylation of receptors not fully phosphorylated by a short

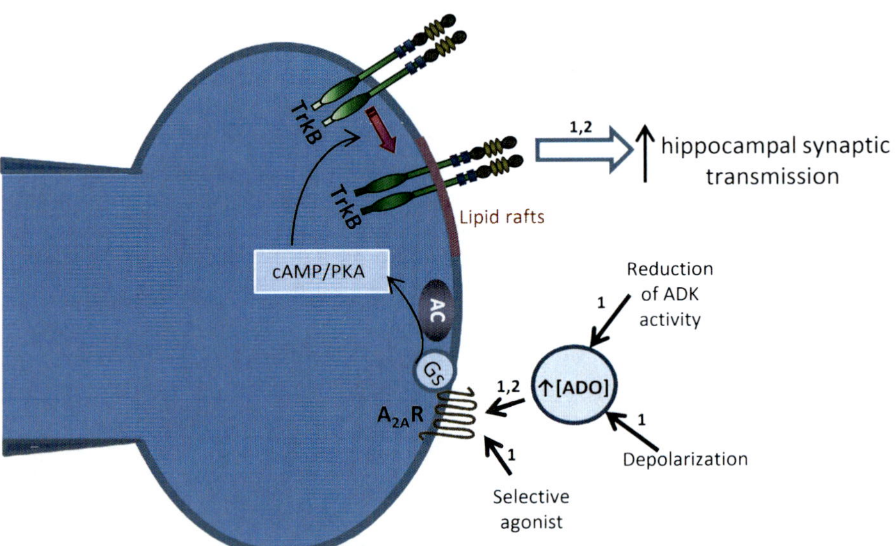

Fig. 4.1 Mechanisms underling facilitatory actions of $A_{2A}R$ activation on BDNF effects upon hippocampal synaptic transmission. In hippocampal slices taken from infant animals (*1*) BDNF excitatory effects upon synaptic transmission are only seen under conditions that favor $A_{2A}R$ activation such as: pre-depolarization; direct pharmacological activation of $A_{2A}R$ by selective agonists or decreased action of enzymes responsible for the degradation of adenosine (*ADO*) such as adenosine kinase (ADK). In these conditions, the activation of $A_{2A}R$, which are positively coupled to Gs proteins, increases the activity of the adenylate cyclase (AC) and consequently promotes formation of cyclic AMP (cAMP) and the activity of the protein kinase A (PKA). The activation of this cAMP/PKA transducing system induces the translocation of TrkB receptors into the lipid rafts. In hippocampal slices taken from older animals (*2*) the activation of $A_{2A}R$ by endogenous adenosine in the synaptic cleft is enough to facilitate the BDNF excitatory action upon synaptic transmission. Abbreviators: *ADO* adenosine; *AC* adenylate cyclase; *ADK* adenosine kinase; *BDNF* brain-derived neurotrophic factor; *cAMP* cyclic AMP; *PKA* protein kinase A

BDNF exposure. In contrast, long-lasting incubations with BDNF (>40 min) cause maximal TrkB translocation to lipid rafts and this no longer requires A$_{2A}$R activation (Assaife-Lopes et al. 2014), again indicating that A$_{2A}$R act to favor TrkB activation but do not exacerbate maximal activation. A$_{2A}$R-induced TrkB translocation to lipid rafts does not seem to require TrkB receptor internalization and involves activation of the cAMP/PKA signaling cascade (Assaife-Lopes et al. 2014), in contrast with the predominant mechanism operated by BDNF to enhance TrkB levels in these microdomains (Assaife-Lopes et al. 2014; Pereira and Chao 2007; Suzuki et al. 2004). Interestingly, relevant actions of BDNF at synapses, such as facilitation of glutamate release and synaptic plasticity, require not only A$_{2A}$R activation but also lipid raft integrity (Assaife-Lopes et al. 2014). Also noteworthy, high frequency stimulation also increases the levels of TrkB receptors in the lipid rafts, and this enhancement is lost when adenosine is not allowed to accumulate extracellularly (Assaife-Lopes et al. 2014). Altogether, the above summarized data suggest that A$_{2A}$R-induced TrkB translocation to lipid rafts plays an important part in the mechanism through which; enhanced neuronal activity; A$_{2A}$R activation; and cAMP elevations facilitate BDNF actions at active synapses.

Levels of Neurotrophic Factors and its Receptors

The influence of A$_{2A}$R on NTF expression was firstly demonstrated in 1997 with a work showing that, in primary glial cultures from cerebral cortices, the activation of A$_{2A}$R increased NGF expression and release through a mechanism dependent on cAMP (Heese et al. 1997). It is now known that tonic activation of A$_{2A}$R is also required to sustain a normal BDNF tone. This was clearly shown by Tebano et al. 2008, who compared BDNF levels present in the hippocampus of A$_{2A}$R KO and wild type (WT) animals, as well as in WT mice pharmacologically treated with an A$_{2A}$R antagonist. In both A$_{2A}$R KO mice and A$_{2A}$R antagonist WT treated animals, the levels of BDNF were significantly reduced (Tebano et al. 2008). Later, by using rat primary cortical neurons, it was demonstrated that activation of A$_{2A}$R enhances BDNF expression and release, through a mechanism that requires activation of Akt-GSK-3β signaling pathway (Jeon et al. 2011).

Oroxylin A is a flavone isolated from a medicinal herb reported to be effective in reducing the inflammatory and oxidative stresses. It also modulates the production of BDNF in cortical neurons by transactivation of the cAMP response element-binding protein (CREB) (Kim et al. 2006). Interestingly, it was shown that oroxylin A-induced increase in BDNF production is prevented by the blockade of A$_{2A}$R (Jeon et al. 2012). Somehow surprisingly on the basis of the above mentioned data, the daily in vivo intraperitoneal administration of the A$_{2A}$R antagonist, SCH 558261, for 14 days did not significantly affect the levels of BDNF evaluated in the anterior cingulate cortex, striatum, parietal cortex and in hippocampus nor the CDNF levels in substancia nigra, striatum, nucleus accumbens, hippocampus, parietal cortex and brainsteam (Gyarfas et al. 2010).

Most of the available data, summarized above, is focused on the effects of $A_{2A}R$ upon NTF levels; however, NTF receptor levels are also a crucial component for proper NTF mediated actions. We recently showed that chronic in vivo administration of an $A_{2A}R$ antagonist, KW6002, severely impairs the levels of mRNA and protein of TrkB-FL receptors in the hippocampus (Jerónimo-Santos et al. 2014), demonstrating that tonic activation of $A_{2A}R$ are not only important for BDNF levels but also for its receptor levels, therefore certainly affecting BDNF mediated actions.

Functional Evidences

Synaptic Actions

Synaptic Transmission

The influence of $A_{2A}R$ activation on BDNF actions has been mostly demonstrated in hippocampal synaptic transmission. It became clear that in young animals (~3 week-old rats) BDNF, *per se,* does not facilitate hippocampal synaptic transmission unless experimental conditions favor $A_{2A}R$ activation such as: (1) pre-depolarization known to increase adenosine release (Pazzagli et al. 1993) and consequently $A_{2A}R$ activation (Diógenes et al. 2004); (2) direct pharmacological activation of $A_{2A}R$ by selective agonists (Diógenes et al. 2004); (3) decreased action of enzymes responsible for the degradation of adenosine such as adenosine kinase (ADK), achieved either by pharmacologically inhibition of ADK (Diógenes et al. 2004) or by using transgenic animals underexpressing ADK (Diógenes et al. 2014) (Fig. 4.1).

In adult animals, BDNF alone is able to enhance hippocampal synaptic transmission (Diógenes et al. 2007; Kang and Schuman 1995; Tebano et al. 2008) but this facilitatory action is lost when $A_{2A}R$ are blocked (Diógenes et al. 2007; Tebano et al. 2008) or when $A_{2A}R$ are deleted (Tebano et al. 2008) (Fig. 4.1). These age-dependent actions of BDNF *per se* upon hippocampal synaptic transmission have been related to the age-related changes in the density of TrkB and of $A_{2A}R$ (Diógenes et al. 2007). Accordingly, in young animals, where $A_{2A}R$ levels are modest, the effect of BDNF upon synaptic transmission is only observed in conditions where $A_{2A}R$ activation is favored. In adult animals, BDNF induces an LTP-like phenomenon dependent on $A_{2A}R$ activation which disappears in old adult rats probably because of the marked decrease in the density of TrkB receptors in this age (Diógenes et al. 2007). Surprisingly, BDNF could enhance synaptic transmission in aged (~80 weeks old) animals. In this age group, it is possible to observe, on one hand, a marked increase in the Bmax value for $A_{2A}R$ binding, indicating higher density in $A_{2A}R$ and, on the other hand, that the effect of BDNF on synaptic transmission also requires $A_{2A}R$ activation (Diógenes et al. 2007). This indicates that the ability of BDNF to enhance synaptic transmission under conditions of low density of TrkB receptors, as it occurs the aged, might be related to the increased levels of $A_{2A}R$, which partially compensates the marked loss of TrkB receptors levels. This relationship between age-related changes in the density of TrkB receptors and A$_{2A}$R, to allow BDNF-induced facilitation of synaptic transmission in the hippocampus, should be taken into consideration whenever designing BDNF-based therapeutic strategies in pathologies of the aged, such as Parkinson's disease (PD).

In interneurons of the hippocampal CA1 stratum radiatum, BDNF was shown to rapidly decrease α7 nicotinic acetylcholine receptor responses. This effect is dependent on the activation of TrkB receptors; involves the actin cytoskeleton and it is compromised when the extracellular levels of the endogenous adenosine are reduced with adenosine deaminase or when A$_{2A}$R are blocked (Fernandes et al. 2008). This interplay between BDNF and A$_{2A}$R upon hippocampal nicotinic mechanisms is of main interest given that nicotinic mechanisms, acting on the hippocampus, influence attention, learning, and memory and constitute a significant therapeutic target for many neurodegenerative disorders such as PD.

Synaptic Plasticity

It is generally accepted that the neurophysiological basis for learning and memory involve modifications in the efficiency of synapses between neurons, that is to say the synaptic adaptations to external stimuli. Experimental paradigms of such plasticity processes are the long-term modifications in synaptic strength induced by specific patterns of stimulation. The forms more commonly studied are those known as long-term potentiation (LTP) (Bliss and Collingridge 1993) and long-term depression (LTD) (Linden 1994).

Ex vivo LTP
The mechanisms underlying the establishment of LTP in the CA1 region of the hippocampus have been the subject of many studies (for a review, see Malenka and Nicoll 1999). There is now clear evidence that LTP is under control by NTF, namely BDNF (Minichiello 2009). The long lasting actions of BDNF upon gene expression and protein-synthesis dependent LTP have also been matter of several reviews (Bramham et al. 2008; Lu et al. 2008), but BDNF also influences earlier phases of LTP and this has been mostly shown at the CA1 area of the hippocampus. LTP at these synapses is deeply impaired in BDNF homozygous and heterozygous mutant mice and can be rescued by exogenous BDNF (Korte et al. 1996; Patterson et al. 1996). In accordance to these results, the application of a BDNF scavenger greatly inhibits hippocampal LTP (Figurov et al. 1996; Kang et al. 1997). Moreover, hippocampal slices taken from TrkB CA1 KO mice have an impaired LTP (Xu et al. 2000), further highlighting the role of the endogenous BDNF upon CA1 LTP.

BDNF expression and release (Balkowiec and Katz 2002; Hartmann et al. 2001) as well as release of adenosine (Pazzagli et al. 1993) and of its precursor ATP (Wieraszko et al. 1989), are much more pronounced upon depolarization and during physiologically relevant patterns of neuronal activity, namely those that induce hippocampal LTP. Consequently, high neuronal activity appears to create ideal physiological conditions for the concomitant increase of both adenosine and BDNF at the synapses and therefore favoring the interplay between $A_{2A}R$ and TrkB receptors.

The first functional evidence for the $A_{2A}R$/TrkB interplay to control hippocampal LTP appeared in 2008 with a work (Fontinha et al. 2008) showing that the facilitatory action of exogenous BDNF upon θ-burst-induced LTP is fully dependent on the co-activation of $A_{2A}R$ through a cAMP/PKA-dependent mechanism. This evidence was further extended while examining the influence of BDNF upon LTP throughout ageing (Diógenes et al. 2011). Interestingly, endogenous BDNF actions upon θ-burst- induced LTP are significantly increased in aged animals (Diógenes et al. 2011) where neuromodulation through $A_{2A}R$ is increased (Costenla et al. 2011; Rebola et al. 2003). Moreover, in vivo chronic blockade of $A_{2A}R$ in adult rats inhibits the facilitatory action of BDNF upon LTP on hippocampal CA1 area and decreases both mRNA and protein levels of the TrkB receptor in hippocampus (Jeronimo-Santos et al. 2014). Whether this relates to the impairment of hippocampal dependent conditional learning caused by i.p. injections of an $A_{2A}R$ antagonists in young animals (Fontinha et al. 2009) is yet unknown. The learning impairment induced by an $A_{2A}R$ antagonist reported by Fontinha et al. (2009) contrasts with the ability of $A_{2A}R$ antagonists to revert learning impairment associated to chronic stress (Batalha et al. 2013). Besides differences in the experimental paradigm (prolonged vs acute, oral vs i.p. administration, rat vs mice, spacial vs conditional learning), the above referred discrepancy may suggest a different role of $A_{2A}R$ in health and disease. A protective role of $A_{2A}R$ activation in neurodegenerative disease models, as Hungtington's disease models has been however shown.

The actions mediated by either endogenous or exogenous BDNF on LTP are more pronounced whenever $A_{2A}R$ are more relevant as in aging (Diógenes et al. 2011) where despite the decrease in TrkB receptor levels (Diógenes et al. 2007), BDNF levels are maintained (Diógenes et al. 2011), and $A_{2A}R$ levels are significantly increased (Diógenes et al. 2007).

Ex vivo LTD

BDNF is thought to display a dual action over synaptic plasticity phenomena mediating opposite events: mature BDNF facilitates LTP through TrkB receptors, whereas the activation of p75NTR by proBDNF has been reported to be necessary for NMDAR-LTD at the CA1 hippocampal area (Rosch et al. 2005; Woo et al. 2005). Moreover, in the visual cortex mature BDNF was shown to impair LTD (Akaneya et al. 1996; Huber et al. 1998; Jiang et al. 2003). In the hippocampus, high concentrations of BDNF (~ 100 ng/ml) attenuate LTD (Ikegaya et al. 2002; Rodrigues et al. 2014a). At this concentration BDNF lacks effect on an adenosine depleted background or under selective $A_{2A}R$ blockade, indicating that it relies on tonic $A_{2A}R$ activation (Rodrigues et al. 2014a). At a lower concentration (~ 20 ng/ml) BDNF can inhibit LTD providing that $A_{2A}Rs$ are activated by either endogenous adenosine or by selective ligands (Rodrigues et al. 2014a).

Altogether, the above evidences indicate that the action of BDNF upon plasticity is under the control of upstream regulators as adenosine, which on one hand boost BDNF facilitation of LTP and on the other hand allow inhibition of LTD even at low BDNF levels, contributing to synaptic reinforcement.

Neuromuscular Transmission

Likewise the hippocampus, the terminals of the motor nerves have both $A_{2A}R$ and TrkB receptors. Data showed that BDNF, even in high concentrations, is not able to affect evoked endplate potentials (EPPs) recorded intracellularly from rat innervated diaphragms paralyzed with tubocurarine. However, when BDNF is applied after a brief depolarizing K^+ pulse or when the $A_{2A}R$ are pharmacologically activated, BDNF can increase EPPs amplitude without influencing the resting membrane potential of the muscle fiber. This action of BDNF is prevented by antagonizing $A_{2A}R$. Experiments preformed using a PKA inhibitor and a PLCγ inhibitor, show that the following sequence of events, in what concerns cooperativity between $A_{2A}R$ and TrkB receptors at the neuromuscular junction, occurs: $A_{2A}R$ activate the PKA pathway, which promotes the action of BDNF through TrkB receptors coupled to PLCγ, leading to enhancement of neuromuscular transmission (Pousinha et al. 2006).

Long-Lasting Phrenic Motor Nerve Facilitation

Long-term changes in respiratory motoneuron function can occur via plasticity of their synaptic inputs (Golder and Mitchell 2005; Mitchell and Johnson 2003). Respiratory synaptic plasticity can be induced by several neuromodulators (Bocchiaro and Feldman 2004; Feldman et al. 2003; Neverova et al. 2007) able to activate downstream signaling mechanisms that often involve the BDNF and TrkB activation (Baker-Herman et al. 2004; Bramham and Messaoudi 2005; Carter et al. 2002). For example, acute intermittent hypoxia (AIH) elicits a long-lasting enhancement of phrenic motor activity known as phrenic long-term facilitation (pLTF) (Mitchell et al. 2001) that requires BDNF synthesis (Baker-Herman et al. 2004).

Similarly of what was described for hippocampal neurons in culture (Lee and Chao 2001), $A_{2A}R$ activation transactivates TrkB receptors in the rat cervical spinal cord near phrenic motoneurons, inducing long-lasting phrenic motor facilitation (Golder et al. 2008). Moreover, $A_{2A}R$ activation increases the synthesis of an immature TrkB protein, induces TrkB signaling through Akt, and strengthens synaptic pathways to phrenic motoneurons. This work (Golder et al. 2008) suggests that adenosine receptor agonists may prove an effective therapeutic strategy in the treatment of patients with ventilator control disorders, such as respiratory insufficiency, after spinal injury or during neurodegenerative diseases.

Neurotransmitter Dynamics

Glial cell line-derived neurotrophic factor (GDNF) was discovered in 1993 as a potent survival-promoting agent for cultured dopaminergic neurons (Lin et al. 1993) and therefore GDNF has been regarded as one of the most promising molecules for NTF based PD therapy (Rodrigues et al. 2014b; Stayte and Vissel 2014). The first evidence for a crosstalk between $A_{2A}R$ and GDNF in the striatum appeared in 2006 in a study showing that this NTF acutely increases evoked dopamine release in rat striatal slices and synaptosomes (presynaptic nerve terminals) and that its action is modulated by $A_{2A}R$ (Gomes et al. 2006). The $A_{2A}R$ that promote the facilitatory action of GDNF upon dopamine release are most probably located presynaptically at dopaminergic nerve terminals, since the $A_{2A}R$/GDNF crosstalk was observed in isolated synaptosomes, where interactions at the circuit level are absent (Gomes et al. 2006). Notably, in rat striatal synaptosomes GDNF was also found to enhance glutamate release, and again, this action requires co-activation of $A_{2A}R$ (Gomes et al. 2009).

BDNF is also an important regulator of neurotransmitter dynamics and the involvement of $A_{2A}R$ being studied. In fact, BDNF through activation of TrkB receptors decreases uptake of GABA mediated by the high-affinity Na^+/Cl^- dependent transporter, GAT-1, in isolated hippocampal rat nerve terminals. In contrast with what has been observed for other actions of BDNF, the inhibition of GABA transport by BDNF does not require tonic activation of $A_{2A}R$ since it is not blocked by a selective $A_{2A}R$ antagonist. However, in synaptosomes depleted of extracellular endogenous adenosine, the pharmacological activation of $A_{2A}R$ enhances the inhibitory effect of BDNF upon GABA transport, an action prevented by blockade of $A_{2A}R$. Therefore, the inhibitory action of BDNF upon GAT-1-mediated GABA transport into nerve endings is not dependent on, but can be enhanced by, TrkB/$A_{2A}R$ receptor cross talk (Vaz et al. 2008). Interestingly, in cultured astrocytes BDNF enhances GAT-1-mediated GABA transport in a mechanism involving the truncated form of the TrkB receptor (TrkB-Tc) coupled to a non-classic PLC-γ/PKC-δ and ERK/MAPK pathway; this action fully requires active $A_{2A}R$ (Vaz et al. 2011).

Regarding GABA and glutamate release, BDNF enhances the release of glutamate and inhibits the release of GABA from rat hippocampal synaptosomes (Canas et al. 2004). These two opposite actions of BDNF involve different mechanisms since BDNF mostly influences the calcium-dependent release of glutamate, while its influence upon the release of GABA does not depend on extracellular calcium and involves GABA transporters (Canas et al. 2004). However, in both situations, enhancement of glutamate release or decrease in GABA release by BDNF are depended on $A_{2A}R$ activation (Parreira 2014).

Neuronal Differentiation and Survival

It is widely accepted that BDNF is a NTF with a central role in neuronal differentiation and survival. Adenosine has also been regarded as a neuromodulator that mediates neuroprotection mainly through A_1R activation. However, there is data showing the involvement of $A_{2A}R$ on cell survival and differentiation mainly by promoting NTF actions or by preventing cellular death induced by NTF withdrawal.

First evidence that adenosine, acting through the $A_{2A}R$, exerts a trophic effect through the engagement of Trk receptors was published in the very first paper showing that $A_{2A}R$ can transactivate TrkB receptors (Lee and Chao 2001). This work showed that $A_{2A}R$ agonists can activate phosphatidylinositol 3-kinase and Akt through a Trk-dependent mechanism increasing survival of PC12 cells or hippocampal neurons after NGF or BDNF withdrawal (Lee and Chao 2001). Moreover, stimulation of the $A_{2A}R$, in PC12 cells, by a selective agonist rescues the blockade of NGF-induced neurite outgrowth when the NGF-evoked MAPK cascade is suppressed. This action of $A_{2A}R$ is dependent on cAMP/PKA transducing system (Cheng et al. 2002). Also in PC12 cells, bacterial nucleoside N6-methyldeoxyadenosine induces an $A_{2A}R$-mediated neurite outgrowth, an action that also depends on MAPK activation by NGF (Charles et al. 2003).

In sympathetic neurons, the available data also provided evidence for neuroprotection mediated by endogenous adenosine via $A_{2A}R$ activation. Rat superior cervical ganglion (SCG) cultures maintained in the continuous presence of NGF but in an environment depleted of endogenous adenosine present a marked increase in cellular apoptosis, to a level that is approximated to what occur as consequence of NGF withdrawal (Ramirez et al. 2004). The addition of exogenous adenosine to NGF-deprived SCG neurons resulted in enhanced cell survival. In addition, a selective $A_{2A}R$ agonist significantly reduced both ADA-induced and NGF-withdrawal-induced neuronal apoptosis. Moreover, the selective $A_{2A}R$ agonist was shown to prevent the induction of early apoptotic events, such as changes in mitochondrial integrity and caspase activation, and to trigger the increase in ERK activation, which is essential for neurotrophin-independent cell survival (Ramirez et al. 2004).

As mentioned above, oroxylin A, regulates BDNF production in cortical neurons through $A_{2A}R$ stimulation. Interestingly the increase on BDNF levels mediated by $A_{2A}R$ activation results in the promotion of cell survival, synapse formation and neurite extension (Jeon et al. 2012).

Regarding motorneurons, it was shown that $A_{2A}R$ contribute to motoneuron survival by transactivating the TrkB receptors (Wiese et al. 2007). On the contrary, there are data showing that the antagonism of $A_{2A}R$ and Trk receptors protects motor neurons from toxic insult (Mojsilovic-Petrovic et al. 2006).

Neuronal Inflammation

Neuroinflammation, as all inflammation in general, is a fundamental immune response engaged to protect the body from damage from internal or external sources. Microglia has been seen as the sentinel immune cell of the brain, being the first responders to tissue injury and initiating the inflammatory response. Microglial activation seems to be a convergence point for diverse stimuli that may promote or compromise neuronal survival; in such a way, the exacerbated or chronic neuroinflammation contributes to cellular injury, therefore participating in the pathophysiology of neurodegenerative diseases (Frank-Cannon et al. 2009). Microglia can broadly exist in two different states (Colton 2009): one is the classical activation, which is typified by the production of inflammatory cytokines and reactive oxygen species, while the second is a state of alternative activation, in which microglia take on an anti-inflammatory phenotype involved in wound repair and debris clearance (Gordon 2003). In neurodegenerative disorders this alternative activation would have a beneficial role, but this field certainly needs a deeper clarification (Cherry et al. 2014).

Adenosine, via $A_{2A}R$ activation, and BDNF, through TrkB receptors, have determinant roles in inflammation. Indeed, it was demonstrated that $A_{2A}R$ are a critical part of the physiological negative feedback mechanism for limitation and termination of both tissue-specific and systemic inflammatory responses (Ohta and Sitkovsky 2001). Moreover, $A_{2A}R$ mediate microglial process retraction (Orr et al. 2009). On the other hand there is data showing that $A_{2A}R$ antagonists prevent neuroinflammation, supporting the hypothesis that $A_{2A}R$ antagonists can control different neurodegenerative diseases through prevention of neuroinflammation (Rebola et al. 2011).

Regarding BDNF, there are data suggesting that intranasal BDNF might protect the brain against an ischemic insult by modulating local inflammation, an action that involves the regulation of the levels of cytokines and transcription factors (Jiang et al. 2011).

Interestingly, there are data showing that the inflammatory trigger lipopolysaccharide (LPS) induces time-dependent changes of the intra- and extracellular levels of BDNF with increased microglial proliferation (Gomes et al. 2013). The maximal LPS-induced BDNF release was shown to be time-coincident with an LPS-induced increase of the $A_{2A}R$ density. Notably, the LPS-mediated increase of BDNF secretion and proliferation, as well as the exogenous BDNF-induced proliferation was prevented by removing endogenous extracellular adenosine or by blocking $A_{2A}R$. These data led the authors to conclude that $A_{2A}R$ activation plays a mandatory role controlling the release of BDNF from activated microglia, as well as the autocrine/paracrine proliferative role of BDNF (Gomes et al. 2013).

Behaviour

The dysfunction of conditioned fear leads to maladaptive fear responses that may underlie neuropsychiatric disorders. Interestingly, forebrain (fb)-specific $A_{2A}R$ knockout (fb-$A_{2A}R$ KO) mice possess altered fear and anxiety-like behaviors associated with a selective reduction of BDNF levels in hippocampus (Wei et al. 2014). Accordingly, the selective deletion of $A_{2A}Rs$ in the striatum increases Pavlovian fear conditioning in striatum-$A_{2A}R$ KO mice, but extending the deletion to the rest of the fb apparently spars context fear conditioning and attenuates tone fear conditioning in fb-$A_{2A}R$ KO mice. Moreover, focal deletion of hippocampal $A_{2A}R$ by AAV5-Cre injection selectively attenuates context (but not tone) fear conditioning. Deletion of $A_{2A}R$ in the entire forebrain in fb-$A_{2A}R$ KO mice also produces an anxiolytic phenotype in both the elevated plus maze and open field tests, and increases the startle response. Whether this extrastriatal forebrain $A_{2A}R$ behavioral effects are caused by a reduction of BDNF levels in the fb-$A_{2A}R$ KO hippocampus remains to be explained. Nevertheless, it is clear, as previously described in this chapter that $A_{2A}R$ are crucial for the preservation of BDNF levels.

Evidence from Animal Models of Diseases

Hungtington's Disease

Huntington's disease (HD) is a genetic neurodegenerative disease caused by a tri-nucleotide expansion in exon 1 of the huntingtin gene without an effective pharmacological treatment. There are evidences from postmortem human samples and from HD mouse model brain cortices that there is an impairment of BDNF signaling in HD (Gines et al. 2006; Zuccato et al. 2001, 2008). Based on striatal gene expression, it was found that both heterozygous and forebrain specific homozygous knock-outs for BDNF, are more like human HD than the other HD models, which strongly implicates reduced trophic support as a major pathway contributing to striatal degeneration in HD (Strand et al. 2007). Moreover, BDNF overexpression in the forebrain rescues HD phenotypes in YAC128 mice (Xie et al. 2010).

Inspired by evidences for an involvement of striatal A$_{2A}$R in HD (Blum et al. 2003; Popoli et al. 2007), Potenza et al. (2007) used two different models of HD, quinolinic acid (QA)-lesioned rats and a transgenic mice model of HD (R6/2 mice) and studied the influence of the pharmacological blockade of A$_{2A}$Rs on BDNF levels. Accordingly to what was described for non-disease animal models (see 2.3 above), in HD animal models the blockade of A$_{2A}$R also significantly reduces striatal BDNF levels (Potenza et al. 2007).

Excitotoxicity mediated by NMDA receptor is thought to play a pivotal role in HD (Levine et al. 1999) and BDNF is known to influence the activity and expression of striatal NMDA receptors (Torres-Peraza et al. 2008). Electrophysiological studies show that in corticostriatal slices from WT mice, NMDA application induces a transient reduction of field potential amplitude while in age-matched symptomatic R6/2 mice (animal model of HD) it induces a permanent (i.e., toxic) reduction of field potential amplitude; interestingly, BDNF potentiates NMDA responses in WT animals, while it protects from NMDA toxicity in R6/2 mice and remarkably, both effects of BDNF were prevented by A$_{2A}$R blockade (Martire et al. 2013).

Summarizing, in HD models it is clear that the blockade of A$_{2A}$R not only impairs the levels of BDNF but also abolishes its neuroprotective actions against NMDA toxicity. Given the recognized role of BDNF in rescuing HD phenotypes in animal mice models (Xie et al. 2010), this data might alert for a prejudicial role of antagonizing A$_{2A}$R in pathologies where the maintenance of BDNF effects is vital.

Amyotrophic Lateral Sclerosis

Amyotrophic lateral sclerosis (ALS) is a late-onset progressive neurodegenerative disease affecting motor neurons. The etiology of most ALS cases remains unknown, but 2 % of the cases are associated to mutations in Cu/Zn superoxide dismutase (SOD1) (Boillee et al. 2006). BDNF mRNA and protein are severely upregulated in the muscle of ALS patients, and total TrkB mRNA is increased in the spinal cord;

however, phosphorylation of the TrkB receptor is impaired (Kust et al. 2002; Mutoh et al. 2000).

Interestingly, it was found that the deletion of truncated form of TrkB receptors, TrkB.T1, which may act as negative modulators of TrkB signaling (Eide et al. 1996), significantly slows the onset of motor neuron degeneration, delays the development of muscle weakness and improves the neurological score at the late stage of the disease. Notably, the treatment with a selective $A_{2A}R$ agonist slowed the onset of motor neuron degeneration and muscle weakness similarly to TrkB.T1 removal (Yanpallewar et al. 2012). $A_{2A}R$ have been considered potential therapeutic targets for several disorders such as ALS (Beghi et al. 2011; Potenza et al. 2013; Yanpallewar et al. 2012), though either agonists (Potenza et al. 2013; Yanpallewar et al. 2012) or antagonists (Beghi et al. 2011) have been regarded as potentially relevant. Detailed information on $A_{2A}R$ changes in ALS is required to better appraise the therapeutic potential of $A_{2A}R$ ligands. A recent work (Nascimento et al. 2014) demonstrates that in a SOD1 mouse model of ALS there is an exacerbation of the $A_{2A}R$-mediated signaling at neuromuscular junctions of pre-symptomatic mice, whereas in the symptomatic phase the $A_{2A}R$ excitatory action disappears (Nascimento et al. 2014). Whether the $A_{2A}R$-mediated exacerbation of neuromuscular transmission in the pre-simptomatic phase acts as a compensatory mechanism, delaying disease progression, is yet unknown.

Taken together this data suggest that $A_{2A}R$ activation and deletion of TrkB.T1 can have a benefic role in ALS. Whether the effect of $A_{2A}R$ activation is mediated by a down regulation of TrkB.T1 receptors or whether the effect of the deletion of TrkB.T1 is mediated by an upregulation of $A_{2A}R$, remains however to be further elucidated.

Relevance for Parkinson's Disease

Parkinson's disease (PD) is a common neurodegenerative disease characterized by a loss of dopaminergic input to the striatum and several motor symptoms as bradykinesia, rigidity, resting tremor and postural instability. The striatum is the brain area with the highest density of $A_{2A}R$, which are mostly, but not exclusively, localized postsynaptically in the medium spiny GABAergic neurons where dopaminergic D_2 receptors co-localize. Non-motor symptoms in PD, as cognitive impairment, may involve other brain areas as the hippocampus (Calabresi et al. 2013). NTF have been frequently regarded as promising therapies for neurodegenerative diseases as Alzheimer's disease and PD (Lu et al. 2013; Rodrigues et al. 2014b; Stayte and Vissel 2014). GDNF is an important survival factor for midbrain dopaminergic neurons and stimulates the growth of processes from immature neurons (Lin et al. 1993) and has proved as beneficial in animal models of PD, including non-human primate models (see Rodrigues et al. 2014b; Sebastião and Ribeiro 2009; Stayte and Vissel 2014). Despite the initial conflicting results in clinical trials using GDNF based therapies (see Rodrigues et al. 2014b; Sebastião and Ribeiro 2009; Stayte and

Vissel 2014), the positive outcomes obtained in animal models of PD encouraged further evaluation of this possibility and gene- therapies aiming to enhance GDNF expression in target brain areas are currently entering Phase I trial (ClinicalTrials. gov NCT01621581) (see Stayte and Vissel 2014).

Direct evidence for the impact of the A$_{2A}$R/NTF cross talk upon PD is still lacking. The commonalities between the different neurodegenerative disorders, together with the evidence already obtained in other neurodegenerative disease models, in particular HD allow to anticipate an impact of this cross-talk also in PD. As mentioned above, GDNF loss of function has been implicated in the etiology of PD (Rodrigues et al. 2014b; Stayte and Vissel 2014) and A$_{2A}$R activation is required for the facilitatory actions of GDNF on dopamine and glutamate release (Gomes et al. 2006, 2009). Boosting the ability of NTF to reinforce synapses may be particularly important at early phases of neurodegenerative diseases if one takes into account the emerging evidence of an early synaptic dysfunction in neurodegenerative disease models. Relevant in this context is the finding that α-synuclein hampers synaptic plasticity even before causing an overt neuronal dysfunction (Diógenes et al. 2012). On the other hand, however, the negative interaction between A$_{2A}$R and D$_2$ receptors in the striatum, together with the ability of A$_{2A}$R antagonists to decrease excitotoxicity phenomena, inspired the possibility that A$_{2A}$R antagonists may prove beneficial in PD. The therapeutic benefit of A$_{2A}$R antagonists have already been evaluated in Phase II and Phase III clinical trials, though the outcome was not as positive as initially expected (see Lopes et al. 2011; Stayte and Vissel 2014).

The evidence pointing out the cross-talk between A$_{2A}$R and NTF highlights to the need of caution about therapies with A$_{2A}$R antagonists in PD. As previously pointed out (Sebastião and Ribeiro 2009), one issue that requires further attention is the optimal time window for combined therapies with NTF and A$_{2A}$R ligands. It is likely that in the early stages of the disease, where neurons and synapses are struggling for life, NTF based therapies are helpful and therefore A$_{2A}$R agonists may be desirable whereas A$_{2A}$R antagonists should be avoided. One may at least anticipate the advantage of not blunting actions of endogenous NTF by using A$_{2A}$R antagonists at early disease states. In latter stages of the disease, dopaminergic replacement therapies are required and in this case, A$_{2A}$R antagonists are most probably desirable to facilitate D$_2$ receptor signaling.

Conclusions

Presently, there are two main views on neurodegenerative disorders in general and on PD in particular: (a) the first one is centered on a specific neuronal function or brain area, in the case of PD on the nigrostriatal neurons, the functioning of dopamine D$_2$ receptors, and the way they are counteracted by striatal A$_{2A}$R, and (b) a second one residing on a sort of "globalization" of the brain, involving dysfunction of circuits across brain regions, that in PD encompass besides the striatum, the hippocampus and the cerebral cortex. Detailed understanding of the first led to the identification of targets for novel therapeutic strategies (e.g. $A_{2A}R$ antagonists). The second will succeed if more holist therapeutic strategies prove useful to fight PD, but this possibility is just emerging nowadays. Particularly promising is the modulation of neurodegeneration and neuroinflamation via microglia directed therapies, but further knowledge of the bidirectional relationship between the actors involved, including the role of the extracellular signaling molecules, their time window of action, and their cellular targets (neurons, microglia, astrocytes, oligodentrocytes), is yet necessary. Among the molecules involved in the communication between different neuronal cell types, the NTF and their "benefactors"—the $A_{2A}R$ agonists, are promising as therapeutic targets especially at early disease states. Clarification of the way the circuits operate and of the function the different regulators at the molecular, cellular and circuitry level, will certainly proportionate an enormous impulse to treat more efficiently PD.

References

Akaneya Y, Tsumoto T, Hatanaka H (1996) Brain-derived neurotrophic factor blocks long-term depression in rat visual cortex. J Neurophysiol 76:4198–4201

Arslan G, Kontny E, Fredholm BB (1997) Down-regulation of adenosine A2A receptors upon NGF-induced differentiation of PC12 cells. Neuropharmacology 36:1319–1326

Assaife-Lopes N, Sousa VC, Pereira DB et al (2014) Regulation of TrkB receptor translocation to lipid rafts by adenosine A(2A) receptors and its functional implications for BDNF-induced regulation of synaptic plasticity. Purinergic Signal 10:251–267

Baker-Herman TL, Fuller DD, Bavis RW et al (2004) BDNF is necessary and sufficient for spinal respiratory plasticity following intermittent hypoxia. Nat Neurosci 7:48–55

Balkowiec A, Katz DM (2002) Cellular mechanisms regulating activity-dependent release of native brain-derived neurotrophic factor from hippocampal neurons. J Neurosci 22:10399–10407

Batalha VL, Pego JM, Fontinha BM et al (2013) Adenosine A(2A) receptor blockade reverts hippocampal stress-induced deficits and restores corticosterone circadian oscillation. Mol Psychiatry 18:320–331

Beghi E, Pupillo E, Messina P et al (2011) Coffee and amyotrophic lateral sclerosis: a possible preventive role. Am J Epidemiol 174:1002–1008

Bliss TV, Collingridge GL (1993) A synaptic model of memory: long-term potentiation in the hippocampus. Nature 361:31–39

Blum D, Hourez R, Galas MC et al (2003) Adenosine receptors and Huntington's disease: implications for pathogenesis and therapeutics. Lancet Neurol 2:366–374

Bocchiaro CM, Feldman JL (2004) Synaptic activity-independent persistent plasticity in endogenously active mammalian motoneurons. Proc Natl Acad Sci U S A 101:4292–4295

Boillee S, Vande Velde C, Cleveland DW (2006) ALS: a disease of motor neurons and their non-neuronal neighbors. Neuron 52:39–59

Boulanger L, Poo MM (1999) Gating of BDNF-induced synaptic potentiation by cAMP. Science 284:1982–1984

Bramham CR, Messaoudi E (2005) BDNF function in adult synaptic plasticity: the synaptic consolidation hypothesis. Prog Neurobiol 76:99–125

Bramham CR, Worley PF, Moore MJ et al (2008) The immediate early gene arc/arg3.1: regulation, mechanisms, and function. J Neurosci 28:11760–11767

Calabresi P, Castrioto A, Di Filippo M et al (2013) New experimental and clinical links between the hippocampus and the dopaminergic system in Parkinson's disease. Lancet Neurol 12:811–821

Canas N, Pereira IT, Ribeiro JA et al (2004) Brain-derived neurotrophic factor facilitates glutamate and inhibits GABA release from hippocampal synaptosomes through different mechanisms. Brain Res 1016:72–78

Carter AR, Chen C, Schwartz PM et al (2002) Brain-derived neurotrophic factor modulates cerebellar plasticity and synaptic ultrastructure. J Neurosci 22:1316–1327

Charles MP, Adamski D, Kholler B et al (2003) Induction of neurite outgrowth in PC12 cells by the bacterial nucleoside N6-methyldeoxyadenosine is mediated through adenosine A2a receptors and via cAMP and MAPK signaling pathways. Biochem Biophys Res Commun 304:795–800

Cheng HC, Shih HM, Chern Y (2002) Essential role of cAMP-response element-binding protein activation by A2A adenosine receptors in rescuing the nerve growth factor-induced neurite outgrowth impaired by blockage of the MAPK cascade. J Biol Chem 277:33930–33942

Cherry JD, Olschowka JA, O'Banion MK (2014) Neuroinflammation and M2 microglia: the good, the bad, and the inflamed. J Neuroinflamm 11:98

Colton CA (2009) Heterogeneity of microglial activation in the innate immune response in the brain. J Neuroimmune Pharmacol 4:399–418

Costenla AR, Diógenes MJ, Canas PM et al (2011) Enhanced role of adenosine A(2A) receptors in the modulation of LTP in the rat hippocampus upon ageing. Eur J Neurosci 34:12–21

Diógenes MJ, Fernandes CC, Sebastião AM et al (2004) Activation of adenosine A2A receptor facilitates brain-derived neurotrophic factor modulation of synaptic transmission in hippocampal slices. J Neurosci 24:2905–2913

Diógenes MJ, Assaife-Lopes N, Pinto-Duarte A et al (2007) Influence of age on BDNF modulation of hippocampal synaptic transmission: interplay with adenosine A2A receptors. Hippocampus 17:577–585

Diógenes MJ, Costenla AR, Lopes LV et al (2011) Enhancement of LTP in aged rats is dependent on endogenous BDNF. Neuropsychopharmacology 36:1823–1836

Diógenes MJ, Dias RB, Rombo DM et al (2012) Extracellular alpha-synuclein oligomers modulate synaptic transmission and impair LTP via NMDA-receptor activation. J Neurosci 32:11750–11762

Diógenes MJ, Neves-Tome R, Fucile S et al (2014) Homeostatic control of synaptic activity by endogenous adenosine is mediated by adenosine kinase. Cereb Cortex 24:67–80

Eide FF, Vining ER, Eide BL et al (1996) Naturally occurring truncated trkB receptors have dominant inhibitory effects on brain-derived neurotrophic factor signaling. J Neurosci 16:3123–3129

Feldman JL, Mitchell GS, Nattie EE (2003) Breathing: rhythmicity, plasticity, chemosensitivity. Annu Rev Neurosci 26:239–266

Fernandes CC, Pinto-Duarte A, Ribeiro JA et al (2008) Postsynaptic action of brain-derived neurotrophic factor attenuates alpha7 nicotinic acetylcholine receptor-mediated responses in hippocampal interneurons. J Neurosci 28:5611–5618

Figurov A, Pozzo-Miller LD, Olafsson P et al (1996) Regulation of synaptic responses to high-frequency stimulation and LTP by neurotrophins in the hippocampus. Nature 381:706–709

Fontinha BM, Diogenes MJ, Ribeiro JA et al (2008) Enhancement of long-term potentiation by brain-derived neurotrophic factor requires adenosine A2A receptor activation by endogenous adenosine. Neuropharmacology 54:924–933

Fontinha BM, Delgado-Garcia JM, Madronal N et al (2009) Adenosine A(2A) receptor modulation of hippocampal CA3-CA1 synapse plasticity during associative learning in behaving mice. Neuropsychopharmacology 34:1865–1874

Frank-Cannon TC, Alto LT, McAlpine FE et al (2009) Does neuroinflammation fan the flame in neurodegenerative diseases? Mol Neurodegener 4:47

Gines S, Bosch M, Marco S et al (2006) Reduced expression of the TrkB receptor in Huntington's disease mouse models and in human brain. Eur J Neurosci 23:649–658

Golder FJ, Mitchell GS (2005) Spinal synaptic enhancement with acute intermittent hypoxia improves respiratory function after chronic cervical spinal cord injury. J Neurosci 25:2925–2932

Golder FJ, Ranganathan L, Satriotomo I et al (2008) Spinal adenosine A2a receptor activation elicits long-lasting phrenic motor facilitation. J Neurosci 28:2033–2042

Gomes CA, Vaz SH, Ribeiro JA et al (2006) Glial cell line-derived neurotrophic factor (GDNF) enhances dopamine release from striatal nerve endings in an adenosine A2A receptor-dependent manner. Brain Res 1113:129–136

Gomes CA, Simões PF, Canas PM et al (2009) GDNF control of the glutamatergic cortico-striatal pathway requires tonic activation of adenosine A receptors. J Neurochem 108:1208–1219

Gomes C, Ferreira R, George J et al (2013) Activation of microglial cells triggers a release of brain-derived neurotrophic factor (BDNF) inducing their proliferation in an adenosine A2A receptor-dependent manner: A2A receptor blockade prevents BDNF release and proliferation of microglia. J Neuroinflamm 10:16

Gordon S (2003) Alternative activation of macrophages. Nat Rev Immunol 3:23–35

Gyarfas T, Knuuttila J, Lindholm P et al (2010) Regulation of brain-derived neurotrophic factor (BDNF) and cerebral dopamine neurotrophic factor (CDNF) by anti-parkinsonian drug therapy in vivo. Cell Mol Neurobiol 30:361–368

Hartmann M, Heumann R, Lessmann V (2001) Synaptic secretion of BDNF after high-frequency stimulation of glutamatergic synapses. EMBO J 20:5887–5897

Heese K, Fiebich BL, Bauer J et al (1997) Nerve growth factor (NGF) expression in rat microglia is induced by adenosine A2a-receptors. Neurosci Lett 231:83–86

Huber KM, Sawtell NB, Bear MF (1998) Brain-derived neurotrophic factor alters the synaptic modification threshold in visual cortex. Neuropharmacology 37:571–579

Ikegaya Y, Ishizaka Y, Matsuki N (2002) BDNF attenuates hippocampal LTD via activation of phospholipase C: implications for a vertical shift in the frequency-response curve of synaptic plasticity. Eur J Neurosci 16:145–148

Jeon SJ, Rhee SY, Ryu JH et al (2011) Activation of adenosine A2A receptor up-regulates BDNF expression in rat primary cortical neurons. Neurochem Res 36:2259–2269

Jeon SJ, Bak H, Seo J et al (2012) Oroxylin A induces BDNF expression on cortical neurons through adenosine A2A receptor stimulation: a possible role in neuroprotection. Biomol Ther (Seoul) 20:27–35

Jerónimo-Santos A, Batalha VL, Muller CE et al (2014) Impact of in vivo chronic blockade of adenosine A2A receptors on the BDNF-mediated facilitation of LTP. Neuropharmacology 83:99–106

Jiang B, Akaneya Y, Hata Y et al (2003) Long-term depression is not induced by low-frequency stimulation in rat visual cortex in vivo: a possible preventing role of endogenous brain-derived neurotrophic factor. J Neurosci 23:3761–3770

Jiang Y, Wei N, Lu T et al (2011) Intranasal brain-derived neurotrophic factor protects brain from ischemic insult via modulating local inflammation in rats. Neuroscience 172:398–405

Kang H, Schuman EM (1995) Long-lasting neurotrophin-induced enhancement of synaptic transmission in the adult hippocampus. Science 267:1658–1662

Kang H, Welcher AA, Shelton D et al (1997) Neurotrophins and time: different roles for TrkB signaling in hippocampal long-term potentiation. Neuron 19:653–664

Kim DH, Jeon SJ, Son KH et al (2006) Effect of the flavonoid, oroxylin A, on transient cerebral hypoperfusion-induced memory impairment in mice. Pharmacol Biochem Behav 85:658–668

Korte M, Griesbeck O, Gravel C et al (1996) Virus-mediated gene transfer into hippocampal CA1 region restores long-term potentiation in brain-derived neurotrophic factor mutant mice. Proc Natl Acad Sci U S A 93:12547–12552

Kust BM, Copray JC, Brouwer N et al (2002) Elevated levels of neurotrophins in human biceps brachii tissue of amyotrophic lateral sclerosis. Exp Neurol 177:419–427

Lee FS, Chao MV (2001) Activation of Trk neurotrophin receptors in the absence of neurotrophins. Proc Natl Acad Sci U S A 98:3555–3560

Levine MS, Klapstein GJ, Koppel A et al (1999) Enhanced sensitivity to N-methyl-D-aspartate receptor activation in transgenic and knockin mouse models of Huntington's disease. J Neurosci Res 58:515–32

Lin LF, Doherty DH, Lile JD et al (1993) GDNF: a glial cell line-derived neurotrophic factor for midbrain dopaminergic neurons. Science 260:1130–1132

Linden DJ (1994) Long-term synaptic depression in the mammalian brain. Neuron 12:457–472

Lopes LV, Sebastião AM, Ribeiro JA (2011) Adenosine and related drugs in brain diseases: present and future in clinical trials. Curr Top Med Chem 11:1087–1101

Lu Y, Christian K, Lu B (2008) BDNF: a key regulator for protein synthesis-dependent LTP and long-term memory? Neurobiol Learn Mem 89:312–323

Lu B, Nagappan G, Guan X et al (2013) BDNF-based synaptic repair as a disease-modifying strategy for neurodegenerative diseases. Nat Rev Neurosci 14:401–416

Malenka RC, Nicoll RA (1999) Long-term potentiation–a decade of progress? Science 285:1870–1874

Martire A, Pepponi R, Domenici MR et al (2013) BDNF prevents NMDA-induced toxicity in models of Huntington's disease: the effects are genotype specific and adenosine A(2A) receptor is involved. J Neurochem 125:225–235

Meyer-Franke A, Kaplan MR, Pfrieger FW et al (1995) Characterization of the signaling interactions that promote the survival and growth of developing retinal ganglion cells in culture. Neuron 15:805–819

Minichiello L (2009) TrkB signalling pathways in LTP and learning. Nat Rev Neurosci 10:850–860

Mitchell GS, Johnson SM (2003) Neuroplasticity in respiratory motor control. J Appl Physiol (1985) 94:358–374

Mitchell GS, Baker TL, Nanda SA et al (2001) Invited review: intermittent hypoxia and respiratory plasticity. J Appl Physiol (1985) 90:2466–2475

Mojsilovic-Petrovic J, Jeong GB, Crocker A et al (2006) Protecting motor neurons from toxic insult by antagonism of adenosine A2a and Trk receptors. J Neurosci 26:9250–9263

Mutoh T, Sobue G, Hamano T et al (2000) Decreased phosphorylation levels of TrkB neurotrophin receptor in the spinal cords from patients with amyotrophic lateral sclerosis. Neurochem Res 25:239–245

Nascimento F, Pousinha PA, Correia AM et al (2014) Adenosine A2A receptors activation facilitates neuromuscular transmission in the pre-symptomatic phase of the SOD1(G93A) ALS mice, but not in the symptomatic phase. Plos One. 9:e104081

Neverova NV, Saywell SA, Nashold LJ et al (2007) Episodic stimulation of alpha1-adrenoreceptors induces protein kinase C-dependent persistent changes in motoneuronal excitability. J Neurosci 27:4435–4442

Ohta A, Sitkovsky M (2001) Role of G-protein-coupled adenosine receptors in downregulation of inflammation and protection from tissue damage. Nature 414:916–920

Orr AG, Orr AL, Li XJ et al (2009) Adenosine A(2A) receptor mediates microglial process retraction. Nat Neurosci 12:872–878

Parreira S (2014) Modulation of GABA and glutamate release by brain-derived neurotrophic factor: role of adenosine A2A receptors. Master Thesis, Faculty of Medicine, University of Lisbon, Portugal

Patterson SL, Abel T, Deuel TA et al (1996) Recombinant BDNF rescues deficits in basal synaptic transmission and hippocampal LTP in BDNF knockout mice. Neuron 16:1137–1145

Pazzagli M, Pedata F, Pepeu G (1993) Effect of K+ depolarization, tetrodotoxin, and NMDA receptor inhibition on extracellular adenosine levels in rat striatum. Eur J Pharmacol 234:61–65

Pereira DB, Chao MV (2007) The tyrosine kinase Fyn determines the localization of TrkB receptors in lipid rafts. J Neurosci 27:4859–4869

Popoli P, Blum D, Martire A et al (2007) Functions, dysfunctions and possible therapeutic relevance of adenosine A2A receptors in Huntington's disease. Prog Neurobiol 81:331–348

Potenza RL, Tebano MT, Martire A et al (2007) Adenosine A(2A) receptors modulate BDNF both in normal conditions and in experimental models of Huntington's disease. Purinergic Signal 3:333–338

Potenza RL, Armida M, Ferrante A et al (2013) Effects of chronic caffeine intake in a mouse model of amyotrophic lateral sclerosis. J Neurosci Res 91:585–592

Pousinha PA, Diógenes MJ, Ribeiro JA et al (2006) Triggering of BDNF facilitatory action on neuromuscular transmission by adenosine A2A receptors. Neurosci Lett 404:143–147

Quiroz C, Gomes C, Pak AC et al (2006) Blockade of adenosine A2A receptors prevents protein phosphorylation in the striatum induced by cortical stimulation. J Neurosci 26:10808–10812

Rajagopal R, Chen ZY, Lee FS et al (2004) Transactivation of Trk neurotrophin receptors by G-protein-coupled receptor ligands occurs on intracellular membranes. J Neurosci 24:6650–6658

Ramirez SH, Fan S, Maguire CA et al (2004) Activation of adenosine A2A receptor protects sympathetic neurons against nerve growth factor withdrawal. J Neurosci Res 77:258–269

Rebola N, Sebastião AM, de Mendonca A et al (2003) Enhanced adenosine A2A receptor facilitation of synaptic transmission in the hippocampus of aged rats. J Neurophysiol 90:1295–1303

Rebola N, Simoes AP, Canas PM et al (2011) Adenosine A2A receptors control neuroinflammation and consequent hippocampal neuronal dysfunction. J Neurochem 117:100–111

Rodrigues TM, Jeronimo-Santos A, Sebastião AM et al (2014a) Adenosine A(2A) receptors as novel upstream regulators of BDNF-mediated attenuation of hippocampal long-term depression (LTD). Neuropharmacology 79:389–398

Rodrigues TM, Jeronimo-Santos A, Outeiro TF et al (2014b) Challenges and promises in the development of neurotrophic factor-based therapies for Parkinson's disease. Drugs Aging 31:239–261

Rosch H, Schweigreiter R, Bonhoeffer T et al (2005) The neurotrophin receptor p75NTR modulates long-term depression and regulates the expression of AMPA receptor subunits in the hippocampus. Proc Natl Acad Sci U S A 102:7362–7367

Sebastião AM, Ribeiro JA (2009) Triggering neurotrophic factor actions through adenosine A2A receptor activation: implications for neuroprotection. Br J Pharmacol 158:15–22

Sebastião AM, Colino-Oliveira M, Assaife-Lopes N et al (2013) Lipid rafts, synaptic transmission and plasticity: impact in age-related neurodegenerative diseases. Neuropharmacology 64:97–107

Stayte S, Vissel B (2014) Advances in non-dopaminergic treatments for Parkinson's disease. Front Neurosci 8:113

Strand AD, Baquet ZC, Aragaki AK et al (2007) Expression profiling of Huntington's disease models suggests that brain-derived neurotrophic factor depletion plays a major role in striatal degeneration. J Neurosci 27:11758–11768

Suzuki S, Numakawa T, Shimazu K et al (2004) BDNF-induced recruitment of TrkB receptor into neuronal lipid rafts: roles in synaptic modulation. J Cell Biol 167:1205–1215

Tebano MT, Martire A, Potenza RL et al (2008) Adenosine A(2A) receptors are required for normal BDNF levels and BDNF-induced potentiation of synaptic transmission in the mouse hippocampus. J Neurochem 104:279–286

Torres-Peraza JF, Giralt A, Garcia-Martinez JM et al (2008) Disruption of striatal glutamatergic transmission induced by mutant huntingtin involves remodeling of both postsynaptic density and NMDA receptor signaling. Neurobiol Dis 29:409–421

Vaz SH, Cristovão-Ferreira S, Ribeiro JA et al (2008) Brain-derived neurotrophic factor inhibits GABA uptake by the rat hippocampal nerve terminals. Brain Res 1219:19–25

Vaz SH, Jorgensen TN, Cristovão-Ferreira S et al (2011) Brain-derived neurotrophic factor (BDNF) enhances GABA transport by modulating the trafficking of GABA transporter-1 (GAT-1) from the plasma membrane of rat cortical astrocytes. J Biol Chem 286:40464–40476

Wei CJ, Augusto E, Gomes CA et al (2014) Regulation of fear responses by striatal and extrastriatal adenosine A2A receptors in forebrain. Biol Psychiatry 75:855–863

Wieraszko A, Goldsmith G, Seyfried TN (1989) Stimulation-dependent release of adenosine triphosphate from hippocampal slices. Brain Res 485:244–250

Wiese S, Jablonka S, Holtmann B et al (2007) Adenosine receptor A2A-R contributes to motoneuron survival by transactivating the tyrosine kinase receptor TrkB. Proc Natl Acad Sci U S A 104:17210–17215

Woo NH, Teng HK, Siao CJ et al (2005) Activation of p75NTR by proBDNF facilitates hippocampal long-term depression. Nat Neurosci 8:1069–1077

Xie Y, Hayden MR, Xu B (2010) BDNF overexpression in the forebrain rescues Huntington's disease phenotypes in YAC128 mice. J Neurosci 30:14708–14718

Xu B, Gottschalk W, Chow A et al (2000) The role of brain-derived neurotrophic factor receptors in the mature hippocampus: modulation of long-term potentiation through a presynaptic mechanism involving TrkB. J Neurosci 20:6888–6897

Yamamoto M, Sobue G, Li M et al (1993) Nerve growth factor (NGF), brain-derived neurotrophic factor (BDNF) and low-affinity nerve growth factor receptor (LNGFR) mRNA levels in cultured rat Schwann cells; differential time- and dose-dependent regulation by cAMP. Neurosci Lett 152:37–40

Yanpallewar SU, Barrick CA, Buckley H et al (2012) Deletion of the BDNF truncated receptor TrkB.T1 delays disease onset in a mouse model of amyotrophic lateral sclerosis. PLoS One 7:e39946

Zuccato C, Ciammola A, Rigamonti D et al (2001) Loss of huntingtin-mediated BDNF gene transcription in Huntington's disease. Science 293:493–498

Zuccato C, Marullo M, Conforti P et al (2008) Systematic assessment of BDNF and its receptor levels in human cortices affected by Huntington's disease. Brain Pathol 18:225–238

Chapter 5
Role of Adenosine A_{2A} Receptors in the Control of Neuroinflammation—Relevance for Parkinson's Disease

Catarina Gomes, Jimmy George, Jiang-Fan Chen and Rodrigo A. Cunha

Abstract The antagonism of adenosine A_{2A} receptors ($A_{2A}R$) is currently a leading non-dopaminergic strategy to delay the onset of Parkinson's disease (PD), but the underlying mechanism of action is still unclear. One prominent feature of PD is the emergence of a neuroinflammation status supported by an increased density of activated microglia in afflicted brain regions, namely the *substantia nigra* and dorsolateral striatum since the onset of PD motor symptoms. This neuroinflammation might contribute for the etiology of PD since anti-inflammatory strategies can attenuate the behavioral and neurochemical changes in both PD patients and PD animal models. We now discuss the possibility that $A_{2A}R$ may control PD features through the control of microgliosis and neuroinflammation since: (1) microglia are endowed with $A_{2A}R$; (2) $A_{2A}R$ are up-regulated in diseased conditions; (3) $A_{2A}R$ can control different facets of microglia function, from proliferation, migration and inflammatory reactivity; (4) $A_{2A}R$ antagonists effectively prevent microgliosis and prevent neuroinflammation, namely in animal models of PD.

Keywords A_{2A} receptor · Adenosine · Microglia · Neuroinflammation

The antagonism of adenosine A_{2A} receptors ($A_{2A}R$) is currently a leading non-dopaminergic strategy to delay the onset of Parkinson's disease (PD), but the underlying mechanism of action is still unclear. One prominent feature of PD is the emergence of a neuroinflammation status supported by an increased density of activated microglia in afflicted brain regions, namely the *substantia nigra* and dorsolateral striatum since the onset of PD motor symptoms. This neuroinflammation might

R. A. Cunha (✉) · C. Gomes · J. George
CNC-Center for Neuroscience and Cell Biology, University of Coimbra, 3004-517 Coimbra, Portugal
e-mail: cunharod@gmail.com

C. Gomes · R. A. Cunha
FMUC-Faculty of Medicine, University of Coimbra, 3004-501 Coimbra, Portugal

J.-F. Chen
Department of Neurology, Boston University School of Medicine, Boston, MA 02129, USA
e-mail: chenjf@.bu.edu

© Springer International Publishing Switzerland 2015
M. Morelli et al. (eds.), *The Adenosinergic System*, Current Topics in Neurotoxicity 10, DOI 10.1007/978-3-319-20273-0_5

contribute for the etiology of PD since anti-inflammatory strategies can attenuate the behavioral and neurochemical changes in both PD patients and PD animal models. We now discuss the possibility that $A_{2A}R$ may control PD features through the control of microgliosis and neuroinflammation since: (1) microglia are endowed with $A_{2A}R$; (2) $A_{2A}R$ are up-regulated in diseased conditions; (3) $A_{2A}R$ can control different facets of microglia function, from proliferation, migration and inflammatory reactivity; (4) $A_{2A}R$ antagonists effectively prevent microgliosis and prevent neuroinflammation, namely in animal models of PD (Fig. 5.1).

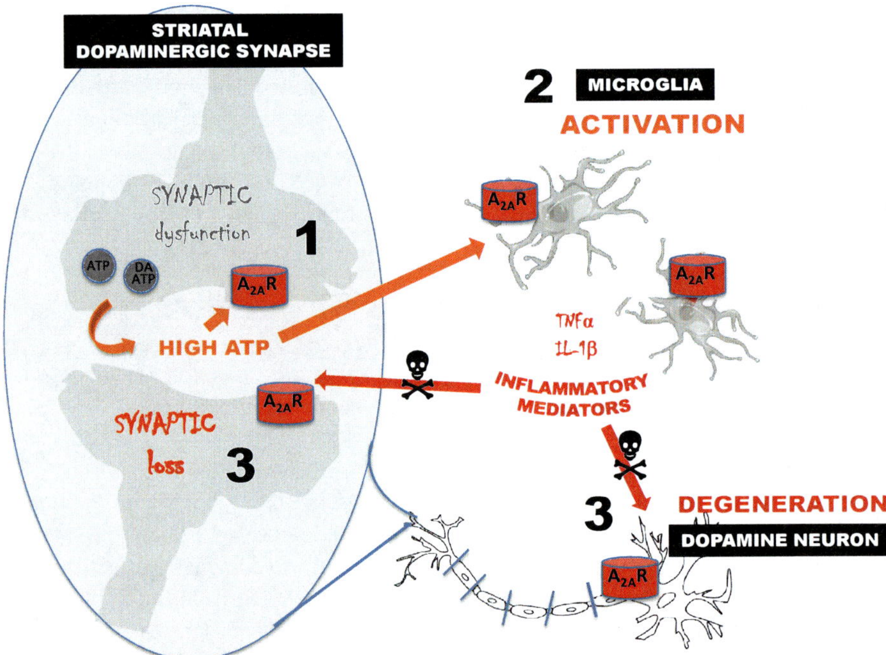

Fig. 5.1 Triple role of adenosine A_{2A} receptors ($A_{2A}R$) in the control of the microglia-associated evolving neurodegeneration in Parkinson's disease (PD). As occurs in most neurodegenerative disorders, it is hypothesized that PD might begin with a dysfunction of dopaminergic synapses controlling cortico-striatal transmission in the dorso-lateral striatum (*1*); notably $A_{2A}R$ are known to participate in the aberrant plasticity in cortico-striatal synapses. Synaptic dysfunction is accompanied by an increased release of ATP, co-stored in synaptic vesicles, which acts as a danger signal, and can act as a chemotaxic signal for microglia and can also trigger a phenotypic modification of microglia (*2*); additionally, ATP is extracellularly converted into adenosine selectively activating $A_{2A}R$, which are known to control microglia dynamics, their proliferation and to assist in the mounting of a neuroinflammatory reaction. The sustained release of pro-inflammatory cytokines can cause deleterious effects both in synapses and in the viability of neurons, causing synaptic loss and neurodegeneration (*3*); as a third level of action of $A_{2A}R$, they are known to control the delirious impact of cytokines (namely interleukin-1β, IL-1β) on synaptic plasticity and on neurodegeneration.

Microglia and Neuroinflammation

Microglial cells are key players of innate immunity in the central nervous system (CNS). They are derived from myeloid cells that migrate early during development into the brain parenchyma (reviewed in Ladeby et al. 2005; Nayak et al. 2014; Saijo and Glass 2011). Thus, microglia are equivalent to a macrophage-like population resident in the CNS expressing chemokine and cytokine receptors that interact with the peripheral immune cells (reviewed in Amor and Woodroofe 2014; Pocock and Kettenmann 2007; Ransohoff and Brown 2012). According to their ontogenic origin, they are endowed with several features characteristic of immune cells able to rapidly expand their population, to chemotaxically migrate to sites of injury and to trigger and sustain inflammatory responses (reviewed in Kettenmann et al. 2011; Lynch 2009; Parkhurst and Gan 2010).

Traditionally, it was assumed that microglia remained quiescent until injury or infection activated them in the brain (Perry and Gordon 1988; Streit et al. 1988). However, it is now recognized that microglia play a key role in supporting the homeostatic functioning of brain function under physiological conditions (Davalos et al. 2005; Nimmerjahn et al. 2005). Thus, what was initially assumed to be a resting phenotype of microglia in the absence of noxious stimuli in fact corresponds to an active surveying state (reviewed in Hanisch and Kettenmann 2007; Raivich 2005; Tremblay et al. 2011; Wake et al. 2013), with a regulatory and supportive role (reviewed in Cherry et al. 2014; Eyo and Wu 2013; Jones and Lynch 2014). The sensor ability of microglia is operated by the constant extension and retraction of cellular processes, requiring moment-to-moment rearrangements of its cytoskeleton (Dailey et al. 2013; Ilschner and Brandt 1996; Janßen et al. 2014). Without displacing the cell body, microglia are able to survey different parts of the nervous system and different subcellular structures of neurons, in particular the synaptic compartment (reviewed in Biber et al. 2007; Tremblay et al. 2011; Wake et al. 2013). In fact, a wealth of recent studies has established a tight association between microglia and synapses (e.g. Kettenmann et al. 2013; Li et al. 2012; Myamoto et al. 2013; Tremblay et al. 2010; Wake et al. 2009) to such as extent that the concept of a quad-partite synapse has been forwarded (Schafer et al. 2013). Microglia are equipped with receptors for neurotransmitters (Murugan et al. 2013; Pocock and Kettenmann 2007), and excitatory transmission (mediated by glutamate) increases whereas inhibitory (mediated by GABA) transmission decreases microglial processes dynamic (Fontainhas et al. 2011; Grinberg et al. 2011; Nimmerjahn et al. 2005; Wong et al. 2011). Conversely, microglia can release a variety of factor that affect synaptic transmission (Antonucci et al. 2012), ranging from chemokines (Piccinin et al. 2010; Schafer et al. 2012), cytokines (Griffin et al. 2006; Rebola et al. 2011), purines (Inoue 2006; Pascual et al. 2012), glutamate and D-serine (Scianni et al. 2013), nitric oxide (NO) (Zhang et al. 2014) or brain-derived neurotrophic factor (BDNF) (Coull et al. 2005; Parkhurst et al. 2013). The importance of this constitutive bi-directional communication between synapses (synaptic activity) and microglia is best heralded by the synaptic dysfunction observed upon manipulation of genetic microglia function (Costello et al. 2011;

Hoshiko et al. 2012; Roumier et al. 2004). Thus, microglia are critical for the dynamic synaptic carving that is essential to entrain the adaptive function of the brain (Cristovão et al. 2014; Ji et al. 2013; Lim et al. 2013; Paolicelli et al. 2011; Ueno et al. 2013; Zhan et al. 2014).

This illustrates that microglia dysfunction can actually act as a trigger of brain disease, because of its critical role on physiological brain function. This should obviously not overshadow the importance of microglia in the adaptive mechanisms associated with noxious brain stimulation. In fact, noxious signals, such as bacteria cell wall fragments, misfolded proteins or intracellular molecules (glutamate, ATP), can trigger a modification of microglia phenotype (Béraud et al. 2013; Doens and Fernández 2014; Färber and Kettenmann 2006; Liu and Bing 2011; Monif et al. 2009; Salminen et al. 2008; Schapansky et al. 2014; Trang et al. 2012; Zielasek and Hartung 1996). This involves a time-coordinated series of processes (Gomez-Nicola and Perry 2015; Santiago et al. 2014): (1) the proliferation of the microglia, which is mostly dependent on the amplification of the brain parenchyma resident population (Ajami et al. 2011; Ladeby et al. 2005; Li et al. 2013; Saijo and Glass 2011); (2) the chemotaxic migration of microglia to the sites of injury, which can be triggered by chemokines, by proteins such as α-synuclein or β-amyloid as well as by purines namely ATP/ADP (Davalos et al. 2005; Färber and Kettenmann 2006; Ohsawa et al. 2007); (3) the re-balance of the pattern of expression and release by microglia of cytokines and other neuroactive substances, corresponding to a re-balance between pro-and anti-inflammatory phenotypes, which is still poorly understood (Aguzzi et al. 2013); (4) the modification of the phagocytic potential of microglia (Fu et al. 2014; Inoue et al. 2009; Sierra et al. 2013), which can promote the elimination of toxic fragments (Neumann et al. 2009) or also promote the elimination of synapses or neurons (Neher et al. 2012; Perry and O'Connor 2010; Rao et al. 2012); (5) the retraction of microglia processes contacting synapses (Orr et al. 2009), thus potentially decreasing the homeostatic support of synapses described above; (6) the apoptosis of microglia (Streit and Xue 2009).

Thus, microglia exist in a variety of flavors, namely surveillant/supportive microglia, phagocytic microglia, pro-inflammatory microglia, anti-inflammatory microglia, proliferating microglia or pro-apoptotic microglia; the classification and analysis of each type of microglia is currently complicated by our inability to ascribe a characteristic molecular profile to each putative type of microglia (Gomez-Nicola and Perry 2015). Furthermore, it is most likely that these different types of microglia co-exist as a continuum during noxious brain conditions, with a dynamic ratio that will depend on the timing of detection, reaction or extinction of the adaptive response of microglia to noxious conditions. This probably contributes for the discussion on the dual ability of microglia to impact on the function and viability of brain function: in fact, eliminating or interfering with microglia function can either aggravate or attenuate brain damage, according to the impact of the noxious stimulus on microglia dynamics and to the timing of intervention of microglia function (Gomez-Nicola and Perry 2015; Santiago et al. 2014).

Neuroinflammation and Microglia in Parkinson's Disease

The possible involvement of neuroinflammation in the pathogenesis of Parkinson's disease (PD) was first prompted by the observation that microglia with an 'activated' morphology were recurrently observed in the afflicted areas of the brain of PD patients, namely in the *substantia nigra pars compacta* (SN) (Banati et al. 1998; Gerhard et al. 2006; Imamura et al. 2003; McGeer et al. 1988; Ouchi et al. 2005; Sawada et al. 2006), as well as in animal models of PD, namely upon acute or chronic administration of 1-methyl-4-phenyl-1,2,3,6-tetrahydropyridine (MPTP) or 6-hydroxydopamine (6-OHDA) (Akiyama and McGeer 1989; Barcia et al. 2013; Depino et al. 2003; He et al. 2001; Kanaan et al. 2008; Maia et al. 2012; McGeer et al. 2003; Walsh et al. 2011). Likewise, increased levels of inflammatory markers, such as nitric oxide synthase (iNOS), cyclooxygenase 2 (COX2) or tumor necrosis factor (TNF-α) receptor 1, are found in the SN of brains of PD patients (Boka et al. 1994; Hunot et al. 1996; Knott et al. 2000; Mogi et al. 2000) and, accordingly, higher levels of pro-inflammatory cytokines such as interleukin-1β, interleukin-6 or TNF-α (tumor necrosis factor) are found in the cerebrospinal fluid of PD patients (Boka et al. 1994; Dobbs et al. 1999; Mogi et al. 1994a, b). Furthermore, different polymorphisms related to inflammation are associated with the risk of developing PD (reviewed in Hirsch and Hunot 2009), as confirmed in genome-wide association studies (Hamza et al. 2010; International Parkinson Disease Genomics Consortium 2011), and inflammatory conditions such as influenza infection and neuroinflammatory conditions such as encephalitis (Ogata et al. 2000; Rail et al. 1981) can trigger PD-like symptoms. In fact, the activation of microglia in the SN with a non-toxic dose of either LPS or interleukin-1β precipitated the loss of dopaminergic neurons and the emergence of motor symptoms in animal models of PD (Ferrari et al. 2006; Godoy et al. 2008; Koprich et al. 2008). The importance of microglia activation-induced neuroinflammation in the emergence of PD is further heralded by the development of animal models of PD based on the intra-nigral injection of LPS or of formylmethionyl-leucyl-phenylalanine, a bacterial-derived chemoattractant (Castano et al. 1998; Gao et al. 2008). This is actually mimicked by the peri-natal or systemic administration of LPS (Ling et al. 2002, 2004; Qin et al. 2007), possibly because the SN has the highest density of microglia in the brain (McGeer et al. 1988) with greater reactivity (Kim et al. 2000). In support of a role of microglia in the genesis of the neuroinflammation that bolsters the risk of PD, it was observed in animal studies that the deletion of key inflammatory mechanisms in microglia cells dampens the PD-associated dysfunction (Hernandes et al. 2013; Pabon et al. 2011), whereas the elimination of anti-inflammatory mechanisms exacerbates PD-associated dysfunction (Zhang et al. 2011). In further accordance with this proposed deleterious impact of microglia-associated neuroinflammation in the evolution of PD, most animal studies showed that different anti-inflammatory drugs (dexamethasone, indomethacin, indomethacin, celecoxib) or inhibitors of microglia activation (minocycline) attenuated different features characteristic of PD (Castano et al. 2002; Du et al. 2001; He et al. 2001; Kurkowska-Jastrzebska et al. 2002,

2004; Quintero et al. 2006; Sanchez-Pernaute et al. 2004). Accordingly, several (but not all) epidemiological studies have reported an association between the intake of anti-inflammatory drugs and the risk to develop PD (reviewed in Gagne and Power 2010; Hirsch and Hunot 2009). In particular, the consumption of ibuprofen displays an inverse dose-response relationship with the risk to develop PD with an odds ratio of 0.62, even after adjusting the data for different possible confounding variables (Gao et al. 2011). Overall, this converging body of animal and human studies has supported the proposal that microglia 'activation' and a neuroinflammation state are associated with the emergence of PD (reviewed in Herrera et al. 2005; Hirsch and Hunot 2009; Kannarkat et al. 2013; Long-Smith et al. 2009; Moehle and West 2014; Qian et al. 2010; Sanchez-Guajardo et al. 2013; Tansey and Goldberg 2010; Whitton 2007; Wilms et al. 2007).

Although the evidence linking neuroinflammation with the emergence of PD is compelling and constitutes a promising opportunity for the development of novel neuroprotective strategies, there are still some open questions. Indeed, although there seems to be a closer association between the onset of alterations of microglia phenotypes especially with early stages of PD (e.g. Ouchi et al. 2005), it is still unclear if neuroinflammation is associated with the onset of PD or instead if it results from dopaminergic damage and is mostly associated with the evolution of PD. Likewise, it is still unclear what might trigger microglia-induced neuroinflammation in PD since several candidates can fulfill this role, namely α-synuclein aggregates (Zhang et al. 2005), ATP (Davalos et al. 2005), MMP-3 (Kim et al. 2005, 2007) or neuromelanin, which is particularly abundant in the SN (Wilms et al. 2003; Zecca et al. 2008). All these substances can be produced upon damage of dopaminergic neurons, which lead to the proposal that microglia-induce neuroinflammation might be an amplification loop to generate PD, whereby an initial dysfunction of dopaminergic neurons would release some of the above mentioned mediators that would trigger microglia-induced release of pro-inflammatory factors to further amplify dopaminergic neuronal loss.

Adenosine A_{2A} Receptor Blockade Prevents Parkinson's Disease

The combined efforts of several groups over the last years have guided adenosine A_{2A} receptor ($A_{2A}R$) antagonists as leading non-dopaminergic therapeutic target in PD (Chen et al. 2013; Ferré et al. 2007; Jenner 2014; Morelli et al 2009; Schwarzschild et al. 2006). $A_{2A}R$ antagonists have demonstrated motor benefits and may have neuroprotective benefits as well. Clinical Phase II–III trials have been completed for the $A_{2A}R$ antagonists KW-6002 (istradefylline, Kyowa, Japan) and SCH420814 (Preladenant, Merck, USA) (Cutler et al. 2012; Hauser 2011), confirming a motor benefit in advanced PD patients. Over the last 5 years, four trials with KW-6002 reported an average reduction in "OFF" time of 1.7 h/day in nearly 1700 patients with advanced PD who were already on optimized L-DOPA regimens. SCH420814 also produced motor benefits, decreasing both OFF time and scores on the unified PD rating scale (UPDRS) in advanced PD patients in a clinical Phase III trial (Hauser 2011). Importantly, both drugs had robust safety profiles in clinical trials. The most exciting prospective role for A$_{2A}$R antagonists as a novel therapy for PD is their potential to attenuate dopaminergic neurodegeneration, as suggested by convergent epidemiological and experimental evidence (reviewed in Prediger 2010): thus, in accordance with the recognition that A$_{2A}$R are the main targets operated by chronic caffeine consumption to generate its psychoactive effects (Cunha and Agostinho 2010; Ferré 2008), three large, long-term (> 30 years follow-up) prospective studies firmly establish a relationship between increased intake of caffeine and decreased risk of developing PD (up to five times lower) in men (compiled in Costa et al. 2010; Palacios et al. 2012; Schwarzschild et al. 2002). However, the mechanism underlying this beneficial neuroprotective effects resulting from the antagonism of A$_{2A}$R in PD still remains to be unraveled.

Adenosine A$_{2A}$ Receptor Control Microglia Reactivity and Neuro-Inflammation

Linked to the role of adenosine as a paracrine signal of distress (Cunha 2001; Newby 1984), A$_{2A}$R are key controllers of immune-inflammatory reactions in the periphery (Sitkovsky et al. 2004). Indeed, A$_{2A}$R are located in all cells of the immune-inflammatory signal and they operate a STOP signal of inflammation (reviewed in Haskó et al. 2008; Sitkovsky et al. 2004). However, the role of A$_{2A}$R in the control of neuroinflammation is less firmly established (Chen and Pedata 2008; Cunha et al. 2007).

The demonstration that A$_{2A}$R are located in microglia cells was first obtained in cultured microglia cells (Saura et al. 2005) and only later in microglia in the brain parenchyma (Yu et al. 2008). A$_{2A}$R are not only present in microglia, but they also control microglia function. The stimulation of A$_{2A}$R triggered the expression and activity of pro-inflammatory mechanisms such as the expression and activity of K$^+$ channels Kv1.3 and ROMK1 (Küst et al. 1999), which control calcium influx and inflammatory cytokine production in activated microglia (Dolga et al. 2012), as well as the expression and activity of iNOS (Saura et al. 2005) and cyclooxygenase-2 (Fiebich et al. 1996). This translated into a functional impact in the control of the production of pro-inflammatory cytokines since the intracerebroventricular injection of a selective A$_{2A}$R antagonist (SCH58261) prevented the lipopolysaccharide (LPS)-induced microglial activation (Yu et al. 2008) and increase of inflammatory mediators like interleukin-1β that causes biochemical changes (p38 and c-jun N-terminal kinase phosphorylation and caspase 3 activation) contributing to neuronal dysfunction typified by decreased long-term potentiation, a form of synaptic plasticity (Rebola et al. 2011). Caffeine also attenuated LPS-induced neuroinflammation (Brothers et al. 2010) and striatal microgliosis induced by the administration of another toxin, 3,4-methylenedioxymethamphetamine (MDMA), a widely used psychostimulant, was abolished in $A_{2A}R$ knockout mice (Ruiz-Medina et al. 2011), as well as by the chronic administration of caffeine (10 mg/kg) (Ruiz-Medina et al. 2013; but see Khairnar et al. 2010). The beneficial effect resulting from $A_{2A}R$-induced increase of microglia-associated neuroinflammation seems to be bolstered by the parallel ability of $A_{2A}R$ blockade to attenuate the interleukin-1β-induced exacerbation of neuronal toxicity (Simões et al. 2012; Stone and Behan 2007).

The control by $A_{2A}R$ of microglia function is not limited to the control of the production of pro-inflammatory factors. Thus, A_2R are required to stimulate microglial proliferation (Gebicke-Haerter et al. 1996) and removing endogenous extracellular adenosine or blocking $A_{2A}R$ prevented the LPS-mediated increase of both BDNF secretion and proliferation, as well as exogenous BDNF-induced proliferation (Gomes et al. 2013). The dynamics of microglia motility is also controlled by $A_{2A}R$, which mediate process retraction in LPS-activated microglia as observed in 3D cultures of primary microglia (Gyoneva et al. 2014a), and $A_{2A}R$ blockade restores the ability of microglia to move towards sites of injury in animal models of neurodegenerative disorders (Gyoneva et al. 2014b).

In apparent contrast to this series of observations that indicate the potential beneficial effects of blocking $A_{2A}R$ to control microglia reactivity, there are several reports supporting that the activation of $A_{2A}R$ might also afford benefits to control brain degeneration associated with neuroinflammation conditions. Thus, low doses of the $A_{2A}R$ agonist CGS 21680 are protective in a rat model of transient cerebral ischemia (Melani et al. 2014), whereas the genetic deletion of $A_{2A}R$ bolstered the mRNA expression and protein levels of pro-inflammatory cytokines (TNF-α, interleukin-1β and interleukin-6) in the corpus callosum upon chronic cerebral hypoperfusion in mice (Duan et al. 2009). Similarly, in animal models of experimental autoimmune encephalomyelitis, the genetic deletion of $A_{2A}R$ also exacerbated demyelination and axonal damage in brainstem, while increasing the levels of pro-inflammatory cytokines and decreasing anti-inflammatory cytokines (Yao et al. 2012); conversely, the increased activation of $A_{2A}R$ with cannabidiol lead to a protective effect of damage and inflammation also in animal models of experimental autoimmune encephalomyelitis (Mecha et al. 2013). Additionally, $A_{2A}R$ agonists also prevent microglia reactivity associated with neuropathic allodynia (Bura et al. 2008; Loram et al. 2009).

This situation seems paradoxical, but is actually understandable in view of the existence of different populations of $A_{2A}R$ in the brain, often with opposite functions (Shen et al. 2008, 2013). Indeed, $A_{2A}R$ in different brain regions have a different overall effect on animal behavior (Wei et al. 2014), in accordance with the observation that the selective $A_{2A}R$ antagonist SCH58261 differently affected the expression of cyclooxygenase-2 (COX-2) in different regions in a rat model of striatal excitotoxicity induced by the intra-cerebral injection of quinolinic acid in cortex and striatum: the $A_{2A}R$ antagonist enhanced COX-2 expression in cortical neurons and prevented it in striatal microglia-like cells (Minghetti et al. 2007). In the case of microglia cells, it has been shown that $A_{2A}R$ actually have a different impact on microglia reactivity according to the environment of microglia cells: thus, $A_{2A}R$ bolster neuroinflammation when extracellular glutamate levels are high and dampen neuroinflammation when extracellular glutamate levels are low (Dai et al. 2010). Since neurodegenerative conditions are associated with an increased glutamatergic excitotoxicity (Lipton and Rosenberg 1994), it is expected that it may be the antagonism of $A_{2A}R$ that may be of greater neuroprotective potential in PD.

Possible Role of $A_{2A}R$-Mediated Control of Neuroinflammation in $A_{2A}R$-Mediated Neuroprotection in Parkinson's Disease

This ability of $A_{2A}R$ antagonists to control neuroinflammation in neurodegenerative disorders (Chen and Pedata 2008; Cunha 2005; Santiago et al. 2014) has also been documented in animal models of Parkinson's disease. Thus, the $A_{2A}R$ antagonist KW-6002 inhibit the nigral microglia activation and prevent the loss of dopaminergic striatal terminals and nigral cell bodies in different MPTP mouse PD models (Pierri et al. 2005). Likewise, we also reported that both the genetic inactivation and the pharmacological blockade of $A_{2A}R$ prevent microgliosis and motor dysfunction in an MPTP mouse model of PD (Yu et al. 2008). Furthermore, it was also shown that $A_{2A}R$ are present in native microglia-like profiles and undergo a robust up-regulation after the MPTP challenge (Yu et al. 2008). The possibility that the effects of $A_{2A}R$ might result from a direct effect of microglial $A_{2A}R$ is supported by another study showing that caffeine (10–20 mg/kg i.p.) and KW6002 (1.5–3 mg/kg i.p.) given once daily for 6 days prevent the changes of different neurochemical features characteristic of PD such as increased extracellular levels of DA, adenosine, glutamate, and hydroxyl radical production, caused by the direct activation of microglial cells by an intra-striatal injection of LPS (Gołembiowska et al. 2013). This was further supported by another study reporting that the $A_{2A}R$ selective antagonist preladenant restore the ability of activated microglia to respond to tissue damage in slices from mice treated for 5 days with MPTP (Gyoneva et al. 2014b).

This hypothesis seems inconsistent with the observation that the selective inactivation of neuronal forebrain $A_{2A}R$, using tissue selective knockout mice, is sufficient to prevent gliosis and the loss of dopaminergic neurons in a MPTP mouse model of PD (Carta et al. 2009). In view of the previously discussed tight interaction between synapses and microglia (reviewed in Biber et al. 2007; Schafer et al. 2013; Tremblay et al. 2011; Wake et al. 2013), this observation prompts the hypothesis that the initial trigger of microgliosis might actually be the initial synaptic dopaminergic neurodegeneration that occurs since the pre-motor phase of PD. In fact, the motor symptoms and the overt degeneration of nigral dopaminergic neurons that are characteristic of PD are preceded by an initial affection of synaptic contacts in the striatum (Day et al. 2006), leading to the loss of striatal dopaminergic nerve terminals (synaptotoxicity), which then evolves to the overt loss of dopaminergic neurons (neurotoxicity) (Berendse et al. 2001; Bézard et al. 2001; Forno et al. 1994). Accordingly, animal models of PD, such as the exposure to

6-hydroxydopamine (6-OHDA) or to mitochondrial toxins (MPTP or rotenone), are based on the destruction of dopaminergic nerve terminals which then evolve to an overt dopaminergic cell loss in the nigra and the emergence of motor symptoms (Simola et al. 2007; Smeyne and Jackson-Lewis 2005). This evolution from striatal synaptotoxicity to nigra dopaminergic cell loss is accompanied by an abnormal function of microglia cells, typified by a microgliosis that is observed both in the striatum and in the nigra in PD (Halliday and Stevens 2011; Teismann and Schulz 2004) and occurs at the onset of motor symptoms in PD patients (Ouchi et al. 2009). Thus, microglia 'activation' would fulfill the role of an amplification system converting the initial synaptotoxicity into an overt damage of dopaminergic neurons that would trigger the emergence of the PD motor symptoms. This hypothetic scenario is further supported by the robust evidence suggesting that synaptic $A_{2A}R$ are prominently up-regulated by noxious stimuli (reviewed in Cunha and Agostinho 2010; Gomes et al. 2011) and seem to play a key role in the control of the neurodegeneration associated with different neurodegenerative conditions (e.g. Coleman et al. 2004; Dadon-Nachum et al. 2011; Gonçalves et al. 2013; Milnerwood and Raymond 2010; Selkoe 2002). Thus, $A_{2A}R$ might have a triple role to control the onset of PD: (1) synaptic $A_{2A}R$ might control the initial synaptic dysfunction that triggers microglia reactivity; (2) microglia $A_{2A}R$ might control different features of microglia reactivity ranging from proliferation to migration to the transformation into a pro-apoptotic phenotype; (3) neuronal $A_{2A}R$ might further control the impact of pro-inflammatory mediators on neuronal viability.

Acknowledgements This work was supported by DARPA, NARSAD, *Santa Casa da Misericórdia de Lisboa* and co-funded by FEDER (QREN), through *Programa Mais Centro* under projects CENTRO-07-ST24-FEDER-002002, CENTRO-07-ST24-FEDER-002006 and CENTRO-07-ST24-FEDER-002008, and through *Programa Operacional Factores de Competitividade*—COMPETE and National funds via FCT—*Fundação para a Ciência e a Tecnologia* under project(s) Pest-C/SAU/LA0001/2013-2014.

References

Aguzzi A, Barres BA, Bennett ML (2013) Microglia: scapegoat, saboteur, or something else? Science 339:156–161

Ajami B, Bennett JL, Krieger C et al (2011) Infiltrating monocytes trigger EAE progression, but do not contribute to the resident microglia pool. Nat Neurosci 14:1142–1149

Akiyama H, McGeer PL (1989) Microglial response to 6-hydroxydopamine- induced substantia nigra lesions. Brain Res 489:247–253

Amor S, Woodroofe MN (2014) Innate and adaptive immune responses in neurodegeneration and repair. Immunology 141:287–291

Antonucci F, Turola E, Riganti L et al (2012) Microvesicles released from microglia stimulate synaptic activity via enhanced sphingolipid metabolism. EMBO J 31:1231–1240

Banati RB, Daniel SE, Blunt SB (1998) Glial pathology but absence of apoptotic nigral neurons in long-standing Parkinson's disease. Mov Disord 13:221–227

Barcia C, Ros CM, Ros-Bernal F et al (2013) Persistent phagocytic characteristics of microglia in the substantia nigra of long-term Parkinsonian macaques. J Neuroimmunol 261:60–66

Béraud D, Hathaway HA, Trecki J et al (2013) Microglial activation and antioxidant responses induced by the Parkinson's disease protein α-synuclein. J Neuroimmune Pharmacol 8:94–117

Berendse HW, Booij J, Francot CM et al (2001) Subclinical dopaminergic dysfunction in asymptomatic Parkinson's disease patients' relatives with a decreased sense of smell. Ann Neurol 50:34–41

Bézard E, Dovero S, Prunier C et al (2001) Relationship between the appearance of symptoms and the level of nigrostriatal degeneration in a progressive 1-methyl-4-phenyl-1,2,3,6-tetrahydro-pyridine-lesioned macaque model of Parkinson's disease. J Neurosci 21:6853–6861

Biber K, Neumann H, Inoue K et al (2007) Neuronal 'On' and 'Off' signals control microglia. Trends Neurosci 30:596–602

Boka G, Anglade P, Wallach D et al (1994) Immunocytochemical analysis of tumor necrosis factor and its receptors in Parkinson's disease. Neurosci Lett 172:151–154

Brothers HM, Marchalant Y, Wenk GL (2010) Caffeine attenuates lipopolysaccharide-induced neuroinflammation. Neurosci Lett 480:97–100

Bura SA, Nadal X, Ledent C et al (2008) A$_{2A}$ adenosine receptor regulates glia proliferation and pain after peripheral nerve injury. Pain 140:95–103

Carta AR, Kachroo A, Schintu N et al (2009) Inactivation of neuronal forebrain A$_{2A}$ receptors protects dopaminergic neurons in a mouse model of Parkinson's disease. J Neurochem 111:1478–1489

Castano A, Herrera AJ, Cano J et al (1998) Lipopolysaccharide intranigral injection induces inflammatory reaction and damage in nigrostriatal dopaminergic system. J Neurochem 70:1584–1592

Castano A, Herrera AJ, Cano J et al (2002) The degenerative effect of a single intranigral injection of LPS on the dopaminergic system is prevented by dexamethasone, and not mimicked by rh-TNF-α, IL-1β and IFN-γ. J Neurochem 81:150–157

Chen JF, Pedata F (2008) Modulation of ischemic brain injury and neuroinflammation by adenosine A$_{2A}$ receptors. Curr Pharm Des 14:1490–1499

Chen JF, Eltzschig HK, Fredholm BB (2013) Adenosine receptors as drug targets—what are the challenges? Nat Rev Drug Discov 12:265–286

Cherry JD, Olschowka JA, O'Banion MK (2014) Are "resting" microglia more "m2"? Front Immunol 5:594

Coleman P, Federoff H, Kurlan R (2004) A focus on the synapse for neuroprotection in Alzheimer disease and other dementias. Neurology 63:1155–1162

Costa J, Lunet N, Santos C et al (2010) Caffeine exposure and the risk of Parkinson's disease: a systematic review and meta-analysis of observational studies. J Alzheimers Dis 20:S221–S238

Costello DA, Lyons A, Denieffe S et al (2011) Long term potentiation is impaired in membrane glycoprotein CD200-deficient mice: a role for Toll-like receptor activation. J Biol Chem 286:34722–34732

Coull JA, Beggs S, Boudreau D et al (2005) BDNF from microglia causes the shift in neuronal anion gradient underlying neuropathic pain. Nature 438:1017–1021

Cristovão G, Pinto MJ, Cunha RA et al (2014) Activation of microglia bolsters synapse formation. Front Cell Neurosci 8:153

Cunha RA (2001) Adenosine as a neuromodulator and as a homeostatic regulator in the nervous system: different roles, different sources and different receptors. Neurochem Int 38:107–125

Cunha RA (2005) Neuroprotection by adenosine in the brain: From A1 receptor activation to A$_{2A}$ receptor blockade. Purinergic Signal 1:111–34

Cunha RA, Agostinho PM (2010) Chronic caffeine consumption prevents memory disturbance in different animal models of memory decline. J Alzheimers Dis 20:S95–S116

Cunha RA, Chen JF, Sitkovsky MV (2007) Opposite modulation of peripheral inflammation and neuroinflammation by adenosine A$_{2A}$ receptors. In: Malva JO, Rego AC, Cunha RA, Oliveira CR (eds) Interaction between neurons and glia in aging and disease. Springer-Verlag, Berlim, pp 53–79

Cutler DL, Tendolkar A, Grachev ID (2012) Safety, tolerability and pharmacokinetics after single and multiple doses of preladenant (SCH420814) administered in healthy subjects. J Clin Pharm Ther 37:578–587

Dadon-Nachum M, Melamed E, Offen D (2011) The "dying-back" phenomenon of motor neurons in ALS. J Mol Neurosci 43:470–477

Dai SS, Zhou YG, Li W et al (2010) Local glutamate level dictates adenosine A_{2A} receptor regulation of neuroinflammation and traumatic brain injury. J Neurosci 30:5802–5810

Dailey ME, Eyo U, Fuller L et al (2013) Imaging microglia in brain slices and slice cultures. Cold Spring Harb Protoc 2013:1142–1148

Davalos D, Grutzendler J, Yang G et al (2005) ATP mediates rapid microglial response to local brain injury in vivo. Nat Neurosci 8:752–758

Day M, Wang Z, Ding J et al (2006) Selective elimination of glutamatergic synapses on striatopallidal neurons in Parkinson disease models. Nat Neurosci 9:251–259

Depino AM, Earl C, Kaczmarczyk E et al (2003) Microglial activation with atypical proinflammatory cytokine expression in a rat model of Parkinson's disease. Eur J Neurosci 18:2731–2742

Dobbs RJ, Charlett A, Purkiss AG et al (1999) Association of circulating TNF-α and IL-6 with ageing and parkinsonism. Acta Neurol Scand 100:34–41

Doens D, Fernández PL (2014) Microglia receptors and their implications in the response to amyloid β for Alzheimer's disease pathogenesis. J Neuroinflammation 11: 48

Dolga AM, Letsche T, Gold M et al (2012) Activation of KCNN3/SK3/K_{Ca}2.3 channels attenuates enhanced calcium influx and inflammatory cytokine production in activated microglia. Glia 60:2050–2064

Du Y, Ma Z, Lin S et al (2001) Minocycline prevents nigrostriatal dopaminergic neurodegeneration in the MPTP model of Parkinson's disease. Proc Natl Acad Sci USA 98:14669–14674

Duan W, Gui L, Zhou Z et al (2009) Adenosine A_{2A} receptor deficiency exacerbates white matter lesions and cognitive deficits induced by chronic cerebral hypoperfusion in mice. J Neurol Sci 285:39–45

Eyo UB, Wu LJ (2013) Bidirectional microglia-neuron communication in the healthy brain. Neural Plast 2013:456857

Färber K, Kettenmann H (2006) Purinergic signaling and microglia. Pflugers Arch 452:615–621

Ferrari CC, Pott Godoy MC et al (2006) Progressive neurodegeneration and motor disabilities induced by chronic expression of IL-1β in the substantia nigra. Neurobiol Dis 24:183–193

Ferré S (2008) An update on the mechanisms of the psychostimulant effects of caffeine. J Neurochem 105:1067–1079

Ferré S, Ciruela F, Quiroz C et al (2007) Adenosine receptor heteromers and their integrative role in striatal function. ScientificWorldJournal 7:74–85

Fiebich BL, Biber K, Lieb K et al (1996) Cyclooxygenase-2 expression in rat microglia is induced by adenosine A_{2a}-receptors. Glia 18:152–160

Fontainhas AM, Wang M, Liang KJ et al (2011) Microglial morphology and dynamic behavior is regulated by ionotropic glutamatergic and GABAergic neurotransmission. PloS One 6:e15973

Forno LS, DeLanney LE, Irwin et al (1994) Evolution of nerve fiber degeneration in the striatum in the MPTP-treated squirrel monkey. Mol Neurobiol 9:163–170

Fu R, Shen Q, Xu P et al (2014) Phagocytosis of microglia in the central nervous system diseases. Mol Neurobiol 49:1422–1434

Gagne JJ, Power MC (2010) Anti-inflammatory drugs and risk of Parkinson disease: a meta-analysis. Neurology 74:995–1002

Gao X, Hu X, Qian L et al (2008) Formyl-methionyl-leucyl-phenylalanine-induced dopaminergic neurotoxicity via microglial activation: a mediator between peripheral infection and neurodegeneration? Environ Health Perspect 116:593–598

Gao X, Chen H, Schwarzschild MA et al (2011) Use of ibuprofen and risk of Parkinson disease. Neurology 76:863–869

Gebicke-Haerter PJ, Christoffel F, Timmer J et al (1996) Both adenosine A_1- and A_2-receptors are required to stimulate microglial proliferation. Neurochem Int 29:37–42

Gerhard A, Pavese N, Hotton G et al (2006) *In vivo* imaging of microglial activation with [^{11}C] (R)-PK11195 PET in idiopathic Parkinson's disease. Neurobiol Dis 21:404–412

Godoy MC, Tarelli R, Ferrari CC et al (2008) Central and systemic IL-1 exacerbates neurodegeneration and motor symptoms in a model of Parkinson's disease. Brain 131:1880–1894

Gołembiowska K, Wardas J, Noworyta-Sokołowska K et al (2013) Effects of adenosine receptor antagonists on the *in vivo* LPS-induced inflammation model of Parkinson's disease. Neurotox Res 24:29–40

Gomes CV, Kaster MP, Tomé AR et al (2011) Adenosine receptors and brain diseases: neuroprotection and neurodegeneration. Biochim Biophys Acta 1808:1380–1399

Gomes C, Ferreira R, George J et al (2013) Activation of microglial cells triggers a release of brain-derived neurotrophic factor (BDNF) inducing their proliferation in an adenosine A$_{2A}$ receptor-dependent manner: A$_{2A}$ receptor blockade prevents BDNF release and proliferation of microglia. J Neuroinflammation 10:16

Gomez-Nicola D, Perry VH (2015) Microglial dynamics and role in the healthy and diseased brain: a paradigm of functional plasticity. Neuroscientist 21:169–184

Gonçalves N, Simões AT, Cunha RA et al (2013) Caffeine and adenosine A$_{2A}$ receptor inactivation decrease striatal neuropathology in a lentiviral-based model of Machado–Joseph disease. Ann Neurol 73:655–666

Griffin R, Nally R, Nolan Y et al (2006) The age-related attenuation in long-term potentiation is associated with microglial activation. J Neurochem 99:1263–1272

Grinberg YY, Milton JG, Kraig RP (2011) Spreading depression sends microglia on Levy flights. PloS One 6:e19294

Gyoneva S, Davalos D, Biswas D et al (2014a) Systemic inflammation regulates microglial responses to tissue damage in vivo. Glia 62:1345–1360

Gyoneva S, Shapiro L, Lazo C et al (2014b) Adenosine A$_{2A}$ receptor antagonism reverses inflammation-induced impairment of microglial process extension in a model of Parkinson's disease. Neurobiol Dis 67:191–202

Halliday GM, Stevens CH (2011) Glia: initiators and progressors of pathology in Parkinson's disease. Mov Disord 26:6–17

Hamza TH, Zabetian CP, Tenesa A et al (2010) Common genetic variation in the HLA region is associated with late-onset sporadic Parkinson's disease. Nat Genet 42:781–785

Hanisch UK, Kettenmann H (2007) Microglia: active sensor and versatile effector cells in the normal and pathologic brain. Nat Neurosci 10:1387–1394

Haskó G, Linden J, Cronstein B et al (2008) Adenosine receptors: therapeutic aspects for inflammatory and immune diseases. Nat Rev Drug Discov 7:759–770

Hauser RA (2011) Future treatments for Parkinson's disease: surfing the PD pipeline. Int J Neurosci 121:53–62

He Y, Appel S, Le W (2001) Minocycline inhibits microglial activation and protects nigral cells after 6-hydroxydopamine injection into mouse striatum. Brain Res 909:187–193

Hernandes MS, Santos GD, Café-Mendes CC et al (2013) Microglial cells are involved in the susceptibility of NADPH oxidase knockout mice to 6-hydroxy-dopamine-induced neurodegeneration. PLoS One 8:e75532

Herrera AJ, Tomás-Camardiel M, Venero JL et al (2005) Inflammatory process as a determinant factor for the degeneration of substantia nigra dopaminergic neurons. J Neural Transm 112:111–119

Hirsch EC, Hunot S (2009) Neuroinflammation in Parkinson's disease: a target for neuroprotection? Lancet Neurol 8:382–397

Hoshiko M, Arnoux I, Avignone E et al (2012) Deficiency of the microglial receptor CX3CR1 impairs postnatal functional development of thalamocortical synapses in the barrel cortex. J Neurosci 32:15106–15111

Hunot S, Boissiere F, Faucheux B et al (1996) Nitric oxide synthase and neuronal vulnerability in Parkinson's disease. Neuroscience 72:355–363

Ilschner S, Brandt R (1996) The transition of microglia to a ramified phenotype is associated with the formation of stable acetylated and detyrosinated microtubules. Glia 18:129–140

Imamura K, Hishikawa N, Sawada M et al (2003) Distribution of major histocompatibility complex class II-positive microglia and cytokine profile of Parkinson's disease brains. Acta Neuropathol 106:518–526

Inoue K (2006) The function of microglia through purinergic receptors: neuropathic pain and cytokine release. Pharmacol Ther 109:210–226

Inoue K, Koizumi S, Kataoka A et al (2009) P2Y$_6$-evoked microglial phagocytosis. Int Rev Neurobiol 85:159–163

International Parkinson Disease Genomics Consortium (2011) Imputation of sequence variants for identification of genetic risks for Parkinson's disease: a meta-analysis of genome-wide association studies. Lancet 377:641–649

Janßen S, Gudi V, Prajeeth CK et al (2014) A pivotal role of nonmuscle myosin II during microglial activation. Exp Neurol 261:666–676

Jenner P (2014) An overview of adenosine A$_{2A}$ receptor antagonists in Parkinson's disease. Int Rev Neurobiol 119:71–86

Ji K, Akgul G, Wollmuth LP et al (2013) Microglia actively regulate the number of functional synapses. PloS One 8:e56293

Jones RS, Lynch MA (2015) How dependent is synaptic plasticity on microglial phenotype? Neuropharmacology. 96:3–10

Kanaan NM, Kordower JH, Collier TJ (2008) Age and region-specific responses of microglia, but not astrocytes, suggest a role in selective vulnerability of dopamine neurons after 1-methyl-4-phenyl-1,2,3,6-tetrahydropyridine exposure in monkeys. Glia 56:1199–1214

Kannarkat GT, Boss JM, Tansey MG (2013) The role of innate and adaptive immunity in Parkinson's disease. J Parkinsons Dis 3:493–514

Kettenmann H, Hanisch UK, Noda M et al (2011) Physiology of microglia. Physiol Rev 91:461–553

Kettenmann H, Kirchhoff F, Verkhratsky A (2013) Microglia: new roles for the synaptic stripper. Neuron 77:10–18

Khairnar A, Plumitallo A, Frau L et al (2010) Caffeine enhances astroglia and microglia reactivity induced by 3,4-methylenedioxymethamphetamine ('ecstasy') in mouse brain. Neurotox Res 17:435–439

Kim WG, Mohney RP, Wilson B et al (2000) Regional difference in susceptibility to lipopolysaccharide-induced neurotoxicity in the rat brain: role of microglia. J Neurosci 20:6309–6316

Kim YS, Kim SS, Cho JJ et al (2005) Matrix metalloproteinase-3: a novel signaling proteinase from apoptotic neuronal cells that activates microglia. J Neurosci 25:3701–3711

Kim YS, Choi DH, Block ML et al (2007) A pivotal role of matrix metalloproteinase-3 activity in dopaminergic neuronal degeneration via microglial activation. FASEB J 21:179–187

Knott C, Stern G, Wilkin GP (2000) Inflammatory regulators in Parkinson's disease: iNOS, lipocortin-1, and cyclooxygenases-1 and -2. Mol Cell Neurosci 16:724–739

Koprich JB, Reske-Nielsen C, Mithal P, Isacson O (2008) Neuroinflammation mediated by IL-1β increases susceptibility of dopamine neurons to degeneration in an animal model of Parkinson's disease. J Neuroinflammation 5:8

Kurkowska-Jastrzebska I, Babiuch M, Joniec I et al (2002) Indomethacin protects against neurodegeneration caused by MPTP intoxication in mice. Int Immunopharmacol 2:1213–1218

Kurkowska-Jastrzebska I, Litwin T, Joniec I et al (2004) Dexamethasone protects against dopaminergic neurons damage in a mouse model of Parkinson's disease. Int Immunopharmacol 4:1307–1318

Küst BM, Biber K, van Calker D et al (1999) Regulation of K+ channel mRNA expression by stimulation of adenosine A$_{2a}$-receptors in cultured rat microglia. Glia 25:120–130

Ladeby R, Wirenfeldt M, Garcia-Ovejero D et al (2005) Microglial cell population dynamics in the injured adult central nervous system. Brain Res Rev 48:196–206

Li Y, Du XF, Liu CS et al (2012) Reciprocal regulation between resting microglial dynamics and neuronal activity in vivo. Dev Cell 23:1189–1202

Li T, Pang S, Yu Y et al (2013) Proliferation of parenchymal microglia is the main source of microgliosis after ischaemic stroke. Brain 136:3578–3588

Lim SH, Park E, You B et al (2013) Neuronal synapse formation induced by microglia and interleukin 10. PloS One 8:e81218

Ling Z, Gayle DA, Ma SY et al (2002) In utero bacterial endotoxin exposure causes loss of tyrosine hydroxylase neurons in the postnatal rat midbrain. Mov Disord 17:116–124

Ling ZD, Chang Q, Lipton JW et al (2004) Combined toxicity of prenatal bacterial endotoxin exposure and postnatal 6- hydroxydopamine in the adult rat midbrain. Neuroscience 124:619–628

Lipton SA, Rosenberg PA (1994) Excitatory amino acids as a final common pathway for neurologic disorders. N Engl J Med 330:613–622

Liu M, Bing G (2011) Lipopolysaccharide animal models for Parkinson's disease. Parkinsons Dis 2011:327089

Long-Smith CM, Sullivan AM, Nolan YM (2009) The influence of microglia on the pathogenesis of Parkinson's disease. Prog Neurobiol 89:277–287

Loram LC, Harrison JA, Sloane EM et al (2009) Enduring reversal of neuropathic pain by a single intrathecal injection of adenosine 2A receptor agonists: a novel therapy for neuropathic pain. J Neurosci 29:14015–14025

Lynch MA (2009) The multifaceted profile of activated microglia. Mol Neurobiol 40:139–156

Maia S, Arlicot N, Vierron E et al (2012) Longitudinal and parallel monitoring of neuroinflammation and neurodegeneration in a 6-hydroxydopamine rat model of Parkinson's disease. Synapse 66:573–583

McGeer PL, Itagaki S, Boyes BE et al (1988) Reactive microglia are positive for HLA-DR in the substantia nigra of Parkinson's and Alzheimer's disease brains. Neurology 38:1285–1291

McGeer PL, Schwab C, Parent A et al (2003) Presence of reactive microglia in monkey substantia nigra years after 1-methyl-4-phenyl-1,2,3,6-tetrahydropyridine administration. Ann Neurol 54:599–604

Mecha M, Feliú A, Iñigo PM et al (2013) Cannabidiol provides long-lasting protection against the deleterious effects of inflammation in a viral model of multiple sclerosis: a role for A_{2A} receptors. Neurobiol Dis 59:141–150

Melani A, Corti F, Cellai L et al (2014) Low doses of the selective adenosine A_{2A} receptor agonist CGS 21680 are protective in a rat model of transient cerebral ischemia. Brain Res 1551:59–72

Milnerwood AJ, Raymond LA (2010) Early synaptic pathophysiology in neurodegeneration: insights from Huntington's disease. Trends Neurosci 33:513–523

Minghetti L, Greco A, Potenza RL et al (2007) Effects of the adenosine A_{2A} receptor antagonist SCH 58621 on cyclooxygenase-2 expression, glial activation, and brain-derived neurotrophic factor availability in a rat model of striatal neurodegeneration. J Neuropathol Exp Neurol 66:363–371

Miyamoto A, Wake H, Moorhouse AJ et al (2013) Microglia and synapse interactions: fine tuning neural circuits and candidate molecules. Front Cell Neurosci 7:70

Moehle MS, West AB (2014) M1 and M2 immune activation in Parkinson's disease: foe and ally? Neuroscience. doi: 10.1016/j.neuroscience.2014.11.018 (in press)

Mogi M, Harada M, Kondo T et al (1994a) Interleukin-1β, interleukin-6, epidermal growth factor and transforming growth factor-α are elevated in the brain from parkinsonian patients. Neurosci Lett 180:147–150

Mogi M, Harada M, Riederer P et al (1994b) Tumor necrosis factor-α (TNF-α) increases both in the brain and in the cerebrospinal fluid from parkinsonian patients. Neurosci Lett 165:208–210

Mogi M, Togari A, Kondo T et al (2000) Caspase activities and tumor necrosis factor receptor R1 (p55) level are elevated in the substantia nigra from parkinsonian brain. J Neural Transm 107:335–341

Monif M, Reid CA, Powell KL et al (2009) The P2X7 receptor drives microglial activation and proliferation: a trophic role for P2X7R pore. J Neurosci 29:3781–3791

Morelli M, Carta AR, Jenner P (2009) Adenosine A_{2A} receptors and Parkinson's disease. Handb Exp Pharmacol 193:589–615

Murugan M, Ling EA, Kaur C (2013) Glutamate receptors in microglia. CNS Neurol Disord Drug Targets 12:773–784

Nayak D, Roth TL, McGavern DB (2014) Microglia development and function. Annu Rev Immunol 32:367–402

Neher JJ, Neniskyte U, Brown GC (2012) Primary phagocytosis of neurons by inflamed microglia: potential roles in neurodegeneration. Front Pharmacol 3:27

Neumann H, Kotter MR, Franklin RJ (2009) Debris clearance by microglia: an essential link between degeneration and regeneration. Brain 132:288–295

Newby AC (1984) Adenosine and the concept of retaliatory metabolites. Trends Biochem Sci 9:42–44

Nimmerjahn A, Kirchhoff F, Helmchen F (2005) Resting microglial cells are highly dynamic surveillants of brain parenchyma in vivo. Science 308:1314–1318

Ogata A, Tashiro K, Pradhan S (2000) Parkinsonism due to predominant involvement of substantia nigra in Japanese encephalitis. Neurology 55:602

Ohsawa K, Irino Y, Nakamura Y et al (2007) Involvement of P2X4 and P2Y12 receptors in ATP-induced microglial chemotaxis. Glia 55:604–616

Orr AG, Orr AL, Li XJ et al (2009) Adenosine A_{2A} receptor mediates microglial process retraction. Nat Neurosci 12:872–878

Ouchi Y, Yoshikawa E, Sekine Y et al (2005) Microglial activation and dopamine terminal loss in early Parkinson's disease. Ann Neurol 57:168–175

Ouchi Y, Yagi S, Yokokura M et al (2009) Neuroinflammation in the living brain of Parkinson's disease. Parkinsonism Relat Disord 15:S200–S204

Pabon MM, Bachstetter AD, Hudson CE et al (2011) CX3CL1 reduces neurotoxicity and microglial activation in a rat model of Parkinson's disease. J Neuroinflammation 8:9

Palacios N, Gao X, McCullough ML et al (2012) Caffeine and risk of Parkinson's disease in a large cohort of men and women. Mov Disord 27:1276–1282

Paolicelli RC, Bolasco G, Pagani F et al (2011) Synaptic pruning by microglia is necessary for normal brain development. Science 333:1456–1458

Parkhurst CN, Gan WB (2010) Microglia dynamics and function in the CNS. Curr Opin Neurobiol 20:595–600

Parkhurst CN, Yang G, Ninan I et al (2013) Microglia promote learning-dependent synapse formation through brain-derived neurotrophic factor. Cell 155:1596–1609

Pascual O, Ben Achour S, Rostaing P et al (2012) Microglia activation triggers astrocyte-mediated modulation of excitatory neurotransmission. Proc Natl Acad Sci USA 109:E197–E205

Perry VH, Gordon S (1988) Macrophages and microglia in the nervous system. Trends Neurosci 11:273–277

Perry VH, O'Connor V (2010) The role of microglia in synaptic stripping and synaptic degeneration: a revised perspective. ASN Neuro 2:e00047

Piccinin S, Di Angelantonio S, Piccioni A et al (2010) CX3CL1-induced modulation at CA1 synapses reveals multiple mechanisms of EPSC modulation involving adenosine receptor subtypes. J Neuroimmunol 224:85–92

Pierri M, Vaudano E, Sager T et al (2005) KW-6002 protects from MPTP induced dopaminergic toxicity in the mouse. Neuropharmacology 48:517–524

Pinna A (2014) Adenosine A_{2A} receptor antagonists in Parkinson's disease: progress in clinical trials from the newly approved istradefylline to drugs in early development and those already discontinued. CNS Drugs 28:455–474

Pocock JM, Kettenmann H (2007) Neurotransmitter receptors on microglia. Trends Neurosci 30:527–535

Prediger RD (2010) Effects of caffeine in Parkinson's disease: from neuroprotection to the management of motor and non-motor symptoms. J Alzheimers Dis 20:S205–S220

Qian L, Flood PM, Hong JS (2010) Neuroinflammation is a key player in Parkinson's disease and a prime target for therapy. J Neural Transm 117:971–979

Qin L, Wu X, Block ML et al (2007) Systemic LPS causes chronic neuroinflammation and progressive neurodegeneration. Glia 55:453–462

Quintero EM, Willis L, Singleton R et al (2006) Behavioral and morphological effects of minocycline in the 6-hydroxydopamine rat model of Parkinson's disease. Brain Res 1093:198–207

Rail D, Scholtz C, Swash M (1981) Post-encephalitic Parkinsonism: current experience. J Neurol Neurosurg Psychiatry 44:670–676

Raivich G (2005) Like cops on the beat: the active role of resting microglia. Trends Neurosci 28:571–573

Ransohoff RM, Brown MA (2012) Innate immunity in the central nervous system. J Clin Invest 122:1164–1171

Rao JS, Kellom M, Kim HW et al (2012) Neuroinflammation and synaptic loss. Neurochem Res 37:903–910

Rebola N, Simões AP, Canas PM et al (2011) Adenosine A$_{2A}$ receptors control neuroinflammation and consequent hippocampal neuronal dysfunction. J Neurochem 117:100–111

Roumier A, Bechade C, Poncer JC et al (2004) Impaired synaptic function in the microglial KARAP/DAP12-deficient mouse. J Neurosci 24:11421–11428

Ruiz-Medina J, Ledent C, Carretón O et al (2011) The A$_{2a}$ adenosine receptor modulates the reinforcement efficacy and neurotoxicity of MDMA. J Psychopharmacol 25:550–564

Ruiz-Medina J, Pinto-Xavier A, Rodríguez-Arias M et al (2013) Influence of chronic caffeine on MDMA-induced behavioral and neuroinflammatory response in mice. Psychopharmacology 226:433–444

Saijo K, Glass CK (2011) Microglial cell origin and phenotypes in health and disease. Nat Rev Immunol 11:775–787

Salminen A, Ojala J, Suuronen T et al (2008) Amyloid-beta oligomers set fire to inflammasomes and induce Alzheimer's pathology. J Cell Mol Med 12:2255–2262

Sanchez-Guajardo V, Barnum CJ, Tansey MG et al (2013) Neuroimmunological processes in Parkinson's disease and their relation to α-synuclein: microglia as the referee between neuronal processes and peripheral immunity. ASN Neuro 5:113–139

Sanchez-Pernaute R, Ferree A, Cooper O et al (2004) Selective COX-2 inhibition prevents progressive dopamine neuron degeneration in a rat model of Parkinson's disease. J Neuroinflammation 1:6

Santiago AR, Baptista FI, Santos PF et al (2014) Role of microglia adenosine A$_{2A}$ receptors in retinal and brain neurodegenerative diseases. Mediators Inflamm 2014:465694

Sawada M, Imamura K, Nagatsu T (2006) Role of cytokines in inflammatory process in Parkinson's disease. J Neural Transm 70:373–381

Saura J, Angulo E, Ejarque A et al (2005) Adenosine A$_{2A}$ receptor stimulation potentiates nitric oxide release by activated microglia. J Neurochem 95:919–929

Schafer DP, Lehrman EK, Kautzman AG et al (2012) Microglia sculpt postnatal neural circuits in an activity and complement-dependent manner. Neuron 74:691–705

Schafer DP, Lehrman EK, Stevens B (2013) The "quad-partite" synapse: microglia-synapse interactions in the developing and mature CNS. Glia 61:24–36

Schapansky J, Nardozzi JD, LaVoie MJ (2014) The complex relationships between microglia, alpha-synuclein, and LRRK2 in Parkinson's disease. Neuroscience. doi:10.1016/j.neuroscience.2014.09.049 (in press)

Schwarzschild MA, Chen JF, Ascherio A (2002) Caffeinated clues and the promise of adenosine A$_{2A}$ antagonists in PD. Neurology 58:1154–1160

Schwarzschild MA, Agnati L, Fuxe K et al (2006) Targeting adenosine A$_{2A}$ receptors in Parkinson's disease. Trends Neurosci 29:647–654

Scianni M, Antonilli L, Chece G et al (2013) Fractalkine (CX3CL1) enhances hippocampal N-methyl-D-aspartate receptor (NMDAR) function via D-serine and adenosine receptor type A$_2$ (A$_{2A}$R) activity. J Neuroinflammation 10:108

Selkoe DJ (2002) Alzheimer's disease is a synaptic failure. Science 298:789–791

Shen HY, Coelho JE, Ohtsuka N et al (2008) A critical role of the adenosine A$_{2A}$ receptor in extrastriatal neurons in modulating psychomotor activity as revealed by opposite phenotypes of striatum and forebrain A$_{2A}$ receptor knock-outs. J Neurosci 28:2970–2975

Shen HY, Canas PM, Garcia-Sanz P et al (2013) Adenosine A$_{2A}$ receptors in striatal glutamatergic terminals and GABAergic neurons oppositely modulate psychostimulant action and DARPP-32 phosphorylation. PLoS One 8:e80902

Sierra A, Abiega O, Shahraz A et al (2013) Janus-faced microglia: beneficial and detrimental consequences of microglial phagocytosis. Fronti Cell Neurosci 7:6

Simões AP, Duarte JA, Agasse F et al (2012) Blockade of adenosine A_{2A} receptors prevents interleukin-1β-induced exacerbation of neuronal toxicity through a p38 mitogen-activated protein kinase pathway. J Neuroinflammation 9:204

Simola N, Morelli M, Carta AR (2007) The 6-hydroxydopamine model of Parkinson's disease. Neurotox Res 11:151–167

Sitkovsky MV, Lukashev D, Apasov S et al (2004) Physiological control of immune response and inflammatory tissue damage by hypoxia-inducible factors and adenosine A_{2A} receptors. Annu Rev Immunol 22:657–682

Smeyne RJ, Jackson-Lewis V (2005) The MPTP model of Parkinson's disease. Mol Brain Res 134:57–66

Stone TW, Behan WM (2007) Interleukin-1β but not tumor necrosis factor-α potentiates neuronal damage by quinolinic acid: protection by an adenosine A_{2A} receptor antagonist. J Neurosci Res 85:1077–1085

Streit WJ, Xue QS (2009) Life and death of microglia. J Neuroimmune Pharmacol 4:371–379

Streit WJ, Graeber MB, Kreutzberg GW (1988) Functional plasticity of microglia: a review. Glia 1:301–307

Tansey MG, Goldberg MS (2010) Neuroinflammation in Parkinson's disease: its role in neuronal death and implications for therapeutic intervention. Neurobiol Dis 37:510–518

Teismann P, Schulz JB (2004) Cellular pathology of Parkinson's disease: astrocytes, microglia and inflammation. Cell Tissue Res 318:149–161

Trang T, Beggs S, Salter MW (2012) ATP receptors gate microglia signaling in neuropathic pain. Exp Neurol 234:354–361

Tremblay ME, Lowery RL, Majewska AK (2010) Microglial interactions with synapses are modulated by visual experience. PLoS Biol 8:e1000527

Tremblay ME, Stevens B, Sierra A et al (2011) The role of microglia in the healthy brain. J Neurosci 31:16064–16069

Ueno M, Fujita Y, Tanaka T et al (2013) Layer V cortical neurons require microglial support for survival during postnatal development. Nat Neurosci 16:543–551

Wake H, Moorhouse AJ, Jinno S et al (2009) Resting microglia directly monitor the functional state of synapses in vivo and determine the fate of ischemic terminals. J Neurosci 29:3974–3980

Wake H, Moorhouse AJ, Miyamoto A et al (2013) Microglia: actively surveying and shaping neuronal circuit structure and function. Trends Neurosci 36:209–217

Walsh S, Finn DP, Dowd E (2011) Time-course of nigrostriatal neurodegeneration and neuroinflammation in the 6-hydroxydopamine-induced axonal and terminal lesion models of Parkinson's disease in the rat. Neuroscience 175:251–261

Wei CJ, Augusto E, Gomes CA et al (2014) Regulation of fear responses by striatal and extrastriatal adenosine A_{2A} receptors in forebrain. Biol Psychiatry 75:855–863

Whitton PS (2007) Inflammation as a causative factor in the aetiology of Parkinson's disease. Br J Pharmacol 150:963–976

Wilms H, Rosenstiel P, Sievers J, Deuschl G, Zecca L, Lucius R (2003) Activation of microglia by human neuromelanin is NF-kappaB dependent and involves p38 mitogen-activated protein kinase: implications for Parkinson's disease. FASEB J 17:500–502

Wilms H, Zecca L, Rosenstiel P et al (2007) Inflammation in Parkinson's diseases and other neurodegenerative diseases: cause and therapeutic implications. Curr Pharm Des 13:1925–1928

Wong WT, Wang M, Li W (2011) Regulation of microglia by ionotropic glutamatergic and GABAergic neurotransmission. Neuron Glia Biol 7:41–46

Yao SQ, Li ZZ, Huang QY et al (2012) Genetic inactivation of the adenosine A_{2A} receptor exacerbates brain damage in mice with experimental autoimmune encephalomyelitis. J Neurochem 123:100–112

Yu L, Shen HY, Coelho JE et al (2008) Adenosine A_{2A} receptor antagonists exert motor and neuroprotective effects by distinct cellular mechanisms. Ann Neurol 63:338–346

Zecca L, Wilms H, Geick S et al (2008) Human neuromelanin induces neuroinflammation and neurodegeneration in the rat substantia nigra: implications for Parkinson's disease. Acta Neuropathol 116:47–55

Zhan Y, Paolicelli RC, Sforazzini F et al (2014) Deficient neuron-microglia signaling results in impaired functional brain connectivity and social behavior. Nat Neurosci 17:400–406

Zhang W, Wang T, Pei Z et al (2005) Aggregated alpha-synuclein activates microglia: a process leading to disease progression in Parkinson's disease. FASEB J 19:533–42

Zhang S, Wang XJ, Tian LP et al (2011) CD200-CD200R dysfunction exacerbates microglial activation and dopaminergic neurodegeneration in a rat model of Parkinson's disease. J Neuroinflammation 8:154

Zhang J, Malik A, Choi HB et al (2014) Microglial CR3 activation triggers long-term synaptic depression in the hippocampus via NADPH oxidase. Neuron 82:195–207

Zielasek J, Hartung HP (1996) Molecular mechanisms of microglial activation. Adv Neuroimmunol 6:191–122

Chapter 6
Purines in Parkinson's: Adenosine A_{2A} Receptors and Urate as Targets for Neuroprotection

Rachit Bakshi, Robert Logan and Michael A. Schwarzschild

Abstract Purines are essential constituents of all living cells. The nucleoside adenosine is not only a precursor of ATP and cyclic AMP but is also released by a wide variety of cells under various physiological and pathological conditions. In mammals, adenosine acts on four subtypes of guanine nucleotide binding protein (G protein)-coupled receptor (GPCR)—A_1, A_{2A}, A_{2B} and A_3. Among these the adenosine A_{2A} receptor has emerged as a particularly attractive target of therapeutics development for Parkinson's disease (PD), in part because it is highly expressed in brain regions innervated by the dopaminergic neurons that degenerate in PD. Urate (also known as uric acid—2,6,8-trioxypurine) is the most abundant antioxidant as well as the end product of purine metabolism in humans. Emerging clinical, epidemiological, and laboratory evidence has identified urate as a potential neuroprotectant for the treatment of PD. The primary intent of this review is to explore the neuroprotective effects of adenosine receptor antagonists and urate and their therapeutic potential in PD with particular attention to epidemiological and preclinical findings linking these purines to PD and other neurodegenerative diseases. This review also summarizes current clinical development of purines as candidate neuroprotectants.

Keywords Adenosine A_{2A} receptor · Urate · Caffeine · Neuroprotection · Parkinson's disease · Purines · Clinical trials · Risk factor · Neurodegenerative disease

Purines are essential constituents of all living cells. In addition to their vital roles in storage and transmission of genetic information (DNA, RNA) and energy reserves (ATP), purines also serve as important molecules for both intracellular and extracellular signaling. The nucleoside purine adenosine is not only a precursor of ATP and

R. Bakshi (✉)
Molecular Neurobiology Laboratory, MassGeneral Institute for Neurodegenerative Disease, Massachusetts General Hospital, 114, 16th street, Rm 3001, Boston, MA 02129, USA
e-mail: Rbakshi1@mgh.harvard.edu

R. Logan · M. A. Schwarzschild
Molecular Neurobiology Laboratory, MassGeneral Institute for Neurodegenerative Disease, Massachusetts General Hospital, Harvard Medical School, Boston, MA 02129, USA

© Springer International Publishing Switzerland 2015

M. Morelli et al. (eds.), *The Adenosinergic System,* Current Topics in Neurotoxicity 10, DOI 10.1007/978-3-319-20273-0_6

cyclic AMP but is also released by a wide variety of cells under various physiological and pathological conditions. In mammals, adenosine acts on four subtypes of guanine nucleotide binding protein (G protein)-coupled receptor (GPCR)—A_1, A_{2A}, A_{2B} and A_3. Among these the adenosine A_{2A} receptor has emerged as a particularly attractive target of therapeutics development for Parkinson's disease (PD), in part because it is highly expressed in brain regions innervated by the dopaminergic neurons that degenerate in PD. Antagonists of the A_{2A} receptor, including purines like caffeine, consistently confer protection in animal models of PD. The end product of purine metabolism in humans is urate, (also known as uric acid– 2,6,8-trioxypurine), the most abundant antioxidant circulating in the plasma. The primary intent of this chapter is to explore the neuroprotective role of urate and adenosine receptor antagonists and their therapeutic potential in PD.

Purine Metabolism: Evolutionary Significance

The purine metabolism pathway is one of the most conserved pathways found among all living things. Loss-of-function gene mutations have played important roles in the adaptive evolution of purine metabolism among vertebrates, leading to functional benefits and diversification between species (Keebaugh and Thomas 2010). Urate is the end product of purine metabolism in human and higher primates in contrast to all other mammals in which urate is readily converted to allantoin by the enzyme urate oxidase (UOx) (Fig. 6.1). This peculiarity in higher primates is a consequence of multiple independent mutations in the urate oxidase gene (*UOx*), which occurred late in primate evolution, approximately 10–15 million years ago, leading to much higher urate concentrations near the limits of its solubility in humans and apes (Christen et al. 1970; Oda et al. 2002; Wu et al. 1992). In addition to reduced catabolism via the loss of UOx, enhanced renal reabsorption of urate via the urate transporter evolution also contributes to higher levels of urate in humans (Hediger et al. 2005).

Accordingly, it has been hypothesized that higher circulating urate in the ancestors of man and apes had evolutionary advantages. The hypothesis that urate possesses antioxidant properties comparable to those of ascorbate (Proctor 1970) was successfully confirmed a decade later (Ames et al. 1981). Urate also accounts for most of the antioxidant capacity in human plasma (Benzie and Strain 1996; Yeum et al. 2004). Several other speculative theories have been advanced to explain the putative benefit of urate elevation. In 1955 physicist Egon Orowan published a theory on "The Origin of Man" (Orowan 1955) in which he posited that hominoids (apes and humans) evolved to have high levels of urate because of its critical role as "catalyzer of mental development". Another, more recent theory suggests that urate may also have had a beneficial hypertensive effect in our primate predecessors at a time when a low-salt diet and resultant hypotension might have posed a survival threat (Watanabe et al. 2002). Despite all the theoretical evolutionary advantages of higher urate levels, their only established health effects on modern day humans are deleterious as increasing urate concentrations can contribute to gout and uric acid kidney stones (Kutzing and Firestein 2008).

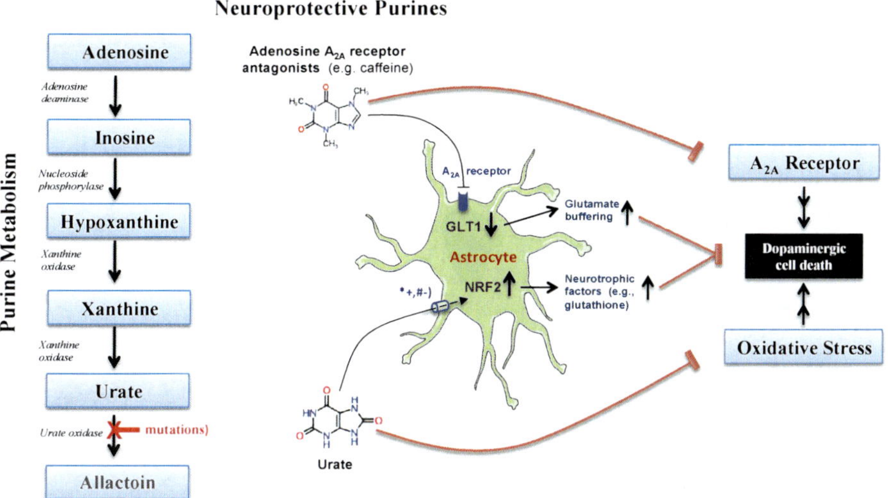

Fig. 6.1 Potential disease modifying actions and astrocytic mechanisms of purines in Parkinson's disease. Urate (or its precursors like Inosine) and adenosine A_{2A} antagonists (including caffeine, a.k.a. tri-methyl-xanthine) have emerged as key neuroprotective purines, which may prevent dopaminergic neuronal death in PD via astrocytes (as shown *above*) or other cell types

Purines in Parkinson's Disease—Epidemiological Clues

There is extensive epidemiological evidence linking greater consumption of caffeine (a non-specific A_{2A} antagonist) with a reduced risk of subsequently developing PD in multiple case-control and longitudinal studies. A full discussion on the epidemiological links between caffeine and PD has been covered separately in this volume (see Chap. 12) and readers are also referred to Morelli et al. 2010 for further information.

Oxidative stress is one of the most prominent pathophysiological processes implicated in dopaminergic cell death in PD (Hauser and Hastings 2013; Jenner 2003). Since urate is one of the most abundant antioxidant in humans, it may be an important determinant of disease susceptibility and progression in PD. The first direct evidence of altered urate in PD demonstrated reduced levels in post-mortem nigrostriatal tissue from PD patients (Church and Ward 1994), and encouraged further studies in humans as well as in laboratory models of PD. The urate antioxidant hypothesis coupled with the initial pathological clue to urate in PD led to a series of epidemiological investigations in case control studies and prospectively followed cohorts. Several case control studies have consistently demonstrated reduced levels of serum urate levels in PD patients compared to controls (Andreadou et al. 2009; Annamaki et al. 2007; Bogdanov et al. 2008; Jesus et al. 2012; Johansen et al. 2009; Larumbe Ilundain et al. 2001). Direct evidence that lower blood urate is a risk factor for PD has come from repeated findings of a reduced rate of developing PD among people with higher blood urate levels in prospectively followed initially healthy cohorts– across regions, races and nationalities (Chen et al. 2009a; Davis et al. 1996;

De Lau et al. 2005; Weisskopf et al. 2007; Winquist et al. 2010). For example, in one of the largest studies conducted at the Harvard School of Public Health, Weisskopf et al. found that men in the top quartile of plasma urate had a 55% lower risk of PD than men in the bottom quartile. The decrease in risk was even stronger (with an 80% PD risk reduction in the highest vs lowest quartile; $p < 0.01$ for trend) in those with blood collected at least 4 years before diagnosis, suggesting that the lower urate in those with PD precedes symptom onset and is thus unlikely to be a consequence of changes in medical treatment, diet or other behaviors early in the course of disease. Interestingly, prospective studies have generally found that lower urate is a risk factor in men, but less so (Chen et al. 2009a) if at all (O'Reilly et al. 2010) in women. Epidemiological findings that people with gout or consuming a urate-elevating diet have a reduced likelihood of PD in prospectively followed men strengthen the urate-PD link (Alonso et al. 2007; De vera et al. 2008; Gao et al. 2008).

This identification of lower urate as a PD risk factor among healthy people led to investigations of whether lower urate in people already diagnosed with PD might be predict slower progression of the disease. This question was initially addressed in two long-term clinical trials known as DATATOP (Parkinson Study Group (PSG) 1989a, b, 1993) and PRECEPT (PSG 2007) together comprising over 1600 early cases of PD. The goal of PRECEPT (Parkinson Research Examination of CEP-1347 Trial) was to determine if the investigational drug CEP-1347 could slow the clinical progression of early PD (PSG 2007). Serum urate was routinely monitored during the trial to assess the safety of CEP 1347, and values were available at enrollment for 804 of the 806 trial participants. It was observed that higher serum urate at baseline was indeed strongly associated with a slower clinical progression. The hazard ratio (HR) of reaching the primary study endpoint (the development of disability warranting dopaminergic therapy) declined with increasing serum urate (p for trend < 0.0001) (Schwarzschild et al. 2008). In the DATATOP (Deprenyl and Tocopherol Antioxidative Therapy of Parkinsonism) trial, conducted nearly two decades earlier, a similar strong association was observed (Ascherio et al. 2009). The HR of progressing to the same primary end point decreased with increasing serum urate concentrations (adjusted HR for highest vs lowest quintile: 0.64; 95% CI: 0.44–0.94; p for trend = 0.002) (Ascherio et al. 2009). These robust findings linking higher serum urate levels with slower clinical decline were paralleled by findings with urate levels in cerebrospinal fluid, which may be particularly relevant to the microenvironment of degenerating neurons. For the 713 subjects with available CSF urate levels at baseline from the DATATOP trial the concentration of CSF urate was inversely related to the primary end point (HR for highest vs lowest quintile: 0.65; 95% CI: 0.44–0.96) and the rate of change in the UPDRS ($p < 0.05$ for trend across quintiles, Ascherio et al. 2009). Moreover, a similarly clear inverse association was observed between baseline serum urate and loss of striatal $^{123}\beta$-CIT uptake, a marker for the presynaptic dopamine (DA) transporter (p for trend $= 0.002$) (Schwarzschild et al. 2008). More recently, serum urate in early PD patients has also been reported to predict a slower rate of worsening for motor and non-motor features (Moccia et al. 2014). Together these data identify urate as a molecular biomarker of the rate of disease progression as well as the risk of PD.

Urate Transporters in Parkinson's Disease: Epidemiological Links

Urate levels in blood depend on an intricate balance between dietary intake of urate precursors, cellular production of urate through the metabolism of purines, and its excretion/reabsorption in the kidneys and secretion in intestine (Lipkowitz 2012). The majority of urate transport is handled by the kidneys. Several specific transporters of urate reside in kidney epithelial cells where they mediate its renal excretion or reabsorption (So and Thorens 2010). GLUT9, a member of the facilitative glucose transporter family, is one of the key urate transporters and is also implicated in PD. The urate transport function of GLUT9 (encoded by the *SLC2A9* gene) was predicted through several genome-wide association studies, which found that the *SLC2A9* sequence variation is the strongest inherited determinant of serum urate levels in humans (Dehghan et al. 2008; Kolz et al. 2009; Li et al. 2007; Vitart et al. 2008) and was subsequently confirmed experimentally (Caulfield et al. 2008). GLUT9 is one of the most ubiquitously expressed urate transporters in brain and hence is relevant to understanding the roles of urate in normal brain function, as well as in neurodegenerative diseases. Single nucleotide polymorphisms (SNPs) in *SLC2A9* which are linked to lower serum urate have also been linked to PD (Gonzalez-Aramburu et al. 2013) and to an earlier age at onset of the disease (Facheris et al. 2011). Recently, *SLC2A9* SNPs were also found to predict faster clinical decline as well as lower serum urate in early PD (Simon et al. 2014), supporting a causal basis for the link between higher urate and favorable outcomes in PD. Although the neurobiology of urate transport is virtually unexplored, its genetic links to PD encourage investigation of GLUT9 and other urate transporters in PD models. Insights into the role of urate transport and metabolism may also help understand PD and other neuro-metabolic disorders involving urate such as Lesch-Nyhan syndrome (Jinnah et al. 2010).

Neuroprotective Potential of Targeting Adenosine Receptors

Adenosine receptors have been implicated in several key physiological processes, ranging from neuro-modulation to immune regulation, and from vascular function to metabolic control. Among the four adenosine receptors, the adenosine A_{2A} receptor has emerged as a particularly attractive non-dopaminergic target and its antagonists have generated considerable attention over their promise as therapeutic agents in PD and related disorders. Although the primary focus of their clinical development has been on short-term antiparkinsonian symptom relief, A_{2A} antagonists may also offer neuroprotection under a range of pathophysiological conditions from PD to Alzheimer's disease to stroke (see Table. 6.1).

Table 6.1 Evidence for the roles of purines in neuroprotection. The epidemiological evidence covers longitudinal studies of healthy cohorts for the risk of disease (Risk), case-control studies for the odds that a factor is associated with disease cases (Odds), and longitudinal studies of diagnosed cohorts for the risk of clinical worsening (Progression). The citation listings in this table are not intended to be exhaustive. The abbreviations used here are as follows Parkinson's disease (*PD*), Alzheimer disease (*AD*), amyotrophic lateral sclerosis (*ALS*), Huntington disease (*HD*), multiple sclerosis (*MS*), multiple system atrophy (*MSA*) and spinal cord injury (*SCI*), *urate oxidase* (*UOx*)

		Evidence	PD	AD	ALS	HD	others
A$_{2A}$ antagonists	Epidemiological	**Risk, Odds** (caffeine)	Strong inverse association [1–9]	Inverse association [10–12]	Inverse association [13] No association [14]	Positive association [15]	No association in MS [16]
		Progression (caffeine)	No association [17,18]				Inverse association in MS [19]
	Laboratory	**Genetic** (A$_{2A}$KO)	Neuroprotective [20–23]			Neurotoxic[24] Neuroprotective [25]	
		Pharmacological (A$_{2A}$ antagonists)	Neuroprotective [20,21,26–28]	Neuroprotective [29–36]	Neurotoxic [37]	Neuroprotective [38–40] Neurotoxic[41]	
Urate	Epidemiological	**Risk, Odds** (urate levels, genes & diet)	Strong inverse association [42–55];except [56]	Inverse association [57–64] No association [65,66]	Inverse association [67,68] No association [69]		Positive association in stroke onset[70,71]
		Progression (urate levels & gene)	Inverse association [72–75]	Trend inverse association [76]	Inverse association [77,78] No association [79,80]	Inverse association [81]	Inverse association in MSA[82]and stroke [83]
	Laboratory	**Genetic** (*UOx*KO)	Neuroprotective [84]				
		Pharmacological (urate) *in vivo*	Neuroprotective [85]				Neuroprotective in MS [93–94], SCI [95,96], Ischemia [97,98] and stroke [99]
		(urate) **cellular**	Neuroprotective [86–91]		(*in vitro* antioxidant [92])		

References cited

[1–9] Hellenbrand et al. (1996), Fall et al. (1999), Benedetti et al. (2000), Ross et al. (2000), Ascherio et al. (2001), Checkoway et al. (2002), Ragonese et al. (2003), Tan et al. (2003), Hu et al. (2007)

[10–12] Lindsay et al. (2002), Eskelinen et al. (2009), Eskelinen and Kivipelto (2010)

[13] Beghi et al. (2011)

[14] Morozova et al. (2008)

[15] Simonin et al. (2013)

[16] Massa et al. (2012)

[17, 18] Schwarzschild et al. (2003), Simon et al. (2008)

[19] D'Hooghe et al. (2012)

[20–23] Chen et al. (2001b), Ikeda et al. (2002), Pierri et al. (2005), Yu et al. (2008), Carta et al. (2009)

[24, 25] Mievis et al. (2011), Fink et al. (2004)

[20, 21, 26–28] Chen et al. (2001b), Ikeda et al. (2002, Joghataie et al. (2004), Kachroo et al. (2010), Sonsalla et al. (2012)

[29–36] Arendash et al. (2006), Arendash et al. (2009), Cao et al. (2009), Chen et al. (2010), Prasanthi et al. (2010), Wostyn et al. (2011), Zeitlin et al. (2011), Laurent et al. (2014)

[37] Potenza et al. (2013)

[38–40] Popoli et al. (2002), Blum et al. (2003), Chou et al. (2005)
[41] Minghetti et al. (2007)
[42–55] Davis et al. (1996), De Lau et al. (2005), Annanmaki et al. (2007), Weisskopf et al. (2007), Bogdanov et al. (2008), De Vera et al. (2008), Gao et al. (2008), Andreadou et al. (2009), Chen et al. (2009a), Johansen et al. (2009), Winquist et al. (2010), Facheris et al. (2011), Jesus et al. (2012), Gonzalez-Aramburu et al. (2013)
[56] O'Reilly et al. (2010)
[57–64] Maesaka et al. (1993), Rinaldi et al. (2003), Polidori et al. (2004), Kim et al. (2006), Euser et al. (2009), Cankurtaran et al. (2012), Can et al. (2013), Al-Khateeb et al. (2015)
[65, 66] Ahlskog et al. (1995), Chen et al. (2014)
[67, 68] Keizman et al. (2009), Ikeda et al. (2012)
[69] Zoccolella et al. (2011)
[70, 71] Bos et al. (2006), Hozawa et al. (2006)
[72–75] Schwarzschild et al. (2008), Ascherio et al. (2009), Moccia (2014), Simon et al. (2014)
[76] Irizarry et al. (2009)
[77, 78] Paganoni et al. (2012), Atassi et al. (2014)
[79, 80] Kataoka et al. (2013), Zheng et al. (2014)
[81] Auigner et al. (2010)
[82] Lee et al. (2011)
[83] Chamorro et al. (2002)
[84] Chen et al. (2013)
[85] Gong et al. (2012)
[86–91] Duan et al. (2002), Haberman et al. (2007), Guerreiro et al. (2009), Cipriani et al. (2012a), Cipriani et al. (2012b), Zhu et al. (2012)
[92] Spasojevic et al. (2010)
[93, 94] Hooper et al. (1998), Hooper et al. (2000)
[95, 96] Scott et al. (2005), Du et al. (2007)
[97, 98] Yu et al. (1998), Haberman et al. (2007)
[99] Romanos et al. (2007)

Neuroprotection and Adenosine A_1 Receptors

A_1 receptors are the most abundant type of adenosine receptor in the brain and exert a major inhibitory effect on neuronal excitability and synaptic transmission (Cunha 2005). Agonists of A_1 receptors confer neuroprotection through reduced calcium influx, hyperpolarizing the cell membrane, inhibiting NMDA receptor activation and attenuating glutamate excitotoxicity (Rudolphi et al. 1992). In a variety of hypoxia/ischemia models using cell culture, brain slices, and in vivo, A_1 receptor agonists are neuroprotective (de Mendonça et al. 2000). A_1 receptor antagonists, however, do not show uniform effects. In models of hypoxia/ischemia A_1 receptor antagonists often exacerbate damage (de Mendonça et al. 2000). In some studies, A_1 receptor antagonists have no effect on ischemic conditions (Lekieffre et al. 1991; Ostwald et al. 1997) while others have shown protective effects (Lekieffre et al. 1991; Seida et al. 1988). There is some evidence to suggest that protection by A_1 receptor antagonists might be a function of time. For example, acute administration of caffeine (a non-selective A_1 and A_{2A} antagonist) was not protective in forebrain ischemia, whereas caffeine taken for 3 weeks prior to ischemic insult was (Sutherland et al. 1991). Similarly, use of the A_1 selective antagonists DPCPX for 2 weeks prior to global forebrain ischemia was neuroprotective (Von Lubitz et al. 1994). In models of PD, A_1 receptor antagonists have not demonstrated meaningful neuroprotective properties, and thus contrast A_{2A} receptor antagonists, which produce robust protection. For example, Chen et al. (2001b) demonstrated that the A_1 receptor antagonist CPX did not mitigate the loss of striatal dopamine in the 1-methyl-4-phenyl-1,2,3,6-tetrahydropyridine (MPTP) mouse model. Further studies are needed to clarify roles of the A_1 receptor and their complexities in the neurodegeneration of PD, and whether they may be targeted therapeutically.

Neuroprotection and A_{2A} Adenosine Receptors

Role of Caffeine and More Selective Antagonists of the Adenosine A_{2A} Receptors in Neuroprotection

Caffeine is the most widely used psychoactive drug in the world (Fredholm et al. 1999). It also has been shown to have neuroprotective properties. For example, when caffeine was administered to rodents at doses that produced blood concentrations comparable to those achieved with typical human consumption, it protected them from neurotoxicity in multiple models of PD (Xu et al. 2005). Intraperitoneal administration of caffeine (10 mg/kg) prevents MPTP-induced loss of striatal dopamine and dopamine transporter binding sites in mice (Chen et al. 2001). In a complementary toxin model, caffeine was protective against unilateral 6-hydroxydopamine (6-OHDA) lesioning in rats (Joghataie et al. 2004; Machado-Filho et al. 2014). Furthermore, caffeine was neuroprotective in mice that were chronically exposed to the neurotoxic pesticides paraquat and maneb. This chronic dual pesticide treatment induced the loss of tyrosine hydroxylase (TH)-positive nigral neurons, which was prevented by caffeine (Kachroo et al. 2010). Caffeine has also been shown to reduce nigral death in mice treated with lipopolysaccharide (Brothers et al. 2010). Additionally, caffeine prevents neurodegeneration-associated disruptions in the blood-brain-barrier in animal models of PD (Chen et al. 2010). Furthermore, it was recently reported that delayed caffeine treatment prevents nigral dopamine neuron loss in a MPTP rat model of PD (Sonsalla et al. 2012). Taken together, with the extensive epidemiological evidence (reviewed in Chap. 12) these studies support the notion that caffeine may offer neuroprotection against PD. Efforts aimed at elucidating the molecular mechanisms of neuroprotection by caffeine have centered on the CNS adenosinergic system because caffeine readily traverses the blood-brain-barrier, and non-specifically but relatively potently antagonizes adenosine receptors including the A_1 and A_{2A} subtypes (Daly et al. 1983; McCall et al. 1982).

Relatively selective A_{2A} receptor antagonists like SCH 58261, DMPX, KW-6002 and 1,3-dipropyl-7-methylxanthine have also been widely shown to be neuroprotective. KW-6002 like caffeine displayed potent neuroprotective properties in the commonly used MPTP and 6-OHDA toxin models of PD (Chen et al. 2001b; Ikeda et al. 2002). Similarly, A_{2A} receptor antagonists have been shown to confer protection against mitochondrial complex inhibitor-induced nigral and striatal damage. For example, DMPX spared the loss of striatal dopamine, TH and GABA in mice and rats exposed to the mitochondrial complex II inhibitor malonate. Interestingly, the A_1 receptor antagonist CPX worsened the damaging effects of malonate (Alfinito et al. 2003). The mitochondrial complex I inhibitor rotenone is also not as neurotoxic to the striatum when co-administered with the A_{2A} receptor antagonists ST1535 or ZM241385 (Belcastro et al. 2009). Additionally, A_{2A} receptor antagonists offer protection against such insults as ischemia, quinolinic acid, 3-nitropropionic acid, malonate, and MPTP across various brain areas including the cortex, hippocampus and striatum (Cunha 2005). Several new A_{2A} antagonists, have been identified using drug screens (Pinna 2014; Scatena et al. 2011) with potent neuroprotective properties. Collectively, these data strongly suggest that in PD models adenosine receptor antagonists, such as caffeine rely on A_{2A} receptors rather than A_1 receptors to exert their neuroprotective effects. The laboratory and epidemiological evidence indicate that A_{2A} receptor antagonists, such as caffeine, make compelling neuroprotective drug candidates.

Genetic Manipulation of A_{2A} Receptor for Neuroprotection

To complement the neuroprotective benefits of pharmacological strategies to disrupt A_{2A} function, targeted mutations in the functionally relevant regions of the A_{2A} receptor gene were done to knockout the receptor in mice. Mice lacking A_{2A} receptor displayed attenuated brain damage in models of ischemic or excitotoxin-induced brain injury (Chen et al. 1999). More recently, using a conditional knockout (Cre/loxP) system to generate mice with a selective postnatal depletion of forebrain neuronal A_{2A} receptors fully prevented neurotoxin-induced degeneration of nigral dopaminergic neurons in a subchronic MPTP model of PD (Carta et al. 2009). However using this conditional knockout method to eliminate neuronal A_{2A} receptors did not protect striatal terminals from a more acute, high-dose MPTP exposure (Yu et al. 2008). The effects of A_{2A} receptor deletion on neuronal damage in the brain are complex and can be paradoxical as A_{2A} knockdown has been shown to exacerbate neuronal toxicity in models of Huntington's disease (HD) and experimental autoimmune encephalomyelitis (Huang et al. 2006; Yao et al. 2012). All together these data, while supporting a neuroprotective outcome of neuronal A_{2A} receptor blockade in PD, highlight the complexities of the roles played by A_{2A} receptors, pointing to distinct actions of cell-type specific receptors in different neurodegenerative conditions.

Glial A$_{2A}$ Receptors and Neuroprotective Mechanisms

The physiological and pathological roles of adenosine receptors have been attributed primarily to their direct action on neurons (Stone et al. 2009), however adenosine receptors are also present in glial cells where they control the metabolism of glucose, astrogliosis, the release of neurotrophic factors and even cell death. Neuroinflammation is thought to play a role in propagating neurodegeneration (Filippo et al. 2010). Consistent with this hypothesis, neuroprotection associates with reduced neuroinflammation, whereas degeneration correlates to increased levels. Resident microglial cells act as the brain's first line of immune defense since the blood-brain-barrier limits its interactions with the peripheral immune system (Banks and Erickson 2010; Erickson et al. 2012; Takeuchi 2013). In response to pathophysiological stressors such as pro-oxidants, hypoxia or necrotic tumors, extracellular adenosine levels rise in the brain dramatically. This increase in adenosine levels precedes a characteristic increase in astrogliosis and reactive microglia, that in PD can be found in the substantia nigra and other brain areas such as the pons, frontal cortex and the temporal cortex (Gerhard et al. 2006; McGeer and McGeer 2008; Niranjan 2013). Astrogliosis is a condition where the astrocyte cell cycle, morphology, and molecular expression patterns are altered, and can occur in response to conditions such as ischemia and neurodegeneration, which dramatically increase the number of astrocytes. In addition to glial activation, lymphocyte infiltration and increased levels of soluble inflammatory factors are seen in human PD as well as in toxin models like 6-OHDA, MPTP and rotenone (Armentero et al. 2011) One such soluble inflammatory factor is the glial-derived cytokine, which in chronic neuroinflammation causes a feed-forward loop that is neurotoxic to dopaminergic neurons (Armentero et al. 2011; Morelli et al. 2010).

A$_{2A}$ receptors are found embedded in glial cell membranes and can modulate neuroinflammation during neurodegeneration. Although there is evidence of neuroprotective adenosinergic signaling in oligodendrocytes, it is not well understood (González-Fernández et al. 2014; Melani et al. 2009; Stevens et al. 2002). The vast majority of what is known about adenosine receptors in glia centers on microglia and astrocytes, in which they have been shown to influence glial secretions and activation states during inflammation. For example, in rats, the adenosine agonist CPCA increased astrogliosis, which was counteracted by the A$_{2A}$ antagonist DPMX (Hindley et al. 1994). Also, Brambilla and colleagues used basic fibroblast growth factor to induce astrogliosis in primary rat striatal astrocytes. They then showed a concentration-dependent anti-astrogliosis effect of the selective A$_{2A}$ antagonists SCH58261 and KW-6002. A$_{2A}$ agonists alone, in the absence of basic fibroblast growth factor, were insufficient to induce astrogliosis (Brambilla et al. 2003). In a model of hippocampal ischemia, the A$_{2A}$ receptor antagonist ZM241385 increased neuronal survival and decreased astrogliosis (Pugliese et al. 2009).

Brain injury or disease is also accompanied by microglia activation, which can release cytotoxic molecules and reactive oxygen intermediates (Dheen et al. 2007). In addition to astrogliosis, the MPTP mouse model has increased levels of activated microglia in the substantia nigra pars compacta and the striatum (Carta et al. 2009).

The selective A_{2A} antagonists KW-6002, ANR-94, and SCH58261 have been shown to reduce the microglial activation seen in subacute and subchronic MPTP mouse models (Carta et al. 2009; Pierri et al. 2005; Pinna et al. 2010). KW-6002 was also shown to reduce gliosis in rats that had unilateral intrastriatal 6-OHDA lesioning (Ikeda et al. 2002). Furthermore, brain inflammation due to lipopolysaccharide or old age was also reduced by caffeine, which concomitantly reduced hippocampal microglia activation in rats (Brothers et al. 2010). Interestingly, Yu et al. provided evidence that in an acute MPTP model A_{2A} receptors in forebrain neurons are responsible for motor stimulation in both normal and dopamine-deficient conditions, but do not offer neuroprotection in an acute MPTP model. Additionally, pretreatment with KW-6002 was equally neuroprotective against MPTP among forebrain A_{2A} knockouts and wild type controls. Yu et al. determined that astrocyte and microglial activation corresponded to neuronal degeneration, both of which were reduced with KW-6002. Therefore, they suggested that the A_{2A} receptors found in astrocytes and microglia are most likely involved in producing neuroprotection, rather than forebrain neuronal A_{2A} receptors (Yu et al. 2008) in contrast to findings by Carta et al. 2009 which showed in subchronic MPTP model of PD, loss of neuronal A_{2A} receptors fully prevented neurotoxin-induced degeneration of nigral dopaminergic neurons (Carta et al. 2009).

A_{2A} antagonists might also mitigate neuroinflammation through reducing toxic secretions from glial cells, such as nitric oxide (NO). In pathological conditions, NO is produced via the inducible nitric oxide synthase (iNOS) enzyme. iNOS is barely present in normal healthy brain, but increases dramatically in the case of inflammation, infectious or ischemic damage, or normal aging (Ladecola et al. 1995; Licinio et al. 1999; Kröncke et al. 1998). NO has many roles in the body including being a pro-inflammatory and pro-oxidant agent in pathological states. Therefore, curbing the excessive release of NO can help ameliorate some of the damage caused by inflammation.

There are some inconsistencies in the literature regarding A_{2A} receptor involvement in modulating NO and iNOS levels. For example, the A_{2A} receptor agonist CGS 21680 was shown by Brodie et al. to inhibit iNOS and NO levels in activated astrocytes in vitro, an effect that was countered by the A_{2A} antagonist CSC (Brodie et al. 1998). However, another group showed that in similarly activated astrocyte cultures, when mixed with activated microglia, CGS 21680 increased NO release, whereas the A_{2A} receptor antagonist ZM-241385 suppressed it (Saura et al. 2005). This apparent discrepancy might be due to the involvement of microglia. Cultured microglia, when activated by lipopolysaccharide, demonstrate an increase in iNOS and NO (Fiebich et al. 1998). It has been reported that microglia express A_{2A} receptor mRNA and demonstrate the presence of A_{2A} receptors with binding studies, whereas astrocytes do not (Fiebich et al. 1996; Saura et al. 2005). CGS 21680 was only shown to potentiate NO production in microglia, when cultured with astrocytes (Saura et al. 2005). Microglia were also shown to inhibit COX-2 expression and PGE2 release via CGS 21680 action on their A_{2A} receptors (Fiebich et al. 1996).

Despite the two aforementioned studies reporting a lack of A_{2A} receptors on astrocytes, there is ample evidence to the contrary. Several studies showed that adenosine acts on A_{2A} receptors in astrocytes to increase intracellular calcium levels and release glutamate (Li et al. 2001; Matos et al. 2013; Nishizaki 2004). A recent study revealed that when adenosine stimulates astrocyte A_{2A} receptors, Na^+/K^+-ATPases are inhibited, which in turn inhibits glutamate transporter-1 and results in reduced glutamate uptake (Matos et al. 2013). When A_{2A} receptors are genetically deleted in mouse astrocytes, astrocyte levels of glutamate increase in both the striatum and the cortex, which might partially explain the neuroprotective effects of A_{2A} antagonists in PD (Matos et al. 2013).

Much work remains to uncover all the molecular underpinnings of neuroprotection by A_{2A} receptor antagonists. They may have complex actions in a variety of brain areas in a multitude of pathological conditions, including neurological disorders such as in PD and AD. Though the complexity of signaling orchestrated by the adenosine A_{2A} receptor poses a challenge to understanding its actions, there is substantial evidence pointing to the neuroprotective properties of A_{2A} antagonists. They have been shown to protect against neurotoxic insults through a variety of mechanisms, likely including the buffering of glutamatergic excitotoxicity or the attenuation of harmful neuroinflammation. A_{2A} antagonists have tremendous promise as therapeutic agents, warranting their continued investigation in CNS pathophysiology and neuropsychiatric disease.

Neuroprotective Potential of Targeting Urate

Based on the evolutionary, antioxidant and epidemiological links between urate and PD, several groups including ours have investigated the protective properties of urate in cellular and rodent models of PD and other neurodegenerative diseases. Interestingly the first direct neuroprotective effect of urate came from cellular and animal models of multiple sclerosis (MS) and its animal correlate experimental autoimmune encephalomyelitis (EAE) (Hooper et al. 1998, 2000; Scott et al. 2002). In these studies the authors show that direct administration of urate or its precursor inosine inhibited CNS inflammation and tissue damage in models of MS or EAE. In rat models of stroke preceding or simultaneous treatment with urate (Romanos et al. 2007; Yu et al. 1998) or its analogs (Haberman et al. 2007) protected against cortical or striatal damage. In models of spinal cord injury treatment with urate protected spinal cord neurons directly (Scott et al. 2005) or through an astrocyte-dependent mechanism (Du et al. 2007).

Urate also confers protection in cellular as well as rodent models of PD. Across a range of models urate has prevented spontaneous degeneration of cultured nigral neurons and dopaminergic cell death induced by oxidative and mitochondrial toxins. For example, in PC12 cells, dopamine-induced apoptosis and oxidative stress was blocked by urate (Jones et al. 2000). In the first report of urate in a rotenone toxicity model of PD, dopaminergic cell death induced by homocysteine plus rotenone or iron was completely prevented by co-treating with urate (Duan et al. 2002). Guerriero et al. 2009 showed that urate at physiologically relevant concentrations can markedly enhance survival of dopaminergic neurons in primary midbrain culture of rat ventral mesencephalon. Recently published studies by our group in primary and immortalized dopaminergic neuronal cells (Cipriani et al. 2012a, b) and others (Zhu et al. 2012) demonstrated protective effects of urate against toxicity induced by hydrogen peroxide (an oxidant), MPP^+ (a mitochondrial toxin) or 6-OHDA. Neuroprotective effects of urate has also been evaluated in vivo in rodent models of PD, and found to attenuate 6-OHDA toxicity (Gong et al. 2012). Similarly, our group has found that mice with a UOx gene knockout, which recapitulates human evolution of purine metabolism (Fig. 6.1), had elevated brain urate levels and were resistant to toxic effects of 6-OHDA on nigral dopaminergic cell counts, striatal dopamine content, and rotational behavior, whereas transgenic overexpression of *UOx* exacerbated these morphological, neurochemical, and functional lesions of the dopaminergic nigrostriatal pathway (Chen et al. 2013). Higher striatal urate levels have been reported with both 6-OHDA treatment (Chen et al. 2013; De Luca et al. 2014) or MPTP treatment (De Luca et al. 2014; Serra et al. 2002) in rodent models of PD possibly due to a compensatory mechanism to combat dopaminergic cell death.

Although considerable evidence indicates that urate is a powerful direct antioxidant few studies have investigated alternative mechanisms of its protective effect. Previous studies (Cipriani et al. 2012a; Du et al. 2007) reported that neuroprotective effects of urate can depend on the presence of astrocytes in cultures. Astrocytes play a critical role in neuroprotection (Brambilla et al. 2012) particularly following an insult due to a toxicant or stress and there is ample precedent for the inducible release of neuroprotectants from astrocytes (Chen et al. 2001a, 2006; Fujishita et al. 2009; Imamura et al. 2008; Li et al. 2009; Rathinam et al. 2012). Moreover, it may help substantiate the growing evidence of critical pathophysiological role for astrocytes in the microenvironment of degenerating neurons in PD (Niranjan 2013).

In addition, a recent study hinted at a role of nuclear factor E2 (erythroid-derived 2)-related factor 2 (Nrf2) in neuroprotective mechanisms of urate in 6-OHDA- and hydrogen peroxide- induced dopaminergic cell death (Zhang et al. 2014). The authors showed that Nrf2 signaling and its downstream targets can be induced by urate treatment in dopaminergic cell lines. Nrf2 is a master regulator of cellular defense against oxidative stress, making it a therapeutic target for neurodegenerative diseases (Joshi and Johnson 2012*)*. Numerous studies have shown that Nrf2 protects different cell types and organ systems from a broad spectrum of toxic and pathogenic treatments (Ellrichmann 2011; Neymotin et al. 2011; Vargas et al. 2008; Yamamoto et al. 2007). Nrf2 KO mice show increased vulnerability to MPTP (Burton et al. 2006; Chen 2009b; Innamorato et al. 2010) and 6-OHDA (Jakel et al. 2007) in toxin models of PD. Conversely, Nrf2 activation protects dopaminergic neurons from 6-OHDA toxicity and MPP^+ toxicity, in vitro (Jakel et al. 2007; Yamamoto et al. 2007) and in vivo (Jakel et al. 2007). It remains to be determined how Nrf2 is activated by urate, and whether the astrocytic Nrf2 pathway explains the astrocyte-dependence of neuroprotection by urate (Fig. 6.1) in cellular and in vivo models of PD.

Purines as Protectants in Other Neurodegenerative Diseases

The use of adenosine receptor antagonists in the control of neuronal damage was first shown by Gao and Phillis 1994 in a model of cerebral ischemic injury. We later confirmed that genetic elimination of A_{2A} receptors also conferred a robust neuroprotection in animal models of brain ischemia (Chen et al. 1999). A_{2A} antagonists like caffeine can confer protection against other neurological diseases like Alzheimer's disease by reducing excitotoxicity, inflammatory responses, sensory and motor deficits and neuronal cell death (Table 6.1 and references within). The role of A_{2A} receptors in HD is also of interest (Lee and Chern 2014), in part because A_{2A} receptors are so densely expressed in the striatal output neurons whose early degeneration in the disease contributes to its characteristic movement disorder. However, the therapeutic potential of targeting these receptors for protection in HD is diminished given evidence for their conflicting actions on striatal neuron survival. Blocking pre-synaptic cortico-striatal or astrocytic A_{2A} receptors can reduce glutamate release or its extracellular levels (Popoli et al. 2002), highlighting the neuroprotective potential of A_{2A} antagonism in HD. On the other hand, blockade of post-synaptic A_{2A} receptor in the striatum may also exacerbate neuronal death (Blum et al. 2003), and A_{2A} agonists have been found to ameliorate neurotoxicity in an HD model (Huang et al. 2011). Similarly, epidemiological evidence that caffeine use (which is an established inverse risk factor for PD, as above), is likely a risk factor of earlier onset in HD (Simonin et al. 2013) supports a predominantly and uniquely deleterious effect of A_{2A} antagonism in the neurodegeneration of HD. The neuroprotective effect of A_{2A} receptor antagonists also correlates with their ability to improve cognitive behavior in mouse models of neurodegenerative disease most likely through its control of neuronal cell death.

Similarly, urate has been associated with other neurodegenerative diseases as a biomarker for favorable CNS outcomes in neurodegeneration (Table 6.1). For example, higher urate levels have also been correlated with slower clinical progression in HD, multiple system atrophy, and possibly mild cognitive impairment (Auigner et al. 2010; Cao et al. 2013; Irizarry et al. 2009; Lee et al. 2011). Similarly urate levels are an independent predictor of progression and survival in Amyotrophic lateral sclerosis (ALS), where higher levels were associated with improved outcomes (Paganoni et al. 2012). Similar results were reported by other groups (Abraham and Drory 2014; Ikeda et al. 2012; Keizman et al. 2009; Zoccolella et al. 2011) including the recent analysis of individuals from PRO-ACT, the largest available ALS clinical trial dataset. In PRO-ACT, urate levels were found to be one of the strongest predictors of favorable ALS progression and survival (Atassi et al. 2014).

Clinical Trials of Purines as Candidate Neuroprotectants

Adenosine A_{2A} Antagonists in PD Trials Sensitive to Neuroprotection

Caffeine Almost all clinical trials conducted on adenosine A_{2A} antagonists for PD have been short-term studies (with less than 3 months of follow-up on study drug) targeting the symptoms of the disease, primarily those affecting its characteristic motor deficits as reviewed in depth elsewhere in this volume (see Chap. 14 see also Pinna 2014). Although no rigorously designed trial has tested a primary hypothesis of disease modification by an A_{2A} antagonist, a couple of large trials have adopted designs capable of addressing the hypothesis. Recently, a long-term (5 year), randomized, placebo-controlled Phase III trial of caffeine (up to 400 mg daily in the active drug arm) was initiated in PD (Postuma access date July 2015). This ambitious study, which is projected to conclude in 2021,has the potential to meaningfully assess the effect of caffeine on the course of clinical progression of PD given its long-term duration,a primary outcome of a well-studied, composite (motor and non-motor; clinician- and patient-reported) scale of parkinsonian features (Movement Disorders Society-Unified Parkinson's Disease Rating Scale; MDS-UPDRS) measured every 6 months, and a 'delayed-start' design (McDermott et al. 2002) (with the placebo group switched after 4.5 years to caffeine for the last 6 months of follow-up). Of note, with its secondary outcomes including measures of dyskinesia (a motor complication of prolonged treatment with standard dopaminergic drug therapy for PD), this trial may also have the potential to test the hypothesis that A_{2A} antagonism delays the development of dyskinesia in PD, a form of neuroplasticity-based disease modification suggested by convergent preclinical (Bibbiani et al. 2003; Xiao et al. 2006) and clinical data (Wills et al. 2013).

Preladenant Although commercial A_{2A} antagonist programs have understandably focused on the lower risk pursuit of a short-term symptomatic indication in PD, the above evidence supporting an additional disease-modifying advantage has encouraged greater investment in this target (Morelli et al. 2009; Schwarzschild et al. 2006). One major randomized, placebo-controlled, Phase III trial of an A_{2A} antagonist (preladenant) followed subjects long-term, and included a secondary analysis of effects of the antagonist over 52 weeks on UPDRS scores (compared to those of placebo for 26 weeks followed by preladenant for an additional 26 weeks) (Merck Sharp & Dohme Corp.ClinicalTrials.gov:NCT01155479). This delayed-start design allows for a separation of potentially confounding, symptomatic benefits of an A_{2A} antagonist from any disease-modifying benefit. Unfortunately, the trial was terminated prematurely (Merck Sharp & Dohme Corp.ClinicalTrials.gov:NCT01155479), possibly due to technical limitations of the study, and only interim primary results having preliminarily reported to date (Stocchi et al. 2014). Nevertheless, the clinical investigation path for preladenant illustrates the feasibility of pursuing a potentially sequential development strategy to obtain regulatory approval of an indication for improved symptomatic treatment through A_{2A} antagonism, which could then be rapidly expanded to fill the major unmet need for disease modification in PD.

Urate-Elevating Treatments in Trials Sensitive to Neuroprotection

Multiple Sclerosis Elevating urate concentrations has been suggested as neuroprotective strategy across a range of neurological disorders from ischemic (stroke) to neuroinfammatory (multiple sclerosis [MS]) to neurodegenerative (PD, ALS, HD and AD). However, urate itself was found to have poor oral bioavailability in humans, apparently due its metabolism by gut flora (Spitsin et al. 2001). This pharmacological limitation prompted an alternative approach to raising endogenous urate levels via metabolic mass action with the urate precursor inosine (see Fig. 6.1), which proved capable of substantially and chronically raising serum and possibly CSF urate levels in a small population of MS patients (Spitsin et al. 2001). Although several long-term trials (Gonsette et al. 2010; Markowitz et al. 2009; Munoz Garcia et al. 2015; Spitsin et al. 2001; Toncev 2006) of urate-elevating inosine treatment in MS have provided inconsistent evidence of improved clinical progression (Yadav et al. 2014), they have helped establish feasibility of using inosine to persistently and relatively safely elevate serum urate, with kidney stones the most consistent serious adverse effect observed thus far (Gonsette et al. 2010; Markowitz et al. 2009).

Parkinson's Disease Informed by this early experience of using oral inosine to elevate urate in MS, and motivated by the strong convergence of laboratory, epidemiological and clinical data suggesting urate as a neuroprotectant in PD, we have pursued the clinical development of inosine as a candidate disease-modifying therapy for PD. A safety-focused, randomized, placebo-controlled, Phase II trial of oral inosine demonstrated its ability to dose-dependently elevate serum and CSF urate levels for months to years (Parkinson Study Group et al. 2014). The Safety of Urate Elevation in PD (SURE-PD) trial enrolled 75 early PD subjects with serum urate below 6 mg/dL and randomized them 1:1:1 to placebo or inosine titrated to mildly or moderately elevate serum urate up to 6–7 or 7–8 mg/dL for up to 24 months. The only serious adverse events that was likely attributable to inosine treatment were kidney stones in three subjects, all of whom recovered fully. Preliminary assessment of long-term clinical outcomes, for which this Phase II study was not powered, nevertheless supported advancing to a Phase III disease modification trial.

Importantly, secondary analyses of SURE-PD suggested inosine produces no short-term effect on parkinsonian features, in contrast, for example, to caffeine, which significantly improved UPDRS scores and their motor component within weeks (Postuma et al. 2012). The presence of such symptomatic benefits can actually complicate the testing of a candidate neuroprotectant because its putative effects on the long-term course of the disease must then be distinguished from its short-term effects, and may necessitate a more complex trial design (such as a 'delayed start' design, which is being employed for caffeine as above). Thus, a simpler trial design may be suitable for Phase III clinical development of inosine as a disease modifying strategy for PD.

Of note, laboratory (Du et al. 2007) and recent clinical (Atassi et al. 2014; Paganoni et al. 2012; Zoccolella et al. 2011) data raising the possibility that higher urate may protect degenerating neurons in ALS as well as PD, have led to a pilot clinical study of urate-elevating inosine treatment in ALS (Paganoni 2014).

Stroke In parallel to its investigation as a therapy for progressive neurodegenerative disorders, urate elevation is being explored as a protective strategy in stroke based on a comparable set of supportive laboratory (Haberman et al. 2007; Romanos et al. 2007) findings and epidemiological data (Chamorro et al. 2002). However, whereas oral inosine is being used to chronically raise urate in neurodegeneration, parenteral (intravenous) administration of urate itself is being tested in stroke to ameliorate the immediate ischemic injury within hours of onset (Amaro et al. 2010). The results of this Phase IIB/III trial (Charmorro et al. 2014) did not demonstrate a statistically significant overall benefit of urate, but were sufficiently suggestive to warrant for fuller Phase III clinical testing. Interestingly a trend toward greater efficacy of urate-elevating therapy for stroke in women, who at baseline have substantially lower serum urate levels than do men, was mirrored for PD in the SURE-PD study (Parkinson Study Group et al. 2014; Schwarzschild et al. 2014) and warrants further attention in future studies.

Acknowledgements The authors would like to thank the Dept of Defense/NETPR program W81XWH-11-1-0150 and NIH R21NS084710 for the funding. The authors declare no competing financial interests.

References

Abraham A, Drory VE (2014) Influence of serum uric acid levels on prognosis and survival in amyotrophic lateral sclerosis: a meta-analysis. J Neurol 261:1133–1138

Ahlskog JE, Uitti RJ, Low PA et al (1995) No evidence for systemic oxidant stress in Parkinson's or Alzheimer's disease. Mov Disord 10:566–573

Alfinito PD, Wang SP, Manzino L et al (2003) Adenosinergic protection of dopaminergic and GABAergic neurons against mitochondrial inhibition through receptors located in the substantia nigra and striatum, respectively. J Neurosci 23:10982–10987

Al-Khateeb E, Althaher A, Al-Khateeb M et al (2015) Relation between uric acid and Alzheimer's disease in elderly Jordanians. J Alzheimers Dis 44:859–865

Alonso A, Rodriguez LA, Logroscino G et al (2007) Gout and risk of Parkinson disease: a prospective study. Neurology 69:1696–1700

Amaro S, Canovas D, Castellanos M et al (2010) The URICO-ICTUS study, a phase 3 study of combined treatment with uric acid and rtPA administered intravenously in acute ischaemic stroke patients within the first 4.5 h of onset of symptoms. Int J Stroke 5:325–328

Ames BN, Cathcart R, Schwiers E et al (1981) Uric acid provides an antioxidant defense in humans against oxidant- and radical-caused aging and cancer: a hypothesis. Proc Natl Acad Sci U S A 78:6858–6862

Andreadou E, Nikolaou C, Gournaras F et al (2009) Serum uric acid levels in patients with Parkinson's disease: their relationship to treatment and disease duration. Clin Neurol Neurosurg 111:724–728

Annanmaki T, Muuronen A, Murros K (2007) Low plasma uric acid level in Parkinson's disease. Mov Disord 22:1133–1137

Arendash GW, Schleif W, Rezai-Zadeh K et al (2006) Caffeine protects Alzheimer's mice against cognitive impairment and reduces brain beta-amyloid production. Neuroscience 142:941–952

Arendash GW, Mori T, Cao C et al (2009) Caffeine reverses cognitive impairment and decreases brain amyloid-beta levels in aged Alzheimer's disease mice. J Alzheimers Dis 17:661–680

Armentero MT, Pinna A, Ferré S et al (2011) Past, present and future of A(2A) adenosine receptor antagonists in the therapy of Parkinson's disease. Pharmacol Ther 132:280–299

Ascherio A, Zhang SM, Hernan MA et al (2001) Prospective study of caffeine consumption and risk of Parkinson's disease in men and women. Ann Neurol 50:56–63

Ascherio A, LeWitt PA, Xu K et al (2009) Urate as a predictor of the rate of clinical decline in Parkinson disease. Arch Neurol 66:1460–1468

Atassi N, Berry J, Shui A et al (2014) The PRO-ACT database: design, initial analyses, and predictive features. Neurology 83:1719–1725

Auinger P, Kieburtz K, McDermott MP (2010) The relationship between uric acid levels and Huntington's disease progression. Mov Disord 25:224–228

Banks WA, Erickson MA (2009) The blood-brain barrier and immune function and dysfunction. Neurobiol Dis 37:26–32

Beghi E, Pupillo E, Messina P et al (2011) Coffee and amyotrophic lateral sclerosis: a possible preventive role. Am J Epidemiol 174:1002–1008

Belcastro V, Tozzi A, Tantucci M et al (2009) A2A adenosine receptor antagonists protect the striatum against rotenone-induced neurotoxicity. Exp Neurol 217:231–234

Benedetti MD, Bower JH, Maraganore DM et al (2000) Smoking, alcohol, and coffee consumption preceding Parkinson's disease: a case-control study. Neurology 55:1350–1358

Benzie I, Strain J (1996) Uric acid: friend or foe? Redox Rep 2:231–234

Bibbiani F, Oh JD, Petzer JP et al (2003) A2A antagonist prevents dopamine agonist-induced motor complications in animal models of Parkinson's disease. Exp Neurol 184:285–294

Blum D, Galas MC, Pintor A et al (2003) A dual role of adenosine A2A receptors in 3-nitropropionic acid-induced striatal lesions: implications for the neuroprotective potential of A2A antagonists. J Neurosci 23:5361–5369

Bogdanov M, Matson WR, Wang L et al (2008) Metabolomic profiling to develop blood biomarkers for Parkinson's disease. Brain 131:389–396

Bos MJ, Koudstaal PJ, Hofman A et al (2006) Uric acid is a risk factor for myocardial infarction and stroke: the Rotterdam study. Stroke 37:1503–1507

Brambilla R, Cottini L, Fumagalli M et al (2003) Blockade of A2A adenosine receptors prevents basic fibroblast growth factor-induced reactive astrogliosis in rat striatal primary astrocytes. Glia 43:190–194

Brambilla L, Martorana F, Rossi D (2012) Astrocyte signaling and neurodegeneration: new insights into CNS disorders. Prion 7:28–36

Brodie C, Blumberg PM, Jacobson KA (1998) Activation of the A2A adenosine receptor inhibits nitric oxide production in glial cells. FEBS Lett 429:139–142

Brothers HM, Marchalant Y, Wenk GL (2010) Caffeine attenuates lipopolysaccharide-induced neuroinflammation. Neurosci Lett 480:97–100

Burton NC, Kensler TW, Guilarte TR (2006) In vivo modulation of the Parkinsonian phenotype by Nrf2. Neurotoxicology 27:1094–1100

Can M, Varlibas F, Guven B et al (2013) Ischemia modified albumin and plasma oxidative stress markers in Alzheimer's disease. Eur Neurol 69:377–380

Cankurtaran M, Yesil Y, Kuyumcu ME et al (2012) Altered levels of homocysteine and serum natural antioxidants links oxidative damage to Alzheimer's disease. J Alzheimers Dis 33:1051–1058

Cao C, Cirrito JR, Lin X et al (2009) Caffeine suppresses amyloid-beta levels in plasma and brain of Alzheimer's disease transgenic mice. J Alzheimers Dis 17:681–697

Cao B, Guo X, Chen K et al (2013) Uric acid is associated with the prevalence but not disease progression of multiple system atrophy in Chinese population. J Neurol 260:2511–2515

Carta AR, Kachroo A, Schintu N et al (2009) Inactivation of neuronal forebrain A receptors protects dopaminergic neurons in a mouse model of Parkinson's disease. J Neurochem 111:1478–1489

Caulfield MJ, Munroe PB, O'Neill D et al (2008) SLC2A9 is a high-capacity urate transporter in humans. PLoS Med 5:e197

Chamorro A, Obach V, Cervera A et al (2002) Prognostic significance of uric acid serum concentration in patients with acute ischemic stroke. Stroke 33:1048–1052

Chamorro A, Amaro S, Castellanos M et al (2014) Safety and efficacy of uric acid in patients with acute stroke (URICO-ICTUS): a randomised, double-blind phase 2b/3 trial. Lancet Neurol 13:453–460

Checkoway H, Powers K, Smith-Weller T et al (2002) Parkinson's disease risks associated with cigarette smoking, alcohol consumption, and caffeine intake. Am J Epidemiol 155:732–738

Chen Y, Vartiainen NE, Ying W et al (2001a) Astrocytes protect neurons from nitric oxide toxicity by a glutathione-dependent mechanism. J Neurochem 77:1601–1610

Chen JF, Xu K, Petzer JP et al (2001b) Neuroprotection by caffeine and A(2A) adenosine receptor inactivation in a model of Parkinson's disease. J Neurosci 21:RC143

Chen PS, Peng GS, Li G et al (2006) Valproate protects dopaminergic neurons in midbrain neuron/glia cultures by stimulating the release of neurotrophic factors from astrocytes. Mol Psychiatry 11:1116–1125

Chen H, Mosley TH, Alonso A et al (2009a) Plasma urate and Parkinson's disease in the Atherosclerosis Risk in Communities (ARIC) study. Am J Epidemiol 169:1064–1069

Chen PC, Vargas MR, Pani AK et al (2009b) Nrf2-mediated neuroprotection in the MPTP mouse model of Parkinson's disease: critical role for the astrocyte. Proc Natl Acad Sci U S A 106:2933–2938

Chen X, Ghribi O, Geiger JD (2010) Caffeine protects against disruptions of the blood-brain barrier in animal models of Alzheimer's and Parkinson's diseases. J Alzheimers Dis 20:S127–S141

Chen X, Burdett TC, Desjardins CA et al (2013) Disrupted and transgenic urate oxidase alter urate and dopaminergic neurodegeneration. Proc Natl Acad Sci U S A 110:300–305

Chen X, Guo X, Huang R et al (2014) Serum uric acid levels in patients with Alzheimer's disease: a meta-analysis. PLoS One 9:e94084

Chou SY, Lee YC, Chen HM et al (2005) CGS 21680 attenuates symptoms of Huntington's disease in a transgenic mouse model. J Neurochem 93:310–320

Christen P, Peacock WC, Christen AE et al (1970) Urate oxidase in primate phylogenesis. Eur J Biochem 12:3–5

Church WH, Ward VL (1994) Uric acid is reduced in the substantia nigra in Parkinson's disease: effect on dopamine oxidation. Brain Res Bull 33:419–425

Cipriani S, Desjardins CA, Burdett TC et al (2012a) Urate and its transgenic depletion modulate neuronal vulnerability in a cellular model of Parkinson's disease. PLoS One 7:e37331

Cipriani S, Desjardins CA, Burdett TC et al (2012b) Protection of dopaminergic cells by urate requires its accumulation in astrocytes. J Neurochem 123:172–181

Cunha RA (2005) Neuroprotection by adenosine in the brain: from A(1) receptor activation to A(2A) receptor blockade. Purinergic Signal 1:111–134

Daly JW, Butts-Lamb P, Padgett W (1983) Subclasses of adenosine receptors in the central nervous system: interaction with caffeine and related methylxanthines. Cell Mol Neurobiol 3:69–80

Davis JW, Grandinetti A, Waslien CI et al (1996) Observations on serum uric acid levels and the risk of idiopathic Parkinson's disease. Am J Epidemiol 144:480–484

de Lau LM, Koudstaal PJ, Hofman A et al (2005) Serum uric acid levels and the risk of Parkinson disease. Ann Neurol 58:797–800

De Luca MA, Cauli O, Morelli M et al (2014) Elevation of striatal urate in experimental models of Parkinson's disease: a compensatory mechanism triggered by dopaminergic nigrostriatal degeneration? J Neurochem 131:284–289

de Mendonça A, Sebastiao AM, Ribeiro JA (2000) Adenosine: does it have a neuroprotective role after all? Brain Res Rev 33:258–274

De Vera M, Rahman MM, Rankin J et al (2008) Gout and the risk of Parkinson's disease: a cohort study. Arthritis Rheum 59:1549–1554

Dehghan A, Kottgen A, Yang Q et al (2008) Association of three genetic loci with uric acid concentration and risk of gout: a genome-wide association study. Lancet 372:1953–1961

Dheen ST, Kaur C, Ling EA (2007) Microglial activation and its implications in the brain diseases. Curr Med Chem 14:1189–1197

D'Hooghe MB, Haentjens P, Nagels G et al (2012) Alcohol, coffee, fish, smoking and disease progression in multiple sclerosis. Eur J Neurol 19:616–624

Du Y, Chen CP, Tseng CY et al (2007) Astroglia-mediated effects of uric acid to protect spinal cord neurons from glutamate toxicity. Glia 55:463–472

Duan W, Ladenheim B, Cutler RG et al (2002) Dietary folate deficiency and elevated homocysteine levels endanger dopaminergic neurons in models of Parkinson's disease. J Neurochem 80:101–110

Ellrichmann G, Petrasch-Parwez E, Lee DH et al (2011) Efficacy of fumaric acid esters in the R6/2 and YAC128 models of Huntington's disease. PLoS One 6:e16172

Erickson MA, Dohi K, Banks WA (2012) Neuroinflammation: a common pathway in CNS diseases as mediated at the blood-brain barrier. Neuroimmunomodulation 19:121–130

Eskelinen MH, Kivipelto M (2010) Caffeine as a protective factor in dementia and Alzheimer's disease. J Alzheimers Dis 20:S167–S174

Eskelinen MH, Ngandu T, Tuomilehto J et al (2009) Midlife coffee and tea drinking and the risk of late-life dementia: a population-based CAIDE study. J Alzheimers Dis 16:85–91

Euser SM, Hofman A, Westendorp RG et al (2009) Serum uric acid and cognitive function and dementia. Brain 132:377–382

Facheris MF, Hicks AA, Minelli C et al (2011) Variation in the uric acid transporter gene SLC2A9 and its association with AAO of Parkinson's disease. J Mol Neurosci 43:246–250

Fall PA, Fredrikson M, Axelson O et al (1999) Nutritional and occupational factors influencing the risk of Parkinson's disease: a case-control study in southeastern Sweden. Mov Disord 14:28–37

Fiebich BL, Biber K, Lieb K et al (1996) Cyclooxygenase-2 expression in rat microglia is induced by adenosine A2a-receptors. Glia 18:152–160

Fiebich BL, Butcher RD, Gebicke-Haerter PJ (1998) Protein kinase C-mediated regulation of inducible nitric oxide synthase expression in cultured microglial cells. J Neuroimmunol 92:170–178

Filippo MD, Chiasserini D, Tozzi A et al (2010) Mitochondria and the link between neuroinflammation and neurodegeneration. J Alzheimers Dis 20:S369–S379

Fink JS, Kalda A, Ryu H et al (2004) Genetic and pharmacological inactivation of the adenosine A2A receptor attenuates 3-nitropropionic acid-induced striatal damage. J Neurochem 88:538–544

Fredholm BB, Battig K, Holmen J et al (1999) Actions of caffeine in the brain with special reference to factors that contribute to its widespread use. Pharmacol Rev 51:83–133

Fujishita K, Ozawa T, Shibata K et al (2009) Grape seed extract acting on astrocytes reveals neuronal protection against oxidative stress via interleukin-6-mediated mechanisms. Cell Mol Neurobiol 29:1121–1129

Gao Y, Phillis JW (1994) CGS 15943, an adenosine A2 receptor antagonist, reduces cerebral ischemic injury in the *Mongolian gerbil*. Life Sci 55:PL61–PL65

Gao X, Chen H, Choi HK et al (2008) Diet, urate, and Parkinson's disease risk in men. Am J Epidemiol 167:831–838

Gerhard A, Pavese N, Hotton G et al (2006) In vivo imaging of microglial activation with [11C](R)-PK11195 PET in idiopathic Parkinson's disease. Neurobiol Dis 21:404–412

Gong L, Zhang QL, Zhang N et al (2012) Neuroprotection by urate on 6-OHDA-lesioned rat model of Parkinson's disease: linking to Akt/GSK3beta signaling pathway. J Neurochem 123:876–885

Gonsette RE, Sindic C, D'Hooghe MB et al (2010) Boosting endogenous neuroprotection in multiple sclerosis: the ASsociation of Inosine and Interferon beta in relapsing- remitting Multiple Sclerosis (ASIIMS) trial. Mult Scler 16:455–462

Gonzalez-Aramburu I, Sanchez-Juan P, Jesus S et al (2013) Genetic variability related to serum uric acid concentration and risk of Parkinson's disease. Mov Disord 28:1737–1740

Gonzalez-Fernandez E, Sanchez-Gomez MV, Perez-Samartin A et al (2014) A3 Adenosine receptors mediate oligodendrocyte death and ischemic damage to optic nerve. Glia 62(2):199–216

Guerreiro S, Ponceau A, Toulorge D et al (2009) Protection of midbrain dopaminergic neurons by the end-product of purine metabolism uric acid: potentiation by low-level depolarization. J Neurochem 109:1118–1128

Haberman F, Tang SC, Arumugam TV et al (2007) Soluble neuroprotective antioxidant uric acid analogs ameliorate ischemic brain injury in mice. Neuromolecular Med 9:315–323

Hauser DN, Hastings TG (2013) Mitochondrial dysfunction and oxidative stress in Parkinson's disease and monogenic parkinsonism. Neurobiol Dis 51:35–42

Hediger MA, Johnson RJ, Miyazaki H et al (2005) Molecular physiology of urate transport. Physiology 20:125–133

Hellenbrand W, Boeing H, Robra BP et al (1996) Diet and Parkinson's disease. II: a possible role for the past intake of specific nutrients. Results from a self-administered food-frequency questionnaire in a case-control study. Neurology 47:644–650

Hindley S, Herman MA, Rathbone MP (1994) Stimulation of reactive astrogliosis in vivo by extracellular adenosine diphosphate or an adenosine A2 receptor agonist. J Neurosci Res 38:399–406

Hooper DC, Spitsin S, Kean RB et al (1998) Uric acid, a natural scavenger of peroxynitrite, in experimental allergic encephalomyelitis and multiple sclerosis. Proc Natl Acad Sci U S A 95:675–680

Hooper DC, Scott GS, Zborek A et al (2000) Uric acid, a peroxynitrite scavenger, inhibits CNS inflammation, blood-CNS barrier permeability changes, and tissue damage in a mouse model of multiple sclerosis. FASEB J 14:691–698

Hozawa A, Folsom AR, Ibrahim H et al (2006) Serum uric acid and risk of ischemic stroke: the ARIC Study. Atherosclerosis 187:401–407

Hu G, Bidel S, Jousilahti P et al (2007) Coffee and tea consumption and the risk of Parkinson's disease. Mov Disord 22:2242–2248

Huang QY, Wei C, Yu L et al (2006) Adenosine A2A receptors in bone marrow-derived cells but not in forebrain neurons are important contributors to 3-nitropropionic acid-induced striatal damage as revealed by cell-type-selective inactivation. J Neurosci 26:11371–11378

Huang NK, Lin JH, Lin JT et al (2011) A new drug design targeting the adenosinergic system for Huntington's disease. PLoS One 6:e20934

Iadecola C, Zhang F, Xu S et al (1995) Inducible nitric oxide synthase gene expression in brain following cerebral ischemia. J Cereb Blood Flow Metab 15:378–384

Ikeda K, Kurokawa M, Aoyama S et al (2002) Neuroprotection by adenosine A2A receptor blockade in experimental models of Parkinson's disease. J Neurochem 80:262–270

Ikeda K, Hirayama T, Takazawa T et al (2012) Relationships between disease progression and serum levels of lipid, urate, creatinine and ferritin in Japanese patients with amyotrophic lateral sclerosis: a cross-sectional study. Intern Med 51:1501–1508

Imamura K, Takeshima T, Nakaso K et al (2008) Pramipexole has astrocyte-mediated neuroprotective effects against lactacystin toxicity. Neurosci Lett 440:97–102

Innamorato NG, Jazwa A, Rojo AI et al (2010) Different susceptibility to the Parkinson's toxin MPTP in mice lacking the redox master regulator Nrf2 or its target gene heme oxygenase-1. PLoS One 5:e11838

Irizarry MC, Raman R, Schwarzschild MA et al (2009) Plasma urate and progression of mild cognitive impairment. Neurodegener Dis 6:23–28

Jakel RJ, Townsend JA, Kraft AD et al (2007) Nrf2-mediated protection against 6-hydroxydopamine. Brain Res 1144:192–201

Jenner P (2003) Oxidative stress in Parkinson's disease. Ann Neurol 53:S26–S36

Jesus S, Perez I, Caceres-Redondo MT et al (2012) Low serum uric acid concentration in Parkinson's disease in southern Spain. Eur J Neurol 20:208–210

Jinnah HA, Ceballos-Picot I, Torres RJ et al (2010) Attenuated variants of Lesch-Nyhan disease. Brain 133:671–689

Joghataie MT, Roghani M, Negahdar F et al (2004) Protective effect of caffeine against neurodegeneration in a model of Parkinson's disease in rat: behavioral and histochemical evidence. Parkinsonism Relat Disord 10:465–468

Johansen KK, Wang L, Aasly JO et al (2009) Metabolomic profiling in LRRK2-related Parkinson's disease. PLoS One 4:e7551

Jones DC, Gunasekar PG, Borowitz JL et al (2000) Dopamine-induced apoptosis is mediated by oxidative stress and Is enhanced by cyanide in differentiated PC12 cells. J Neurochem 74:2296–2304

Joshi G, Johnson JA (2012) The Nrf2-ARE pathway: a valuable therapeutic target for the treatment of neurodegenerative diseases. Recent Pat CNS Drug Discov 7:218–229

Kachroo A, Irizarry MC, Schwarzschild MA (2010) Caffeine protects against combined paraquat and maneb-induced dopaminergic neuron degeneration. Exp Neurol 223:657–661

Kataoka H, Kiriyama T, Kobayashi Y et al (2013) Clinical outcomes and serum uric acid levels in elderly patients with amyotrophic lateral sclerosis aged ≥70 years. Am J Neurodegener Dis 2:140–144

Keebaugh AC, Thomas JW (2010) The evolutionary fate of the genes encoding the purine catabolic enzymes in hominoids, birds, and reptiles. Mol Biol Evol 27:1359–1369

Keizman D, Ish-Shalom M, Berliner S et al (2009) Low uric acid levels in serum of patients with ALS: further evidence for oxidative stress? J Neurol Sci 285:95–99

Kim TS, Pae CU, Yoon SJ et al (2006) Decreased plasma antioxidants in patients with Alzheimer's disease. Int J Geriatr Psychiatry 21:344–348

Kolz M, Johnson T, Sanna S et al (2009) Meta-analysis of 28,141 individuals identifies common variants within five new loci that influence uric acid concentrations. PLoS Genet 5:e1000504

Kroncke KD, Fehsel K, Kolb-Bachofen V (1998) Inducible nitric oxide synthase in human diseases. Clin Exp Immunol 113:147–156

Kutzing MK, Firestein BL (2008) Altered uric acid levels and disease states. J Pharmacol Exp Ther 324:1–7

Ladecola C, Zhang F, Xu S et al (1995) Inducible nitric oxide synthase gene expression in brain following cerebral ischemia. J Cereb Blood Flow Metab 15:378–384

Larumbe Ilundain R, Ferrer Valls JV, Vines Rueda JJ et al (2001) Case-control study of markers of oxidative stress and metabolism of blood iron in Parkinson's disease. Rev Esp Salud Publica 75:43–53

Laurent C, Eddarkaoui S, Derisbourg M et al (2014) Beneficial effects of caffeine in a transgenic model of Alzheimer's disease-like tau pathology. Neurobiol Aging 35:2079–2090

Lee CF, Chern Y (2014) Adenosine receptors and Huntington's disease. Int Rev Neurobiol 119:195–232

Lee JE, Song SK, Sohn YH et al (2011) Uric acid as a potential disease modifier in patients with multiple system atrophy. Mov Disord 26:1533–1536

Lekieffre D, Callebert J, Plotkine M et al (1991) Enhancement of endogenous excitatory amino acids by theophylline does not modify the behavioral and histological consequences of forebrain ischemia. Brain Res 565:353–357

Li XX, Nomura T, Aihara H et al (2001) Adenosine enhances glial glutamate efflux via A2a adenosine receptors. Life Sci 68:1343–1350

Li S, Sanna S, Maschio A et al (2007) The GLUT9 gene is associated with serum uric acid levels in Sardinia and Chianti cohorts. PLoS Genet 3:e194

Li XZ, Bai LM, Yang YP et al (2009) Effects of IL-6 secreted from astrocytes on the survival of dopaminergic neurons in lipopolysaccharide-induced inflammation. Neurosci Res 65:252–258

Licinio J, Prolo P, McCann SM et al (1999) Brain iNOS: current understanding and clinical implications. Mol Med Today 5:225–232

Lindsay J, Laurin D, Verreault R et al (2002) Risk factors for Alzheimer's disease: a prospective analysis from the Canadian Study of Health and Aging. Am J Epidemiol 156:445–453

Lipkowitz MS (2012) Regulation of uric acid excretion by the kidney. Curr Rheumatol Rep 14:179–188

Machado-Filho JA, Correia AO, Montenegro AB et al (2014) Caffeine neuroprotective effects on 6-OHDA-lesioned rats are mediated by several factors, including pro-inflammatory cytokines and histone deacetylase inhibitions. Behav Brain Res 264:116–125

Maesaka JK, Wolf-Klein G, Piccione JM et al (1993) Hypouricemia, abnormal renal tubular urate transport, and plasma natriuretic factor(s) in patients with Alzheimer's disease. J Am Geriatr Soc 41:501–506

Markowitz CE, Spitsin S, Zimmerman V et al (2009) The treatment of multiple sclerosis with inosine. J Altern Complement Med 15:619–625

Massa J, O'Reilly EJ, Munger KL et al (2012) Caffeine and alcohol intakes have no association with risk of multiple sclerosis. Mult Scler 19:53–58

Matos M, Augusto E, Agostinho P et al (2013) Antagonistic interaction between adenosine A2A receptors and Na+/K+-ATPase-α2 controlling glutamate uptake in astrocytes. J Neurosci 33:18492–18502

McCall AL, Millington WR, Wurtman RJ (1982) Blood-brain barrier transport of caffeine: dose-related restriction of adenine transport. Life Sci 31:2709–2715

McDermott MP, Hall WJ, Oakes D et al (2002) Design and analysis of two-period studies of potentially disease-modifying treatments. Control Clin Trials 23:635–649

McGeer PL, McGeer EG (2008) Glial reactions in Parkinson's disease. Mov Disord 23:474–483

Melani A, Cipriani S, Vannucchi MG et al (2009) Selective adenosine A2a receptor antagonism reduces JNK activation in oligodendrocytes after cerebral ischaemia. Brain 132:1480–1495

Merck Sharp & Dohme Corp. (2015) A placebo- and active-controlled study of preladenant in early Parkinson's disease. http://clinicaltrials.gov/show/NCT01155479

Mievis S, Blum D, Ledent C (2011) A2A receptor knockout worsens survival and motor behaviour in a transgenic mouse model of Huntington's disease. Neurobiol Dis 41:570–576

Minghetti L, Greco A, Potenza RL et al (2007) Effects of the adenosine A2A receptor antagonist SCH 58621 on cyclooxygenase-2 expression, glial activation, and brain-derived neurotrophic factor availability in a rat model of striatal neurodegeneration. J Neuropathol Exp Neurol 66:363–371

Moccia M, Picillo M, Erro R et al (2014) Presence and progression of non-motor symptoms in relation to uric acid in de novo Parkinson's disease. Eur J Neurol 22:93–98

Morelli M, Carta AR, Jenner P (2009) Adenosine A2A receptors and Parkinson's disease. Handb Exp Pharmacol 193:589–615

Morelli M, Carta AR, Kachroo A et al (2010) Pathophysiological roles for purines: adenosine, caffeine and urate. Prog Brain Res 183:183–208

Morozova N, Weisskopf MG, McCullough ML et al (2008) Diet and amyotrophic lateral sclerosis. Epidemiology 19:324–337

Munoz Garcia D, Midaglia L, Martinez Vilela J et al (2015) Associated Inosine to interferon: results of a clinical trial in multiple sclerosis. Acta Neurol Scand. 131:405–410

Neymotin A, Calingasan NY, Wille E et al (2011) Neuroprotective effect of Nrf2/ARE activators, CDDO ethylamide and CDDO trifluoroethylamide, in a mouse model of amyotrophic lateral sclerosis. Free Radic Biol Med 51:88–96

Niranjan R (2013) The role of inflammatory and oxidative stress mechanisms in the pathogenesis of Parkinson's disease: focus on astrocytes. Mol Neurobiol 49:28–38

Nishizaki T (2004) ATP- and adenosine-mediated signaling in the central nervous system: adenosine stimulates glutamate release from astrocytes via A2a adenosine receptors. J Pharmacol Sci 94:100–102

Oda M, Satta Y, Takenaka O et al (2002) Loss of urate oxidase activity in hominoids and its evolutionary implications. Mol Biol Evol 19:640–653

O'Reilly EJ, Gao X, Weisskopf MG et al (2010) Plasma urate and Parkinson's disease in women. Am J Epidemiol 172:666–670

Orowan E (1955) The origin of man. Nature 175:683–684

Ostwald P, Park SS, Toledano AY et al (1997) Adenosine receptor blockade and nitric oxide synthase inhibition in the retina: impact upon post-ischemic hyperemia and the electroretinogram. Vision Res 37:3453–3461

Paganoni S (2014) A Pilot Study of Inosine in ALS. http://clinicaltrials.gov/NCT02288091. Accessed July 2015

Paganoni S, Zhang M, Quiroz Zarate A et al (2012) Uric acid levels predict survival in men with amyotrophic lateral sclerosis. J Neurol 259:1923–1928

Parkinson's-Study-Group, Schwarzschild MA, Ascherio A et al (2014) Inosine to increase serum and cerebrospinal fluid urate in Parkinson disease: a randomized clinical trial. JAMA Neurol 71:141–150

Pierri M, Vaudano E, Sager T et al (2005) KW-6002 protects from MPTP induced dopaminergic toxicity in the mouse. Neuropharmacology 48:517–524

Pinna A (2014) Adenosine A2A receptor antagonists in Parkinson's disease: progress in clinical trials from the newly approved istradefylline to drugs in early development and those already discontinued. CNS Drugs 28:455–474

Pinna A, Tronci E, Schintu N et al (2010) A new ethyladenine antagonist of adenosine A(2A) receptors: behavioral and biochemical characterization as an antiparkinsonian drug. Neuropharmacology 58:613–623

Polidori MC, Mattioli P, Aldred S et al (2004) Plasma antioxidant status, immunoglobulin g oxidation and lipid peroxidation in demented patients: relevance to Alzheimer disease and vascular dementia. Dement Geriatr Cogn Disord 18:265–270

Popoli P, Pintor A, Domenici MR et al (2002) Blockade of striatal adenosine A2A receptor reduces, through a presynaptic mechanism, quinolinic acid-induced excitotoxicity: possible relevance to neuroprotective interventions in neurodegenerative diseases of the striatum. J Neurosci 22:1967–1975

Postuma R (2015) Caffeine as a therapy for Parkinson's disease. http://clinicaltrials.gov/show/NCT01738178. Accessed July 2015

Postuma RB, Lang AE, Munhoz RP et al (2012) Caffeine for treatment of Parkinson disease: a randomized controlled trial. Neurology 79:651–658

Potenza RL, Armida M, Ferrante A et al (2013) Effects of chronic caffeine intake in a mouse model of amyotrophic lateral sclerosis. J Neurosci Res 91:585–592

Prasanthi JR, Dasari B, Marwarha G et al (2010) Caffeine protects against oxidative stress and Alzheimer's disease-like pathology in rabbit hippocampus induced by cholesterol-enriched diet. Free Radic Biol Med 49:1212–1220

Proctor P (1970) Similar functions of uric acid and ascorbate in man? Nature 228:868

PSG (1989a) Effect of deprenyl on the progression of disability in early Parkinson's disease. The Parkinson Study Group. N Engl J Med 321:1364–1371

PSG (1989b) DATATOP: a multicenter controlled clinical trial in early Parkinson's disease. Parkinson Study Group. Arch Neurol 46:1052–1060

PSG (1993) Effects of tocopherol and deprenyl on the progression of disability in early Parkinson's disease. N Engl J Med 328:176–183

PSG (2007) Mixed lineage kinase inhibitor CEP-1347 fails to delay disability in early Parkinson disease. Neurology 69:1480–1490

Pugliese AM, Traini C, Cipriani S et al (2009) The adenosine A2A receptor antagonist ZM241385 enhances neuronal survival after oxygen-glucose deprivation in rat CA1 hippocampal slices. Br J Pharmacol 157:818–830

Ragonese P, Salemi G, Morgante L et al (2003) A case-control study on cigarette, alcohol, and coffee consumption preceding Parkinson's disease. Neuroepidemiology 22:297–304

Rathinam ML, Watts LT, Narasimhan M et al (2012) Astrocyte mediated protection of fetal cerebral cortical neurons from rotenone and paraquat. Environ Toxicol Pharmacol 33:353–360

Rinaldi P, Polidori MC, Metastasio A et al (2003) Plasma antioxidants are similarly depleted in mild cognitive impairment and in Alzheimer's disease. Neurobiol Aging 24:915–919

Romanos E, Planas AM, Amaro S et al (2007) Uric acid reduces brain damage and improves the benefits of rt-PA in a rat model of thromboembolic stroke. J Cereb Blood Flow Metab 27:14–20

Ross GW, Abbott RD, Petrovitch H et al (2000) Association of coffee and caffeine intake with the risk of Parkinson disease. JAMA 283:2674–2679

Rudolphi KA, Schubert P, Parkinson FE et al (1992) Adenosine and brain ischemia. Cerebrovasc Brain Metab Rev 4:346–369

Saura J, Angulo E, Ejarque A et al (2005) Adenosine A2A receptor stimulation potentiates nitric oxide release by activated microglia. J Neurochem 95:919–929

Scatena A, Fornai F, Trincavelli ML et al (2011) 3-(Fur-2-yl)-10-(2-phenylethyl)-[1,2,4] triazino[4,3-a]benzimidazol-4(10H)-one, a novel adenosine receptor antagonist with A(2A)-mediated neuroprotective effects. ACS Chem Neurosci 2:526–535

Schwarzschild MA, Xu K, Oztas E et al (2003) Neuroprotection by caffeine and more specific A2A receptor antagonists in animal models of Parkinson's disease. Neurology 61:S55–S61

Schwarzschild MA, Agnati L, Fuxe K et al (2006) Targeting adenosine A2A receptors in Parkinson's disease. Trends Neurosci 29:647–654

Schwarzschild MA, Schwid SR, Marek K et al (2008) Serum urate as a predictor of clinical and radiographic progression in Parkinson disease. Arch Neurol 65:716–723

Schwarzschild MA, Macklin EA, Ascherio A (2014) Urate and neuroprotection trials. Lancet Neurol 13:758

Scott GS, Spitsin SV, Kean RB et al (2002) Therapeutic intervention in experimental allergic encephalomyelitis by administration of uric acid precursors. Proc Natl Acad Sci U S A 99:16303–16308

Scott GS, Cuzzocrea S, Genovese T et al (2005) Uric acid protects against secondary damage after spinal cord injury. Proc Natl Acad Sci U S A 102:3483–3488

Seida M, Wagner HG, Vass K et al (1988) Effect of aminophylline on postischemic edema and brain damage in cats. Stroke 19:1275–1282

Serra PA, Sciola L, Delogu MR et al (2002) The neurotoxin 1-methyl-4-phenyl-1,2,3,6-tetrahydropyridine induces apoptosis in mouse nigrostriatal glia. Relevance to nigral neuronal death and striatal neurochemical changes. J Biol Chem 277:34451–34461

Simon DK, Swearingen CJ, Hauser RA et al (2008) Caffeine and progression of Parkinson disease. Clin Neuropharmacol 31:189–196

Simon KC, Eberly S, Gao X et al (2014) Mendelian randomization of serum urate and parkinson disease progression. Ann Neurol 76:862–868

Simonin C, Duru C, Salleron J et al (2013) Association between caffeine intake and age at onset in Huntington's disease. Neurobiol Dis 58:179–182

So A, Thorens B (2010) Uric acid transport and disease. J Clin Invest 120:1791–1799

Sonsalla PK, Wong LY, Harris SL et al (2012) Delayed caffeine treatment prevents nigral dopamine neuron loss in a progressive rat model of Parkinson's disease. Exp Neurol 234:482–487

Spasojevic I, Stevic Z, Nikolic-Kokic A et al (2010) Different roles of radical scavengers—ascorbate and urate in the cerebrospinal fluid of amyotrophic lateral sclerosis patients. Redox Rep 15:81–86

Spitsin S, Hooper DC, Leist T et al (2001) Inactivation of peroxynitrite in multiple sclerosis patients after oral administration of inosine may suggest possible approaches to therapy of the disease. Mult Scler 7:313–319

Stevens B, Porta S, Haak LL et al (2002) Adenosine: a neuron-glial transmitter promoting myelination in the CNS in response to action potentials. Neuron 36:855–868

Stocchi F, Rascol O, Hauser R et al (2014) Phase-3 clinical trial of the adenosine 2a antagonist preladenant, given as monotherapy, in patients with Parkinson's disease. Neurology 82:S7.004

Stone TW, Ceruti S, Abbracchio MP (2009) Adenosine receptors and neurological disease: neuroprotection and neurodegeneration. Handb Exp Pharmacol 193:535–587

Sutherland GR, Peeling J, Lesiuk HJ et al (1991) The effects of caffeine on ischemic neuronal injury as determined by magnetic resonance imaging and histopathology. Neuroscience 42:171–182

Takeuchi H (2013) Roles of glial cells in neuroinflammation and neurodegeneration. Clin Exp Neuroimmunol 4:2–16

Tan EK, Tan C, Fook-Chong SM et al (2003) Dose-dependent protective effect of coffee, tea, and smoking in Parkinson's disease: a study in ethnic Chinese. J Neurol Sci 216:163–167

Toncev G (2006) Therapeutic value of serum uric acid levels increasing in the treatment of multiple sclerosis. Vojnosanit Pregl 63:879–882

Vargas MR, Johnson DA, Sirkis DW et al (2008) Nrf2 activation in astrocytes protects against neurodegeneration in mouse models of familial amyotrophic lateral sclerosis. J Neurosci 28:13574–13581

Vitart V, Rudan I, Hayward C et al (2008) SLC2A9 is a newly identified urate transporter influencing serum urate concentration, urate excretion and gout. Nat Genet 40:437–442

Von Lubitz DK, Lin RC, Melman N et al (1994) Chronic administration of selective adenosine A1 receptor agonist or antagonist in cerebral ischemia. Eur J Pharmacol 256:161–167

Watanabe S, Kang DH, Feng L et al (2002) Uric acid, hominoid evolution, and the pathogenesis of salt-sensitivity. Hypertension 40:355–360

Weisskopf MG, O'Reilly E, Chen H et al (2007) Plasma urate and risk of Parkinson's disease. Am J Epidemiol 166:561–567

Wills AM, Eberly S, Tennis M et al (2013) Caffeine consumption and risk of dyskinesia in CALM-PD. Mov Disord 28:380–383

Winquist A, Steenland K, Shankar A (2010) Higher serum uric acid associated with decreased Parkinson's disease prevalence in a large community-based survey. Mov Disord 25:932–936

Wostyn P, Van Dam D, Audenaert K et al (2011) Increased cerebrospinal fluid production as a possible mechanism underlying caffeine's protective effect against Alzheimer's disease. Int J Alzheimers Dis 2011:617420

Wu XW, Muzny DM, Lee CC et al (1992) Two independent mutational events in the loss of urate oxidase during hominoid evolution. J Mol Evol 34:78–84

Xiao D, Bastia E, Xu YH et al (2006) Forebrain adenosine A2A receptors contribute to L-3,4-dihydroxyphenylalanine-induced dyskinesia in hemiparkinsonian mice. J Neurosci 26:13548–13555

Xu K, Bastia E, Schwarzschild M (2005) Therapeutic potential of adenosine A(2A) receptor antagonists in Parkinson's disease. Pharmacol Ther 105:267–310

Yadav V, Bever C Jr, Bowen J et al (2014) Summary of evidence-based guideline: complementary and alternative medicine in multiple sclerosis: report of the guideline development subcommittee of the American Academy of Neurology. Neurology 82:1083–1092

Yamamoto N, Sawada H, Izumi Y et al (2007) Proteasome inhibition induces glutathione synthesis and protects cells from oxidative stress: relevance to Parkinson disease. J Biol Chem 282:4364–4372

Yao SQ, Li ZZ, Huang QY et al (2012) Genetic inactivation of the adenosine A(2A) receptor exacerbates brain damage in mice with experimental autoimmune encephalomyelitis. J Neurochem 123:100–112

Yeum KJ, Russell RM, Krinsky NI et al (2004) Biomarkers of antioxidant capacity in the hydrophilic and lipophilic compartments of human plasma. Arch Biochem Biophys 430:97–103

Yu ZF, Bruce-Keller AJ, Goodman Y et al (1998) Uric acid protects neurons against excitotoxic and metabolic insults in cell culture, and against focal ischemic brain injury in vivo. J Neurosci Res 53:613–625

Yu L, Shen HY, Coelho JE et al (2008) Adenosine A2A receptor antagonists exert motor and neuroprotective effects by distinct cellular mechanisms. Ann Neurol 63:338–346

Zeitlin R, Patel S, Burgess S et al (2011) Caffeine induces beneficial changes in PKA signaling and JNK and ERK activities in the striatum and cortex of Alzheimer's transgenic mice. Brain Res 1417:127–136

Zhang N, Shu HY, Huang T et al (2014) Nrf2 signaling contributes to the neuroprotective effects of urate against 6-OHDA toxicity. PLoS One 9:e100286

Zheng Z, Guo X, Wei Q et al (2014) Serum uric acid level is associated with the prevalence but not with survival of amyotrophic lateral sclerosis in a Chinese population. Metab Brain Dis 29:771–775

Zhu TG, Wang XX, Luo WF et al (2012) Protective effects of urate against 6-OHDA-induced cell injury in PC12 cells through antioxidant action. Neurosci Lett 506:175–179

Zoccolella S, Simone IL, Capozzo R et al (2011) An exploratory study of serum urate levels in patients with amyotrophic lateral sclerosis. J Neurol 258:238–243

Chapter 7
Adenosine A_{2A} Receptor Antagonists as Drugs for Symptomatic Control of Parkinson's Disease in Preclinical Studies

Annalisa Pinna

Abstract Parkinson's disease (PD) is primarily a neurological basal ganglia (BG)-related disorder caused by progressive degeneration of the nigrostriatal dopaminergic neurons, which results in the cardinal motor symptoms of PD, including bradykinesia (slow movement and difficulty in initiation movement), resting tremor, muscle tone rigidity, postural instability, and sensorimotor integration deficits. The gold standard of PD therapy is characterized by the dopamine precursor L-DOPA however, after several years, this therapy leads to neuropsychiatric and motor complications, including fluctuations in motor response and dyskinesias, which develop in the majority of patients. Consequently, one of the main targets of research in PD is to identify alternative therapeutic approaches to ameliorate PD symptoms without inducing motor complications. Among the non-dopaminergic strategies for PD, one of the most promising is represented by adenosine A_{2A} receptor antagonists, due to the colocalization of these receptors and dopamine D_2 receptors in the striatopallidal neurons of the BG, which provides the anatomical basis for the existence of a functional antagonistic interaction between these receptors. Thus, extensive preclinical studies have been performed to prove the effectiveness of adenosine A_{2A} receptor blockade in counteracting the cardinal motor symptoms of PD.

This chapter describes the effects of A_{2A} antagonists alone or in combination with L-DOPA against the cardinal motor symptoms of PD, using rodent and primate models of PD, and the main mechanisms responsible for these anti-parkinsonian effects. In addition, findings suggesting the potential utilization of A_{2A} antagonists, as adjunctive treatments to L-DOPA to reduce the L-DOPA induced *wearing-off* without modifying dyskinetic movements, have been reviewed.

Keywords Parkinson's disease · Rodent models · Non-human primate models · Catalepsy · Rigidity · Tremor · 6-Hydroxydopamine lesion · MPTP lesion · A_{2A} receptor antagonists · Adenosine

A. Pinna (✉)
Institute of Neuroscience, National Research Council of Italy (CNR), Cagliari, Italy
e-mail: apinna@unica.it

© Springer International Publishing Switzerland 2015
M. Morelli et al. (eds.), *The Adenosinergic System,* Current Topics in Neurotoxicity 10,
DOI 10.1007/978-3-319-20273-0_7

Parkinson's Disease

Parkinson's disease (PD) is the second most common chronic neurodegenerative disease, with a progressive course, affecting over 5 million individuals worldwide (Van Den Eeden et al. 2003). Age is the greatest risk factor for PD, with an average age of onset of approximately 55–65 years (Obeso et al. 2000; Van Den Eeden et al. 2003). The prevalence of PD is expected to rise dramatically over the next 20 years as the population ages (Dorsey et al. 2007).

Symptomatically, PD is characterized by debilitating motor impairment, including akinesia, bradykinesia, muscle rigidity, resting tremor, gait disorders, and postural instability (Marsden 1994; Obeso et al. 2000). Additionally, PD patients are affected by a variety of non-motor symptoms, including cognitive dysfunction, autonomic abnormalities, sleep disturbance, and depression (Chaudhuri et al. 2006).

Pathologically, PD is characterized by degeneration of the nigrostriatal dopaminergic system, which is responsible for many of the motor symptoms observed in the disease. The principal effect of dopaminergic neurodegeneration in the striatum or caudate-putamen (CPu) of parkinsonian patients leads to a disruption of processing in the basal ganglia (BG) circuitry, which is responsible for the integration of sensorimotor information that controls the planning and initiation of voluntary movement (Obeso et al. 2000) (Fig. 7.1). However, neuronal loss has also been observed in brain areas other than the BG, producing changes in neurotransmitters, such as noradrenaline, serotonin, glutamate, acetylcholine, and adenosine, which contribute to the symptomatology of PD (Jellinger 2002). Additionally, widespread Lewy body pathology is observed in both the central and peripheral nervous systems (Braak et al. 2003).

The prime cause of dopaminergic neurodegeneration in PD has not yet been identified, but a large amount of data suggest that, from an aetiological and pathogenetic perspective, it might depend on a combination of environmental and genetic factors, such as toxins, genetic susceptibility, and the aging process (Alves et al. 2008). In particular, several known factors causing PD pathogenesis are mitochondrial dysfunction, oxidative damage, anomalous protein aggregation, and neuroinflammation (Schapira 2006). These processes, once started, persist to cause dopaminergic neuron injury, and have a negative impact on the effectiveness of the current PD therapy.

Since the finding of nigrostriatal dopamine depletion in the BG of parkinsonian patients, the dopaminergic neurotransmitter system has been the main focus of pharmacological therapies for the cardinal features of PD. Dopamine replacement with the dopamine precursor L-DOPA (in combination with a peripheral decarboxylase inhibitor), remains the most efficacious treatment to counteract PD motor symptoms (Olanow et al. 2009). Although L-DOPA is of substantial benefit to antagonize the main motor symptoms in parkinsonian patients, its loses effectiveness over time; specifically, after several years of treatment, the duration of L-DOPA effect shortens (known as *wearing-off*), responses become less predictable (with rapid switching between time spent by patients in a state of mobility *on-time* and immobility *off-time*). Moreover, patients affected by these motor response swings

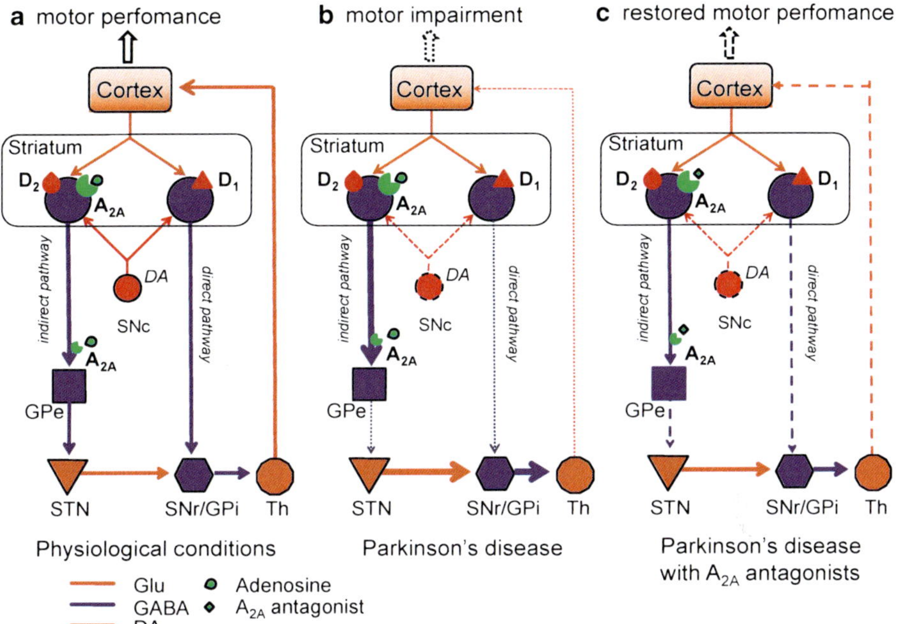

Fig. 7.1 Changes in the function of basal ganglia circuits during Parkinson's disease and effect of adenosine A$_{2A}$ receptor blockade. Under physiological conditions (**a**), the SNc sends dopaminergic inputs to striatal neurons. Endogenous DA then activates the neurons belonging to the striatonigral, or "direct", pathway. These neurons send GABAergic projections to the substantia nigra pars reticulata/globus pallidus pars interna *(SNr/GPi)*, and express stimulatory DA D$_1$ receptors. At the same time, endogenous DA inhibits the neurons belonging to the striato-pallidal, or "indirect", pathway. These neurons send GABAergic projections to the *SNr/GPi* via globus pallidus pars externa *(GPe)* and subthalamic nucleus *(STN)*, and express inhibitory DA D$_2$ receptors. The balanced activity of the two striatal efferent pathways underlies the correct execution of movement. Degeneration of SNc neurons during PD removes DA input to the striatum (**b**). This causes the disinhibition of the neurons in the striato-pallidal pathway, and boosts the inhibitory influence these neurons exert on the *GPe*, in turn leading to an overactivation of the *STN* glutamatergic output neurons. The reduction in striatal DA inputs from the SNc also causes a decreased activation of neurons in the striatonigral pathway. Taken together, these modifications in basal ganglia circuits result in an imbalanced activity of the striatal efferent pathways, and in an increased inhibitory output from the *SNr–GPi* complex to the thalamus *(Th)*. As a consequence, the excessive inhibition of thalamocortical neurons causes the motor deficits associated with PD. Adenosine A$_{2A}$ receptors are selectively located in the striato-pallidal pathway, both on striatal medium-sized spiny neurons and on their terminals projecting to GPe (a-b-c). Blockade of these receptors, by counteracting the function of D$_2$ receptors, alleviates the excessive inhibition of the striato-pallidal pathway, in turn restoring a certain degree of functional balance between the "direct" and "indirect" striatal efferent pathways, and favoring the performance of movement (c)

often show a range of types of choreic or dystonic drug-induced involuntary movements which, in themselves, could become a major source of disability (Olanow et al. 2004) Beside motor fluctuations, neuropsychiatric complications can develop (Obeso et al. 2000; Olanow et al. 2004). Even though, a few pharmacological and surgical strategies exist to ameliorate L-DOPA induced motor complications, they do not completely solve this problem (Horstink et al. 2006). As a result, the therapy of PD will remain an urgent healthcare issue, and requires an alternative approach to pharmacological intervention that can improve the symptomatology of parkinsonian patients, while, at the same time, offering a lower incidence of adverse effects.

Basal Ganglia Circuitry

The motor BG circuitry involved in the pathophysiology of movement disorders, consist of several subcortical structures, including the striatum or CPu, the globus pallidus (internal [GPi] and external [GPe] divisions), the substantia nigra (pars reticulata [SNr], pars compacta [SNc]), and the subthalamic nucleus (STN) (Delong and Wichmann 2007; Galvan and Wichmann 2008) (Fig. 7.1). All BG-related nuclei are connected through well-established neurochemical circuits, and with the specific cortical areas from which they originate. Briefly, the striatum, under tonic dopaminergic conditions, receives and integrates glutamatergic input from the thalamus and cerebral cortex, and this information is transmitted to the output nuclei, such as the SNr and GPi, which then provide BG projections to the thalamus and cerebral cortex (Delong and Wichmann 2007) (Fig. 7.1). Other BG output nuclei connect with the tegmental pedunculopontine nucleus as well as with the caudal intralaminar nuclei (Delong and Wichmann 2007). The neural population of the striatum is characterized by 95% of medium-sized spiny GABAergic neurons and by 5% of aspiny interneurons, including GABAergic and cholinergic interneurons. The striatal population of medium spiny GABAergic neurons is divided into two neuronal pathways: the monosynaptic "striatonigral direct projection" which connects the striatum with the SNr or GPi and the polysynaptic "striatopallidal indirect projection" which connects the striatum with the GP or GPe (Fig. 7.1). The striatonigral neurons mainly express dopamine D_1 receptors and the neuropeptides substance P and dynorphin, whereas the striatopallidal neurons express predominantly dopamine D_2 receptors and the neuropeptide enkephalin. Dopaminergic input to the striatum arises primarily from the mesencephalon, either from the SNc or the ventral tegmental area, and plays a critical modulatory role in neuronal signalling at this level, exerting a dual effect, depending on the type of post-synaptic dopaminergic receptor stimulated. Specifically, dopamine modulates motor coordination and fine movements by facilitating the direct pathway activity acting on the excitatory dopamine D_1 receptors and by inhibiting the indirect pathway function acting on inhibitory dopamine D_2 receptors (Gerfen and Bolam 2010). In PD, the nigrostriatal dopaminergic neurodegeneration causes dopamine depletion in the striatum, consequently reducing activation of both dopamine D_1 and D_2 receptors (Fig. 7.1). This lack of striatal dopamine generates an imbalance in the activity of striatal output pathways, characterized by reduced excitation of the striatonigral direct pathway, which leads to a decrease in inhibitory control of the GPi/SNr and a concomitant disinhibition of the striatopallidal indirect pathway to the STN and increases stimulation of the GPi/SNr neurons (Fig. 7.1). Taken together, this sequence of events exacerbates the activation of GABAergic BG output neurons, finally leading to excessive inhibition of thalamocortical projections of the motor systems, causing parkinsonian motor symptoms (Albin et al. 1989; Delong 1990; Obeso et al. 2000) (Fig. 7.1). Although, this proposed model of BG function and dysfunction provides an excellent starting point (Albin et al. 1989; Delong 1990), it is important to highlight that BG organization is far more sophisticated than supposed in this model (reviewed in Bar-Gad and Bergman 2001; Obeso et al. 2000). Thus, recent findings detailing neurotransmission throughout the BG networks should be taken into consideration for a more complete understanding of its organization and activity (Armentero et al. 2011; see also Chap. 2).

Adenosine A$_{2A}$ Receptor Antagonists

The discovery of the restricted expression of adenosine A$_{2A}$ receptors in the BG circuitry and of their close interaction with dopamine, especially with dopamine D$_2$ receptors, rendered adenosine A$_{2A}$ receptors very attractive as a non-dopaminergic target to be explored for PD therapy. Indeed, adenosine A$_{2A}$ receptors are localized in areas of the BG associated with the dopaminergic nigrostrial and mesolimbic neuronal pathways, including the striatum, GP, nucleus accumbens, and olfactory tubercle (Rosin et al. 1998). Specifically, in the striatum, adenosine A$_{2A}$ receptors are predominantly restricted on the dendritic spines of GABAergic striatopallidal neurons, where they are colocalized with dopamine D$_2$ receptors (Hettinger et al. 2001), whereas striatonigral neurons do not contain appreciable levels of adenosine A$_{2A}$ receptors (Hettinger et al. 2001) . This colocalization of adenosine A$_{2A}$ and dopamine D$_2$ receptors in the striatopallidal neurons leads to a functional antagonistic interaction between these receptors (Ferré et al. 1997; Hettinger et al. 2001; Svenningsson et al. 1999). Specifically, stimulation of the dopamine D$_2$ receptors by dopamine or dopamine D$_2$ receptor agonists enhances motor activity, whereas stimulation of the adenosine A$_{2A}$ receptors reduces this effect (Ferré et al. 1997). At the biochemical level, this antagonistic functional interaction between adenosine A$_{2A}$ and dopamine D$_2$ receptors takes place both directly, with an intramembrane receptor–receptor interaction, in which the activation of adenosine A$_{2A}$ receptors decreases the binding affinity of D$_2$ receptors for dopamine (Ferré et al. 1991) and at the level of second messengers, such as adenylyl cyclase, in which stimulation of adenosine A$_{2A}$ receptors counteracts the dopamine D$_2$ receptor-mediated inhibition of 3′,5′-cyclic adenosine monophosphate (cAMP) formation and D$_2$ receptor-induced intracellular Ca^{2+} responses (Hillion et al. 2002; Olah and Stiles 2000; for more details please see also Chaps. 1 and 2).

Modulation of Adenosine A_{2A} Receptors Located in the BG Circuitry

To better understand the anti-parkinsonian efficacy of A_{2A} receptor antagonists on the cardinal symptoms of PD, it is necessary to illustrate the main role played by adenosine A_{2A} receptors in the motor BG circuitry involved in the pathophysiology of movement disorders (Fig. 7.1).

As described above, the colocalization of adenosine A_{2A} and dopamine D_2 receptors in the striatopallidal neurons provides the anatomical basis for the existence of a functional antagonistic interaction between these receptors (Ferré et al. 1997) (Fig. 7.1). Adenosine A_{2A} receptor blockade leads to motor activity by reducing the excessive inhibitory output of the BG indirect pathway, similar to dopamine D_2 receptor activation (Ferré et al. 1997) (Fig. 7.1). In addition, activation or blockade of the adenosine A_{2A} receptors in the indirect striatopallidal pathway, impairs or facilitates dopaminergic D_1-mediated responses as well (Ferré et al. 1997; Pinna et al. 1996; Pollack and Fink 1996). Thus, A_{2A} receptor antagonists seem to restore some balance between the striatonigral and striatopallidal neurons, consequently relieving thalamocortical activity (Fig. 7.1). Moreover, an important function of adenosine A_{2A} receptors has been showed in the GP (Fig. 7.1). Indeed, in PD, the blockade of pallidal adenosine A_{2A} receptors, by reducing extracellular GABA, may contribute to restoring GP activity, and, in turn, STN activity, leading to a more balanced activation of the direct and indirect pathways and, when associated with dopaminergic receptor agonists, an enhancement of their motor-stimulating effects (Ochi et al. 2004; Shindou et al. 2003; Simola et al. 2004, 2008). Furthermore, stimulation of the postsynaptic adenosine A_{2A} receptors antagonizes the inhibitory modulation of the N-methyl-D-aspartate (NMDA) receptor activity mediated by dopamine D_2 receptors (Azdad et al. 2009; Higley and Sabatini 2010). This interaction appears to be responsible for most of the locomotor activation and depression induced by A_{2A} receptor antagonists and agonists, respectively (Ferré et al. 2008). Additionally, further contribution to the anti-parkinsonian effects, in particular the anti-tremorigenic effect, of adenosine A_{2A} receptor antagonists may be related to a cholinergic mechanism (Armentero et al. 2011; Kurokawa et al. 1996; Simola et al. 2004). Indeed, functional antagonism between the adenosine A_{2A} and dopamine D_2 receptors was recently reported in striatal cholinergic interneurons (Tozzi et al. 2011). Moreover, adenosine A_{2A} receptors have been shown to interact either directly or indirectly with various receptors, such as the dopamine D_3, NMDA, cannabinoid CB_1, serotonin 5-HT$_{1A}$, metabotropic glutamate 4 (mGlu4) and 5 (mGlu5), receptors and to form heteromeric complexes with some of them, suggesting a more complex explanation of their influence on PD motor deficits (Armentero et al. 2011; Bogenpohl et al. 2012; Gerevich et al. 2002; Jones et al. 2012; Łukasiewicz et al. 2007) (see more details in Chap. 2).

The functionally opposing roles of the adenosine A_{2A} and dopamine D_2 receptors on the indirect pathway neurons offers a rationale for the extensive investigation of the activity of A_{2A} receptor antagonists on counteracting motor deficits in pharmacological and toxicological animal models of PD.

The following sections of this chapter illustrate, in detail, the modulatory role played by A_{2A} receptor antagonists on each cardinal PD motor symptom, such as akinesia/bradykinesia, gait impairments, sensorimotor integration deficit, muscle rigidity, and tremor, demonstrated in rodent and primate models of PD.

Effect of A_{2A} Receptor Antagonists on Akinesia, Bradykinesia, and Motor Activity

Akinesia strictly means absence of movement, but in PD, it usually refers to slowness of movement execution (bradykinesia) or lack of spontaneous movements (hypokinesia) (Obeso et al. 2000). In PD, there is a decrease in the amplitude and rate of movements. Bradykinesia may significantly impair the quality of life of PD patients, because it takes much longer to perform everyday tasks, such as eating or dressing. Automatic movements, such as step length or arm swings when walking, or more complex voluntary movements, such as writing or drinking, can all be involved. The effects of A_{2A} receptor antagonists against the symptomatic parkinsonian akinesia, bradykinesia, and motor activity impairment have been demonstrated using a wide range of pharmacological and/or toxicological rodent and primate models of PD, including counteraction of hypomotility or catalepsy induced by haloperidol or reserpine, and modulation of rotational behaviour in rodents, as well as a reduction of motor impairment in non-human primates (Table 7.1 and Fig. 7.2) (Pinna and Morelli 2014; Simola et al. 2008; Xu et al. 2005). In rodents, the monoamine-depleting agent reserpine or the typical neuroleptic haloperidol induces a dramatic reduction of motor activity, principally characterized by akinesia, hypokinesia and catalepsy, which are representative of parkinsonian symptoms and caused by hypofunctionality of the striatum (Duty and Jenner 2011; Gerlach and Riederer 1996). Moreover, rodents administered a range of different doses of reserpine or haloperidol showed other parkinsonian-like symptoms, such as hindlimb rigidity and tremor (as described in the following sections of this chapter) (Duty and Jenner 2011; Gerlach and Riederer 1996; Lorenc-Koci et al. 1996; Salamone et al. 2008). Drugs commonly used in PD treatment are known to counteract the catalepsy induced by haloperidol or reserpine (for review see Duty and Jenner 2011). Moreover, specifically, the catalepsy test induced by haloperidol is useful to underline the pharmacokinetic differences of the compounds tested (Gillespie et al. 2009; Neustadt et al. 2007; Pinna et al. 2005; Weiss et al. 2003).

The majority of A_{2A} receptor antagonists were able to counteract, in a dose-dependent manner, catalepsy and/or hypolocomotion induced by haloperidol or reserpine in rodents, reducing their duration and severity, thereby demonstrating an improvement of parkinsonian motor impairment by these drugs (Table 7.1 and Fig. 7.2) (Drabczyńska et al. 2011; Gillespie et al. 2009; Hodgson et al. 2009; Jones et al. 2013; Kanda et al. 1994; Mandhane et al. 1997; Pinna et al. 2005; Shiozaki et al. 1999; Shook et al. 2010, 2013; Stasi et al. 2006; Villanueva-Toledo 2003; Wardas et al. 2003; Weiss et al. 2003). Furthermore, the co-administration of several

Table 7.1 Summary of the effects exerted by A_{2A} antagonists on cardinal parkinsonian-like symptoms in rodent and primate models of PD

Parkinsonian-like symptoms	Effects of A_{2A} antagonists in rodent models of PD
Parkinsonian-like akinesia, bradykinesia and motor impairment	Reversal of hypolocomotion and catalepsy induced by haloperidol or reserpine (alone or in combination with L-DOPA) [1–14]
	Potentiation of contralateral rotational behaviour induced by DAergic antiparkinsonian drugs in the hemiparkinsonian rats [7, 15–24]
	Restoration of the impaired adjusting steps and initiation time of stepping of the forelimb contralateral to the lesion [19, 25, 26]
	Restoration the lost functionality of hindlimb bradykinesia and in rotarod test of MitoPark mice [27]
Parkinsonian-like sensorimotor integration deficit	Restoration of the placement of the contralateral forelimb after vibrissae brushing lesion [19, 25, 26]
Parkinsonian-like muscle rigidity	Amelioration of parkinsonian-like muscle rigidity produced by either reserpine or haloperidol (alone or in combination with L-DOPA) [14, 28]
Parkinsonian-like tremor	Counteraction of parkinsonian-like tremor elicited by several tremorigenic agents in rodents (for review see Chap. 13 by Salamone)
Parkinsonian-like symptoms	Effects of A_{2A} antagonists in primate models of PD
Parkinsonian-like akinesia, bradykinesia and motor impairment	Reversal of catalepsy induced by haloperidol [29]
	Relieve of motor impairment in MPTP-treated primates (alone or in combination with DAergic antiparkinsonian drugs) [30–34]
Parkinsonian-like muscle rigidity	Amelioration of parkinsonian-like muscle rigidity produced by haloperidol [29]

[1] Drabczyńska et al. 2011; [2] Gillespie et al. 2009; [3] Hodgson et al. 2009; [4] Kanda et al. 1994; [5] Jones et al. 2013; [6] Mandhane et al. 1997; [7] Pinna et al. 2005; [8] Shiozaki et al. 1999; [9] Shook et al. 2010; [10] Shook et al. 2013; [11] Stasi et al. 2006; [12] Villanueva-Toledo 2003; [13] Weiss et al. 2003; [14] Wardas et al. 2003; [15] Fenu et al. 1997; [16] Hodgson et al. 2009; [17] Koga et al. 2000; [18] Pinna et al. 1996; [19] Pinna et al. 2010; [20] Pollack and Fink 1996; [21] Rose et al. 2007; [22] Tronci et al. 2007; [23] Vellucci et al. 1993; [24] Weiss et al. 2003; [25] Pinna et al. 2007; [26] Pinna and Morelli 2014; [27] Smith et al. 2014; [28] Wardas et al. 2001; [29] Varty et al. 2008; [30] Grondin et al. 1999; [31] Hodgson et al. 2010; [32] Kanda et al. 1998; [33] Kanda et al. 2000; [34] Rose et al. 2006

A_{2A} receptor antagonists with L-DOPA has been demonstrated to strengthen the anti-cataleptic effect induced by L-DOPA suggesting that there may be a synergism between the adenosine A_{2A} receptor antagonists and the dopaminergic agents (Table 7.1; Kanda et al. 1994; Shiozaki et al. 1999; Stasi et al. 2006) . Interestingly, Varty and collaborators have also evaluated the efficacy of A_{2A} receptor antagonists against catalepsy induced by haloperidol in primates (Table 7.1 and Fig. 7.2; Varty et al. 2008). The catalepsy induced by haloperidol in primates is characterized by immobility with open eyes, usually accompanied by unusual postures, including rigid limb extensions and/or a twisted torso. Consistent with rodent studies, adenosine A_{2A} receptor blockade can attenuate haloperidol-induced cataleptic motor impairment in monkeys (Table 7.1 and Fig. 7.2; Varty et al. 2008).

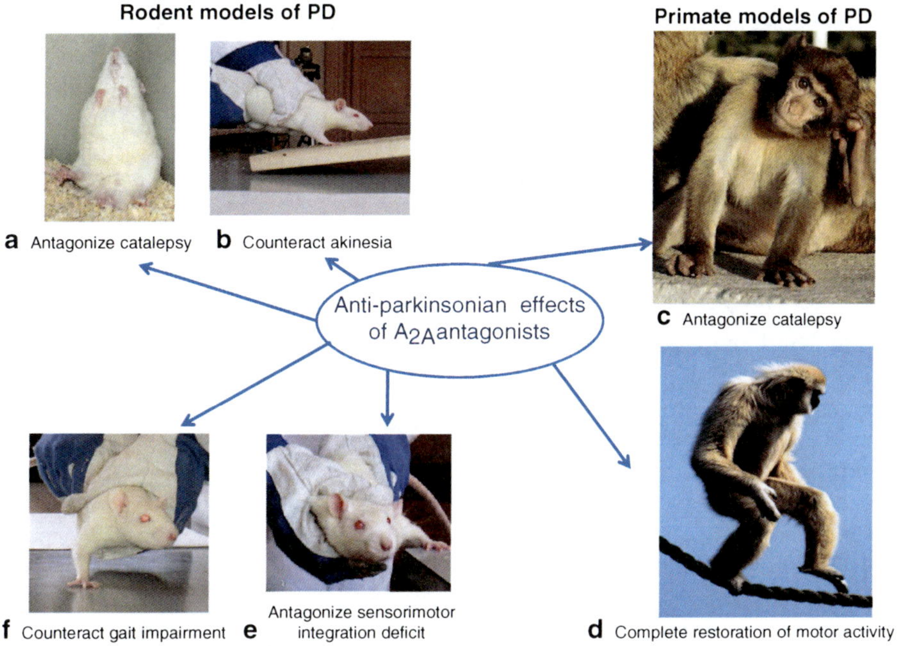

Fig. 7.2 Illustration of diverse rodent and primate models utilized for behavioural evaluation of A_{2A} receptor antagonists. Clockwise shows test performed in rodents *(left panel)* and primates *(right panel)* for **a** catalepsy in rodent, **b** akinesia in rodent, **c** catalepsy in primate, **d** akinesia in primate, **e** sensorimotor integration deficit in rodent, and **f** gait impairment in rodent

To better verify the anti-parkinsonian effects of A_{2A} receptor antagonists, these compounds have been evaluated in the most frequently used PD model of hemiparkinsonian rats, characterized by a unilateral intracerebral infusion of the dopaminergic neurotoxin 6-hydroxydopamine (6-OHDA), which produces massive degeneration of the nigrostriatal dopaminergic neurons, similar to that occurring in idiopathic PD (Schwarting and Huston 1996; Simola et al. 2007; Ungerstedt 1968). In this model, the ability of a specific drug to induce contralateral rotational behaviour, as well as to potentiate the rotational behaviour stimulated by dopamine receptor agonists, can be assumed as a parameter reflecting its anti-parkinsonian activity (Schwarting and Huston 1996; Simola et al. 2007).

A_{2A} receptor antagonists clearly showed a motor facilitative activity in hemiparkinsonian rats. Specifically, acute administration of several adenosine A_{2A} receptor antagonists induced no contralateral rotations *per se*, but significantly potentiated rotational behaviour induced by L-DOPA or apomorphine and by either dopamine D_1 or D_2 receptor agonists, in hemiparkinsonian rodents (Table 7.1; Fenu et al. 1997; Hodgson et al. 2009; Koga et al. 2000; Pinna et al. 1996, 2005, 2010; Pollack and Fink 1996; Rose et al. 2007; Tronci et al. 2007; Vellucci et al. 1993; Weiss et al. 2003).

Furthermore, in hemiparkinsonian rats, more sophisticated measurements of akinesia, bradykinesia, and gait impairment have been assessed. Indeed, as a consequence of unilateral 6-OHDA lesion, rats develop gait impairment and forelimb akinesia considered to be analogous to PD symptoms in humans. Different strategies, such as adjusting step counting and initiation time of stepping have been developed in order to evaluate and quantify these symptoms and their relief by drugs (Chang et al. 1999; Meredith and Kang 2006; Olsson et al. 1995).

A few weeks after unilateral lesioning of the nigrostriatal pathway in 6-OHDA in rats, the motor performance of the forelimb contralateral to the lesion is significantly and progressively impaired compared with the motor performance of the same forelimb before the lesion. Indeed, hemiparkinsonian rats made less steps with the forelimb contralateral to the lesion, compared with their ipsilateral forelimb, showing a marked reduction of movements defined as hypokinesia (Chang et al. 1999; Olsson et al. 1995; Pinna et al. 2007, 2010). Moreover, hemiparkinsonian rats show marked and long-lasting impairment in the initiation time of stepping movement of the contralateral to the lesioned side, an impairment considered to be of symptomatic validity for the initiation of movement deficit present in parkinsonian patients (Meredith and Kang 2006; Olsson et al. 1995; Pinna et al. 2007, 2010). Both deficits described were effectively counteracted by a dose of L-DOPA at sub-threshold levels for induction of rotation. Administration of the A_{2A} receptor antagonists, similarly to L-DOPA significantly counteracted forelimb akinesia/hypokinesia and motor initiation deficit, as demonstrated by their effect in increasing the number of steps performed in both a forward and backward direction and in improving initiation time of stepping by the forelimb contralateral to the lesion, with different intensity, depending on the A_{2A} receptor antagonists tested (Table 7.1 and Fig. 7.2; Pinna et al. 2007, 2010; Pinna and Morelli 2014). Notably, hemiparkinsonian rats did not show any spontaneous recovery in the adjusting test and in initiation time in the stepping test during the period in which the drug test was performed (Pinna et al. 2007, 2010).

It is important to underline that even though A_{2A} receptor antagonists do not *per se* induce contralateral rotations in drug-naïve hemiparkinsonian rats, but only potentiate contralateral rotation induced by L-DOPA (Fenu et al. 1997; Koga et al. 2000), they appear, as shown by the above-mentioned results, to be effective in counteracting specific motor deficits associated with dopamine neuron degeneration, such as akinesia/hypokinesia and initiation of movement deficits.

Recently, the anti-akinetic/bradykinetic effects of A_{2A} receptor antagonists have been evaluated in a genetic mouse model of PD that displays a progressive loss of dopamine neurons, such as in the MitoPark mouse (Table 7.1; Smith et al. 2014). The dopamine cell loss in these mice is associated with a deep akinetic phenotype that is sensitive to L-DOPA (Smith et al. 2014). In this PD genetic mouse model, blockade of adenosine A_{2A} receptors increased locomotor activity in a dose-dependent way, completely restored the lost functionality in a measure of hindlimb bradykinesia, and partially restored functionality in a rotarod test, confirming the efficacy of A_{2A} receptor antagonists against these motor deficits (Table 7.1; Smith et al. 2014).

The anti-parkinsonian activity of A$_{2A}$ receptor antagonists against bradykinesia, akinesia, and motor disability shown in the rodent model of PD have been confirmed in a neurotoxic primate model of PD (Table 7.1 and Fig. 7.2).

Primates treated with the neurotoxin 1-methyl-4-phenyl-1,2,3,6-tetrahydropyridine (MPTP) is the model of PD, which closely mimics the clinical features of PD in humans, and in which all currently used anti-parkinsonian medications have been shown to be effective; thus, this model is undoubtedly the most clinically relevant of all the available models (Duty and Jenner 2011). Indeed, the MPTP intoxication induced a parkinsonian syndrome, characterized by all of the cardinal symptoms of PD and similar anatomical and functional characteristics of dopaminergic neurodegeneration observed in idiopathic PD (Duty and Jenner 2011). Moreover, the MPTP-treated primate develops clear dyskinesia when repeatedly exposed to L-DOPA and these parkinsonian animals have shown responses to novel dopaminergic agents that are highly predictive of their effect in humans.

Acute administration of A$_{2A}$ receptor antagonists increased locomotor activity and reversed motor disability in a dose-dependent manner in primates previously rendered parkinsonian with MPTP (Table 7.1 and Fig. 7.2; Grondin et al. 1999; Hodgson et al. 2010; Kanda et al. 1998, 2000; Rose et al. 2006). Furthermore, when co-administered with L-DOPA A$_{2A}$ receptor antagonists enhanced the intensity and duration of the efficacy of L-DOPA in reversing motor disabilities and increasing locomotor activity in parkinsonian monkeys (Table 7.1; Hodgson et al. 2010; Kanda et al. 2000; Rose et al. 2006). Similar results have been obtained with the combined administration of A$_{2A}$ receptor antagonists with dopamine D$_1$ and D$_2$ receptor agonists (Table 7.1; Kanda et al. 2000). Interestingly, despite producing an enhanced anti-parkinsonian response, acute A$_{2A}$ receptor antagonists did not exacerbate the dyskinesia induced by L-DOPA in MPTP-treated primates previously rendered dyskinetic by L-DOPA exposure (Grondin et al. 1999; Hodgson et al. 2010; Kanda et al. 1998).

Effect of A$_{2A}$ Receptor Antagonists on Sensorimotor Integration Deficit

Similar to parkinsonian patients, hemiparkinsonian rats showed marked sensorimotor integration deficits correlated with a unilateral lesion of the dopaminergic nigrostriatal pathway (Schallert et al. 2000). These sensorimotor deficits, assessed by means of the vibrissae-elicited forelimb placing test, hampered the hemiparkinsonian rats when placing their forelimb contralateral to the lesion on the table surface after brushing of the vibrissae on the same side, whereas the ipsilateral forelimb was not affected by the lesion (Meredith and Kang 2006; Pinna et al. 2007, 2010; Schallert et al. 2000). A few A$_{2A}$ receptor antagonists, similarly to L-DOPA completely restored placement of the contralateral forelimb by rats, with different intensity depending on the different A$_{2A}$ receptor antagonists tested, suggesting a potential efficacy of these compounds to ameliorate the sensorimotor integration deficits in PD patients (Table 7.1 and Fig. 7.2; Pinna and Morelli 2014; Pinna et al. 2007, 2010). This effect was not due to spontaneous recovery of sensorimotor integration deficits by hemiparkinsonian rats (Pinna and Morelli 2014; Pinna et al. 2007, 2010).

Effect of A_{2A} Receptor Antagonist on Muscle Rigidity

The other cardinal symptom of PD, as frequently disabling for patients as brady-kinesia and akinesia, is muscle rigidity, which is clinically defined as a sustained increase in resistance to passive movement of a joint throughout its range (Del-waide 2001). The most common clinical characteristic of rigidity is an increased resistance to passive movement of the PD patient's limbs, usually associated with a cogwheel phenomenon, and which could be reproduced in rodents by administra-tion of adequate doses of haloperidol, reserpine, or bilateral 6-OHDA into the SN (Lorenc-Koci et al. 1995, 1996). Both haloperidol and reserpine evoke a muscle ri-gidity with a lot of electromyographic (EMG) and mechanographic (MMG, muscle resistance) peculiarities similar to those observed in PD patients (Lorenc-Koci et al. 1995, 1996; Wolfarth et al. 1996). Specifically, such rigidity develops in response to passive movements and is characterized by increased resistance of rodent hindlimbs to passive displacement, potentiation of EMG components, and co-activation of an-tagonistic muscles in response to passive movements. Moreover, as in parkinsonian patients, a tonic EMG activity develops at rest, which reflects some difficulty in relaxing the muscles (Lee 1989).

This combined EMG and MMG method to measure haloperidol or reserpine-induced muscular rigidity has been validated by the fact that muscle rigidity can be reduced by anti-parkinsonian dopaminomimetic agents, including L-DOPA (War-das et al. 2001).

Although a generic effect of A_{2A} receptor antagonists on counteracting postural rigidity, one of the components of catalepsy induced by haloperidol or reserpine, has been shown in the above section of this chapter, a more precise evaluation of the anti-parkinsonian-like muscular rigidity exerted by these compounds has been made by means of this combined EMG and MMG method (Table 7.1; Wardas 2003; Wardas et al. 2001).

Blockade of adenosine A_{2A} receptors counteracted both components (EMG and MMG) of muscle rigidity induced by haloperidol or reserpine plus alpha-methyl-p-tyrosine in rodents (Table 7.1; Wardas 2003; Wardas et al. 2001). Furthermore, the blockade of adenosine A_{2A} receptors potentiated the alleviating effect of a low dose of L-DOPA which alone did not affect the reserpine- or haloperidol-induced muscular rigidity (Table 7.1; Wardas 2003; Wardas et al. 2001). These beneficial ef-fects on parkinsonian-like muscular rigidity of A_{2A} receptor antagonists have been suggested to be mediated by the facilitation of dopamine transmission at the post-synaptic level, as described above (Wardas 2003; Wardas et al. 2001). Moreover, although, the study by Varty et al. (2008) did not perform a specific measure of haloperidol-induced muscle rigidity by means of suitable equipment, counteraction of this symptom was observed in primates after combined administration of A$_{2A}$ receptor antagonists (Table 7.1 and Fig. 7.2; Varty et al. 2008). These findings regarding the effectiveness of A$_{2A}$ receptor antagonists on muscle rigidity in rodent and primate PD models indicate that these drugs might be particularly effective in counteracting parkinsonian-like muscle rigidity in PD patients, which is often resistant to common anti-parkinsonian drugs.

Effect of A$_{2A}$ Receptor Antagonist on Tremor Model of PD

Another important anti-parkinsonian effect exerted by A$_{2A}$ receptor antagonists is the anti-tremorigenic effect (Table 7.1). Indeed, tremor is one of the cardinal symptoms of parkinsonism, which is experienced by more than 70 % of PD patients (Deuschl et al. 2012). Tremor in PD is typically resting and disappears when voluntary movement is performed. The distal joints of the limbs are preferentially affected. Tremor can be intermittent and is increased by stress (Deuschl et al. 2012). In addition, resting tremor has been shown to respond poorly to traditional anti-parkinsonian medications, including L-DOPA (Jiménez and Vingerhoets 2012). Therefore, research addresses improving the management of this disturbance. To date, experimental models of parkinsonian tremor characterized by tremulous jaw movements (TJMs) induced by several drugs, such as acetylcholinesterase inhibitors, muscarinic agonists, DA receptor antagonists, and neurotoxic degeneration of DA neurons, have been validated for evaluating the anti-tremorigenic effects of drugs (Cousins et al. 1997; Ishiwari et al. 2005; Salamone et al. 1998; Simola et al. 2004; for review see Chap. 8). These tremorigenic compounds produced TJMs which possess many of the pharmacological and electromyographic characteristics of the parkinsonian tremor in humans (for review see Collins-Praino et al. 2011; Chap. 8). Acute administration of several A$_{2A}$ receptor antagonists significantly reversed jaw tremor induced by tacrine, pilocarpine, haloperidol, reserpine, and pimozide in rats, suggesting a beneficial use of these compounds as specific drugs against this parkinsonian symptom (Table 7.1; Betz et al. 2009; Collins et al. 2010, 2012; Collins-Praino et al. 2011; Correa et al. 2004; Pinna et al. 2010; Salamone et al. 2008; Simola et al. 2004, 2006; Tronci et al. 2007). Consistent with these findings, A$_{2A}$ receptor antagonism or genetic deletion of the adenosine A$_{2A}$ receptor significantly attenuated the TJMs induced by pilocarpine in mice (Table 7.1; Salamone et al. 2013). The anti-tremorigenic effect of A$_{2A}$ receptor antagonists appears to be focused particularly on the ventrolateral portion of the striatum, in which a specific increase in adenosine A$_{2A}$ receptor mRNA expression was detected following dopamine denervation in hemiparkinsonian rats (Pinna et al. 2002; Simola et al. 2004). Considering the important role played by an increase in striatal acetylcholine in tremor development, and the reduction of the evoked release of this neurotransmitter exerted by A$_{2A}$ receptor antagonists, it might be suggested that modulation of this anticholinergic effect by blockade of the adenosine A$_{2A}$ receptors may explain its anti-tremorigenic effect (Simola et al. 2004, 2006). Recently, it has been suggested that A_{2A} receptor antagonists might also be used to modulate the anti-tremorigenic effect of STN deep brain stimulation in PD patients (for details see Chap. 8) (Collins-Praino et al. 2013). Taken together, these data concerning the efficacy of A_{2A} receptor antagonists achieved in rodent models of parkinsonian tremor show that A_{2A} receptor antagonism might be useful to attenuate parkinsonian-like resting tremor, a symptom hardly managed (for details see Chap. 8).

Persistent Efficacy of A_{2A} Receptor Antagonists on Cardinal Symptoms of PD

Specific studies have thus been performed to verify whether the symptomatic anti-parkinsonian acute effects of A_{2A} receptor antagonists persist over time during repeated treatment in animal models of PD, as required by their utilization in a chronic pathology, such as PD. Indeed, discouraging results have been provided by the non-specific adenosine receptor antagonist caffeine, which loses its motor-stimulant effect during chronic exposure (Fredholm et al. 1999; Halldner et al. 2000). In contrast to caffeine, chronic administration of A_{2A} receptor antagonists has been demonstrated to effectively improve motor deficits in rodent and primate models of PD, and not to produce tolerance to their motor-stimulant effects (Kanda et al. 1998; Koga et al. 2000; Pinna et al. 2001). Interestingly, repeated treatment with A_{2A} receptor antagonists for 1 and 2 weeks produced not merely tolerance, but also led to an enhancement of the intensity of the L-DOPA induced rotational behaviour compared with that observed after acute administration of A_{2A} receptor antagonists in hemiparkinsonian rats (Pinna et al. 2001). Similarly, combined administration of A_{2A} receptor antagonists with apomorphine produced a specific increase in duration rather than in intensity of rotational behaviour in hemiparkinsonian rats (Koga et al. 2000). Moreover, the long-lasting efficacy of A_{2A} receptor antagonists in preventing the reduction of spontaneous locomotor activity has recently been demonstrated in both early (8 weeks of age) and mild to severe (12–22 weeks of age) parkinsonian genetic MitoPark mice (Marcellino et al. 2010; Smith et al. 2014).

Effects of A_{2A} Receptor Antagonists on L-DOPA Induced Motor Fluctuations Like Wearing-off and on–off Phenomena

The main limitation of long-term therapy with L-DOPA in PD patients is characterized by motor fluctuations consistent with the progressive reduction of the drug's efficacy in preventing parkinsonian motor symptoms, usually known as *"wearing-off"* and *"on–off"* phenomena (Olanow et al. 2004). During *wearing-off*, L-DOPA counteracts PD motor deficits for a shorter period of time, after which akinesia and rigidity become manifest again. In the *on–off* phenomenon, the patient fluctuates from "*on*" periods in which the parkinsonian impairments are counteracted, to "*off*" periods in which the patient shows bradykinesia and rigidity. In hemiparkinsonian rats, the duration of rotational behaviour induced by L-DOPA progressively decreases during the long-term treatment with this drug, a phenomenon that mimics the *wearing-off* of L-DOPA observed in parkinsonian patients (Marin et al. 2005; Oh and Chase 2002). Consistent with the acute effect of A$_{2A}$ receptor antagonists producing an increased duration of rotational behaviour induced by L-DOPA or apomorphine (Koga et al. 2000; Pinna and Morelli 2014), the co-administration of the A$_{2A}$ receptor antagonists with L-DOPA reversed the shortening of rotational behaviour, supporting a potential beneficial influence of adenosine A$_{2A}$ receptor blockade on L-DOPA induced *wearing-off* (Bibbiani et al. 2003; Bové et al. 2002; 2006; Koga et al. 2000; Pinna and Morelli 2014). However, when the A$_{2A}$ receptor antagonist 8-(3-chlorostyryl)caffeine was chronically administered in combination with L-DOPA it seems to reverse, but not to prevent, the shortening response of rotational behaviour induced by repeated treatment with L-DOPA (Bové et al. 2002) . Despite this controversial single study, numerous clinical trials in advanced PD patients have demonstrated the efficacy of A$_{2A}$ receptor antagonists in reducing the *wearing off* phenomenon and in increasing the *on period* (for review see Chap. 14), providing effort to approve the commercialization of the A$_{2A}$ receptor antagonist istradefylline as a drug to counteract *wearing-off* in PD patients (for details see Chap. 13).

Effects of A$_{2A}$ Receptor Antagonists on L-DOPA Induced Dyskinesia

Chronic therapy with L-DOPA is associated with the development of dyskinesia, characterized by abnormal involuntary movements (AIMs), such as dystonia (a painful, involuntary spasm of muscles in various parts of the body) and chorea (brief semi-directed, irregular movements that are not repetitive or rhythmic, but appear to flow from one muscle to the next), which are highly disabling for parkinsonian patients (Olanow et al. 2004). As described extensively in the Chap. 9 by Morelli, the influence of adenosine A$_{2A}$ receptor blockade on dyskinesia has been investigated by means of validated experimental paradigms in which dyskinetic movements induced by chronic L-DOPA are expressed both in hemiparkinsonian rodents (sensitization of rotational behaviour and/or AIMs affecting parts of the body contralateral to the lesion) (Henry et al. 1998; Lundblad et al. 2003; Pinna et al. 2001; Tronci et al. 2007), and in parkinsonian MPTP-treated primates (dyskinetic movements affecting several parts of the body, similar to those observed in parkinsonian patients) (Bibbiani et al. 2005). Summarizing the main findings concerning dyskinesia, it has been demonstrated that A$_{2A}$ receptor antagonists, when administered alone, did not induce dyskinesia in both rodents and primates previously rendered dyskinetic by chronic L-DOPA (Grondin et al. 1999; Hodgson et al. 2010; Jones et al. 2013; Kanda et al. 1998; Lundblad et al. 2002). Moreover, in hemiparkinsonian rats, long-term treatment with a combination of an A_{2A} receptor antagonist and low doses of L-DOPA induced a stable response in both rotational behaviour and AIMs, suggesting that this association between the two drugs represents a treatment with a low dyskinetic potential (Hodgson et al. 2009; Pinna et al. 2001; Tronci et al. 2007). Conversely, blockade of the adenosine A_{2A} receptors did not produce any effect on the severity of the AIMs induced by repeated L-DOPA at full dose, when the two drugs were chronically co-administered in hemiparkinsonian rats (Jones et al. 2013; Lundblad et al. 2003). Interestingly, this hypothesis has been supported by studies showing that genetic deletion of the adenosine A_{2A} receptor prevents the sensitization of rotational behaviour and AIMs stimulated by L-DOPA in hemiparkinsonian mice (Fredduzzi et al. 2002; Xiao et al. 2006). Findings in dyskinetic parkinsonian primates confirmed that A_{2A} receptor antagonists associated with a low non-dyskinetic dose of L-DOPA may ameliorate satisfactory motor deficits, limiting the severity of L-DOPA induced dyskinesia (Hodgson et al. 2010; Kanda et al. 2000). Taken together, these results suggested that although no study has yet demonstrated the ability of A_{2A} receptor antagonists to revert an already established dyskinesia in both rodents and primates, the association of A_{2A} receptor antagonists with a low non-dyskinetic dose of L-DOPA might produce an efficient improvement of motor symptoms, with a concomitant reduction of the intensity of dyskinetic movements (for details see Chap. 9).

Conclusions

In conclusion, data reported in the present chapter describe A_{2A} receptor antagonists as being extremely promising compounds to be used in the therapy of PD. Their potential is largely represented by the marked efficacy demonstrated in alleviating every cardinal PD motor symptom observed in pharmacological and toxicological animal models of PD. The findings achieved in both rodent and primate models of PD suggested that A_{2A} receptor antagonist agents might have symptomatic therapeutic effectiveness in the early stages of PD, when motor complications have not yet appeared, because A_{2A} receptor antagonists do not counteract dyskinesia. Specifically, they suggested that A_{2A} receptor antagonists, *per se*, may improve akinesia/bradykinesia, initiation of movement and gait impairments, and muscle rigidity, and, at the same time, ameliorate the sensorimotor integration deficits and tremor that characterize PD. Moreover, experiments performed in hemiparkinsonian rodents demonstrated that the combined administration of A_{2A} receptor antagonists with a low sub-threshold dose of L-DOPA potentiated the effect of L-DOPA Moreover, the persistent anti-parkinsonian efficacy of A_{2A} receptor antagonists during chronic treatment is of greatest interest in a condition requiring long-term pharmacological management, such as PD, in which drugs should retain their motor-facilitating properties over a chronic regimen. In addition, experimental data show the efficacy of A_{2A} receptor antagonists in reducing the *wearing off* phenomenon and in increasing the *"on periods"* with no exacerbation of dyskinesia.

Altogether, these preclinical studies demonstrate the need to investigate, through clinical trials, the potential utilization of A$_{2A}$ receptor antagonists in the management of the cardinal symptoms of parkinsonian patients, both as monotherapies and in combination with low doses of L-DOPA.

References

Albin RL, Young AB, Penney JB (1989) The functional anatomy of basal ganglia disorders. Trends Neurosci 12:366–375

Alves G, Forsaa EB, Pedersen KF et al (2008) Epidemiology of Parkinson's disease. J Neurol 255:18–32

Armentero MT, Pinna A, Ferré S et al (2011) Past, present and future of A(2A) adenosine receptor antagonists in the therapy of Parkinson's disease. Pharmacol Ther 132:280–299

Azdad K, Gall D, Woods AS et al (2009) Dopamine D2 and adenosine A$_{2A}$ receptors regulate NMDA-mediated excitation in accumbens neurons through A$_{2A}$-D2 receptor heteromerization. Neuropsychopharmacology 34:972–986

Bar-Gad I, Bergman H (2001) Stepping out of the box: information processing in the neural networks of the basal ganglia. Curr Opin Neurobiol 11:689–695

Betz AJ, Vontell R, Valenta J et al (2009) Effects of the adenosine A$_{2A}$ antagonist KW 6002 (istradefylline) on pimozide-induced oral tremor and striatal c-Fos expression: comparisons with the muscarinic antagonist tropicamide. Neuroscience 163:97–108

Bibbiani F, Oh JD, Petzer JP et al (2003) A$_{2A}$ antagonist prevents dopamine agonist-induced motor complications in animal models of Parkinson's disease. Exp Neurol 184:285–294

Bibbiani F, Costantini LC, Patel R et al (2005) Continuous dopaminergic stimulation reduces risk of motor complications in parkinsonian primates. Exp Neurol 192:73–78

Bogenpohl JW, Ritter SL, Hall RA et al (2012) Adenosine A$_{2A}$ receptor in the monkey basal ganglia: ultrastructural localization and colocalization with the metabotropic glutamate receptor 5 in the striatum. J Comp Neurol 520:570–589

Bové J, Marin C, Bonastre M et al (2002) Adenosine A$_{2A}$ antagonism reverses levodopa-induced motor alterations in hemiparkinsonian rats. Synapse 15:251–257

Bové J, Serrats J, Mengod G et al (2006) Reversion of levodopa-induced motor fluctuations by the A$_{2A}$ antagonist CSC is associated with an increase in striatal preprodynorphin mRNA expression in 6-OHDA-lesioned rats. Synapse 59:435–444

Braak H, Del Tredici K, Rüb U et al (2003) Staging of brain pathology related to sporadic Parkinson's disease. Neurobiol Aging 24:197–211

Chang JW, Wachtel SR, Young D et al (1999) Biochemical and anatomical characterization of foreforelimb adjusting steps in rat models of Parkinson's disease: studies on medial forebrain bundle and striatal lesions. Neuroscience 88:617–628

Chaudhuri KR, Healy DG, Schapira AH et al (2006) Non-motor symptoms of Parkinson's disease: diagnosis and management. Lancet Neurol 5:235–245

Collins LE, Galtieri DJ, Brennum LT et al (2010) Oral tremor induced by the muscarinic agonist pilocarpine is suppressed by the adenosine A$_{2A}$ antagonists MSX-3 and SCH58261, but not the adenosine A1 antagonist DPCPX. Pharmacol Biochem Behav 94:561–569

Collins LE, Sager TN, Sams AG et al (2012) The novel adenosine A$_{2A}$ antagonist Lu AA47070 reverses the motor and motivational effects produced by dopamine D2 receptor blockade. Pharmacol Biochem Behav 100:498–505

Collins-Praino LE, Paul NE, Rychalsky KL et al (2011) Pharmacological and physiological characterization of the tremulous jaw movement model of parkinsonian tremor: potential insights into the pathophysiology of tremor. Front Syst Neurosci 5:49

Collins-Praino LE, Paul NE, Ledgard F et al (2013) Deep brain stimulation of the subthalamic nucleus reverses oral tremor in pharmacological models of parkinsonism: interaction with the effects of adenosine A_{2A} antagonism. Eur J Neurosci 38:2183–2191

Correa M, Wisniecki A, Betz A et al (2004) The adenosine A_{2A} antagonist KF17837 reverses the locomotor suppression and tremulous jaw movements induced by haloperidol in rats: possible relevance to parkinsonism. Behav Brain Res 148:47–54

Cousins MS, Carriero DL, Salamone JD (1997) Tremulous jaw movements induced by the acetylcholinesterase inhibitor tacrine: effects of antiparkinsonian drugs. Eur J Pharmacol 322:137–145

DeLong M (1990) Primate models of movement disorders of basal ganglia origin. Trends Neurosci 13:281–285

DeLong MR, Wichmann T (2007) Circuits and circuit disorders of the basal ganglia. Arch Neurol 64:20–24

Delwaide PJ (2001) Parkinsonian rigidity. Funct Neurol 16:147–156

Deuschl G, Papengut F, Hellriegel H (2012) The phenomenology of parkinsonian tremor. Parkinsonism Relat Disord 18:S87–S89

Dorsey ER, Constantinescu R, Thompson JP et al (2007) Projected number of people with Parkinson disease in the most populous nations, 2005 through 2030. Neurology 68:384–386

Drabczyńska A, Zygmunt M, Sapa J et al (2011) Antiparkinsonian effects of novel adenosine A(2A) receptor antagonists. Arch Pharm (Weinheim) 344:20–27

Duty S, Jenner P (2011) Animal models of Parkinson's disease: a source of novel treatments and clues to the cause of the disease. Br J Pharmacol 164:1357–1391

Fenu S, Pinna A, Ongini E et al (1997) Adenosine A_{2A} receptor antagonism potentiates L-DOPA-induced turning behaviour and c-fos expression in 6-hydroxydopamine-lesioned rats. Eur J Pharmacol 321:143–147

Ferré S, von Euler G, Johansson B et al (1991) Stimulation of high-affinity adenosine A_2 receptors decreases the affinity of dopamine D2 receptors in rat striatal membranes. Proc Natl Acad Sci USA 88:7238–7241

Ferré S, Fredholm BB, Morelli M et al (1997) Adenosine dopamine receptor-receptor interactions as an integrative mechanism in the basal ganglia. Trends Neurosci 20:482–487

Ferré S, Quiroz C, Woods AS et al (2008) An update on adenosine A_{2A}-dopamine D2 receptor interactions. Implications for the function of G protein-coupled receptors. Curr Pharm Des 14:1468–1474

Fredduzzi S, Moratalla R, Monopoli A et al (2002) Persistent behavioral sensitization to chronic L-DOPA requires A_{2A} adenosine receptors. J Neurosci 22:1054–1062

Fredholm BB, Battig K, Holmen J et al (1999) Actions of caffeine in the brain with special reference to factors that contribute to its widespread use. Pharmacol Rev 51:83–133

Galvan A, Wichmann T (2008) Pathophysiology of parkinsonism. Clin Neurophysiol 119:1459–1474

Gerevich Z, Wirkner K, Illes P (2002) Adenosine A_{2A} receptors inhibit the N-methyl-D-aspartate component of excitatory synaptic currents in rat striatal neurons. Eur J Pharmacol 451:161–164

Gerfen CR, Bolam JP (2010) The neuroanatomical organization of the basal ganglia part A. In: Handbook of basal ganglia structure and function, vol 1. Elsevier Science, pp 3–28

Gerlach M, Riederer P (1996) Animal models of Parkinson's disease: an empirical comparison with the phenomenology of the disease in man. J Neural Transm 103:987–1041

Gillespie RJ, Bamford SJ, Botting R et al (2009) Antagonists of the human A_{2A} adenosine receptor. 4. Design, synthesis, and preclinical evaluation of 7-Aryltriazolo[4,5-d]pyrimidines. J Med Chem 52:33–47

Grondin R, Bedard PJ, Hadj Tahar A et al (1999) Antiparkinsonian effect of a new selective adenosine A_{2A} receptor antagonist in MPTP-treated monkeys. Neurology 52:1673–1677

Halldner L, Lozza G, Lindström K et al (2000) Lack of tolerance to motor stimulant effects of a selective adenosine A(2A) receptor antagonist. Eur J Pharmacol 406:345–354

Henry B, Crossman AR, Brotchie JM (1998) Characterization of enhanced behavioral responses to L-DOPA following repeated administration in the 6-hydroxydopamine-lesioned rat model of Parkinson's disease. Exp Neurol 151:334–342

Hettinger BD, Lee A, Linden J et al (2001) Ultrastructural localization of the adenosine A$_{2A}$ receptors suggests multiple cellular sites for modulation of GABAergic neurons in rat striatum. J Comp Neurol 431:331–346

Higley MJ, Sabatini BL (2010) Competitive regulation of synaptic Ca2+ influx by D2 dopamine and A$_{2A}$ adenosine receptors. Nat Neurosci 13:958–966

Hillion J, Canals M, Torvinen M et al (2002) Coaggregation, cointernalization, and codesensitization of adenosine A$_{2A}$ receptors and dopamine D2 receptors. J Biol Chem 277:18091–18097

Hodgson RA, Bertorelli R, Varty GB et al (2009) Characterization of the potent and highly selective A$_{2A}$ receptor antagonists preladenant and SCH 412348 in rodent models of movement disorders and depression. J Pharmacol Exp Ther 33:294–303

Hodgson RA, Bedard PJ, Varty GB et al (2010) Preladenant, a selective A(2A) receptor antagonist, is active in primate models of movement disorders. Exp Neurol 225:384–390

Horstink M, Tolosa E, Bonuccelli U et al (2006) Review of the therapeutic management of Parkinson's disease. Report of a joint task force of the European Federation of Neurological Societies (EFNS) and the Movement Disorder Society-European Section (MDS-ES). Part II: late (complicated) Parkinson's disease. Eur J Neurol 13(11):1186–1202

Ishiwari K, Betz A, Weber S et al (2005) Validation of the tremulous jaw movement model for assessment of the motor effects of typical and atypical antipychotics: effects of pimozide (Orap) in rats. Pharmacol Biochem Behav 80:351–362

Jellinger KA (2002) Recent developments in the pathology of Parkinson's disease. J Neural Transm 62:347–376

Jiménez MC, Vingerhoets FJ (2012) Tremor revisited: treatment of PD tremor. Parkinsonism Relat Disord 18:S93–S95

Jones CK, Bubser M, Thompson AD et al (2012) The metabotropic glutamate receptor 4-positive allosteric modulator VU0364770 produces efficacy alone and in combination with L-DOPA or an adenosine 2A antagonist in preclinical rodent models of Parkinson's disease. J Pharmacol Exp Ther 340:404–421

Jones N, Bleickardt C, Mullins D et al (2013) A$_{2A}$ receptor antagonists do not induce dyskinesias in drug-naive or L-dopa sensitized rats. Brain Res Bull 98:163–169

Kanda T, Shiozaki S, Shimada J et al (1994) KF17837: a novel selective adenosine A$_{2A}$ receptor antagonist with anticataleptic activity. Eur J Pharmacol 256:263–268

Kanda T, Jackson MJ, Smith LA et al (1998) Adenosine A$_{2A}$ antagonist: a novel antiparkinsonian agent that does not provoke dyskinesia in parkinsonian monkeys. Ann Neurol 43:507–513

Kanda T, Jackson MJ, Smith LA et al (2000) Combined use of the adenosine A(2A) antagonist KW-6002 with L-DOPA or with selective D1 or D2 dopamine agonists increases antiparkinsonian activity but not dyskinesia in MPTP-treated monkeys. Exp Neurol 162:321–327

Koga K, Kurokawa M, Ochi M et al (2000) Adenosine A(2A) receptor antagonists KF17837 and KW-6002 potentiate rotation induced by dopaminergic drugs in hemi-Parkinsonian rats. Eur J Pharmacol 408:249–255

Kurokawa M, Koga K, Kase H et al (1996) Adenosine A2a receptor-mediated modulation of striatal acetylcholine release in vivo. J Neurochem 66:1882–1888

Lee RG (1989) Pathophysiology of rigidity and akinesia in Parkinson's disease. Eur Neurol 29:13–18

Lorenc-Koci E, Ossowska K, Wardas J et al (1995) Does reserpine induce parkinsonian rigidity? J Neural Transm Park Dis Dement Sect 9:211–223

Lorenc-Koci E, Wolfarth S, Ossowska K (1996) Haloperidol-increased muscle tone in rats as a model of parkinsonian rigidity. Exp Brain Res 109:268–276

Łukasiewicz S, Błasiak E, Faron-Górecka A et al (2007) Fluorescence studies of homooligomerization of adenosine A$_{2A}$ and serotonin 5-HT1A receptors reveal the specificity of receptor interactions in the plasma membrane. Pharmacol Rep 59:379–392

Lundblad M, Andersson M, Winkler C et al (2002) Pharmacological validation of behavioural measures of akinesia and dyskinesia in a rat model of Parkinson's disease. Eur J Neurosci 15:120–132

Lundblad M, Vaudano E, Cenci MA (2003) Cellular and behavioural effects of the adenosine A2a receptor antagonist KW-6002 in a rat model of l-DOPA-induced dyskinesia. J Neurochem 84:1398–1410

Mandhane SN, Chopde CT, Ghosh AK (1997) Adenosine A_2 receptors modulate haloperidol-induced catalepsy in rats. Eur J Pharmacol 328:135–141

Marcellino D, Lindqvist E, Schneider M et al (2010) Chronic A_{2A} antagonist treatment alleviates parkinsonian locomotor deficiency in MitoPark mice. Neurobiol Dis 40:460–466

Marin C, Aguilar E, Bonastre M et al (2005) Early administration of entacapone prevents levodopa-induced motor fluctuations in hemiparkinsonian rats. Exp Neurol 192:184–193

Marsden CD (1994) Parkinson's disease. J Neurol Neurosurg Psychiatry 57:672–681

Meredith GE, Kang UJ (2006) Behavioral models of Parkinson's disease in rodents: a new look at an old problem. Mov Disord 21:1595–1606

Neustadt BR, Hao J, Lindo N et al (2007) Potent, selective, and orally active adenosine A_{2A} receptor antagonists: arylpiperazine derivatives of pyrazolo[4,3-e]-1,2,4-triazolo[1,5-c]pyrimidines. Bioorg Med Chem Lett 17:1376–1380

Obeso JA, Rodriguez-Oroz MC, Rodriguez M et al (2000) Pathophysiology of the basal ganglia in Parkinson's disease. Trends Neurosci 23:S8–S19

Ochi M, Shiozaki S, Kase H (2004) Adenosine A_{2A} receptor-mediated modulation of GABA and glutamate release in the output regions of the basal ganglia in a rodent model of Parkinson's disease. Neuroscience 127:223–231

Oh JD, Chase TN (2002) Glutamate-mediated striatal dysregulation and the pathogenesis of motor response complications in Parkinson's disease. Amino Acids 23:133–139

Olah ME, Stiles GL (2000) The role of receptor structure in determining adenosine receptor activity. Pharmacol Ther 85:55–75

Olanow CW, Agid Y, Mizuno Y et al (2004) Levodopa in the treatment of Parkinson's disease: current controversies. Mov Disord 19:997–1005

Olanow CW, Stern MB, Sethi K (2009) The scientific and clinical basis for the treatment of Parkinson disease. Neurology 72:S1–S136

Olsson M, Nikkhah G, Bentlage C et al (1995) Forelimb akinesia in the rat Parkinson model: differential effects of dopamine agonists and nigral transplants as assessed by a new stepping test. J Neurosci 15:3863–3875

Pinna A, Morelli M (2014) A critical evaluation of behavioral rodent models of motor impairment used for screening of antiparkinsonian activity: the case of adenosine A(2A) receptor antagonists. Neurotox Res 25:392–401

Pinna A, Di Chiara G, Wardas J et al (1996) Blockade of A2a adenosine receptors positively modulates turning behaviour and c-Fos expression induced by D1 agonists in dopamine-denervated rats. Eur J Neurosci 8:1176–1181

Pinna A, Fenu S, Morelli M (2001) Motor stimulant effects of the adenosine A(2A) receptor antagonist SCH 58261 do not develop tolerance after repeated treatments in 6-hydroxydopamine-lesioned rats. Synapse 39:233–238

Pinna A, Corsi C, Carta AR et al (2002) Modification of adenosine extracellular levels and adenosine A(2A) receptor mRNA by dopamine denervation. Eur J Pharmacol 446:75–82

Pinna A, Volpini R, Cristalli G et al (2005) New adenosine A_{2A} receptor antagonists: actions on Parkinson's disease models. Eur J Pharmacol 512:157–164

Pinna A, Pontis S, Borsini F et al (2007) Adenosine A(2A) receptor antagonists improve deficits in initiation of movement and sensory motor integration in the unilateral 6-hydroxydopamine rat model of Parkinson's disease. Synapse 61:606–614

Pinna A, Tronci E, Schintu N et al (2010) A new ethyladenine antagonist of adenosine A(2A) receptors: behavioral and biochemical characterization as an antiparkinsonian drug. Neuropharmacology 58:613–623

Pollack AE, Fink JS (1996) Synergistic interaction between an adenosine antagonist and a D1 dopamine agonist on rotational behaviour and striatal c-Fos induction in 6-hydroxydopamine-lesioned rats. Brain Res 743:124–130

Rose S, Jackson MJ, Smith LA et al (2006) The novel adenosine A2a receptor antagonist ST1535 potentiates the effects of a threshold dose of L-DOPA in MPTP treated common marmosets. Eur J Pharmacol 546:82–87

Rose S, Ramsay Croft N, Jenner P (2007) The novel adenosine A2a antagonist ST1535 potentiates the effects of a threshold dose of l-dopa in unilaterally 6-OHDA-lesioned rats. Brain Res 1133:110–114

Rosin DL, Robeva A, Woodard RL et al (1998) Immunohistochemical localization of adenosine A$_{2A}$ receptors in the rat central nervous system. J Comp Neurol 401:163–186

Salamone JD, Mayorga AJ, Trevitt JT et al (1998) Tremulous jaw movements in rats: a model of parkinsonian tremor. Prog Neurobiol 56:591–611

Salamone JD, Betz AJ, Ishiwari K et al (2008) Tremorolytic effects of adenosine A$_{2A}$ antagonists: implications for parkinsonism. Front Biosci 13:3594–3605

Salamone JD, Collins-Praino LE, Pardo M et al (2013) Conditional neural knockout of the adenosine A(2A) receptor and pharmacological A(2A) antagonism reduce pilocarpine-induced tremulous jaw movements: studies with a mouse model of parkinsonian tremor. Eur Neuropsychopharmacol 23:972–977

Schallert T, Fleming SM, Leasure JL et al (2000) CNS plasticity and assessment of forelimb sensirimotor outcome in unilateral rat model of stroke, cortical ablation, parkinsonism, and spinal cord injury. Neuropharmacology 39:777–787

Schapira AH (2006) Etiology of Parkinson's disease. Neurology 66:S10–S23

Schwarting RK, Huston JP (1996) The unilateral 6-hydroxydopamine lesion model in behavioral brain research. Analysis of functional deficits, recovery and treatments. Prog Neurobiol 50:275–331

Shindou T, Richardson PJ, Mori A et al (2003) Adenosine modulates the striatal GABAergic inputs to the globus pallidus via adenosine A$_{2A}$ receptors in rats. Neurosci Lett 352:167–170

Shiozaki S, Ichikawa S, Nakamura J et al (1999) Actions of adenosine A$_{2A}$ receptor antagonist KW-6002 on drug-induced catalepsy and hypokinesia caused by reserpine or MPTP. Psychopharmacology 147:90–95

Shook BC, Rassnick S, Osborne MC et al (2010) In vivo characterization of a dual adenosine A$_{2A}$/A1 receptor antagonist in animal models of Parkinson's disease. J Med Chem 53:8104–8115

Shook BC, Chakravarty D, Barbay JK et al (2013) Substituted thieno[2,3-d]pyrimidines as adenosine A$_{2A}$ receptor antagonists. Bioorg Med Chem Lett 23:2688–2691

Simola N, Fenu S, Baraldi PG et al (2004) Blockade of adenosine A$_{2A}$ receptors antagonizes parkinsonian tremor in the rat tacrine model by an action on specific striatal regions. Exp Neurol 189:182–188

Simola N, Fenu S, Baraldi PG et al (2006) Dopamine and adenosine receptor interaction as basis for the treatment of Parkinson's disease. J Neurol Sci 248:48–52

Simola N, Morelli M, Carta AR (2007) The 6-hydroxydopamine model of Parkinson's disease. Neurotox Res 11:151–167

Simola N, Morelli M, Pinna A (2008) Adenosine A$_{2A}$ receptor antagonists and Parkinson's disease: state of the art and future directions. Curr Pharm Des 14:1475–1489

Smith KM, Browne SE, Jayaraman S et al (2014) Effects of the selective adenosine A$_{2A}$ receptor antagonist, SCH 412348, on the parkinsonian phenotype of MitoPark mice. Eur J Pharmacol 728:31–38

Stasi MA, Borsini F, Varani K et al (2006) ST 1535: a preferential A$_{2A}$ adenosine receptor antagonist. Int J Neuropsychopharmacol 9:575–584

Svenningsson P, Moine C L, Fisone G et al (1999) Distribution, biochemistry and function of striatal adenosine A$_{2A}$ receptors. Prog Neurobiol 59:355–396

Tozzi A, de Iure A, Di Filippo M et al (2011) The distinct role of medium spiny neurons and cholinergic interneurons in the D$_2$/A$_2$A receptor interaction in the striatum: implications for Parkinson's disease. J Neurosci 31:1850–1862

Tronci E, Simola N, Borsini F et al (2007) Characterization of the antiparkinsonian effects of the new adenosine A$_{2A}$ receptor antagonist ST1535: acute and subchronic studies in rats. Eur J Pharmacol 566:94–102

Ungerstedt U (1968) 6-hydroxy-dopamine induced degeneration of central monoamine neurons. Eur J Pharmacol 5:107–110

Van Den Eeden SK, Tanner CM, Bernstein AL et al (2003) Incidence of Parkinson's disease: variation by age, gender, and race/ethnicity. Am J Epidemiol 157:1015–1022

Varty GB, Hodgson RA, Pond AJ et al (2008) The effects of adenosine A_{2A} receptor antagonists on haloperidol-induced movement disorders in primates. Psychopharmacology (Berl) 200:393–401

Vellucci SV, Sirinathsinghji DJ, Richardson PJ (1993) Adenosine A_2 receptor regulation of apomorphine-induced turning in rats with unilateral striatal dopamine denervation. Psychopharmacology 111:383–388

Villanueva-Toledo J, Moo-Puc RE, Góngora-Alfaro JL (2003) Selective A_{2A}, but not A1 adenosine antagonists enhance the anticataleptic action of trihexyphenidyl in rats. Neurosci Lett 346:1–4

Wardas J (2003) Synergistic effect of SCH 58261, an adenosine A_{2A} receptor antagonist, and L-DOPA on the reserpine-induced muscle rigidity in rats. Pol J Pharmacol 55:155–164

Wardas J, Konieczny J, Lorenc-Koci E (2001) SCH 58261, an A(2A) adenosine receptor antagonist, counteracts parkinsonian-like muscle rigidity in rats. Synapse 41:160–171

Wardas J, Pietraszek M, Dziedzicka-Wasylewska M (2003) SCH 58261, a selective adenosine A_{2A} receptor antagonist, decreases the haloperidol-enhanced proenkephalin mRNA expression in the rat striatum. Brain Res 977:270–277

Weiss SM, Benwell K, Cliffe IA et al (2003) Discovery of nonxanthine adenosine A_{2A} receptor antagonists for the treatment of Parkinson's disease. Neurology 61:S101–S106

Wolfarth S, Konieczny J, Smiałowska M et al (1996) Influence of 6-hydroxydopamine lesion of the dopaminergic nigrostriatal pathway on the muscle tone and electromyographic activity measured during passive movements. Neuroscience 74:985–996

Xiao D, Bastia E, Xu YH (2006) Forebrain adenosine A_{2A} receptors contribute to L-3,4-dihydroxyphenylalanine-induced dyskinesia in hemiparkinsonian mice. J Neurosci 26:13548–13555

Xu K, Bastia E, Schwarzschild M (2005) Therapeutic potential of adenosine A(2A) receptor antagonists in Parkinson's disease. Pharmacol Ther 105:267–310

Chapter 8
Dopamine/Adenosine Interactions Related to Tremor in Animal Models of Parkinsonism

John D. Salamone, Samantha J. Podurgiel, Lauren L. Long,
Eric J. Nunes and Mercè Correa

Abstract Adenosine A_{2A} receptor antagonists have been shown to exert antiparkinsonian effects in human clinical studies and animal models. The present chapter reviews experiments that were conducted to study the role of adenosine A_{2A} receptors in the regulation of tremor. In particular, these studies have focused on the tremulous jaw movement model of Parkinsonian tremor. Systemic and intrastriatal injections of adenosine A_{2A} receptor antagonists have been shown to reduce the oral tremor induced by dopamine antagonists, dopamine depletion, and cholinomimetic stimulation. Adenosine A_{2A} receptor knockout mice are resistant to the pharmacological induction of tremulous jaw movements. Moreover, stimulation of adenosine A_{2A} receptors with CGS 21680 was capable of inducing tremulous jaw movements. These results demonstrate that adenosine A_{2A} antagonists can exert anti-tremor effects in animal models, which supports their use as antiparkinsonian agents in humans.

Keywords Tremulous jaw movements · Parkinson's disease · Caudate putamen · Striatum · DARPP-32 · D_2 · A_{2A} · Receptor · Electromyography

Introduction

Neurotransmitter interactions in the basal ganglia are thought to regulate normal and pathological aspects of motor processes, including motor dysfunctions related to Parkinsonism (Collins-Praino et al. 2011). Though much work has focused upon the role of caudate/putamen dopamine (DA), a substantial body of research has implicated several other basal ganglia neurotransmitters, including acetylcholine, serotonin, glutamate, and GABA, in aspects of basal ganglia-related motor function

J. D. Salamone (✉) · S. J. Podurgiel · L. L. Long · E. J. Nunes · M. Correa
Behavioral Neuroscience Division, Department of Psychology, University of Connecticut,
06269-1020 Storrs, CT, USA
e-mail: john.salamone@uconn.edu

M. Correa
Area de Psicobiologia, Universitat Jaume I, Castello, Spain

© Springer International Publishing Switzerland 2015
M. Morelli et al. (eds.), *The Adenosinergic System,* Current Topics in Neurotoxicity 10,
DOI 10.1007/978-3-319-20273-0_8

and dysfunction. Over the last several years, evidence has accumulated indicating that the purine neuromodulator adenosine also plays an important role in regulating the motor functions of the striatal complex, including both nucleus accumbens and neostriatum (Chen et al. 2001; Correa et al. 2004; Ferré et al. 1997; Hauber et al. 1998, 2001; Ishiwari et al. 2007; Kanda et al. 1994; Pinna et al. 1997, 1999, 2007; Salamone et al. 2008a, b; Simola et al. 2004, 2006; Svenningsson et al. 1999). Much of this work has focused upon the functions of adenosine A_{2A} receptors. There are four G-protein coupled adenosine receptors, but the adenosine A_{2A} receptor subtype is expressed to a very high degree in DA-rich striatal regions (Cieslak et al. 2008; Ferré et al. 1997, 2001; Rosin et al. 1998). Adenosine A_{2A} receptors in the striatum are largely expressed on enkephalin-positive striatopallidal neurons that co-localize DA D_2 receptors; these adenosine and DA receptors interact by forming hetero-meric complexes and converging onto the same signal transduction mechanisms (Ferré et al. 1997, 2001, 2008; Fuxe et al. 2003; Hauber et al. 2001). Because of the functional interactions between DA D_2 and adenosine A_{2A} receptors, and the movement-related effects of adenosine A_{2A} receptor antagonists in animal models, it has been widely suggested that adenosine A_{2A} antagonists could be used as non-dopaminergic treatments for Parkinsonian symptoms (Ferré et al. 1997, 2001; Fox 2013; Morelli and Pinna 2001; Morelli et al. 2010; Pinna 2009; Salamone 2010). Several adenosine A_{2A} antagonists have been developed and assessed at various stages of human clinical trials, with variable results. While positive results have generally been shown with initial studies and Phase II clinical trials (LeWitt et al. 2008), there have been mixed results upon further investigation and Phase III clini-cal trials in the US for drugs such as istradefylline, vipadenant, and preladenant (Barkhoudarian and Schwarzschild 2011; Jenner 2014). Nevertheless, istradefylline (NOURIAST) was recently approved for clinical use in Japan after the results of a clinical trial showing significant decreases in OFF time in patients also treated with L-DOPA (Mizuno and Kondo 2013). These inconsistent results, with some promising indications of positive effects, point to the need for further assessment of adenosine A_{2A} antagonists in animal models.

A number of tests related to motor function are used in rodent models of Par-kinsonism, and several of these procedures have been employed for the assessment of adenosine A_{2A} antagonists. The A_{2A} antagonist SCH 58261 reversed the rigidity induced by the DA antagonist haloperidol in rats (Wardas et al. 2001), and the cata-lepsy induced by DA antagonists was shown to be attenuated by MSX-3 (Hauber et al. 1998, 2001; Salamone et al. 2008a). Several studies have focused upon the effects of adenosine A_{2A} antagonists on locomotion in rodents. The adenosine A_{2A} antagonist KW-6002 reversed the suppression of locomotor activity induced by the monoamine depleting agent reserpine (Shiozaki et al. 1999). The reduced activity seen in D_2 receptor deficient mice was rescued by istradefylline (Aoyama et al. 2000). Systemic and intra-accumbens injections of adenosine A_{2A} antagonists re-versed the suppression of locomotion induced by acute or subchronic injections of the D_2 antagonists haloperidol and eticlopride (Collins et al. 2010b; Correa et al. 2004; Ishiwari et al. 2007; Salamone et al. 2008a).

The Tremulous Jaw Movement Model

Tremor is defined as a "periodic oscillation of a body member" (Findley and Gresty 1981), and can be a feature of several different movement disorders. Tremors are classified in various ways, including their local frequency, the state under which they occur (e.g., resting tremor, action tremor), and the pathological conditions associated with the tremor. Although resting tremor is one of the cardinal symptoms of idiopathic and drug-induced Parkinsonism, relatively few clinical studies have specifically emphasized the pharmacology of tremor (e.g. Schneider and Deuschl 2014; Schrag et al. 1999; Sung et al. 2008), and there is considerable uncertainty about the neurochemical mechanisms that underlie tremorogenesis (Bergman and Deuschl 2002; Deuschl et al. 2000; Sung et al. 2008). Thus, it is important to focus attention on the neurochemistry and physiology of tremor (Muthuraman et al. 2008; Schneider and Deuschl 2014), and studies employing animal models are a critical aspect of this strategy.

Drug-induced tremulous jaw movements are a well validated rodent model of Parkinsonian tremor (Collins-Praino et al. 2011; Salamone et al. 1998, 2005, 2008a, b). Tremulous jaw movements are rapid vertical deflections of the lower jaw that are not directed at any stimulus (Salamone et al. 1998). Tremulous jaw movements in rats can be induced by several dopaminergic conditions that are known to be associated with Parkinsonism in humans, including neurotoxic depletion of striatal DA (Delattre et al. 2010; Finn et al. 1997; Jicha and Salamone 1991; Rodriguez-Diaz et al. 2001), DA depleting agents such as reserpine (Baskin and Salamone 1993; Salamone and Baskin 1996; Salamone et al. 2008a, b; Steinpreis and Salamone 1993) and tetrabenazine (Podurgiel et al. 2013a), and DA antagonists (i.e., Betz et al. 2007, 2009; Ishiwari et al. 2005; Jicha and Salamone 1991; Steinpreis and Salamone 1993; Steinpreis et al. 1993; Trevitt et al. 1998). The tremulous jaw movements induced by DA antagonists do not require chronic administration, and can be induced by either acute or subchronic treatments (Jicha and Salamone 1991; Steinpreis and Salamone 1993; Steinpreis et al. 1993; Trevitt et al. 1998); thus, they are not strictly speaking a model of tardive dyskinesia (Collins-Praino et al. 2011). Furthermore, although "typical" antipsychotics such as haloperidol and pimozide readily induce tremulous jaw movements, "atypical" antipsychotics such as clozapine, olanzapine and quetiapine do not (Betz et al. 2005, 2009; Ishiwari et al. 2005; Trevitt et al. 1998, 1999). Tremulous jaw movements also are induced by cholinomimetic drugs such as muscarinic agonists (Baskin et al. 1994; Salamone et al. 1986, 1990; Stewart et al. 1988) and the anticholinesterases physostigmine, tacrine and galantamine (Collins et al. 2011; Kelley et al. 1989; Mayorga et al. 1997).

Considerable evidence indicates that tremulous jaw movements share many characteristics with Parkinsonian tremor. As measured by analysis of freeze-frame video, as well as electromyographic (EMG) methods, these movements occur in phasic bursts of repetitive jaw movement activity in the 3–7 Hz local frequency range, which resembles the local frequency of Parkinsonian resting tremor (Collins et al. 2010a; Cousins et al. 1998; Finn et al. 1997; Ishiwari et al. 2005; Mayorga et al. 1997; Podurgiel et al. 2013a; Salamone and Baskin 1996; Salamone et al. 1998 see Fig. 8.1). Tremulous jaw movements can be reduced by both established and

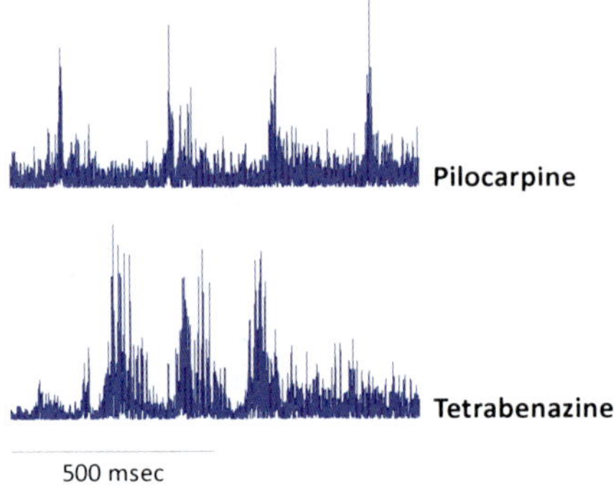

Fig. 8.1 Full-wave rectified EMG traces (1.0 s) from the lateral temporalis muscle (i.e. a jaw closing muscle) of two rats showing drug-induced tremulous jaw movements. Bipolar fine wire tungsten electrodes were implanted into the muscle prior to drug treatment. *Top*: The rat received an IP injection of 4.0 mg/kg of the muscarinic agonist pilocarpine, and showed four tremulous jaw movements during this sweep. *Bottom*: The rat received an IP injection of 2.0 mg/kg of the DA depleting agent tetrabenazine, and showed three tremulous jaw movements during the middle part of this trace

putative antiparkinsonian drugs from several different classes, including DAergic agents such as apomorphine, L-DOPA, bromocriptine, ropinerole and pergolide (Cousins et al. 1997; Salamone et al. 2005), muscarinic antagonists (i.e., benztropine, scopolamine, atropine and tropicamide; Betz et al. 2007, 2009; Cousins et al. 1997; Steinpreis et al. 1993), the T-type calcium channel blocker zonisamide (Miwa et al. 2008, 2009), and the MAO-B inhibitor safinamide (Podurgiel et al. 2013b). Furthermore, tremulous jaw movements can be attenuated by deep brain stimulation of the subthalamic nucleus (Collins-Praino et al. 2013), which is a major brain target in human deep brain stimulation treatments for Parkinsonian patients.

Consistent with the known involvement of neostriatal mechanisms in human Parkinsonism, several lines of evidence indicate that the ventrolateral neostriatum (VLS) is a critical striatal subregion at which DA and acetylcholine receptor mechanisms interact to regulate tremulous jaw movements (see Salamone et al. 1998, 2008a, b). Depletions of DA in the VLS by local injections of 6-hydroxydopamine were shown to induce tremulous jaw movements, while injections in other striatal regions were ineffective (Jicha and Salamone 1991). Local injections of the cholinomimetics physostigmine (Kelley et al. 1989) and pilocarpine (Salamone et al. 1990) into the VLS induced tremulous jaw movements, while injections into other striatal areas did not. Extracellular levels of ACh in VLS as measured by microdialysis were significantly correlated with the jaw movements induced by the anticholinesterases tacrine and physostigmine (Cousins et al. 1999). Cholinomimetic-induced tremulous jaw movements were suppressed by local injections of the muscarinic antagonist scopolamine into the VLS (Mayorga et al. 1997; Salamone et al. 1990). Hemicholinium, which reduces ACh synthesis by blocking high affinity choline uptake, was shown to suppress tacrine-induced jaw movements when injected into the VLS, but not into overlying cortex (Cousins et al. 1999). The suppression of pilocarpine-induced jaw movements that was produced by the DA D_1 agonist SKF 82958 was reversed by injections of the D_1 antagonist SCH 23390 into VLS, but not overlying cortex (Mayorga et al.

1999a). In addition, injections of the c-AMP analogue 8-bromo-c-AMP into the VLS suppressed pilocarpine-induced jaw movements, while injections into cortex were ineffective (Mayorga et al. 1999b). Anatomical evidence indicates that the VLS is the homologue of the ventral region of putamen, and that this region of striatum receives input from sensory and motor cortices related to head, orofacial and forepaw areas (Salamone et al. 1998). The lateral striatum of rodents, like the putamen of primates, is somatotopically organized, and the putamen is a striatal region that is associated with tremor in Parkinson's disease (Salamone et al. 1998).

One important striatal output pathway that appears to be important for the tremulous jaw movements induced by DA antagonism is the GABAergic striatopallidal system. Recent studies were undertaken to determine if extracellular levels of GABA in globus pallidus are associated with the induction of tremulous jaw movements by the DA D_2 antagonist haloperidol (Collins-Praino et al. 2012). Both acute and repeated haloperidol administration induced tremulous jaw movements, and also significantly increased extracellular GABA in globus pallidus as measured by microdialysis. Pooling across the different treatment conditions, there was a significant positive correlation between pallidal GABA levels and the number of tremulous jaw movements induced during the first three samples collected after haloperidol injection. Interestingly, administration of 4.0 mg/kg pilocarpine had no effect on pallidal GABA release, despite the ability of this drug to induce tremulous jaw movements. These results indicate that the tremulous jaw movements induced by DA D_2 receptor antagonism and those induced through muscarinic receptor stimulation appear to be generated via distinct mechanisms.

Adenosine A_{2A} Receptor Regulation of Tremulous Jaw Movements

As reviewed above, considerable evidence indicates that tremulous jaw movements are a useful model for investigating the anatomy, pathophysiology, neurochemistry and pharmacology of tremor. Within the last decade, this model has been used to study the potential tremorolytic effects of adenosine A_{2A} antagonists. Correa et al. (2004) reported that KF 17837 could suppress the tremulous jaw movements induced by repeated administration of haloperidol. Simola et al. (2004) observed that the tremulous jaw movements induced by systemic administration of the anticholinesterase tacrine could be suppressed by the adenosine A_{2A} antagonist SCH 58261. Since those initial reports, a wide variety of adenosine A_{2A} antagonists, including istradefylline, SCH BT2, ST1535, Lu AA47070, MSX-3 and MSX-4, have all been shown to suppress the tremulous jaw movements induced by DA antagonists or cholinomimetics (Betz et al. 2009; Collins et al. 2010a, 2012; Correa et al. 2004; Salamone et al. 2008a; Santerre et al. 2012; Simola et al. 2004, 2006; Tronci et al. 2007). MSX-3 also was shown to suppress the tremulous jaw movements induced by tetrabenazine, which depletes DA via antagonism of vesicular storage (Podurgiel et al. 2013a). In contrast, the selective adenosine A_1 antagonist DPCPX (8-cyclopentyl-1,3-dipropylxanthine) failed to suppress the tremulous jaw movements that were induced by either the muscarinic agonist pilocarpine or the DA antagonist pimozide (Collins et al. 2010a). The anti-tremor effects of adenosine A_{2A} antagonists also have been shown to be induced by local administration of adenosine antagonists directly into the VLS (Salamone et al. 2008a; Simola et al. 2004, 2006; Tronci et al. 2007), which is consistent with the known involvement of this striatal subregion in tremorogenesis.

Additional lines of evidence support the involvement of adenosine A_{2A} receptors in the generation of tremulous jaw movement activity. Systemic administration of sub-sedative doses of the selective adenosine A_{2A} agonist CGS 21680 was able to induce tremulous jaw movements (Collins-Praino et al. 2011). Furthermore, tremulous jaw movements can be induced in mice as well as rats, and some studies have investigated the effects of adenosine A_{2A} receptor knockout on these movements. Conditional neural knockout of adenosine A_{2A} receptors in mice with a C57/Bl6 background suppressed the tremulous jaw movements induced by the muscarinic agonist pilocarpine (Salamone et al. 2013). In addition, the tremulous jaw movements induced by the DA depleting agent tetrabenazine were suppressed in adenosine A_{2A} receptor knockout mice with a CD1 background (Podurgiel et al. 2013a). Together with the research focusing on the effects of adenosine A_{2A} receptor antagonists, these experiments with knockout mice provide substantial support for the hypothesis that adenosine A_{2A} receptors participate in the regulation of tremorogenesis.

Recent studies assessed the effects of deep brain stimulation of the subthalamic nucleus on tremulous jaw movements in rats, and also investigated the interaction between brain stimulation and administration of an adenosine A_{2A} antagonist (Collins-Praino et al. 2013). Subthalamic deep brain stimulation reduced the tremulous jaw movements induced by the DA D_2 family antagonist pimozide and the cholinomimetics pilocarpine and galantamine. The effectiveness of the anti-tremor actions of deep brain stimulation was dependent upon the neuroanatomical locus being stimulated (i.e., subthalamic nucleus vs. a striatal control site), and also upon the frequency and intensity of stimulation used. Importantly, administration of the adenosine A_{2A} receptor antagonist MSX-3 reduced the frequency and intensity parameters needed to attenuate tremulous jaw movements, making the animals more sensitive to the tremor suppression produced by brain stimulation (Collins-Praino et al. 2013). These results have implications for the clinical use of deep brain stimulation combined with adenosine A_{2A} receptor antagonism in human patients.

D_2/A_{2A} Interactions and Markers of Signal Transduction

As described above, adenosine A_{2A} receptors in striatal areas are co-localized with DA D_2 receptors on enkephalin-positive medium spiny neurons. These receptors can form heteromers, and also converge onto the same cAMP/protein kinase A related signal transduction cascade. D_2 receptors are linked to G_i, which results in an inhibition of adenylate cyclase activity, while A_{2A} receptor stimulation increases adenylate cyclase activity via actions on $G_{s/olf}$ (Ferré et al. 2008). Because of these opposing effects on signal transduction mechanisms, one of the plausible mechanisms for the neural basis of A_{2A}/D_2 interactions is that the signal transduction effects of reduced D_2

receptor transmission are reversed by blockade of A_{2A} receptors. Based upon this idea, several studies have used markers of signal transduction activity to characterize the pharmacological interaction between drugs acting on A_{2A} and D_2 receptors. One useful marker of the cellular effects of D_2 receptor antagonism is the induction of c-Fos immunoreactivity in neostriatum. Pinna et al. (1999) reported that the induction of neostriatal c-Fos immunoreactivity by haloperidol was reduced by co-administration of the adenosine A_{2A} antagonist SCH 58261. Betz et al. (2009) studied the effect of istradefylline on pimozide-induced tremulous jaw movements, and in parallel, measured the expression of c-Fos in the striatum. A dose of istradefylline that reduced pimozide-induced tremulous jaw movements also suppressed the induction of VLS c-Fos expression in pimozide-treated rats (Betz et al. 2009). Farrar et al. (2010) studied A_{2A}/D_2 interactions in the ventral striatum, and found that intracranial injections of a dose of MSX-3 that reversed the behavioral effects of the D_2 antagonist eticlopride also reversed eticlopride-induced increases in c-Fos immunoreactive cells. Santerre et al. (2012) reported that eticlopride-induced increases in ventral striatal c-Fos immunoreactivity were suppressed by behaviorally active doses of systemically administered MSX-3 and MSX-4. Furthermore, the induction of ventral and dorsal striatal c-Fos expression by administration of the DA depleting agent tetrabenazine also was suppressed by MSX-3 (Nunes et al. 2013; Podurgiel et al. 2013a).

Another critical marker of striatal signal transduction activity is DA and c-AMP-related phosphoprotein (DARPP-32; Bateup et al. 2008). D_1 receptor stimulation increases c-AMP production and protein kinase A (PKA) activity, which phosphorylates DARPP-32 to yield pDARPP-32(Thr34). This effect is thought to take place predominantly in substance P-positive neurons that mainly express D_1 receptors. D_2 receptor stimulation decreases c-AMP production and PKA activity, which decreases the dephosphorylation of pDARPP-32(Thr75) by protein phosphatase 2A (PP-2A), and therefore increases pDARPP-32(Thr75) expression and decreases pDARPP-32(Thr34) expression in enkephalin-positive neurons (Bateup et al. 2008). In turn, blockade of D_2 receptors is thought to have the opposite effect, which would lead to an increase in the expression of pDARPP-32(Thr34) in enkephalin-positive medium spiny striatal neurons. Recently it was shown that administration of the selective D_2 receptor antagonist eticlopride increased expression of ventral striatal pDARPP-32(Thr34) (Santerre et al. 2012). Consistent with the hypothesized interaction between adenosine A_{2A} and DA D_2 receptors, this effect of eticlopride was attenuated by co-administration of behaviorally effective doses of the A_{2A} antagonists MSX-3 and MSX-4 (Santerre et al. 2012).

More recent studies have focused on the effects of tetrabenazine, which depletes striatal DA (Nunes et al. 2013). Immunocytochemical analyses of different forms of phosphorylated DARPP-32 indicated that tetrabenazine significantly increased ventral striatal expression of both pDARPP-32(Thr34) and pDARPP-32(Thr75). Based upon previous studies, these results suggested that tetrabenazine-induced increases in pDARPP-32(Thr75) would reflect reduced transmission at DA D_1 family receptors in substance P positive neurons, while the increases in pDARPP-32(Thr34) would mark reduced transmission at DA D_2 family receptors located on enkephalin-positive neurons (Bateup et al. 2008; Santerre et al. 2012; Svenningsson et al. 2004; Yger and Girault 2011; see Fig. 8.2, top). Indeed, immunofluorescence

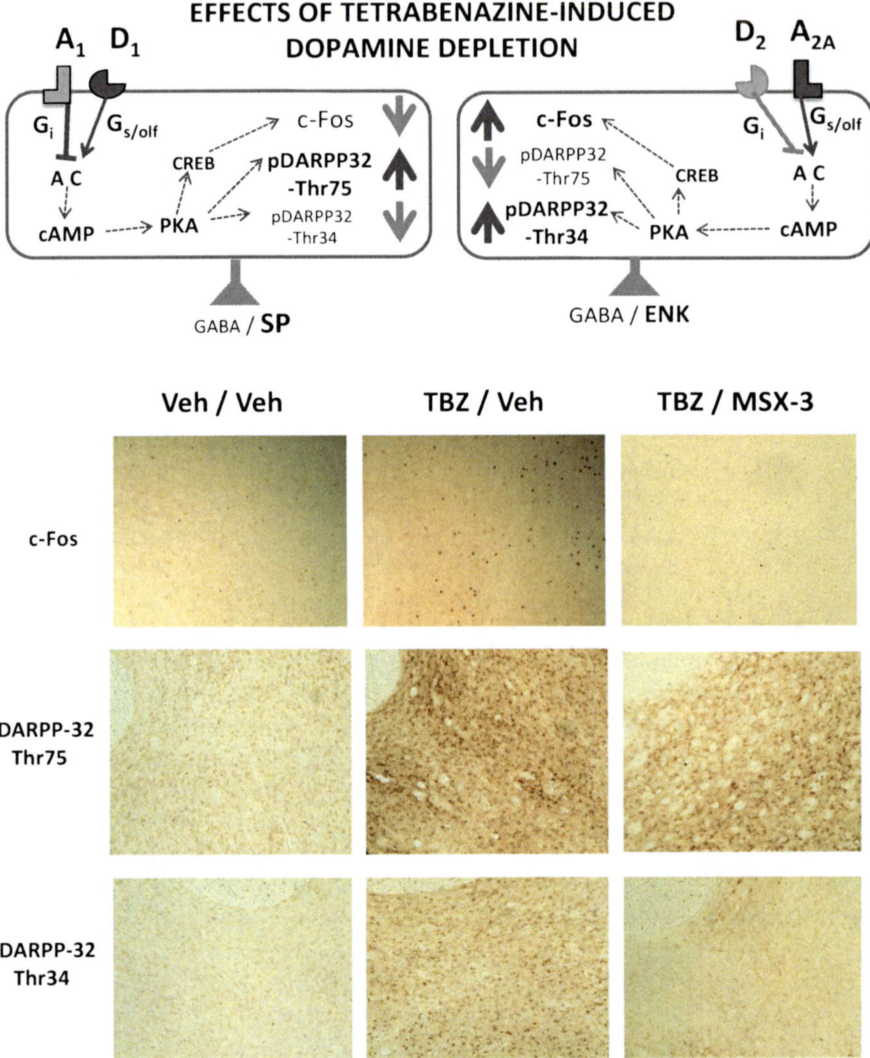

Fig. 8.2 *Top*: This diagram shows the localization of DA and adenosine receptor subtypes on striatal medium spiny neurons, and depicts the effects of tetrabenazine, which depletes DA, on markers of striatal signal transduction (based upon data from Nunes et al. 2013). Tetrabenazine increased expression of pDARPP-32(Thr75), which reflects reduced transmission at DA D₁ family receptors in substance P positive neurons. Tetrabenazine also increased expression of c-Fos and pDARPP-32(Thr34), which marked reduced transmission at DA D₂ family receptors located on enkephalin-positive neurons. *Bottom*: Expression of c-Fos, pDARPP-32(Thr34), and pDARPP-32(Thr75) immunoreactivity in ventral striatum after injection of vehicle plus vehicle (Veh/Veh), 0.75 mg/kg tetrabenazine plus vehicle (TBZ/Veh), or tetrabenazine plus 2.0 mg/kg MSX-3 (Photomicrographs of individual animals; group data are described in Nunes et al. 2013). As described in text, MSX-3 reduced expression of pDARPP-32(Thr34), but did not affect pDARPP-32(Thr75) immunoreactivity. *AC* adenylate cyclase; *PKA* protein kinase A; *SP* substance P; *ENK* enkephalin

double-labeling for different forms of phosphorylated DARPP, as well as the peptides substance P and enkephalin, confirmed this hypothesized effect of tetrabenazine (Nunes et al. 2013). Interestingly, the adenosine A_{2A} antagonist MSX-3 attenuated the effects of tetrabenazine on pDARPP-32(Thr34) expression, but not pDARPP-32(Thr75) expression (Nunes et al. 2013; see Fig. 8.2, bottom). This pattern of effects is consistent with studies demonstrating that adenosine A_{2A} receptors are co-localized with D_2 receptors on enkephalin-positive neurons, but not with D_1 receptors on substance-P positive neurons (Svenningsson et al. 1999), and that A_{2A} and D_2 receptors can form heteromers, and interact via convergence onto c-AMP signal transduction cascades (Ferré et al. 2008).

Taken together, these studies on signal transduction pathways provide valuable neural markers of the interactions between A_{2A} and D_2 receptors, which can offer comparisons with the behavioral measures that are used to characterize this interaction. Furthermore, they provide insights into the cellular mechanisms underlying the antiparkinsonian effects of A_{2A} antagonists.

Summary and Conclusions

In summary, a number of studies have shown that adenosine A_{2A} antagonists can attenuate the tremulous jaw movements induced by DA antagonists, DA depletion, and cholinomimetic drugs (see review by Collins-Praino et al. 2011). Furthermore, adenosine A_{2A} antagonism appears to enhance sensitivity to the tremor suppression induced by subthalamic nucleus deep brain stimulation (Collins-Praino et al. 2013). These findings are broadly consistent with the hypothesis that adenosine A_{2A} antagonists could be useful as treatments for idiopathic Parkinson's disease, as well as the drug-induced Parkinsonism resulting from administration of antipsychotic drugs. Moreover, studies showing tremorolytic effects of adenosine A_{2A} antagonists in animal models are consistent with the results of Bara-Jimenez et al. (2003), who reported that istradefylline was particularly effective at suppressing tremor. Although clinical studies often do not provide direct or objective measures of specific symptoms such as tremor, it may be useful for future clinical studies to provide such information. For example, it is possible that adenosine A_{2A} antagonists would be particularly effective at suppressing tremor relative to other symptoms, or for the treatment of tremor-dominant Parkinson's disease.

In addition to being characterized by cardinal motor symptoms such as akinesia and tremor, Parkinson's Disease patients also can show a variety of non-motor symptoms, including motivational or depression-related symptoms such as anergia, fatigue, or lack of exertion of effort (Salamone et al. 2010). Moreover, similar psychomotor/motivational dysfunctions are frequently seen in patients with major depression or related disorders (Salamone et al. 2007, 2010; Treadway et al. 2012). A large body of data from studies involving animal models of behavioral activation and effort-related functions indicates that adenosine A_{2A} antagonists can reverse the motivational dysfunctions induced by D_2 antagonists and tetrabenazine

(Farrar et al. 2007, 2010; Mott et al. 2009; Nunes et al. 2010, 2013; Pardo et al. 2012; Salamone et al. 2009), as well as the pro-inflammatory cytokine interleukin 1β (Nunes et al. 2014). Taken together with the research involving motor dysfunctions such as akinesia and tremor, these studies on effort-related motivational functions strongly suggest that adenosine A_{2A} antagonists offer much in the way of therapeutic utility.

Acknowledgements This research was supported by grants to JDS from the United States NIH/ NIMH and the University of Connecticut Research Foundation. Many thanks to Emily Errante for assisting with the EMG analysis, and to Dr. James Chrobak for the use of his recording facilities.

References

Aoyama S, Kase H, Borrelli E (2000) Rescue of locomotor impairment in dopamine D2 receptor-deficient mice by an adenosine A2A receptor antagonist. J Neurosci 20:5848–5852

Bara-Jimenez W, Sherzai A, Dimitrova T et al (2003) Adenosine A(2A) receptor antagonist treatment of Parkinson's disease. Neurology 61:293–296

Barkhoudarian MT, Schwarzschild MA (2011) Preclinical jockeying on the translational track of adenosine A2A receptors. Exp Neurol 228:160–164

Baskin P, Salamone JD (1993) Vacuous jaw movements in rats induced by acute reserpine administration: interactions with different doses of apomorphine. Pharmacol Biochem Behav 46:793–797

Baskin PP, Gianutsos G, Salamone JD (1994) Repeated scopolamine injections sensitize rats to pilocarpine-induced vacuous jaw movements and enhance striatal muscarinic receptor binding. Pharmacol Biochem Behav 49:437–442

Bateup HS, Svenningsson P, Kuroiwa M et al (2008) Cell type-specific regulation of DARPP-32 phosphorylation by psychostimulant and antipsychotic drugs. Nat Neurosci 11:932–939

Bergman H, Deuschl G (2002) Pathophysiology of Parkinson's disease: from clinical neurology to basic neuroscience and back. Mov Disord 17:S28–S40

Betz A, Ishiwari K, Wisniecki A et al (2005) Quetiapine (Seroquel) shows a pattern of behavioral effects similar to the atypical antipsychotics clozapine and olanzapine: studies with tremulous jaw movements in rats. Psychopharmacology 179:383–392

Betz AJ, McLaughlin PJ, Burgos M et al (2007) The muscarinic receptor antagonist tropicamide suppresses tremulous jaw movements in a rodent model of Parkinsonian tremor: possible role of M4 receptors. Psychopharmacology 194:347–359

Betz AJ, Vontell R, Valenta J et al (2009) Effects of the adenosine A_{2A} antagonist KW-6002 (istradefylline) on pimozide-induced oral tremor and striatal c-Fos expression: comparisons with the muscarinic antagonist tropicamide. Neuroscience 163:97–108

Chen JF, Moratalla R, Impagnatiello F et al (2001) The role of the D2 dopamine receptor (D2R) in A2a adenenosine-receptor (A2aR) mediated behavioral and cellular responses as revealed by A2a and D2 receptor knockout mice. Proc Natl Acad Sci U S A 98:1970–1975

Cieslak M, Komoszynski M, Wojtczak A (2008) Adenosine A2A receptors in Parkinson's disease treatment. Purinergic Signal 4:305–312

Collins LE, Galtieri DJ, Brennum LT et al (2010a) Cholinomimetic-induced tremulous jaw movements are suppressed by the adenosine A2A antagonists MSX-3 and SCH58261, but not the adenosine A1 antagonist DPCPX: possible relevance for drug-induced Parkinsonism. Pharmacol Biochem Behav 94:561–569

Collins LE, Galtieri DJ, Collins P et al (2010b) Interactions between adenosine and dopamine receptor antagonists with different selectivity profiles: effects on locomotor activity. Behav Brain Res 211:148–155

Collins LE, Paul NE, Abbas SF et al (2011) Oral tremor induced by galantamine in rats: a model of the Parkinsonian side effects of cholinomimetics used to treat Alzheimer's disease. Pharmacol Biochem Behav 99:414–422

Collins LE, Sager TN, Sams AG et al (2012) The novel adenosine A2A antagonist Lu AA47070 reverses the motor and motivational effects produced by dopamine D2 receptor blockade. Pharmacol Biochem Behav 100:498–505

Collins-Praino LE, Paul NE, Rychalsky KL et al (2011) Pharmacological and physiological characterization of the tremulous jaw movement model of Parkinsonian tremor: potential insights into the pathophysiology of tremor. Front Syst Neurosci 5:49

Collins-Praino LE, Podurgiel SJ, Kovner R et al (2012) Extracellular GABA in globus pallidus increases during the induction of oral tremor by haloperidol but not by muscarinic receptor stimulation. Behav Brain Res 234:129–135

Collins-Praino LE, Paul NE, Ledgard F et al (2013) Deep brain stimulation of the subthalamic nucleus reverses oral tremor in pharmacological models of Parkinsonism: interaction with the effects of adenosine A2A antagonism. Eur J Neurosci 38:2183–2191

Correa M, Wisniecki A, Betz A et al (2004) The adenosine A2A antagonist KF17837 reverses locomotor suppression and tremulous jaw movements induced by haloperidol in rats: possible relevance to Parkinsonism. Behav Brain Res 148:47–54

Cousins MS, Carriero DL, Salamone JD (1997) Tremulous jaw movements induced by the acetylcholinesterase inhibitor tacrine; effects of antiparkinsonian drugs. Eur J Pharmacol 322:137–145

Cousins MS, Atherton A, Salamone JD (1998) Behavioral and electromyographic characterization of the local frequency of tacrine-induced tremulous jaw movements. Physiol Behav 64:153–158

Cousins MS, Finn M, Trevitt J et al (1999) The role of ventrolateral striatal acetylcholine in the production of tacrine-induced jaw movements. Pharmacol Biochem Behav 62:439–447

Delattre AM, Kiss A, Szawka RE et al (2010) Evaluation of chronic omega-3 fatty acids supplementation on behavioral and neurochemical alterations in 6-hydroxydopamine-lesion model of Parkinson's disease. Neurosci Res 66:256–264

Deuschl G, Raethjen J, Baron R et al (2000) The pathophysiology of Parkinsonian tremor: a review. J Neurol 247:V33–V48

Farrar AM, Pereira M, Velasco F et al (2007) Adenosine A(2A) receptor antagonism reverses the effects of (DA) receptor antagonism on instrumental output and effort-related choice in the rat: implications for studies of psychomotor slowing. Psychopharmacology 191:579–586

Farrar AM, Segovia KN, Randall PA et al (2010) Nucleus accumbens and effort-related functions: behavioral and neural markers of the interactions between adenosine A_{2A} and dopamine D_2 receptors. Neuroscience 166:1056–1067

Ferré S, Fredholm B, Morelli M et al (1997) Adenosine-dopamine receptor-receptor interactions as an integrative mechanism in the basal ganglia. Trends Neurosci 20:482–486

Ferré S, Popoli P, Gimenez-Llort L et al (2001) Adenosine/dopamine ineteraction: implications for the treatment of Parkinson's disease. Parkinsonism Relat Disord 7:235–241

Ferré S, Quiroz C, Woods AS et al (2008) An update on adenosine A2A-dopamine D2 receptor interactions: implications for the function of G-protein coupled receptors. Curr Pharm Des 14:1468–1474

Findley LJ, Gresty MA (1981) Tremor. Br J Hosp Med 26:16–32

Finn M, Jassen A, Baskin P et al (1997) Tremulous characteristic of vacuous jaw movements induced by pilocarpine and ventrolateral striatal dopamine depletions. Pharmacol Biochem Behav 57:243–249

Fox SH (2013) Non-dopaminergic treatments for motor control in Parkinson's disease. Drugs 73:1405–1415

Fuxe K, Agnati LF, Jacobsen K et al (2003) Receptor heteromerization in adenosine A2A receptor signaling: relevance for striatal function and Parkinson's disease. Neurology 61:S19–S23

Hauber W, Nagel J, Sauer R et al (1998) Motor effects induced by a blockade of adenosine A2A receptors in the caudate-putamen. Neuroreport 9:1803–1806

Hauber W, Neuscheler P, Nagel J et al (2001) Catalepsy induced by a blockade of dopamine D1 or D2 receptors was reversed by a concomitant blockade of adenosine A2A receptors in the caudate putamen of rats. Eur J Neurosci 14:1287–1293

Ishiwari K, Betz A, Weber S et al (2005) Validation of the tremulous jaw movement model for assessment of the motor effects of typical and atypical antipychotics: effects of pimozide (Orap) in rats. Pharmacol Biochem Behav 80:351–362

Ishiwari K, Madson LJ, Farrar AM et al (2007) Injections of the selective adenosine A2A antagonist MSX-3 into the nucleus accumbens core attenuate the locomotor suppression induced by haloperidol in rats. Behav Brain Res 178:190–199

Jenner P (2014) An overview of adenosine A2A receptor antagonists in Parkinson's disease. Int Rev Neurobiol 119:71–86

Jicha G, Salamone JD (1991) Vacuous jaw movements and feeding deficits in rats with ventrolateral striatal dopamine depletions: possible model of Parkinsonian symptoms. J Neurosci 11:3822–3829

Kanda T, Shiozaki S, Shimada J et al (1994) KF17837: a novel selective adenosine A2A receptor antagonist with anticataleptic activity. Eur J Pharmacol 256:263–268

Kelley AE, Bakshi VP, Delfs JM et al (1989) Cholinergic stimulation of the ventrolateral striatum elicits mouth movements in rats: pharmacological and regional specificity. Psychopharmacology 99:542–549

LeWitt PA, Guttman M, Tetrud JW et al (2008) Adenosine A2A receptor antagonist istradefylline (KW-6002) reduces "off" time in Parkinson's disease: a double-blind, randomized, multicenter clinical trial (6002-US-005). Ann Neurol 63:295–302

Mayorga AJ, Carriero DL, Cousins MS et al (1997) Tremulous jaw movements produced by acute tacrine administration: possible relation to Parkinsonian side effects. Pharmacol Biochem Behav 56:273–279

Mayorga AJ, Trevitt JT, Conlan A et al (1999a) Striatal and nigral D_1 mechanisms involved in the antiparkinsonian effects of SKF 82958 (APB): studies of tremulous jaw movements in rats. Psychopharmacology 143:72–81

Mayorga AJ, Gianutsos G, Salamone JD (1999b) Effects of striatal injections of 8-bromo-cyclic-AMP on pilocarpine-induced tremulous jaw movements in rats. Brain Res 829:180–184

Miwa H, Hama K, Kajimoto Y et al (2008) Effects of zonisamide on experimental tremors in rats. Parkinsonism Relat Disord 14:33–36

Miwa H, Kubo T, Suzuki A et al (2009) Effects of zonisamide on c-Fos expression under conditions of tacrine-induced tremulous jaw movements in rats: a potential mechanism underlying its anti-parkinsonian tremor effect. Parkinsonism Relat Disord 15:30–35

Mizuno Y, Kondo T (2013) Japanese Istradefylline study group. Adenosine A2A receptor antagonist istradefylline reduces daily OFF time in Parkinson's disease. Mov Disord 28:1138–1141

Morelli M, Pinna A (2001) Interaction between dopamine and adenosine A2A receptors as a basis for the treatment of Parkinson's disease. Neurol Sci 22:71–72

Morelli M, Carta AR, Kachroo A et al (2010) Pathophysiological roles for purines: adenosine, caffeine and urate. Prog Brain Res 183:183–208

Mott AM, Nunes EJ, Collins LE et al (2009) The adenosine A2A antagonist MSX-3 reverses the effects of the (DA) antagonist haloperidol on effort-related decision making in a T-maze cost/benefit procedure. Psychopharmacology 204:103–112

Muthuraman M, Raethjen J, Hellriegel H et al (2008) Imaging coherent sources of tremor related EEG activity in patients with Parkinson's disease. Conf Proc IEEE Eng Med Biol Soc 2008:4716–4719

Nunes EJ, Randall PA, Santerre JL et al (2010) Differential effects of selective adenosine antagonists on the effort-related impairments induced by (DA) D1 and D2 antagonism. Neuroscience 170:268–280

Nunes EJ, Randall PA, Hart EE et al (2013) Effort-related motivational effects of the VMAT-2 inhibitor tetrabenazine: implications for animal models of the motivational symptoms of depression. J Neurosci 33:19120–19130

Nunes EJ, Randall PA, Estrada A et al (2014) Effort-related motivational effects of the pro-inflammatory cytokine interleukin 1-beta: studies with the concurrent fixed ratio 5/chow feeding choice task. Psychopharmacology 231:727–736

Pardo M, López-Cruz L, Valverde O et al (2012) Adenosine A2A receptor antagonism and genetic deletion attenuate the effects of dopamine D2 antagonism on effort-related decision making in mice. Neuropharmacology 62:2068–2077

Pinna A (2009) Novel investigational adenosine A2A receptor antagonists for Parkinson's disease. Expert Opin Investig Drugs 18:1619–1631

Pinna A, Wardas J, Cristalli G et al (1997) Adenosine A2A receptor agonists increase Fos-like immunoreactivity in mesolimbic areas. Brain Res 759:41–49

Pinna A, Wardas J, Cozzolino A et al (1999) Involvement of adenosine A2A receptors in the induction of c-fos expression by clozapine and haloperidol. Neuropsychopharmacology 20:44–51

Pinna A, Pontis S, Borsini F et al (2007) Adenosine A2A receptor antagonists improve deficits in initiation of movement and sensory motor integration in the unilateral 6-hydroxydopamine rat model of Parkinson's disease. Synapse 61:606–614

Podurgiel SJ, Nunes EJ, Yohn SE et al (2013a) The vesicular monoamine transporter (VMAT-2) inhibitor tetrabenazine induces tremulous jaw movements in rodents: implications for pharmacological models of Parkinsonian tremor. Neuroscience 250:507–519

Podurgiel S, Collins-Praino LE, Yohn S et al (2013b) Tremorolytic effects of safinamide in animal models of drug-induced Parkinsonian tremor. Pharmacol Biochem Behav 105:105–111

Rodriguez Diaz M, Abdala P, Barroso-Chinea P et al (2001) Motor behavioural changes after intracerebroventricular injection of 6-hydroxydopamine in the rat: an animal model of Parkinson's disease. Behav Brain Res 122:79–92

Rosin DL, Robeva A, Woodard RL et al (1998) Immunohistochemical localization of adenosine A2A receptors in the rat central nervous system. J Comp Neurol 401:163–186

Salamone JD (2010) Preladenant, a novel adenosine A(2A) receptor antagonist for the potential treatment of Parkinsonism and other disorders. IDrugs 13:723–731

Salamone JD, Baskin P (1996) Vacuous jaw movements induced by acute reserpine and low-dose apomorphine: possible model of parkinsonian tremor. Pharmacol Biochem Behav 53:179–183

Salamone JD, Lalies MD, Channell SL et al (1986) Behavioural and pharmacological characterization of the mouth movements induced by muscarinic agonists in the rat. Psychopharmacology 88:467–471

Salamone JD, Johnson CJ, McCullough LD et al (1990) Lateral striatal cholinergic mechanisms involved in oral motor activities in the rat. Psychopharmacology 102:529–534

Salamone JD, Mayorga AJ, Trevitt JT et al (1998) Tremulous jaw movements in rats: a model of Parkinsonian tremor. Prog Neurobiol 56:591–611

Salamone JD, Carlson BB, Rios C et al (2005) Dopamine agonists suppress cholinomimetic-induced tremulous jaw movements in an animal model of Parkinsonism: tremorolytic effects of pergolide, ropinirole and CY 208–243. Behav Brain Res 156:173–179

Salamone JD, Correa M, Farrar A et al (2007) Effort-related functions of nucleus accumbens (DA) and associated forebrain circuits. Psychopharmacology 191:461–482

Salamone JD, Betz AJ, Ishiwari K et al (2008a) Tremorolytic effects of adenosine A2A antagonists: implications for Parkinsonism. Front Biosci 13:3594–3605

Salamone JD, Ishiwari K, Betz AJ et al (2008b) Dopamine/adenosine interactions related to locomotion and tremor in animal models: possible relevance to Parkinsonism. Parkinsonism Relat Disord 14:S130–S134

Salamone JD, Farrar AM, Font L et al (2009) Differential actions of adenosine A1 and A2A antagonists on the effort-related effects of dopamine D2 antagonism. Behav Brain Res 201:216–222

Salamone JD, Correa M, Farrar AM et al (2010) Role of dopamine-adenosine interactions in the brain circuitry regulating effort-related decision making: insights into pathological aspects of motivation. Future Neurol 5:377–392

Salamone JD, Collins-Praino LE, Pardo M et al (2013) Conditional neural knockout of the adenosine A(2A) receptor and pharmacological A(2A) antagonism reduce pilocarpine-induced tremulous jaw movements: studies with a mouse model of Parkinsonian tremor. Eur Neuropsychopharmacol 23:972–977

Santerre JL, Nunes EJ, Kovner R et al (2012) The novel adenosine A(2A) antagonist prodrug MSX-4 is effective in animal models related to motivational and motor functions. Pharmacol Biochem Behav 102:477–487

Schneider SA, Deuschl G (2014) The treatment of tremor. Neurotherapeutics 11:128–138

Schrag A, Schelosky L, Scholz U et al (1999) Reduction of Parkinsonian signs in patients with Parkinson's disease by dopaminergic versus anticholinergic single-dose challenges. Mov Disord 14:252–255

Shiozaki S, Ichikawa S, Nakamura J et al (1999) Actions of adenosine A2A receptor antagonist KW-6002 on drug-induced catalepsy and hypokinesia caused by reserpine or MPTP. Psychopharmacology 147:90–95

Simola N, Fenu S, Baraldi PG et al (2004) Blockade of adenosine A2A receptors antagonizes Parkinsonian tremor in the rat tacrine model by an action on specific striatal regions. Exp Neurol 189:182–188

Simola N, Fenu S, Baraldi PG et al (2006) Dopamine and adenosine receptor interaction as basis for the treatment of Parkinson's disease. J Neurol Sci 248:48–52

Steinpreis RE, Salamone JD (1993) Effects of acute haloperidol and reserpine administration on vacuous jaw movements in three different age groups of rats. Pharmacol Biochem Behav 46:405–409

Steinpreis RE, Baskin P, Salamone JD (1993) Vacuous jaw movements induced by sub-chronic administration of haloperidol: interactions with scopolamine. Psychopharmacology 111:99–105

Stewart BR, Jenner P, Marsden CD (1988) The pharmacological characterization of pilocarpine-induced chewing in the rat. Psychopharmacology 96:55–62

Sung YH, Chung SJ, Kim SR et al (2008) Factors predicting response to dopaminergic treatments for resting tremor of Parkinson's disease. Mov Disord 23:137–140

Svenningsson P, Le Moine CL, Fisone G et al (1999) Distribution, biochemistry and function of striatal adenosine A2A receptors. Prog Neurobiol 59:355–396

Svenningsson P, Nishi A, Fisone G et al (2004) DARPP32: an integrator of neurotransmission. Annu Rev Pharmacol Toxicol 44:269–296

Treadway MT, Bossaller NA, Shelton RC et al (2012) Effort-based decision making in major depressive disorder: a translational model of motivational anhedonia. J Abnorm Psychol 121:553–555

Trevitt J, Atherton A, Aberman J et al (1998) Effects of subchronic administration of clozapine, thioridazine and haloperidol on tests related to extrapyramidal motor function in the rat. Psychopharmacology 137:61–66

Trevitt JT, Carlson BB, Salamone JD (1999) Behavioral assessment of atypical antipsychotics in rats: studies of the effects of olanzapine (Zyprexa). Psychopharmacology 145:309–316

Tronci E, Simola N, Borsini F et al (2007) Characterization of the antiparkinsonian effects of the new adenosine A2A receptor antagonist ST1535: acute and subchronic studies in rats. Eur J Pharmacol 566:94–102

Wardas J, Konieczny J, Lorenc-Koci E (2001) SCH 58261, an A2A adenosine receptor antagonist, counteracts Parkinsonian-like muscle rigidity in rats. Synapse 41:160–171

Yger M, Girault JA (2011) DARPP-32, jack of all trades… master of which? Front Behav Neurosci 5:56

Chapter 9
Adenosine A_{2A} Receptor Antagonists in L-DOPA-Induced Motor Fluctuations

Giulia Costa and Micaela Morelli

Abstract Motor fluctuations, and in particular dyskinesia, affect a large percentage of parkinsonian patients under dopamine replacement therapy. Adenosine A_{2A} receptor antagonists may be a new strategy for the treatment of Parkinson's disease (PD) since they potentiate L-DOPA efficacy without worsening dyskinesia. By discussing recent studies in rodents, non-human primates and humans, this chapter summarizes the pharmacology of adenosine A_{2A} receptor antagonist and their interaction with dopaminergic, glutamatergic and cannabinoid receptors, with specific relevance to motor fluctuations and dyskinesia.

Keywords 6-OHDA · AIMs · Basal ganglia · Dopamine replacement therapy · Dyskinesia · Istradefylline · MPTP · Preladenant · Primate · Rat

Parkinson's Disease-Linked Motor Impairment and L-DOPA-Associated Motor Complications

The motor symptoms that characterize Parkinson's disease (PD), i.e. bradykinesia, rigidity, and postural instability, are primarily due to the degeneration of the dopaminergic nigrostriatal neurons; therefore, PD therapy is mainly based on the replacement of impaired dopaminergic transmission (Marsden 1994; Olanow and Tatton 1999).

Dopamine (DA) replacement therapy (DRT), such as the DA precursor L-3,4-dihydroxyphenylalanine (L-DOPA), has significantly advanced the pharmacological treatment of PD, improving survival (Lloyd et al. 1975). However, a major limiting factor in chronic and pulsatile L-DOPA therapy is the development of motor fluctuations that appear after several years of treatment (Jenner 2008; Obeso et al. 2000). These motor disturbances are characterized by "ON" periods in which the patient fully responds to the treatment, and "OFF" periods in which the patient has severe immobility, the duration of the effect of L-DOPA is decreased (*wearing-off*), and

M. Morelli (✉) · G. Costa
Department of Biomedical Sciences, Section of Neuropsychopharmacology,
University of Cagliari, Cagliari, Italy
e-mail: morelli@unica.it

© Springer International Publishing Switzerland 2015   163
M. Morelli et al. (eds.), *The Adenosinergic System*, Current Topics in Neurotoxicity 10,
DOI 10.1007/978-3-319-20273-0_9

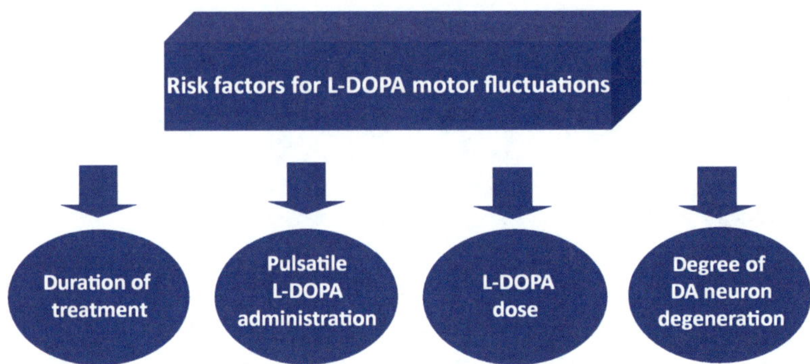

Fig. 9.1 Summary of key points to be considered at the origin of motor fluctuations

dyskinesia is present. Dyskinesia affects about 40 % of chronically treated patients (Jankovic 2005; Nutt 1987; Obeso et al. 2004) and consists of involuntary choreic-dystonic movements that are usually very severe, including twitches, jerking, twisting, or restlessness, when the drug produces its maximal effect. Motor fluctuations and dyskinesia, however, are not only due to the duration, dosage, and pulsatile administration of the pharmacologic treatment, but are also linked to intrinsic factors, such as the degree of loss of dopaminergic innervation in the basal ganglia (BG) (Papa et al. 1994) (Fig. 9.1).

While in the early stages of PD L-DOPA is transformed into DA, released, and reuptaked steadily in the remaining presynaptic dopaminergic terminals, in the later stages, when dopaminergic innervation is totally lost, L-DOPA is mainly transformed into DA in non-dopaminergic neurons, such as the serotoninergic neurons, where no DA reuptake occurs or autoreceptors are present, and, therefore, the duration of the effect of L-DOPA seems to reflect its plasma half-life rise and fall (Carta and Tronci 2014; Sohn et al. 1994). Continuous administration of L-DOPA, in contrast, causes less dyskinesia (Bezard 2013; Jenner 2004; Xie et al. 2014).

L-DOPA-associated motor fluctuations, and in particular dyskinesia, are very difficult to treat since they can only be improved by reducing the L-DOPA dosage, but this reduction worsens the motor symptoms of PD.

Prolonged and pulsatile L-DOPA treatment induces long-term neuronal changes that are at the basis of motor fluctuations. At the same time, high doses of L-DOPA are more dyskinetic than low doses, and high dopaminergic neuron degeneration underlies more severe dyskinesia.

DRT Inadequacy and Long-Term Outcome

At the beginning of the treatment, DRT is very effective; however, as mentioned above, DRT is also at the origin of a number of motor complications, which appear several years after starting treatment (Jenner 2008; Obeso et al. 2000).

Pulsatile stimulation of the denervated DA receptors by L-DOPA or short-acting dopaminergic agents exposes the BG to non-physiologically high (at peak of dose) or low (at the end of each treatment period) stimulation (Bezard 2013; Chase 1998; Jenner 2004; Olanow and Obeso 2000; Xie et al. 2014), leading to irreversible maladaptive neuroplasticity, which underlies the development of dyskinesia (Cenci and Lundblad 2006; Picconi et al. 2003). The oscillations in DA receptor stimulation are known to activate early genes (Canales and Graybiel 2000; Carta et al. 2002; Pavón et al. 2006; Westin et al. 2001), alter neuronal firing activity (Boraud et al. 2001; Calabresi et al. 2000), and induce the loss of low-frequency stimulation, a form of plasticity that facilitates information storage in the neuronal networks of the BG (Picconi et al. 2003).

These shortfalls have prompted research to find non-dopaminergic adjunctive treatments that modulate dopaminergic transmission, rather than stimulating it directly, in order to reduce the above-mentioned side effects. Among the non-dopaminergic pharmacologic approaches to minimize motor dysfunction in PD, antagonists of adenosine A$_{2A}$ receptors have emerged as leading candidates (Kanda and Uchida 2014; Morelli et al. 2007; Schwarzschild et al. 2006; Xu et al. 2005). The basis of this proposal, arising from research in experimental animals, in which A$_{2A}$ receptor antagonists, by potentiating the effects of L-DOPA, allowed the reduction of the L-DOPA dosage, decreasing, in turn, the long-term consequences of its chronic administration.

Studies with A$_{2A}$ Receptor Antagonists or A$_{2A}$ Receptor Knockout (KO) Rodents in PD Models

The unilaterally 6-hydroxydopamine (6-OHDA)-lesioned rat is the most utilized model to mimic PD. In this model, degeneration of the nigrostriatal dopaminergic neurons is caused by 6-OHDA, producing a unilateral parkinsonism (Ungerstedt 1968). Upon DA receptor agonists administration in this model, it is possible to evaluate both the contralateral rotational behavior, which indicates the therapeutic response to a drug used in PD, and the abnormal involuntary movements (AIMs), which indicate the propensity of the drug to induce dyskinesia (Carta et al. 2006a, b; Lundblad et al. 2002; Ungerstedt 1971). AIMs consist of forelimb dyskinesia, axial dystonia, and oral dyskinesia, and have a predictive validity as a model of clinical dyskinesia (Fig. 9.2). Moreover, administration of L-DOPA or DA receptor agonists, either once or repeatedly, leads to sensitization of contralateral rotational behavior (Carta et al. 2006a) that correlates with biochemical changes that are similar to those observed in AIMs and to those correlated with dyskinesia in humans (Frau et al. 2013; Henry et al. 1998; Lindgren et al. 2007).

Chronic administration of A$_{2A}$ receptor antagonists together with L-DOPA at dosages that induced the same number of contralateral rotations of a full effective dose of L-DOPA alone, did not lead to a modification of the intensity of rotational behavior or AIMs during treatment, while L-DOPA alone produced a sensitization in rotational behavior intensity (index of dyskinesia), and increased the intensity of AIMs (Pinna et al. 2001; Tronci et al. 2007), strongly indicating the lack of dyskinetic potential of A$_{2A}$ receptor antagonists.

Fig. 9.2 Rat affected by forelimb, axial, and orolingual AIMs. The purposeless up-and-down movement of the forelimb is called forelimb AIMs. The twisting movement of the neck and upper trunk towards the side contralateralto the lesion is called axial AIMs. Orolingual AIMs comprise opening and closing of the jaw and tongue protrusion towards the side contralateral to the lesion.

Interesting results were obtained in a different experimental setting by Lundblad et al. (2003), in which the authors showed that in unilaterally 6-OHDA-lesioned rats, rendered dyskinetic by chronic administration of L-DOPA, different A_{2A} receptor antagonists, given in combination with a full effective dose of L-DOPA, did not affect the severity of AIMs. Most recent studies with the new A_{2A} receptor antagonist SCH 412348, showed that A_{2A} receptor antagonists neither exacerbated nor prevented the induction of AIMs when administered with chronic L-DOPA (Hodgson et al. 2009; Jones et al. 2013). Altogether, these results suggest that A_{2A} receptor antagonists have a reduced dyskinetic potential and, when administered with L-DOPA, do not worsen L-DOPA-induced dyskinesia. The same issue was evaluated in A_{2A} KO mice, providing preclinical evidence that sensitization of contralateral rotational behavior and AIMs were prevented in these mice (Fredduzzi et al. 2002; Xiao et al. 2006).

Although limitations in the models used to evaluate the efficacy of A_{2A} receptor antagonists exists (Pinna and Morelli 2014), these studies clarified the role of the A_{2A} receptor in the development/expression of sensitized responses to repeated L-DOPA administration. They collectively indicate that A_{2A} receptor antagonists, besides having a low therapeutic activity and reduced dyskinetic liability relative to L-DOPA, do not block established dyskinesia or dyskinesia induced by L-DOPA when the two drugs are coadministered chronically. Moreover, in these studies, a prolongation of L-DOPA efficacy was observed, suggesting a potential therapeutic efficacy of A_{2A} receptor antagonists on *wearing off* and ON/OFF when administered in association with L-DOPA (Hodgson et al. 2009; Pinna et al. 2001).

Studies with A_{2A} Receptor Antagonists in Primate Models of PD

The issue of A_{2A} receptor antagonists and motor fluctuations/dyskinesia was also evaluated in 1-methyl-4-phenyl-1,2,3,6-tetrahydropyridine (MPTP)-treated non-human primates, the best model so far available to reproduce the motor symptoms and complications of PD. Studies in non-human primates are essential for the investigation of new drugs in clinical trials. In these studies, similar to unilaterally 6-OHDA-lesioned rats, the acute administration of the A$_{2A}$ receptor antagonist istradefylline (KW-6002), in chronically L-DOPA-treated animals showing motor complications, enhanced the antiparkinsonian action of a low dose of L-DOPA with no exacerbation of the existing dyskinesia and, very importantly, istradefylline did not lose its efficacy when it was administered chronically (Grondin et al. 1999; Kanda et al. 2000). A further interesting study was performed by the research group of Bibbiani et al. (2003) who showed that istradefylline delayed the shortening of the motor response after chronic administration of the mixed DA D$_1$–D$_2$ receptor agonist apomorphine, suggesting that A$_{2A}$ receptors may play an important role in the development of dyskinesia, rather than its expression, in non-human primates.

Moreover, a study by Hodgson et al. (2010) showed that in MPTP-treated cynomolgus monkeys rendered dyskinetic by chronic L-DOPA, administration of the A$_{2A}$ receptor antagonist preladenant (SCH 420814) alone produced no evidence of dyskinesia at a dose effective in inducing motor activation. Similarly, in association with L-DOPA, both preladenant and istradefylline while decreasing the parkinsonian deficits caused by MPTP and potentiating locomotor activity induced by L-DOPA, did not increase any dyskinesia or motor complications induced by L-DOPA (Hodgson et al. 2010; Uchida et al. 2014).

Clinical Studies on A$_{2A}$ Receptor Antagonists

Several clinical trials with A$_{2A}$ receptor antagonists have so far been performed in PD patients with advanced disease and motor complications. In the majority of those studies, A$_{2A}$ receptor antagonists were utilized as an adjunct to L-DOPA, and results generally showed that the A$_{2A}$ receptor antagonists tested were safe and well tolerated (Kanda and Uchida 2014; Pinna 2014). Regarding antiparkinsonian efficacy, the drugs were, in a consistent way, effective in reducing the waking time spent in the OFF state and increasing the ON state. During the ON state, an increase in dyskinesia classified as non-troublesome was observed (Kanda and Uchida 2014; Pinna 2014).

The A$_{2A}$ receptor antagonists so far tested in Phase II, IIB or III trials are istradefylline (manufactured by Kyowa Hakko Kyogo, now Kyowa Hakko Kirin), vipadenant (BIIB014; manufactured by Vernalis and commercialized in collaboration with Biogen Idec), preladenant (manufactured by Schering-Plough Corp, and now manufactured by Merck) and tozadenant (SYN115; manufactured by Biotie) (Factor et al. 2013; Hauser et al. 2003, 2011, 2014; Kase et al. 2003; LeWitt et al. 2008; Papapetropoulos et al. 2010; Stacy et al. 2008; Tao and Liang 2015; Zhu et al. 2014).

The current clinical results show that istradefylline, the first A$_{2A}$ receptor antagonists entered in clinical trials, produced mixed results, although the most consistent results demonstrated that the drug had a beneficial effect on the *wearing off* and on motor fluctuations (Hauser et al. 2008; Knebel et al. 2012; Mizuno and Kondo

2013). In an initial trial by Bara-Jimenez et al. (2003), istradefylline had no effect when added to an optimal dose of L-DOPA, while it improved PD motor scores when added to a low-dose L-DOPA. The antiparkinsonian response, when it was added to low-dose L-DOPA, was similar to an optimal dose of L-DOPA, while dyskinesia was lower than that observed with L-DOPA. The suggestion originating from this study was that a low dose of L-DOPA plus istradefylline might produce an antiparkinsonian benefit with reduced dyskinesia compared with a full dose of L-DOPA. However, for practical reasons, this experimental setting was not tested in further clinical Phase II or IIB trials, which were instead performed by adding istradefylline to an optimal dose of L-DOPA (Hauser et al. 2003; Kase et al. 2003; LeWitt et al. 2008; Stacy et al. 2008; Tao and Liang 2015; Zhu et al. 2014). In those studies, the drug increased the ON time, but dyskinesia, although classified as non-troublesome, increased more than in placebo-treated subjects.

Clinical use of istradefylline was approved in Japan in 2013, but is not approved in the United States of America (Kyowa Hakko Kirin Co. Ltd. Approval for manufacturing and marketing of NOURIAST® tablets 20 mg, a novel antiparkinsonian agent (2013) News release [available at http://www.kyowa-kirin.com/news_releases/2013/e20130325_04.html]).

Concerning preladenant, early studies revealed promising effects of this compound on OFF periods since OFF time was significantly reduced compared with placebo, while ON time with non-troublesome dyskinesia was at the same time increased. Results were similar to those obtained with istradefylline, and it was concluded that A_{2A} receptor antagonists do not reduce dyskinesia and much of the reduction in OFF time was replaced by ON time with non-troublesome dyskinesia (Factor et al. 2013; Hauser et al. 2011). Preladenant was generally well tolerated (Cutler et al. 2012); however, in May 2013, Merck announced that the Phase III trials did not provide evidence of efficacy over placebo, and the clinical trials on preladenant were terminated (Merck. Newsroom. News releases—research and development news (May 2013). Merck provides update on Phase III clinical program for preladenant, the company's investigational Parkinson's disease medicine [available at http://www.mercknewsroom.com/press-release/research-and-development-news/merck-provides-update-phase-iii-clinical-program-prelade]).

Finally, although positive results were achieved with vipadenant in Phase II, clinical studies were discontinued on the basis of toxicological studies (Papapetropoulos et al. 2010) (Vernalis 2010. Media Centre. Vernalis announces A_{2A} receptor antagonist programme for Parkinson's disease continues with next generation compound [available at http://www.vernalis.com/media-centre/latest-releases/2010-releases/584]).

The latest clinical report published on A_{2A} receptor antagonists was on the investigation of a multicenter, Phase IIB, randomized, double-blind study on tozadenant in which the drug was tested, similar to the above-mentioned studies, in L-DOPA-treated patients who had motor fluctuations (Hauser et al. 2014). The drug was well tolerated and showed efficacy in reducing OFF time and its efficacy will be investigated in Phase III.

Of interest, a recent study by Wills et al. (2013) suggested the possibility that caffeine, which antagonizes A_1 and A_{2A} receptors, may reduce the likelihood of developing dyskinesia.

Based on the preclinical results and on the first findings by Bara-Jimenez et al. (2003), it remains to be evaluated whether A_{2A} receptor antagonists may reduce the development of dyskinesia if administered when DRT is started, and whether lowering the L-DOPA dose and adding an A_{2A} receptor antagonist will maintain antiparkinsonian activity with reduced dyskinesia.

In view of the results of clinical trials, modifications in the A_{2A} receptors were evaluated in parkinsonian patients utilizing positron emission tomography (PET) or magnetic resonance imaging (MRI). In line with older studies of brain slices of PD patients (Calon et al. 2004) showing an increase in A_{2A} receptor density correlated with the onset of dyskinesia, these image studies reported modifications in the A_{2A} receptors of parkinsonian patients with dyskinesia.

Mishina et al. (2011), using PET with the A_{2A} receptor antagonists [7-methyl-11C]-(E)-8-(3,4,5-trimethoxystyryl)-1,3,7-trimethylxanthine ([^{11}C]TMSX), found that the distribution volume ratio of the A_{2A} receptors in the putamen was larger in patients with L-DOPA-induced dyskinesias than in controls and that L-DOPA treatment tended to increase the presence of the A_{2A} receptors in the putamen. Further studies by Ramlackhansingh et al. (2011) using PET and [^{11}C]SCH-442416, in line with the previous study, found that A_{2A} receptor binding was higher in the caudate and putamen of PD patients with L-DOPA-induced dyskinesia than in PD patients without dyskinesia. These studies supported the view that A_{2A} receptor antagonists may prove beneficial in the treatment of motor complications associated with L-DOPA treatment. A further study using MRI conducted during a clinical trial with the A_{2A} receptor antagonists tozadenant in PD patients, showed that patients treated with tozadenant displayed a dose-dependent decrease in thalamic blood flow, indicating a reduced thalamic inhibition via the striatonigral pathway by the drug (Black et al. 2010).

Mechanisms of A_{2A} Receptor Antagonists: Interaction with DA Receptors

The mechanisms at the basis of A_{2A} receptor antagonist action in PD are firstly related to their interaction with DA receptors. However, besides dopaminergic transmission, other neurotransmitters play an important role in the effects of A_{2A} receptors antagonists. These include interaction with glutamatergic and cannabinoid receptors, to quote the most relevant for DRT-induced motor complications.

Interest in A_{2A} receptor interactions has increased with the discovery that these receptors can form heteromeric complexes with other receptors in the striatum (Fuxe et al. 2003). The receptor heteromer concept postulates that receptors from different families combine to generate complexes with distinctive biochemical and functional characteristics, thus generating unique functional entities and novel potential targets for therapy (Ferré et al. 2007, 2009; Franco et al. 2008).

The localization of the A_{2A} and DA receptors in the BG and their signal transduction mechanisms should be taken into account when considering the relationship between these receptors. In the BG, the direct and indirect striatal efferent pathways regulate movement via opposing direct excitatory (D_1) and indirect inhibitory (D_2) inputs to the substantia nigra, which, through the thalamic nuclei, project to the motor cortex (Fig. 9.3). Since the two striatal efferent pathways are GABAergic, stimulation of the direct pathway through the D_1 receptors by inhibiting the substantia nigra pars reticulata (SNr), facilitates movement through disinhibition of the thalamocortical projection; similarly, inhibition of the activity of the indirect pathway through the D_2 receptors contributes to the disinhibition of the thalamocortical projection (Fig. 9.3).

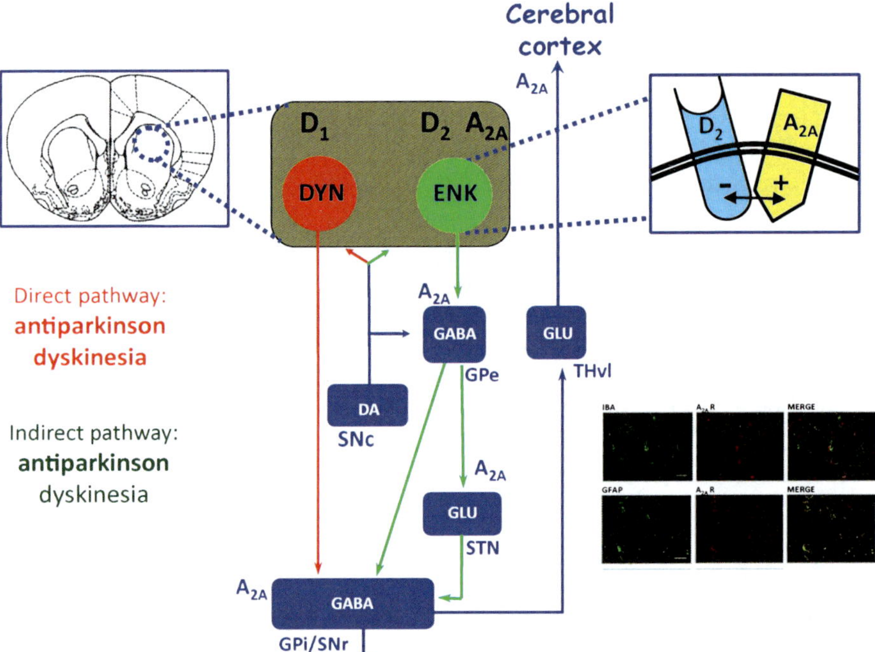

Fig. 9.3 Schematic diagram of the direct and indirect pathways of the basal ganglia. Adenosine A_{2A} receptors are expressed on striatal GABAergic efferent neurons. Direct pathway (*red*) neurons express the neuropeptide DYN together with D_1 receptors, whereas indirect pathway (*green*) neurons contain the neuropeptide ENK and mainly express D_2 receptors. A_{2A} receptor antagonists coadministered with L-DOPA facilitate the inhibitory action of the D_2 receptors on striatopallidal GABA neurons (see insert on the right) and, through the basal circuit, may indirectly facilitate activation of the striatonigral GABA pathway stimulated by the D_1 receptors. A_{2A} receptors are also physiologically expressed in the glial cells, the SNc, both the GPe and GPi, the STN, and the cerebral cortex. *A_{2A}*, adenosine 2A receptor; *DA*, dopamine; *D_1*, dopamine 1 receptor; *D_2*, dopamine 2 receptor; *DYN*, dynorphin; *ENK*, enkephalin; *GABA*, γ-aminobutyric acid *GFAP*, glial fibrillary acidic protein; *GLU*, glutamate; *GPe*, globus pallidus, external segment; *GPi*, globus pallidus, internal segment; *IBA*, ionized calcium binding adapter molecule; *SNc*, substantia nigra pars compacta; *SNr*, substantia nigra pars reticulata; *STN*, subthalamic nucleus; *THvl*, ventrolateral thalamic nucleus. The picture of glia-A_{2A} receptor immunofluorescence is adapted from Paterniti et al. (2011).

In the striatum, the A_{2A} receptors have a peculiar localization with the D_2 receptors on the neurons of the indirect pathway, and have the highest concentration in this area (Fig. 9.3; Fuxe et al. 2007; Hillion et al. 2002; Svenningsson et al. 1999). Activation of the A_{2A} receptors directly opposes the effect of D_2 receptor activation through both a receptor–receptor and second messenger interaction (Figs. 9.3 and 9.4a).

In line with this evidence, A_{2A} receptor agonist treatment reduces the binding affinity of the D_2 receptors in the rat striatum (Aoyama et al. 2000; Ferrè et al. 1991; Svenningsson et al. 1999). An indirect interaction, through the BG loop, instead takes place between the A_{2A} and D_1 receptors, similar to the synergism between the D_1 and D_2 receptors (Fig. 9.3; Pinna et al. 1996; Pollack and Fink 1996; Robertson and Robertson 1986). Regulation of the striatal efferent pathways is therefore mediated by a reciprocal inhibitory interaction between the DA and adenosine receptors, and therefore by blocking adenosine tone, dopaminergic transmission is facilitated.

As mentioned above, A_{2A} receptor antagonists do not counteract dyskinesia; however, when administered chronically with a low dose of L-DOPA, they potentiate the motor efficacy of this drug and extend its duration without exacerbating dyskinesia compared with a full dose of L-DOPA (Pinna et al. 2001; Tronci et al. 2007).

In agreement with behavioral studies, biochemical evidence demonstrates that in 6-OHDA-lesioned rats, the antagonistic interaction between the A_{2A} and DA receptors modulates the expression of striatal genes that may be involved in the pathophysiology of dyskinesia. A_{2A} receptor antagonists administered chronically with L-DOPA at dosages that produced the same acute motor activation of a full dose of L-DOPA, induce a lower activation of striatal enkephalin (ENK), dynorphin (DYN) and glutamic acid decarboxylase-67 (GAD-67) mRNA compared with a full dose of L-DOPA (Carta et al. 2002). In particular, the repeated L-DOPA treatment alone increases the expression of DYN in the striatonigral pathway, which is highly involved in the development dyskinesia (Fig. 9.3; Cenci et al. 1998; Engber et al. 1991; Henry et al. 1999). Overexpression of DYN mRNA induced by chronic L-DOPA in 6-OHDA-lesioned mice was also counteracted in A_{2A} KO mice (Fredduzzi et al. 2002). Attenuation of the L-DOPA-induced modification in DYN expression may help to avoid the maladaptive striatal changes that underlie dyskinesia. However, even if it is not clear whether modifications in opioid peptides may underlie AIMs, they effectively reflect neuroplasticity of the striatal efferent neurons and represent good markers for striatal efferent neuron activity. Besides DYN, chronic intermittent L-DOPA treatment at full dosage increased the striatal levels of GAD-67 and ENK mRNA in 6-OHDA-lesioned rat striatum compared with vehicle treatment, whereas chronic intermittent SCH 58261 plus a lower dose of L-DOPA did not produce any significant modification, although a similar behavioral effect was seen. These results show that the combination SCH 58261 and L-DOPA did not produce long-term changes in markers of striatal efferent neuron activity, confirming the notion that administration of the two drugs have lower dyskinetic potential than a full dose of L-DOPA given alone (Carta et al. 2002, 2003a). All together, these results suggest that through modulation of peptide expression, A_{2A} receptor antagonists, when given with a low dose of L-DOPA, by preventing the development of long-term changes, might restore the balance between indirect and direct pathways, preventing AIMs.

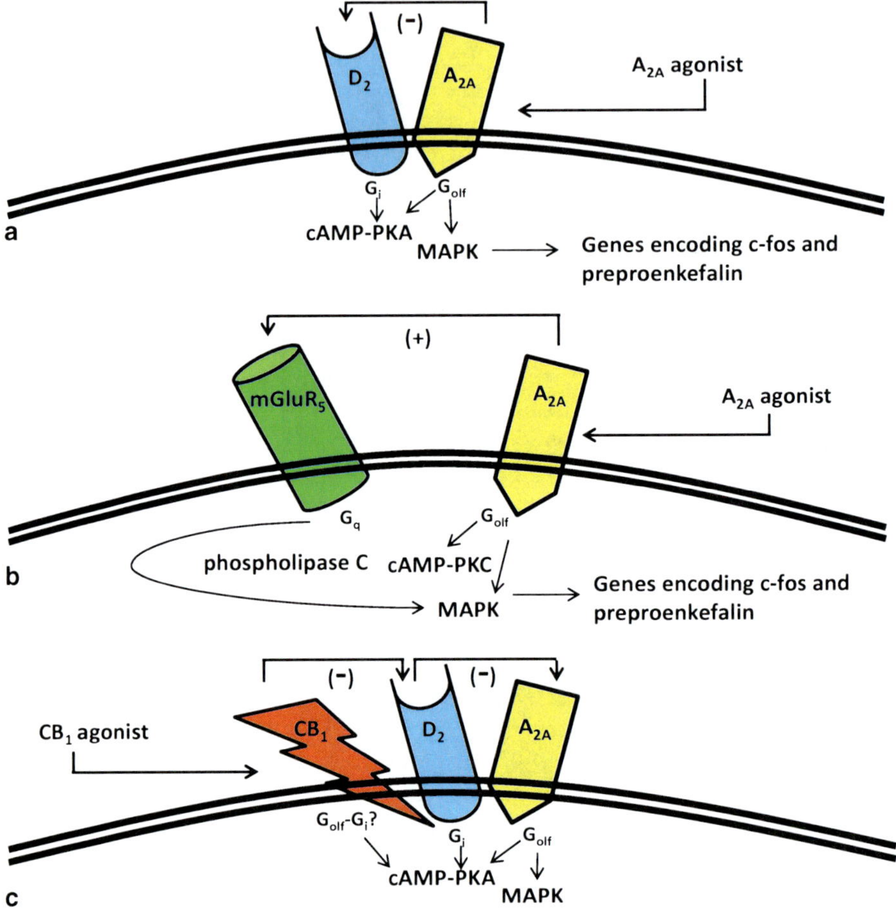

Fig. 9.4 Antagonistic intramembrane A_{2A}–D_2, mGluR5–A_{2A} and A_{2A}–D_2–CB_1 receptor interactions in striatum.

a Activation of the A_{2A} receptor stimulates AC, with subsequent activation of the PKA signaling pathway and induction of the expression of different genes, such as those encoding c-fos and preproenkephalin, by the transcription factor CREB.

b The signal transduction pathways used by the mGluR5 depend on the activation of phospholipase C. Induction of c-fos by the A_{2A} receptors is markedly increased when the mGluR5 is also activated.

c Stimulation of the CB_1 receptor results in an antagonistic CB_1–D_2 interaction that leads to a removal of the D_2 brake on A_{2A} signaling. Thus, the A_{2A} receptor activated AC increases intracellular cAMP and leads to excitation and PKA-mediated changes in gene expression. *A_{2A}* adenosine 2A receptor, *AC* adenylate cyclase, *cAMP* 3′,5′-cyclic adenosine monophosphate, *CB_1* cannabinoid 1 receptor type, *CREB* cAMP-responsive element binding protein, *D_1* dopamine 1 receptor, *D_2* dopamine 2 receptor, *MAPK* mitogen-activated protein kinase, *mGluR5* metabotropic glutamate receptor type 5, *PKA* protein kinase A

When the A$_{2A}$ receptor antagonist was given in combination with a full dose of L-DOPA, this combined treatment induced a dyskinesia similar to a full dose of L-DOPA, and did not significantly alter the expression of DYN and ENK mRNA induced by the 6-OHDA-lesioned striatum (Lundblad et al. 2003). A similar result was obtained on the early gene FosB/ΔFosB-like immunoreactivity (Lundblad et al. 2003). In contrast, it was observed that while high levels of the early gene *zif-268*, together with a persistent hyperresponsiveness of the striatonigral dynorphinergic neurons and hyporesponsiveness of the striatopallidal neurons, were associated with a chronic high dosage of L-DOPA, a low dosage of L-DOPA plus an A$_{2A}$ receptor antagonist did not induce these changes (Carta et al. 2005; Pinna et al. 2010).

In addition to studies in the striatum, results from chronic L-DOPA show that an increase in GAD-67 mRNA in the globus pallidus (GP) and a decrease in the SNr underlie dyskinetic movements induced by L-DOPA (Nielsen and Soghomonian 2003). In contrast a lack of GAD-67 mRNA changes in the GP and a less marked inhibition of the SNr might correlate with the absence of dyskinetic potential observed after the A$_{2A}$ receptor antagonist SCH-58261 plus L-DOPA (Carta et al. 2002, 2003b). These effects could be due to the opposite functional interactions between the A$_{2A}$ and the D$_1$ and D$_2$ receptors mentioned above (Morelli et al. 2007); the first occurring via a polysynaptic interaction at a different BG level, the second, as a direct interaction of the A$_{2A}$–D$_2$ receptors on the striatopallidal neurons. These interactions might contribute to the amplification of DA signaling, but not to DA-induced long-term effects, contributing to the lack of induction of the long-term changes induced by full doses of L-DOPA, which are known to be dyskinetic. Therefore, the sparing of dopaminomimetic drugs in combination with an A$_{2A}$ receptor antagonists may contribute to counteract the long-term aberrant modifications that underlie dyskinesia.

Mechanisms of A$_{2A}$ Receptor Antagonists: Interaction with Glutamate Receptors

While the dopaminergic nigrostriatal projection, which modulates efferent medium spiny neurons, is at the basis of the DA–A$_{2A}$ receptor interaction, the cortical glutamatergic projection modulates these projecting neurons through multiple glutamate receptor subtypes. Neuroanatomical ultrastructural studies have shown that striatal A$_{2A}$ receptors are highly expressed in the dendrites and dendritic spines of GABAergic postsynaptic neurons in asymmetric excitatory synapses (Hettinger et al. 2001; Rosin et al. 2003) and, therefore, these receptors may not only interact with the DA receptors, but also with the glutamatergic ionotropic and metabotropic receptors. In addition, A$_{2A}$ receptors modulate glutamatergic transmission at the extracellular level and the excessive increase in glutamate plays an important role in the neuroplasticity taking place in the BG (Popoli et al. 2003). Therefore, as several lines of preclinical evidence have demonstrated, glutamate plays a most important role in neuroplasticity and in the abnormal modifications related to DRT, in particular dyskinesia (Morin and Di Paolo 2014). Consequently, multiple presynaptic as well as postsynaptic mechanisms could contribute to the modulatory role played by the A_{2A} receptors on glutamatergic transmission.

Several glutamatergic receptors have been involved in these events, the most important being the α-amino-3-hydroxy-5-methyl-4-isoxazole-propionic acid (AMPA), N-methyl-D aspartate (NMDA) and the metabotropic glutamate receptors (mGluR).

In unilaterally 6-OHDA-lesioned rats, chronic administration of L-DOPA was reported to induce a hyperphosphorylation of the AMPA receptor, an effect that was significantly attenuated when L-DOPA was administered in combination with an A_{2A} receptor antagonist (Bibbiani et al. 2003; Chase et al. 2003). Since A_{2A} receptors activate protein kinase A (PKA) and C (PKC) (Cheng et al. 2002; Shindou et al. 2002), it is possible that A_{2A} receptor antagonists, by inhibiting these kinases, might attenuate hyperphosphorylation of these glutamatergic receptors.

In addition, A_{2A} receptors regulate the conductance (Nörenberg et al. 1998; Wirkner et al. 2004) and phosphorylation of NMDA receptors (Köles et al. 2001). A role of the striatal A_{2A} and NMDA receptors in dyskinesia is also suggested by a study by Ekonomou et al. (2004) who showed that A_{2A} receptor stimulation upregulates a subunit of the NMDA receptor of weaver mice, in which overactivity of these receptors worsens dyskinesia.

Therefore, through these mechanisms, both the AMPA and NMDA receptors may play a major role in the long-term changes that underlie motor fluctuations induced by DRT. Since the inactivation of the A_{2A} receptors may modify AIMs by impairing long-term potentiation (LTP) processes in the striatum (Schiffmann et al. 2003), and since LTP partially depends on the A_{2A} receptors (D'Alcantara et al. 2001), A_{2A} modulation of the ionotropic glutamate receptors, may possibly attenuate AIMs.

Besides the ionotropic receptors, the mGluRs are present in the BG and interaction with the A_{2A} receptors and some of them has been described (Fig. 9.4b) (Bogenpohl et al. 2012; Lopez et al. 2008).

The mGlu5 receptor (mGluR5) antagonists were the first to be considered as a therapeutic approach for PD. It was, in fact, reported that chronic treatment with mGluR5 antagonists normalized glutamate neurotransmission and reduced the development of dyskinesia (Morin and Di Paolo 2014; Vallano et al. 2013). Moreover, a positive interaction between the A_{2A} receptors and the mGluR5 antagonists was described in models of PD (Coccurello et al. 2004; Kachroo et al. 2005), the basis of which might be the modulatory role of the A_{2A} receptors and the mGluR5 on striatal electrical activity (Domenici et al. 2004).

A different example of an A_{2A} and mGluR interaction in preclinical models of PD, is the mGlu4 receptor (mGluR4). Agonists and positive allosteric modulators (PAM) of the mGluR4 have been suggested to be efficacious in reducing L-DOPA dose, while maintaining the same benefit on PD motor impairment (Amalric et al. 2013; Bennouar et al. 2013; Jones et al. 2012; Lopez et al. 2011). Interestingly, the selective PAM of mGluR4, VU0364770, produces synergistic effects when administered with either L-DOPA or the A_{2A} receptor antagonist preladenant (Jones et al. 2012).

Mechanisms of A_{2A} Receptor Antagonists: Interaction with Cannabinoid Receptors

Additional mechanisms involved in the modulation of DRT-induced abnormal motor responses include the cannabinoid CB_1 receptors (Fernández-Ruiz 2009).

Selective CB_1 antagonists may enhance the antiparkinsonian action of DRT and allow the use of lower doses of DA receptor agonists, possibly reducing side effects (Cao et al. 2007), whereas dyskinesias may be alleviated by activation of the CB_1 receptors (Ferrer et al. 2003; Morgese et al. 2007).

A_{2A}–CB_1 receptor heteromeric complexes in co-transfected HEK-293T cells and rat striatum have been shown (Carriba et al. 2007; Marcellino et al. 2008). Moreover, A_{2A}, cannabinoid CB_1 and D_2 receptors may interact to form A_{2A}–CB_1–D_2 receptor heteromers in co-transfected cells as well as in the rat striatum (Carriba et al. 2007; Marcellino et al. 2008). These heteromers are post-synaptically located in the spines of GABAergic enkephalinergic neurons (Carriba et al. 2007; Pickel et al. 2006).

In rats, the motor effects induced by the intrastriatal administration of CB_1 receptor agonists could be counteracted by A_{2A} receptor antagonists (Carriba et al. 2007), whereas the inhibitory effect of CB_1 receptor agonists on D_2 receptor agonist-induced hyperlocomotion could be counteracted by CB_1 receptor antagonists and A_{2A} receptor antagonists (Marcellino et al. 2008), providing evidence for the existence of functional receptor–receptor interactions. A_{2A}–CB_1–D_2 receptor heteromers are also present in the striatum of 6-OHDA-lesioned rats; however, following acute or chronic treatment with L-DOPA, the heteromer cross-talk is lost (Pinna et al. 2014; Fig. 9.4c).

Similar results were obtained in non-human primates in which the expression of A_{2A}–CB_1, A_{2A}–D_2, and CB_1–D_2 heteromers was reduced in the caudate nucleus of monkeys that received chronic L-DOPA treatment compared with control, supporting the evidence that DRT alters heteromer expression in models of PD (Bonaventura et al. 2014). As reported in a paper by Bonaventura et al. (2014), L-DOPA-induced disruption of A_{2A}–CB_1–D_2 receptor heteromers contributed to the alteration of the balance between striatal direct and indirect efferent pathways by eliminating the "brake" that A_{2A} or CB_1 receptor activation exerts on D_2 receptor-mediated motor behavior. Therefore, drugs targeting A_{2A}–CB_1–D_2 receptor heteromers might balance striatal efferent pathways and prevent DRT-induced motor complications.

Conclusions

A_{2A} receptor antagonists are emerging as leading non-dopaminergic candidates for symptom- and disease-modifying therapy in PD. Our knowledge of the mechanisms underlying their effect, including their interaction with other neurotransmitter systems, has grown over the last few years. Although other neurotransmitters besides

DA, glutamate, and cannabinoids interact with the A_{2A} receptors in the mediation of motor behavior, not all of them have been implicated in the motor complications induced by DRT in PD. In this chapter, therefore, we have reviewed the interaction of the A_{2A} and dopaminergic, glutamatergic, and cannabinoid receptors since, these are the only receptors for which substantial results have been obtained.

Hope of an active translation to clinical trials in PD is growing, as well as the expectation of potential antidepressant cognitive enhancement and neuroprotective effects of A_{2A} receptor antagonists in PD.

References

Amalric M, Lopez S, Goudet C et al (2013) Group III and subtype 4 metabotropic glutamate receptor agonists: discovery and pathophysiological applications in Parkinson's disease. Neuropharmacology 66:53–64

Aoyama S, Kase H, Borrelli E (2000) Rescue of locomotor impairment in dopamine D2 receptor-deficient mice by an adenosine A_{2A} receptor antagonist. J Neurosci 20:5848–5852

Bara-Jimenez W, Sherzai A, Dimitrova T et al (2003) Adenosine A(2A) receptor antagonist treatment of Parkinson's disease. Neurology 61:293–296

Bennouar KE, Uberti MA, Melon C et al (2013) Synergy between L-DOPA and a novel positive allosteric modulator of metabotropic glutamate receptor 4: implications for Parkinson's disease treatment and dyskinesia. Neuropharmacology 66:158–169

Bezard E (2013) Experimental reappraisal of continuous dopaminergic stimulation against L-dopa-induced dyskinesia. Mov Disord 28:1021–1022

Bibbiani F, Oh JD, Petzer JP et al (2003) A_{2A} antagonist prevents dopamine agonist-induced motor complications in animal models of Parkinson's disease. Exp Neurol 184:285–294

Black KJ, Koller JM, Campbell MC et al (2010) Quantification of indirect pathway inhibition by the adenosine A_{2a} antagonist SYN115 in Parkinson disease. J Neurosci 30:16284–16292

Bogenpohl JW, Ritter SL, Hall RA et al (2012) Adenosine A_{2A} receptor in the monkey basal ganglia: ultrastructural localization and colocalization with the metabotropic glutamate receptor 5 in the striatum. J Comp Neurol 520:570–589

Bonaventura J, Rico AJ, Moreno E et al (2014) L-DOPA-treatment in primates disrupts the expression of A(2A) adenosine-CB(1) cannabinoid-D(2) dopamine receptor heteromers in the caudate nucleus. Neuropharmacology 79:90–100

Boraud T, Bezard E, Bioulac B et al (2001) Dopamine agonist-induced dyskinesias are correlated to both firing pattern and frequency alterations of pallidal neurones in the MPTP-treated monkey. Brain 124:546–557

Calabresi P, Centonze D, Bernardi G (2000) Electrophysiology of dopamine in normal and denervated striatal neurons. Trends Neurosci 23:S57–S63

Calon F, Dridi M, Hornykiewicz O et al (2004) Increased adenosine A_{2A} receptors in the brain of Parkinson's disease patients with dyskinesias. Brain 127:1075–1084

Canales JJ, Graybiel AM (2000) Patterns of gene expression and behavior induced by chronic dopamine treatments. Ann Neurol 47:S53–S59

Cao X, Liang L, Hadcock JR et al (2007) Blockade of cannabinoid type 1 receptors augments the antiparkinsonian action of levodopa without affecting dyskinesias in 1-methyl-4-phenyl-1,2,3,6-tetrahydropyridine-treated rhesus monkeys. J Pharmacol Exp Ther 323:318–326

Carriba P, Ortiz O, Patkar K et al (2007) Striatal adenosine A_{2A} and cannabinoid CB1 receptors form functional heteromeric complexes that mediate the motor effects of cannabinoids. Neuropsychopharmacology 32:2249–2259

Carta M, Tronci E (2014) Serotonin system implication in l-DOPA-induced dyskinesia: from animal models to clinical investigations. Front Neurol 5:78

Carta AR, Pinna A, Cauli O et al (2002) Differential regulation of GAD67, enkephalin and dynorphin mRNAs by chronic-intermittent L-dopa and A$_{2A}$ receptor blockade plus L-dopa in dopamine-denervated rats. Synapse 44:166–174

Carta AR, Pinna A, Tronci E et al (2003a) Adenosine A$_{2A}$ and dopamine receptor interactions in basal ganglia of dopamine denervated rats. Neurology 61:S39–S43

Carta AR, Tabrizi MA, Baraldi PG et al (2003b) Blockade of A$_{2A}$ receptors plus l-DOPA after nigrostriatal lesion results in GAD67 mRNA changes different from l-DOPA alone in the rat globus pallidus and substantia nigra reticulata. Exp Neurol 184:679–687

Carta AR, Tronci E, Pinna A et al (2005) Different responsiveness of striatonigral and striatopallidal neurons to L-DOPA after a subchronic intermittent L-DOPA treatment. Eur J Neurosci 21:1196–1204

Carta AR, Pinna A, Morelli M (2006a) How reliable is the behavioural evaluation of dyskinesia in animal models of Parkinson's disease? Behav Pharmacol 17:393–402

Carta M, Lindgren HS, Lundblad M et al (2006b) Role of striatal L-DOPA in the production of dyskinesia in 6-hydroxydopamine lesioned rats. J Neurochem 96:1718–1727

Cenci MA, Lundblad M (2006) Post- versus presynaptic plasticity in L-DOPA-induced dyskinesia. J Neurochem 99:381–392

Cenci MA, Lee CS, Bjorklund A (1998) L-dopa-induced dyskinesia in the rat is associated with striatal overexpression of prodynorphinand glutamic acid decarboxylase mRNA. Eur J Neurosci 10:2694–2706

Chase TN (1998) The significance of continuous dopaminergic stimulation in the treatment of Parkinson's disease. Drugs 55:1–9

Chase TN, Bibbiani F, Oh JD (2003) Striatal glutamatergic mechanisms and extrapyramidal movement disorders. Neurotox Res 5:139–146

Cheng HC, Shih HM, Chern Y (2002) Essential role of cAMP-response element-binding protein activation by A$_{2A}$ adenosine receptors in rescuing the nerve growth factor-induced neurite outgrowth impaired by blockage of the MAPK cascade. J Biol Chem 277:33930–33942

Coccurello R, Breysse N, Amalric M (2004) Simultaneous blockade of adenosine A$_{2A}$ and metabotropic glutamate mGlu5 receptors increase their efficacy in reversing Parkinsonian deficits in rats. Neuropsychopharmacology 29:1451–1461

Cutler DL, Tendolkar A, Grachev ID (2012) Safety, tolerability and pharmacokinetics after single and multiple doses of preladenant (SCH420814) administered in healthy subjects. J Clin Pharm Ther 37:578–587

D'Alcantara P, Ledent C, Swillens S et al (2001) Inactivation of adenosine A$_{2A}$ receptor impairs long term potentiation in the accumbens nucleus without altering basal synaptic transmission. Neuroscience 107:455–464

Domenici MR, Pepponi R, Martire A et al (2004) Permissive role of adenosine A$_{2A}$ receptors on metabotropic glutamate receptor 5 (mGluR5)-mediated effects in the striatum. J Neurochem 90:1276–1279

Ekonomou A, Poulou PD, Matsokis N et al (2004) Stimulation of adenosine A$_{2A}$ receptors elicits zif/268 and NMDA epsilon2 subunit mRNA expression in cortex and striatum of the "weaver" mutant mouse, a genetic model of nigrostriatal dopamine deficiency. Neuroscience 123:1025–1036

Engber TM, Susel Z, Kuo S et al (1991) Levodopa replacement therapy alters enzyme activities in striatum and neuropeptide content in striatal output regions of 6-hydroxydopamine lesioned rats. Brain Res 552:113–118

Factor SA, Wolski K, Togasaki DM et al (2013) Long-term safety and efficacy of preladenant in subjects with fluctuating Parkinson's disease. Mov Disord 28:817–820

Fernández-Ruiz J (2009) The endocannabinoid system as a target for the treatment of motor dysfunction. Br J Pharmacol 156:1029–1040

Ferré S, von Euler G, Johansson B et al (1991) Stimulation of high-affinity adenosine A2 receptors decreases the affinity of dopamine D2 receptors in rat striatal membranes. Proc Natl Acad Sci U S A 88:7238–7241

Ferré S, Agnati LF, Ciruela F et al (2007) Neurotransmitter receptor heteromers and their integrative role in 'local modules': the striatal spine module. Brain Res Rev 55:55–67

Ferré S, Baler R, Bouvier M et al (2009) Building a new conceptual framework for receptor heteromers. Nat Chem Biol 5:131–134

Ferrer B, Asbrock N, Kathuria S et al (2003) Effects of levodopa on endocannabinoid levels in rat basal ganglia: implications for the treatment of levodopa-induced dyskinesias. Eur J Neurosci 18:1607–1614

Franco R, Casadó V, Cortés A et al (2008) Novel pharmacological targets based on receptor heteromers. Brain Res Rev 58:475–482

Frau L, Morelli M, Simola N (2013) Performance of movement in hemiparkinsonian rats influences the modifications induced by dopamine agonists in striatal efferent dynorphinergic neurons. Exp Neurol 247:663–672

Fredduzzi S, Moratalla R, Monopoli A et al (2002) Persistent behavioral sensitization to chronic L-DOPA requires A_{2A} adenosine receptors. J Neurosci 22:1054–1062

Fuxe K, Agnati LF, Jacobsen K et al (2003) Receptor heteromerization in adenosine A_{2A} receptor signaling: relevance for striatal function and Parkinson's disease. Neurology 61:S19–S23

Fuxe K, Ferré S, Genedani S et al (2007) Adenosine receptor-dopamine receptor interactions in the basal ganglia and their relevance for brain function. Physiol Behav 92:210–217

Grondin R, Bédard PJ, Hadj Tahar A et al (1999) Antiparkinsonian effect of a new selective adenosine A_{2A} receptor antagonist in MPTP-treated monkeys. Neurology 52:1673–1677

Hauser RA, Hubble JP, Truong DD (2003) Istradefylline US-001 study group. Randomized trial of the adenosine A(2A) receptor antagonist istradefylline in advanced PD. Neurology 61:297–303

Hauser RA, Shulman LM, Trugman JM et al (2008) Istradefylline 6002-US-013 study group. Study of istradefylline in patients with Parkinson's disease on levodopa with motor fluctuations. Mov Disord 23:2177–2185

Hauser RA, Cantillon M, Pourcher E et al (2011) Preladenant in patients with Parkinson's disease and motor fluctuations: a phase 2, double-blind, randomised trial. Lancet Neurol 10:221–229

Hauser RA, Olanow CW, Kieburtz KD et al (2014) Tozadenant (SYN115) in patients with Parkinson's disease who have motor fluctuations on levodopa: a phase 2b, double-blind, randomised trial. Lancet Neurol 13:767–776

Henry B, Crossman AR, Brotchie JM (1998) Characterization of enhanced behavioral responses to L-DOPA following repeated administration in the 6-hydroxydopamine-lesioned rat model of Parkinson's disease. Exp Neurol 151:334–342

Henry B, Crossman AR, Brotchie JM (1999) Effect of repeated L-dopa, bromocriptine, or lisuride administration on preproenkephalin-A and preproenkephalin-B mRNA levels in striatum of the 6-hidroxydopamine-lesioned rat. Exp Neurol 155:204–220

Hettinger BD, Lee A, Linden J et al (2001) Ultrastructural localization of adenosine A_{2A} receptors suggests multiple cellular sites for modulation of GABAergic neurons in rat striatum. J Comp Neurol 431:331–346

Hillion J, Canals M, Torvinen M et al (2002) Coaggregation, cointernalization, and codesensitization of adenosine A_{2A} receptors and dopamine D2 receptors. J Biol Chem 277:18091–18097

Hodgson RA, Bertorelli R, Varty GB et al (2009) Characterization of the potent and highly selective A_{2A} receptor antagonists preladenant and SCH 412348 [7-[2-[4-2,4-difluorophenyl]-1-piperazinyl]ethyl]-2-(2-furanyl)-7H-pyrazolo[4,3-e][1,2,4]triazolo[1,5-c]pyrimidin-5-amine] in rodent models of movement disorders and depression. J Pharmacol Exp Ther 330:294–303

Hodgson RA, Bedard PJ, Varty GB et al (2010) Preladenant, a selective A(2A) receptor antagonist, is active in primate models of movement disorders. Exp Neurol 225:384–390

Jankovic J (2005) Motor fluctuations and dyskinesias in Parkinson's disease: clinical manifestations. Mov Disord 20:S11–S16

Jenner P (2004) Avoidance of dyskinesia: preclinical evidence for continuous dopaminergic stimulation. Neurology 62:S47–S55

Jenner P (2008) Molecular mechanisms of L-DOPA-induced dyskinesia. Nat Rev Neurosci 9:665–677

Jones CK, Bubser M, Thompson AD et al (2012) The metabotropic glutamate receptor 4-positive allosteric modulator VU0364770 produces efficacy alone and in combination with L-DOPA or an adenosine 2A antagonist in preclinical models of Parkinson's disease. J Pharmacol Exp Ther 340:404–421

Jones N, Bleickardt C, Mullins D et al (2013) A$_{2A}$ antagonists do not induce dyskinesias in drug-naive or L-dopa sensitized rats. Brain Res 163–169

Kachroo A, Orlando LR, Grandy DK et al (2005) Interaction between metabotropic glutamate 5 and adenosine A$_{2A}$ receptors in normal and parkinsonian. J Neurosci 25:10414–10419

Kanda T, Uchida S (2014) Clinical/Pharmacological aspect of adenosine A$_{2A}$ receptor antagonist for dyskinesia. Int Rev Neurobiol 119:127–150

Kanda T, Jackson MJ, Smith LA et al (2000) Combined use of the adenosine A(2A) antagonist KW-6002 with L-DOPA or with selective D1 or D2 dopamine agonist increases antiparkinsonian activity but not dyskinesia in MPTP-treated monkeys. Exp Neurol 162:321–327

Kase H, Aoyama S, Ichimura M et al (2003) KW-6002 US-001 study group. Progress in pursuit of therapeutic A$_{2A}$ antagonists: the adenosine A$_{2A}$ receptor selective antagonist KW6002: research and development toward a novel nondopaminergic therapy for Parkinson's disease. Neurology 61:S97–S100

Knebel W, Rao N, Uchimura T et al (2012) Population pharmacokinetic-pharmacodynamic analysis of istradefylline in patients with Parkinson disease. J Clin Pharmacol 52:1468–14

Köles L, Wirkner K, Illes P (2001) Modulation of ionotropic glutamate receptor channels. Neurochem Res 26:925–932

Kyowa Hakko Kirin Co. Ltd. (2013) Approval for manufacturing and marketing of NOURIAST© tablets 20 mg, a novel antiparkinsonian agent. News releases. 2013. http://www.kyowa-kirin.com/news_releases/2013/e20130325_04.html. Accessed 15 Feb 2014

LeWitt PA, Guttman M, Tetrud JW et al (2008) 6002-US-005 study group. Adenosine A$_{2A}$ receptor antagonist istradefylline (KW-6002) reduces "off" time in Parkinson's disease: a double-blind, randomized, multicenter clinical trial (6002-US-005). Ann Neurol 63:295–302

Lindgren HS, Rylander D, Ohlin KE et al (2007) The "motor complication syndrome" in rats with 6-OHDA lesions treated chronically with L-DOPA: relation to dose and route of administration. Behav Brain Res 177:150–159

Lloyd KG, Davidson L, Hornykiewicz O (1975) The neurochemistry of Parkinson's disease: effect of L-dopa therapy. J Pharmacol Exp Ther 195:453–464

Lopez S, Turle-Lorenzo N, Johnston TH et al (2008) Functional interaction between adenosine A$_{2A}$ and group III metabotropic glutamate receptors to reduce parkinsonian symptoms in rats. Neuropharmacology 55:483–490

Lopez S, Bonito-Oliva A, Pallottino S et al (2011) Activation of metabotropic glutamate 4 receptors decreases L-DOPA-induced dyskinesia in a mouse model of Parkinson's disease. J Parkinsons Dis 1:339–346

Lundblad M, Andersson M, Winkler C et al (2002) Pharmacological validation of behavioural measures of akinesia and dyskinesia in a rat model of Parkinson's disease. Eur J Neurosci 15:120–132

Lundblad M, Vaudano E, Cenci MA (2003) Cellular and behavioural effects of the adenosine A$_{2A}$ receptor antagonist KW-6002 in a rat model of l-DOPA-induced dyskinesia. J Neurochem 84:1398–1410

Marcellino D, Carriba P, Filip M et al (2008) Antagonistic cannabinoid CB1/dopamine D2 receptor interactions in striatal CB1/D2 heteromers. A combined neurochemical and behavioral analysis. Neuropharmacology 54:815–823

Marsden CD (1994) Parkinson's disease. J Neurol Neurosurg Psychiatry 57:672–681

Merck. Newsroom. News releases-research and development news (May 2013) Merck provides update on Phase III clinical program for preladenant, the company's investigational Parkinson's disease medicine. http://www.mercknewsroom.com/press-release/research-and-development-news/merck-providesupdate-phase-iii-clinical-program-prelade. Accessed 15 Feb 2014

Mishina M, Ishiwata K, Naganawa M et al (2011) Adenosine A(2A) receptors measured with [C] TMSX PET in the striata of Parkinson's disease patients. PLoS One 6:e17338

Mizuno Y, Kondo T (2013) Japanese istradefylline study group. Adenosine A$_{2A}$ receptor antagonist istradefylline reduces daily OFF time in Parkinson's disease. Mov Disord 28:1138–1141

Morelli M, Di Paolo T, Wardas J et al (2007) ... of adenosine A_{2A} receptors in parkinsonian motor impairment and l-DOPA-induced ... complications. Prog Neurobiol 83:293–309

Morgese MG, Cassano T, Cuomo V et al (2007) Anti-dyskinetic effects of cannabinoids in a rat model of Parkinson's disease: ... B(1) and TRPV1 receptors. Exp Neurol 208:110–119

Morin N, Di Paolo T (2014) Interaction of adenosine receptors with other receptors from therapeutic perspective in Parkinson's disease. Int Rev Neurobiol 119:151–167

Nielsen KM, Soghomonian J (...) Dual effects of intermittent or continuous L-DOPA administration on gene expression in the globus pallidus and subthalamic nucleus of adult rats with a unilateral 6-OHDA lesion. Synapse 49:246–260

Nörenberg W, Wirkner K, Assmann H et al (1998) Adenosine A_{2A} receptors inhibit the conductance of NMDA receptor channels in rat neostriatal neurons. Amino Acids 14:33–39

Nutt JG (1987) On-off phenomenon: relation to levodopa pharmacokinetics and pharmacodynamics. Ann Neurol ...535–540

Obeso JA, Rodriguez-Oroz MC, Chana P et al (2000) The evolution and origin of motor complications in Parkinson's disease. Neurology 55:S13–S23

Obeso JA, Rodriguez-Oroz M, Marin C et al (2004) The origin of motor fluctuations in Parkinson's disease: importance of dopaminergic innervation and basal ganglia circuits. Neurology 62:S17–S30

Olanow CW, Obeso JA (2000) Pulsatile stimulation of dopamine receptors and levodopa-induced motor complications in Parkinson's disease: implications for the early use of COMT inhibitors. Neurology 55:72–77

Olanow CW, Tatton WG (1999) Etiology and pathogenesis of Parkinson's disease. Annu Rev Neurosci 22:123–144

Papa SM, Engber TM, Kask AM et al (1994) Motor fluctuations in levodopa treated parkinsonian rats: relation to lesion extent and treatment duration. Brain Res 662:69–74

Papapetropoulos S, Borgohain R, Kellet M et al (2010) The adenosine A_{2A} receptor antagonist BIIB014 is effective in improving ON-time in Parkinson's disease (PD) patients with motor fluctuations. Mov Disord 25:S305

Paterniti I, Melani A, Cipriani S et al (2011) Selective adenosine A_{2A} receptor agonists and antagonists protect against spinal cord injury through peripheral and central effects. J Neuroinflammation 8:31

Pavón N, Martín AB, Mendialdua A et al (2006) ERK phosphorylation and FosB expression are associated with L-DOPA-induced dyskinesia in hemiparkinsonian mice. Biol Psychiatry 59:64–74

Picconi B, Centonze D, Håkansson K et al (2003) Loss of bidirectional striatal synaptic plasticity in L-DOPA-induced dyskinesia. Nat Neurosci 6:501–506

Pickel VM, Chan J, Kearn CS et al (2006) Targeting dopamine D2 and cannabinoid-1 (CB1) receptors in rat nucleus accumbens. J Comp Neurol 495:299–313

Pinna A (2014) Adenosine A_{2A} receptor antagonists in Parkinson's disease: progress in clinical trials from the newly approved istradefylline to drugs in early development and those already discontinued. CNS Drugs 28:455–474

Pinna A, Morelli M (2014) A critical evaluation of behavioral rodent models of motor impairment used for screening of antiparkinsonian activity: The case of adenosine A(2A) receptor antagonists. Neurotox Res 25:392–401

Pinna A, Chiara G D, Wardas J et al (1996) Blockade of A_{2A} adenosine receptors positively modulates turning behaviour and c-Fos expression induced by D1 agonists in dopamine-denervated rats. Eur J Neurosci 8:1176–1181

Pinna A, Fenu S, Morelli M (2001) Motor stimulant effects of the adenosine A_{2A} receptor antagonist SCH 58261 do not develop tolerance after repeated treatments in 6-hydroxydopamine-lesioned rats. Synapse 39:233–238

Pinna A, Tronci E, Schintu N et al (2010) A new ethyladenine antagonist of adenosine A(2A) receptors: behavioral and biochemical haracterization as an antiparkinsonian drug. Neuropharmacology 58:613–623

Pinna A, Bonaventura J, Farré D et al (2014) L-DOPA disrupts adenosine A(2A)-cannabinoid CB(1)-dopamine D(2) receptor heteromer cross-talk in the striatum of hemiparkinsonian rats: biochemical and behavioral studies. Exp Neurol 253:180–191

Pollack AE, Fink JS (1996) Synergistic interaction between an adenosine antagonist and a D1 dopamine agonist on rotational behavior and striatal c-Fos induction in 6-hydroxydopamine-lesioned rats. Brain Res 743:124–130

Popoli P, Frank C, Tebano MT et al (2003) Modulation of glutamate release and excitotoxicity by adenosine A_{2A} receptors. Neurology 61:S69–S71

Ramlackhansingh AF, Bose SK, Ahmed I et al (2011) Adenosine 2A receptor availability in dyskinetic and nondyskinetic patients with Parkinson disease. Neurology 76:1811–1816

Robertson GS, Robertson HA (1986) Synergistic effects of D1 and D2 dopamine agonists on turning behaviour in rats. Brain Res 384:387–390

Rosin DL, Hettinger BD, Lee A et al (2003) Anatomy of adenosine A_{2A} receptors in brain: morphological substrates for integration of striatal function. Neurology 61:S12–S18

Schiffmann SN, Dassesse D, D'Alcantara P et al (2003) A_{2A} receptor and striatal cellular functions: regulation of gene expression, currents, and synaptic transmission. Neurology 61:S24–S29

Schwarzschild MA, Agnati L, Fuxe K et al (2006) Targeting adenosine A_{2A} receptors in Parkinson's disease. Trends Neurosci 29:647–654

Shindou T, Nonaka H, Richardson PJ et al (2002) Presynaptic adenosine A_{2A} receptors enhance GABAergic synaptic transmission via a cyclic AMP dependent mechanism in the rat globus pallidus. Br J Pharmacol 136:296–302

Sohn YH, Metman LV, Bravi D et al (1994) Levodopa peak response time reflects severity of dopamine neuron loss in Parkinson's disease. Neurology 44:755–757

Stacy M, Silver D, Mendis T et al (2008) A 12-week, placebo-controlled study (6002-US-006) of istradefylline in Parkinson disease. Neurology 70:2233–2240

Svenningsson P, Le Moine C, Fisone G et al (1999) Distribution, biochemistry and function of striatal adenosine A_{2A} receptors. Prog Neurobiol 59:355–396

Tao Y, Liang G (2015) Efficacy of adenosine A_{2A} receptor antagonist istradefylline as augmentation for Parkinson's disease: a meta-analysis of randomized controlled trials. Cell Biochem Biophys. 71:57–62

Tronci E, Simola N, Borsini F et al (2007) Characterization of the antiparkinsonian effects of the new adenosine A_{2A} receptor antagonist ST1535: acute and subchronic studies in rats. Eur J Pharmacol 566:94–102

Uchida S, Tashiro T, Kawai-Uchida M et al (2014) Adenosine A_2A-receptor antagonist istradefylline enhances the motor response of L-DOPA without worsening dyskinesia in MPTP-treated common marmosets. J Pharmacol Sci 124:480–485

Ungerstedt U (1968) 6-hydroxy-dopamine induced degeneration of central monoamine neurons. Eur J Pharmacol 5:107–110

Ungerstedt U (1971) Striatal dopamine release after amphetamine or nerve degeneration revealed by rotational behaviour. Acta Physiol Scand Suppl 367:49–68

Vallano A, Fernandez-Duenas V, Garcia-Negredo G et al (2013) Targeting striatal metabotropic glutamate receptor type 5 in Parkinson's disease: bridging molecular studies and clinical trials. CNS Neurol Disord Drug Targets 12:1128–1142

Vernalis (2010) Media centre. Vernalis announces A_{2A} receptor antagonist programme for Parkinson's disease continues with next generation compound. http://www.vernalis.com/media-centre/latest-releases/2010-releases/584

Westin JE, Andersson M, Lundblad M et al (2001) Persistent changes in striatal gene expression induced by long-term L-DOPA treatment in a rat model of Parkinson's disease. Eur J Neurosci 14:1171–1176

Wills AM, Eberly S, Tennis M et al (2013) Parkinson study group. Caffeine consumption and risk of dyskinesia in CALM-PD. Mov Disord 28:380–383

Wirkner K, Gerevich Z, Krause T et al (2004) Adenosine A_{2A} receptor-induced inhibition of NMDA and GABAA receptor-mediated synaptic currents in a subpopulation of rat striatal neurons. Neuropharmacology 46:994–1007

Xiao D, Bastia E, Xu YH et al (2006) Forebrain adenosine A_{2A} receptors contribute to L-3,4-dihydroxyphenylalanine-induced dyskinesia in hemiparkinsonian mice. J Neurosci 26:13548–13555

Xie CL, Wang WW, Zhang SF et al (2014) Continuous dopaminergic stimulation (CDS)-based treatment in Parkinson's disease patients with motor complications: a systematic review and meta-analysis. Sci Rep 4:6027

Xu K, Bastia E, Schwarzschild M (2005) Therapeutic potential of adenosine A(2A) receptor antagonists in Parkinson's disease. Pharmacol Ther 105:267–310

Zhu C, Wang G, Li J et al (2014) Adenosine A_{2A} receptor antagonist istradefylline 20 versus 40 mg/day as augmentation for Parkinson's disease: a meta-analysis. Neurol Res 36:1028–1034

Chapter 10
Adenosine A_{2A} Receptor-Mediated Control of Non-Motor Functions in Parkinson's Disease

Rui Daniel Prediger, Filipe Carvalho Matheus, Paulo Alexandre de Oliveira, Daniel Rial, Morgana Moretti, Ana Cristina Guerra de Souza, Aderbal Silva Aguiar and Rodrigo A. Cunha

Abstract Parkinson's disease (PD) is traditionally recognized as a motor disease. However, non-motor symptoms associated with PD are frequent and currently difficult to manage, being reported by patients to represent a significant burden. We now review the ability of adenosine A_{2A} receptor ($A_{2A}R$) antagonist to attenuate several non-motor PD symptoms including olfactory impairments, anxiety, depression and cognitive deficits. This paves the way to consider $A_{2A}R$ antagonists as novel holistic drugs for PD patients since they not only ameliorate the efficacy of L-DOPA and attenuate its dyskinetic effects, but also afford neuroprotection and attenuate mood and cognitive dysfunctions associated with PD. This clearly prompts the need to detail the underlying mechanisms to understand when and how $A_{2A}R$ should be exploited to maximize benefits for PD patients.

Keywords Parkinson's disease · Adenosine A_{2A} receptors · Non-motor symptoms · Depression · Memory · Olfaction

R. D. Prediger (✉) · F. C. Matheus · P. A. de Oliveira · D. Rial · M. Moretti · A. C. G. de Souza · A. S. Aguiar
Laboratório Experimental de Doenças Neurodegenerativas (LEXDON), Centro de Ciências Biológicas (CCB), Universidade Federal de Santa Catarina (UFSC), Florianópolis, SC, 88040-900, Brazil
e-mail: ruidsp@hotmail.com

D. Rial · R. A. Cunha
CNC-Center for Neuroscience and Cell Biology, University of Coimbra, 3004-517 Coimbra, Portugal

A. S. Aguiar
Laboratório de Neurobiologia do Exercício Físico, Centro de Ciências Biológicas (CCB), Universidade Federal de Santa Catarina (UFSC), Florianópolis, SC, 88040-900, Brazil

R. A. Cunha
FMUC-Faculty of Medicine, University of Coimbra, 3004-501 Coimbra, Portugal

© Springer International Publishing Switzerland 2015

M. Morelli et al. (eds.), *The Adenosinergic System,* Current Topics in Neurotoxicity 10, DOI 10.1007/978-3-319-20273-0_10

Parkinson's Disease Encompasses Non-Motor Symptoms

Classically, Parkinson's disease (PD) is considered to be a motor system disease and its diagnosis is based on the presence of a set of cardinal motor signs (e.g., rigidity, bradykinesia, rest tremor and postural reflex disturbance). These symptoms of PD mainly result from the progressive degeneration of dopamine neurons of the *substantia nigra pars compacta* (SNc), which causes a consequent reduction of dopamine levels in the striatum (Hirsch et al. 1988). Dopamine-replacement therapy has dominated the treatment of PD since the early 1960s and, although the currently approved anti-parkinsonian agents offer an effective relief of the motor deficits during the early stages of the disease, but they have not been found to alleviate the underlying dopaminergic neuron degeneration (Allain et al. 2008). Another major limitation of chronic dopaminergic therapy is the numerous adverse effects such as the development of abnormal involuntary movements (namely dyskinesia), psychosis and behavioral disturbance (e.g., compulsive gambling, hypersexuality) (Ahlskog and Muenter 2001).

The diagnosis of a PD patient is classically based on the motor dysfunction and the clinical criteria are only fulfilled when approximately 70% of the neurons of the *substantia nigra* (SN) are already degenerated and the striatal dopamine (DA) content is reduced by 80% (Braak et al. 2004; Riederer and Wuketich 1976). The more detailed clinical evaluation of patients developing PD, coupled to the advent of different image techniques (like the PET-SCAN), have contributed to consolidate the initial neuropathological evidence suggesting that extra-striatal deficits are also present (Braak et al. 2003, 2004); these are designated as non-motor symptoms and often pre-date the motor symptoms corresponding to overt PD. These early non-motor symptoms are increasingly used to provide an early diagnose of PD and are increasingly recognized as a major burden associated with PD. Among these pre-motor symptoms, particular emphasis is nowadays given to olfactory deficits, cognitive deficits, and alterations of mood typified by anxiety, depression, anhedonia and apathy (Chaudhuri et al. 2006). Notably, these non-motor features of PD invariably respond poorly to dopaminergic medication and are probably the major current challenge faced in the clinical management of PD (Chaudhuri et al. 2006).

The Adenosine Neuromodulation System—Focus on Adenosine $A_{2A}R$ and Motor Control in Parkinson's Disease

Adenosine is an extracellular signaling molecule, acting through four different G-protein coupled receptors out of which the A_1 and A_{2A} receptors ($A_{2A}R$) are the predominant players in the control of brain function (Fredholm et al. 2005). The A_1R is mostly located in synapses (Rebola et al. 2003; Tetzlaff et al. 1987), namely in glutamatergic synapses (Lambert and Teyler 1991; Yoon and Rothman 1991), where they inhibit the release of glutamate and trigger a hyperpolarization that decreases post-synaptic responsiveness (Dunwiddie and Masino 2001; Thompson et al. 1992). The A$_{2A}$R is present at a lower density in the brain, mainly with a synaptic localization (Rebola et al. 2005), although they are also located in glia cells (e.g. Matos et al. 2012; Orr et al. 2009). The combined synaptic action of A$_1$R, controlling excessive noise and heterosynaptic depression (Manzoni et al. 1994; Serrano et al. 2006), and A$_{2A}$R, bolstering N-methyl-D aspartate (NMDA) receptor activation (Rebola et al. 2008) and synaptic plasticity (Costenla et al. 2011; d'Alcantara et al. 2001; Flajolet et al. 2008; Rebola et al. 2008; Shen et al. 2008), constitutes an integrated neuromodulation system assisting the implementation of information salience in brain networks (Cunha 2008). Accordingly, a modification of the efficiency of this adenosine system is expected to play a critical role in the mal-adaptive functioning of neuronal networks underlying brain diseases (Gomes et al. 2011; Lopes et al. 2011).

The relevance of the adenosine neuromodulation system is further heralded by the evidence that adenosine receptors are the main molecular of non-toxic doses of caffeine (Fredholm et al. 1999; Yang et al. 2009). And over the last decade, several lines of evidence have suggested the potential of caffeine in the prevention of PD as well as the ability of caffeine to attenuate of motor deficits in different animal models of PD (Chen et al. 2001; Costa et al. 2010; Palacios et al. 2012; Prediger 2010; Schwarzschild et al. 2002). Notably, A$_{2A}$R are now recognized as the main targets operated by chronic caffeine consumption to generate its psychoactive effects (Cunha and Agostinho 2010; Ferré 2008; Prediger 2010). Their interest in the control of motor dysfunction in PD is further heralded by the particular enrichment of A$_{2A}$R in the striatum.

The striatum is mainly composed of projection neurons (medium spiny neurons, MSNs), which use γ-aminobutyric acid (GABA) as their neurotransmitter. These MSNs are driven by glutamatergic cortico-thalamic inputs and are traditionally divided into two populations based on their function and neurochemical phenotype: striatonigral or striatopallidal pathways (Alexander and Crutcher 1990; Gerfen et al. 1990); these populations give rise to the direct (expressing dopamine D$_1$ receptors, D$_1$R) and indirect pathways (expressing dopamine D$_2$R), respectively, which are differentially modulated by dopamine through its action on D$_1$R (G$_s$-coupled) or D$_2$R (G$_{i/o}$-coupled). A simplified view of the striatal circuitry identifies the direct pathway with an enhanced locomotion and prompting actions, whereas the indirect pathway acts is associated with behavioral inhibition (Eagle and Baunez 2010; Jahfari et al. 2011). The crucial impact of dopamine on striatal circuits is best summarized by its ability to inhibit the brake (inhibitory D$_2$R-mediated action in MSNs of the indirect pathway) and facilitate the accelerator (facilitatory D$_1$R-mediated actions in MSNs of the direct pathway).

Notably, A$_{2A}$R are particularly enriched post-synaptically in striatopallidal medium spiny neurons of the indirect pathway (Schiffmann and Vanderhaeghen 1993; Svenningsson et al. 1999). Here, A$_{2A}$R interact and heterodimerize with D$_2$R, which constitute the basis of the antagonistic interaction between A$_{2A}$R and D$_2$R (Canals et al. 2003; Hillion et al. 2002). Thus, A$_{2A}$R agonists inhibit striatal D$_2$R binding, D$_2$R-mediated neurotransmitter release and immediate early gene expression; conversely, A$_{2A}$R antagonists mimic the molecular and neurochemical effects of D$_2$R agonists (Fredholm et al. 2005). Given that reduced D$_2$R mediated signaling is thought to be partly responsible for PD symptoms, this $A_{2A}R$-D_2R antagonistic interaction provides a strong anatomical and molecular basis for the motor benefits of $A_{2A}R$ antagonists in PD (Ferré et al. 2007; Schwarzschild et al. 2006). Furthermore, $A_{2A}R$ also control post-synaptic NMDA receptors (Gerevich et al. 2002; Higley and Sabatini 2010) and dimerize with metabotropic group 5 glutamate receptors (mGluR5) (Bogenpohl et al. 2012; Ferré et al. 2002; Rodrigues et al. 2005; Tebano et al. 2005), as well as with presynaptic CB1 (Ferreira et al. 2015; Martire et al. 2011) and A_1R (Ciruela et al. 2006), in accordance with the localization of $A_{2A}R$ in glutamatergic nerve terminals driving the firing of medium spiny neurons (Rebola et al. 2005; Rosin et al. 2003). This combined pre- and post-synaptic localization of $A_{2A}R$, together with their integrative interactions of the main transmitter systems at cortico-striatal synapses provide a molecular basis for the ability of $A_{2A}R$ to fine-tune MSN signal processing (Schiffmann et al. 2007). Accordingly, striatal $A_{2A}R$ are critical regulators of cortico-striatal synaptic plasticity, as typified by a control of long-term potentiation (LTP) (d'Alcantara et al. 2001; Flajolet et al. 2008), long-term depression (LTD, Lerner and Kreitzer 2012) and spike-timing-dependent plasticity at cortico-striatal synapses (Shen et al. 2008). Thus, $A_{2A}R$ are uniquely positioned to integrate incoming information (glutamate signals) and neuronal sensitivity to this incoming information (dopamine signals) to control striatal synaptic plasticity and behavior (reviewed in Schiffmann et al. 2007).

Not surprisingly, $A_{2A}R$ have recently emerged as a leading non-dopaminergic therapeutic target in PD (Ferré et al. 2007; Schwarzschild et al. 2006). $A_{2A}R$ antagonists have demonstrated motor benefits and may have neuroprotective benefits as well. Clinical Phase II-III trials have been completed for the $A_{2A}R$ antagonists KW-6002 (istradefylline, Kyowa, Japan) and SCH 420814 (Preladenant, Merck, USA) (Cutler et al. 2012; Hauser 2011), confirming a motor benefit in advanced PD patients. Over the last 5 years, four trials with KW-6002 reported an average reduction in "OFF" time of 1.7 h/day in nearly 1700 patients with advanced PD who were already on optimized L-DOPA regimens. SCH420814 also produced motor benefits, decreasing both OFF time and scores on the unified PD rating scale (UPDRS) in advanced PD patients in a clinical Phase III trial (Hauser 2011). Importantly, both drugs had robust safety profiles in clinical trials. The most exciting prospective role for $A_{2A}R$ antagonists as a novel therapy for PD is their potential to attenuate dopaminergic neurodegeneration, as suggested by convergent epidemiological and experimental evidence (reviewed in Prediger 2010). Three large, long-term (> 30 years follow-up) prospective studies firmly establish a relationship between increased intake of caffeine (an $A_{2A}R$ antagonist) and decreased risk of developing PD (up to five times lower) in men.

Adenosine Impacts Psychiatric Symptoms in Parkinson's Disease

Anxiety is a trait often found in PD patients, with an incidence varying between 30.7 and 55.8 % (Leentjens et al. 2011; Todorova et al. 2014). These anxiety symptoms form a constellation of alterations independent of motor symptoms but may also arise as a psychological reaction to the development of the cardinal motor symptoms characteristic of PD. An anxious parkinsonian patient displays panic attacks comparable to these evidenced by patients with primary psychiatric panic disorder; likewise, PD animal models also display increased anxiety (Prediger et al. 2012b).

There is increased evidence that the adenosinergic system can modulate anxiety disorders both through A$_1$R (Bruns et al. 1983; Marangos and Boulanger 1985) and A$_{2A}$R (reviewed in Correa and Font 2008; Cunha et al. 2008a). The ability of acutely administered methylxanthines to trigger anxiety and panic attacks, particularly in panic disorder patients (Charney et al. 1985), has been associated with the pharmacological blockade of A$_1$R (Bruns et al. 1983; Florio et al. 1998; Maximino et al. 2011; Prediger et al. 2006; Snyder et al. 1981), since the anxiogenic effect of acute caffeine administration is not mimicked by selective A$_{2A}$R antagonists (El Yacoubi et al. 2000). However, the genetic evidence is less clear: in fact both A$_1$R as well as A$_{2A}$R knockout mice display an aggressive and anxious behavior (Giménez-Llort et al. 2002; Johansson et al. 2001; Ledent et al. 1997); also, there is an association of A$_{2A}$R polymorphisms with anxious personalities (Hohoff et al. 2010), autism spectrum disorders (Freitag et al. 2010), the incidence of panic attacks (Alsene et al. 2003; Deckert et al. 1998; Hamilton et al. 2004) and anxious behavior after acute consumption of coffee (Alsene et al. 2003; Childs et al. 2008) or amphetamine (Hohoff et al. 2005). This association between A$_{2A}$R and anxiety (reviewed in Correa and Font 2008; Cunha et al. 2008a) is further re-enforced by the observation that cortical and striatal A$_{2A}$R trigger opposite modifications of startling and conditioned fear (Wei et al. 2014) and A$_{2A}$R over-expression in the forebrain triggers anxiogenic responses (Coelho et al. 2014). This highlights the putative role of A$_{2A}$R in the development of anxiety-related non-motor symptoms of PD, which are currently managed with drugs such as benzodiazepines, buspirone and tricyclic antidepressants (Prediger et al. 2012b).

Depression is another frequent non-motor symptom of PD, with a prevalence approaching 45 %, which often precedes the motor signs of PD (Lemke 2008). The most commonly used antiparkinsonian drugs do not affect depressive symptoms of PD patients (Lemke 2008), prompting the hypothesis that PD-associated depression mainly involves non-dopaminergic systems (Kano et al. 2011; Ongini 2003).

As summarized in Fig. 10.1, there is increasing evidence implicating the adenosine system in the control of depression (reviewed in Cunha et al. 2008a). In fact, a case-control study showed that poor lifetime caffeine consumption increases the risk of Lewy body dementia and depression (Boot et al. 2013), while a 10-year prospective follow-up study showed that increased lifetime caffeine consumption decreased risk of depression (Lucas et al. 2011) and also strongly correlated inversely with the incidence of depression in retired individuals (Smith 2009) and with the risk of suicide (Kawachi et al. 1996; Lucas et al. 2013). The regular (not acute) caffeine consumption also has the ability to prevent different alterations caused by repeated stress (Alzoubi et al. 2013; Haskell et al. 2005; Kale and Addepalli 2014; Pechlivanova et al. 2012), which is a major risk factor for the development of different neuropsychiatric disorders, namely depression, in both humans and animal models (Kim and Diamond 2002; McEwen 2007). Interestingly, caffeine consumption increases in individuals experiencing stressful conditions (Harris et al. 2007) and the consumption of caffeine correlates inversely with enhanced plasma cortisol

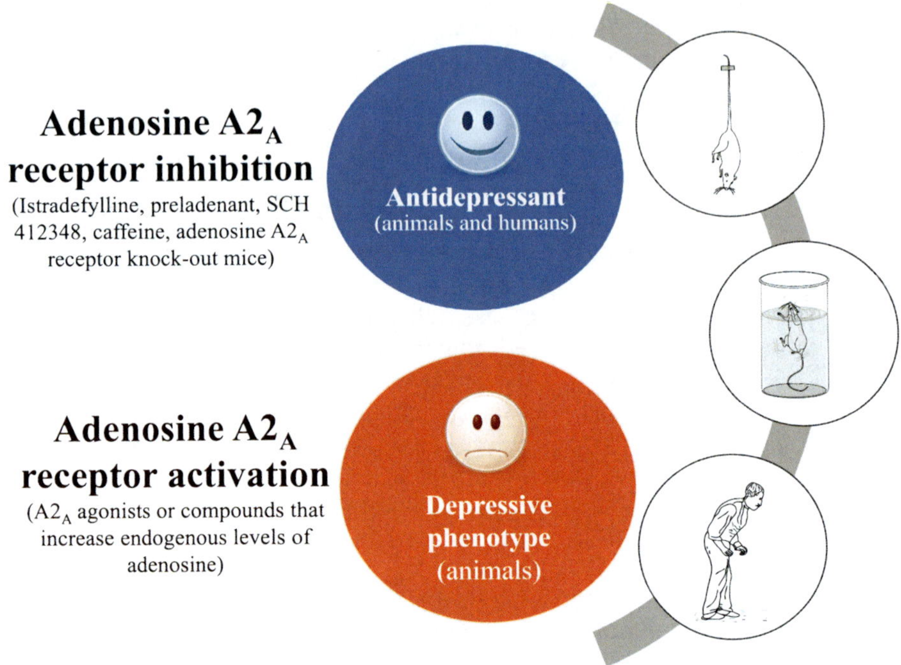

Fig. 10.1 The role of adenosine A_{2A} receptors on mood and depression

levels (Harris et al. 2007). $A_{2A}R$ emerge as the most likely molecular targets of caffeine in the control of mood disorders (reviewed in Cunha et al. 2008a) since hippocampal $A_{2A}R$, but not A_1R, are up-regulated upon chronic repeated stress (Batalha et al. 2013; Cunha et al. 2006).

Accordingly, preclinical studies revealed that $A_{2A}R$ agonists induce depressive-like behaviors (Woodson et al. 1998), while $A_{2A}R$ knockout mice displays increased resistance to 'depressogenic' challenges (El Yacoubi et al. 2001); likewise, $A_{2A}R$ antagonists are antidepressants in animals (Batalha et al. 2013; El Yacoubi et al. 2001; Minor et al. 1994; Yamada et al. 2014), with an efficacy at least similar to desipramine and fluoxetine (Yamada et al. 2014), prolonging escape directed behavior in two screening tests for antidepressant activity (El Yacoubi et al. 2003; Minor et al. 2008; Yamada et al. 2014) and preventing maternal separation-induced long-term cognitive consequences (Batalha et al. 2013). Although caffeine consumption does not seem to change depression scores of PD patients (Altman et al. 2011; Postuma et al. 2012), there seems to be an association between $A_{2A}R$ and depression in PD, as summarized in Fig. 10.1.

The $A_{2A}R$ antagonist istradefylline reduces the daily OFF time in PD patients without troublesome dyskinesia (Hauser et al. 2008; Mizuno et al. 2010; Stacy et al. 2008; Uchida et al. 2014) and the same effective istradefylline dose range attenuated helpless responses in rodents (Yamada et al. 2013). Likewise, other selective $A_{2A}R$ antagonists namely preladenant and SCH 412348 also showed antiparkinsonian outcomes in PD animals models, including primates (Hodgson et al. 2009;

Smith et al. 2014; Varty et al. 2008) and simultaneously reduced the depressive-like behavior in a manner similar to the tricyclic antidepressant desipramine (Hodgson et al. 2009).

Altogether, these data support that an A$_{2A}$R antagonist, such as istradefylline, may be a useful alternative for treating depression in PD. Indeed, a Phase II, double-blind randomized clinical trial demonstrated that preladenant improves motor function, motivation/initiative, thought disorder and depression of PD patients (Hauser et al. 2011). These evidences support a promising antidepressant potential of A$_{2A}$R antagonists to manage PD-associated depression, which has a high prevalence and is not adequately managed by conventionally used pharmacotherapy.

Memory Normalizing Effects of Adenosine A$_{2A}$R Antagonists in Parkinson's Disease

Cognitive dysfunction is a common feature of PD occurring in all PD stages. These cognitive impairments are characterized by subtle changes that are difficult to detect and diagnose (Appollonio et al. 1994) and include impairments in set-shifting (Monchi et al. 2004), in task-switching (Cameron et al. 2010), in probabilistic reversal learning (Peterson et al. 2009), in a delayed win–stay task related to both striatal and prefrontal cortex dysfunction (Partiot et al. 1996), in recognition memory (Higginson et al. 2005; Whittington et al. 2006), and in implicit memory (Knowlton et al. 1996; Wang et al. 2009). At earlier stages, cognitive impairment affects PD patients with prevalence around 25 % (Foltynie et al. 2004; Hely et al. 2008) being a major burden reported by patient (Cools et al. 2010; Klepac et al. 2008). These minor cognitive deficits can appear prior to the onset of motor symptoms, as shown in siblings of patients with familial PD (Kéri et al. 2010) and around 20–40 % of PD patients exhibit cognitive impairments at disease onset (Aarsland et al. 2007). Approximately 20–40 % of these patients will progress to dementia in advanced stages of PD (Aarsland et al. 2001; Williams-Gray et al. 2009), when dementia occurs in 90 % of PD patients (Aarsland et al. 2005). This incidence of dementia emerges as a result of the progressive atrophy of the limbic system in demented PD patients (Beyer and Aarsland 2008; Bouchard et al. 2008; Ibarretxe-Bilbao et al. 2009) and the parallel lesion of the dopaminergic nigrostriatal and mesolimbic pathways further increases the spectrum of cognitive disorders in PD (Owen et al. 1997; Scatton et al. 1983). Thus, dopamine denervation diminishes the dopamine phasic firing that provides the "error" prediction signal in basal ganglia-dependent learning, and also impairs executive functions (Kehagia et al. 2010; Sawamoto et al. 2008; Shohamy et al. 2004).

Convergent evidence from human and animal studies supports the existence of dopamine-dependent cognitive deficits in PD (Lewis et al. 2003) and dopaminergic-based treatment enhances patient performance on tasks sensitive to frontal lobe dysfunction (Gotham et al. 1988; Lange et al. 1992; Taylor et al. 1987). In PD patients, dopaminergic drugs mainly improve aspects of cognition that involve cognitive flexibility, including planning on the Tower of London test, task switching, response inhibition and working memory (Cools et al. 2003). Furthermore, PET studies show that, in many cases, the amelioration of executive function deficits by L-DOPA is associated with a normalization of blood flow in the right PFC, sensorimotor cortex, and premotor cortex (Sawamoto et al. 2008; Shohamy et al. 2004), supporting an extra-striatal main focus for these cognitive alterations. These cognitive impairments are also present in animal models of PD, such as reserpine, 6-hydroxydopamine (6-OHDA), 1-methyl-4-phenyl-1,2,3,6-tetrahydropyridine (MPTP), and α-synuclein overexpression (Aguiar et al. 2009; Chesselet et al. 2012; da Cunha et al. 2002; Dauer and Przedborski 2003; Prediger et al. 2005a, b, 2009, 2010, 2011, 2012a).

The potential of the adenosine modulation system to control memory dysfunction is best heralded by the evidence gathered in both human and animal that concurs in the conclusion that the chronic consumption of moderate doses of caffeine prevents memory impairment, namely the onset, evolution and neuropathology of Alzheimer's disease (reviewed in a special issue of *Journal of Alzheimers Disease,* volume 20, Supplement 1, 2010, at http://iospress.metapress.com/content/t13614762731/). Thus, six key large and prospective epidemiological studies have identified that the regular consumption of moderate doses of caffeine correlates inversely with memory impairments associated both with aging as well as a reduced risk of developing Alzheimer's disease (AD, Cao et al. 2012; Eskelinen et al. 2009; Gelber et al. 2011; Hameleers et al. 2000; Lindsay et al. 2002; Ritchie et al. 2007; van Boxtel et al. 2003; van Gelder et al. 2007).

Despite the considerable strength of the correlation, epidemiological investigation cannot definitively isolate caffeine intake from other lifestyle choices that potentially affect cognition. Therefore, it is of uppermost importance to show that chronic consumption of caffeine can also prevent and counteract memory impairment in animal models of aging and AD: our group has previously found that the chronic consumption of caffeine abrogates memory impairment in animal models of AD (Dall'Igna et al. 2007; Espinosa et al. 2013) and in other brain conditions leading to memory impairment (Cognato et al. 2010; Duarte et al. 2012; Pandolfo et al. 2013), the same occurring in aged rodents (Costa et al. 2008; Leite et al. 2011; Prediger et al. 2005; Vila-Luna et al. 2012). Furthermore, studies of aged AD transgenic (APPsw, Swedish mutation) mice found that long-term (between early 4–9 months as well as aged 18–19 months old) administration of a 1.5 mg daily dose of caffeine (equivalent to 500 mg in human) to APPsw mice reduced brain Aβ levels and protected against certain cognitive impairments (Arendash et al. 2006, 2009; Cao et al. 2009), the same occurring in an animal model of tauopathy (Laurent et al. 2014a). This is in accordance with the ability of caffeine to control synaptic plasticity phenomena in the hippocampus (Alhaider et al. 2010; Alzoubi et al. 2013; Costenla et al. 2010), an effect that becomes more evident upon aging (Costenla et al. 2011); it is also in agreement with recent evidence showing an ability of caffeine to critically modulate the efficiency of memory retrieval in humans (Borota et al. 2014).

As previously discussed for mood-related disorders, the evidence available also implicates the antagonism of $A_{2A}R$ as the likely mechanism by which caffeine attenuates memory deterioration (Cunha and Agostinho 2010; Takahashi et al. 2008). In fact, both the pharmacological and the genetic blockade of $A_{2A}R$ can prevent or reverse cognitive impairments in aging (Prediger et al. 2005a) and in brain disease models, including murine AD models (Canas et al. 2009; Cunha et al. 2008b; Dall'Igna et al. 2007; Laurent et al. 2014b), early life convulsion (Cognato et al. 2010) or attention deficit and hyperactivity disorder (Pandolfo et al. 2013). Silencing $A_{2A}R$ also enhanced working memory (Wei et al. 2011; Zhou et al. 2009), spatial recognition memory (Wang et al. 2006), reversal learning and goal-directed behavior (Mott et al. 2009; Nam et al. 2013; Pardo et al. 2012; Yu et al. 2009). Conversely, the activation and overexpression of $A_{2A}R$ receptors impairs memory retrieval in rats (Gimenez-Llort et al. 2007; Pereira et al. 2005) and the optical recruitment of $A_{2A}R$ signaling was actually sufficient to impair spatial reference memory in naive animals (Li et al. 2015). Overall, these studies provide strong evidence for $A_{2A}R$ to be necessary and sufficient for the expression of memory deficits.

This ability of caffeine and $A_{2A}R$ to prevent cognitive deficits seems extensible to PD. In fact, in animal models of PD, caffeine consumption attenuated cognitive impairment in MPTP-lesioned rats (Gevaerd et al. 2001). These effects were corroborated in several studies from our laboratory using different animal models (Prediger 2010). We demonstrated that the $A_{2A}R$ antagonist ZM 241385 reversed the amnesic effects of reserpine (Prediger et al. 2005c) and another $A_{2A}R$ antagonist istradefylline improved dopamine amount and cognitive performance of 6-OHDA-lesioned rats (Kadowaki Horita et al. 2013). By contrast, the A_1R antagonist DPCPX failed to attenuate cognitive benefits in reserpinized rats (Prediger et al. 2010, 2011).

Adenosine Receptors: A Promising Target for Olfactory Dysfunction in Parkinson's Disease

Hyposmia is a prominent prodromal feature of PD, with increased usefulness to assist the diagnosis of early PD (Morley and Duda 2014; Ponsen et al. 2004; Xiao et al. 2014). There is an increasing interest in the role purinergic receptor signaling in sensory transduction and information coding in sense organs. Purinergic receptors mediate fast transmission of sensory signals and have modulatory roles in the regulation of synaptic transmitter release, namely in the adaptation to sensory overstimulation, apart from their role in the regulation of neuron-glia bidirectional communication and in the fine-tuning of the turnover of sensory epithelia by modulating apoptosis and progenitor proliferation (Housley et al. 2009). In the odor detection and processing, ATP has been well characterized as a mediator of odor sensitivity and information processing in taste buds (Finger et al. 2005; Hegg et al. 2003; Kinnamon and Finger 2013), whereas a role for adenosine receptors has been ascribed to sweet buds in particular (Dando et al. 2012; Kataoka et al. 2012) and in the control of astrocytes (Doengi et al. 2008). In support of a role of adenosine as extracellular neuromodulators for smelling, it is observed that many proteins and mRNA of the adenosinergic system were found in the olfactory bulb. Thus, A_1R and $A_{2A}R$ mRNA is highly expressed in the granule cell layer of adult and developing olfactory bulb of rodents (Kaelin-Lang et al. 1999). The interferon-inducible RNA-specific adenosine deaminase (ADA) is an RNA editing enzyme implicated in the site-selective deamination of adenosine to inosine in cellular pre-mRNAs (Liu et al.

1999; Senba et al. 1987) and CD73 (ecto-5′-nucleotidase) is a cell surface enzyme that regulates purinergic signaling by dephosphorylating extracellular AMP to adenosine (Kulesskaya et al. 2013) and both display an high expression and activity in the olfactory bulb (Kulesskaya et al. 2013; Liu et al. 1999; Senba et al. 1987). Caffeine modulates dopamine release (Hadfield 1997), which plays a critical role in olfaction trough adenosine receptors, similar to that adenosine-dopamine interaction in the striatum (Fink et al. 1992). Altogether, these findings suggest a relevant role of adenosine in olfactory function.

Adenosine A_{2A} receptors may also participate in olfactory disorders. For instance, caffeine and ZM 241385 (a selective adenosine A_{2A}R antagonist) improved the odor recognition deficits displayed by aging rats (Prediger et al. 2005a) and ethanol-intoxicated rats (Prediger and Takahashi 2003; Prediger et al. 2005a; Spinetta et al. 2008). PD is associated with olfactory bulb damage and about 90 % of PD patients present olfactory dysfunction at early (pre-clinical) stages of disease (Doty et al. 1988).

Our group investigated the effects of the pharmacological modulation of adenosine receptors in the olfactory function of reserpine-treated rats, an animal model of PD (Gerlach and Riederer 1996). Reserpine depletes monoamine storage in catecholaminergic neurons and high reserpine doses (3–5 mg/kg, i.p.) induce transient bradykinesia and muscular rigidity. On the other hand, a single administration of a low reserpine dose (1 mg/kg, i.p.) disrupts olfactory short-term social memory without alter locomotor activity (Prediger et al. 2005b). Similar to that observed in aged rats, caffeine and ZM 241385 reversed reserpine-induced olfactory and social memory dysfunction (Prediger et al. 2005b). Interestingly, recent findings from our group have indicated that genetic deletion of A_{2A}R did not prevent the olfactory dysfunction in MPTP-treated mice (Prediger et al., unpublished data) (Fig. 10.2).

In addition, small lifetime caffeine consumption was associated with olfactory dysfunction in first-degree relatives of PD patients (Siderowf et al. 2007). These evidences suggest that antagonism of A_{2A}R may benefit olfaction of PD (Fig. 10.2). Furthermore, long-term neuroprotective effects of the purinergic system might be an additional benefit since purines display neuroprotective and neuroproliferative effects in mouse olfactory epithelium (Jia et al. 2011), in accordance with the abundant presence of the ecto-nucleotidase NTPDase2 in the germinal zones of the developing and adult rat brain (Braun et al. 2003), namely in olfactory pathways (Vandenbeuch et al. 2013).

Adenosine A_{2A} Receptors and Gastrointestinal Dysfunction in Parkinson's Disease

Gastrointestinal dysfunction affects up to 80 % of PD patients over the course of the disease (Jost 2010; Pfeiffer 2003). The most commonly reported gastrointestinal sign in PD is related to constipation and difficulty in defecation, with prevalence rates in over 60 % of patients (Azmin et al. 2014). Prospective follow-up and case-control studies found evidence for increased risk of PD in individuals exhibiting a reduced frequency of bowel movements (Abbott et al. 2001; Savica et al. 2009),

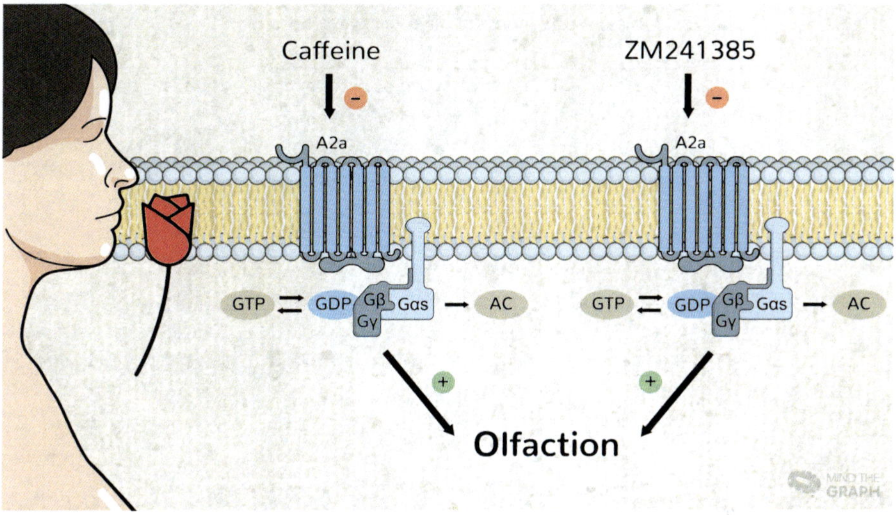

Fig. 10.2 Adenosine A$_{2A}$ receptor antagonists as putative treatments to improve olfactory dysfunction in Parkinson's disease

suggesting that constipation may be one of the earliest symptoms of degenerative process in PD. Corroborating these clinical findings, premotor gastrointestinal dysfunction has also been demonstrated in animal models of PD (Drolet et al. 2009; Greene et al. 2009; Kuo et al. 2010).

The neuropathology underlying constipation in PD is not entirely clear, but possibly involves α-synuclein accumulation in the gastrointestinal tract (Pouclet et al. 2012; Hilton et al. 2014) and cell loss in the dorsal motor nucleus of vagus nerve and enteric nervous system (Bloch et al. 2006), both pathological changes emerging at early stages of PD (Braak et al. 2003). Since adenosine A$_{2A}$R antagonists are a promising new non-dopaminergic approach to manage PD, it would be of interest to exploring the potential involvement of A$_{2A}$R in PD-related gastrointestinal dysfunction. Although at this moment there are no preclinical studies investigating the involvement of A$_{2A}$R in gastrointestinal dysfunction in premotor PD, two double-blind placebo-controlled pilot studies of preladenant in the treatment of fluctuating PD reported constipation as adverse effect (Factor et al. 2013; Hauser et al. 2011). PD patients treated with istradefylline also complained of constipation (Hauser et al. 2008). Certainly further preclinical and clinical research studies are required to establish the impact and possible mechanism of action of A$_{2A}$R in gastrointestinal dysfunction since these non-motor PD symptoms impose a significant burden on patients and their caregivers.

Role of Adenosine A$_{2A}$ Receptors on Pain in Parkinson's Disease

Modification of pain perception in PD patients were described by James Parkinson in 1817 (Parkinson 1817). The high prevalence of pain in PD exceeds that of the general population. Pain is considered underestimated non-motor PD symptom that

Table 10.1 Current pharmacological treatments in different types of pain in PD

Treatment	Category of pain
Opioid analgesics	Central or primary pain, Akathisia
Non-opioid analgesics	Musculoskeletal pain
Dopaminergic drugs	Dystonia, central or primary pain, Akathisia
Anticonvulsants and antidepressants	Central or primary pain

reduces quality of life of patients. These symptoms begin at PD stage 2, according to Braak classification, with increasing sensitivity and excitability of central pain control system (Braak et al. 2004). The scenario worsens with other comorbidities of aging and PD, like other non-motor and motor PD symptoms, osteoarthritis, rheumatoid arthritis, and osteoporosis (Wasner and Deuschl 2012).

Dopaminergic drugs are ineffective, and the prescribed analgesic strategies are inappropriate to manage the particular characteristics of pain in PD patients (Granovsky et al. 2013; see Table 10.1). In general, the treatment offered for pain does not satisfy PD patients (Beiske et al. 2009; Granovsky et al. 2013; Rana et al. 2014). Pain rating for PD covers a broad spectrum, ranging from musculoskeletal, radicular/neuropathic, dystonia-related, central PD pain and akathitic discomfort/pain (Fil et al. 2013; Ford 2010). The correct pain classification improves outcomes and direct the choice of the different drug classes that are most adequate, as summarized in the Table 10.1.

Adenosine stands out in modern pathophysiology of pain associated with PD. Thus, caffeine is an analgesic adjuvant with a favorable risk-benefit (Petersen 2014); in a recent Cochrane review (19 studies, 7238 patients), caffeine improved efficacy of paracetamol, ibuprofen and aspirin (Derry et al. 2012), with effect size resembling the doubling of the dose of the primary analgesic. Analgesia by caffeine is explained by antagonism at adenosine receptors (Derry et al. 2012; Petersen 2014). Adenosine receptors are widely expressed in the central and peripheral nervous system (CNS and PNS), including fibers driving pain (Dixon et al. 1996). Notably, mice lacking $A_{2A}R$ are hypoalgesic, and have altered analgesic responses to receptor-selective opioid agonists (Bailey et al. 2002; Ledent et al. 1997) . There is evidence suggesting a role for the $A_{2A}R$ in sensitizing afferent fibers projecting to the spinal cord (Bura et al. 2008; Hussey et al. 2007). ATL 313[1] or CGS 21680[2] alleviates neuropathic nociception in animals (Loram et al. 2009). A double-blind, placebo-controlled, Phase II, randomized study showed efficacy and tolerability of BVT 115959[3] in reducing neuropathic pain of diabetic patients[4]. Therefore, it will be of interest to analyze data pertinent to the effect of $A_{2A}R$ antagonist on pain perception in PD patients and this is a relevant issue to be addressed in future studies with animal models of PD.

---

[1] ATL 313: a potent and selective adenosine A_{2A} receptor agonist.
[2] CGS 21680: a selective adenosine A_{2A} receptor agonist.
[3] BVT 115959: an adenosine A_{2A} receptor agonist.
[4] ClinicalTrials.gov identifier: NCT00452777.

Conclusion

Non-motor symptoms associated with PD are frequent and currently difficult to manage, being reported by patients to represent a significant burden. The present observation that adenosine A_{2A} receptor antagonist seem efficacious to attenuate several non-motor PD symptoms, paves the way to consider these drugs novel holistic drugs for PD patients: in fact, $A_{2A}R$ antagonists not only ameliorate the efficacy of L-DOPA and attenuate its dyskinetic effects, but also afford neuroprotection and attenuate anxiety, depression and cognitive deficits associated with PD. This clearly prompts the need to detail the underlying mechanisms to understand when and how $A_{2A}R$ should be exploited to maximize benefits for PD patients.

References

Aarsland D, Andersen K, Larsen JP et al (2001) Risk of dementia in Parkinson's disease: a community-based, prospective study. Neurology 56:730–736

Aarsland D, Zaccai J, Brayne C (2005) A systematic review of prevalence studies of dementia in Parkinson's disease. Mov Disord 20:1255–1263

Aarsland D, Kvaløy JT, Andersen K et al (2007) The effect of age of onset of PD on risk of dementia. J Neurol 254:38–45

Abbott RD, Petrovitch H, White LR et al (2001) Frequency of bowel movements and the future risk of Parkinson's disease. Neurology 57:456–462

Aguiar AS Jr, Araujo AL, da-Cunha TR et al (2009) Physical exercise improves motor and short-term social memory deficits in reserpinized rats. Brain Res Bull 79:452–457

Ahlskog JE, Muenter MD (2001) Frequency of levodopa-related dyskinesias and motor fluctuations as estimated from the cumulative literature. Mov Disord 16:448–458

Alexander GE, Crutcher MD (1990) Functional architecture of basal ganglia circuits: neural substrates of parallel processing. Trends Neurosci 13:266–271

Alhaider IA, Aleisa AM, Tran TT et al (2010) Chronic caffeine treatment prevents sleep deprivation-induced impairment of cognitive function and synaptic plasticity. Sleep 33:437–444

Allain H, Bentue-Ferrer D, Akwa Y (2008) Disease-modifying drugs and Parkinson's disease. Progr Neurobiol 84:25–39

Alsene K, Deckert J, Sand P et al (2003) Association between A_{2A} receptor gene polymorphisms and caffeine-induced anxiety. Neuropsychopharmacology 28:1694–16702

Altman RD, Lang AE, Postuma RB (2011) Caffeine in Parkinson's disease: a pilot open-label, dose-escalation study. Mov Disord 26:2427–2431

Alzoubi KH, Abdul-Razzak KK, Khabour OF et al (2013) Caffeine prevents cognitive impairment induced by chronic psychosocial stress and/or high fat-high carbohydrate diet. Behav Brain Res 237:7–14

Appollonio I, Grafman J, Clark K et al (1994) Implicit and explicit memory in patients with Parkinson's disease with and without dementia. Arch Neurol 51:359–367

Arendash GW, Schleif W, Rezai-Zadeh K et al (2006) Caffeine protects Alzheimer's mice against cognitive impairment and reduces brain beta-amyloid production. Neuroscience 142:941–952

Arendash GW, Mori T, Cao C et al (2009) Caffeine reverses cognitive impairment and decreases brain amyloid-beta levels in aged Alzheimer's disease mice. J Alzheimers Dis 17:661–680

Azmin S, Khairul Anuar AM, Tan HJ et al (2014) Nonmotor symptoms in a malaysian Parkinson's disease population. Parkinson's Dis 2014:472157

Bailey A, Ledent C, Kelly M et al (2002) Changes in spinal delta and kappa opioid systems in mice deficient in the A_{2A} receptor gene. J Neurosci 22:9210–9220

Batalha VL, Pego JM, Fontinha BM et al (2013) Adenosine A_{2A} receptor blockade reverts hippo-campal stress-induced deficits and restores corticosterone circadian oscillation. Mol Psychiatry 18:320–331

Beiske AG, Loge JH, Ronningen A et al (2009) Pain in Parkinson's disease: prevalence and characteristics. Pain 141:173–177

Beyer MK, Aarsland D (2008) Grey matter atrophy in early versus late dementia in Parkinson's disease. Parkinsonism Relat Disord 14:620–625

Bloch A, Probst A, Bissig H et al (2006) Alpha-synuclein pathology of the spinal and peripheral autonomic nervous system in neurologically unimpaired elderly subjects. Neuropathol Appl Neurobiol 32:284–295

Bogenpohl JW, Ritter SL, Hall RA et al (2012) Adenosine A_{2A} receptor in the monkey basal ganglia: ultrastructural localization and colocalization with the metabotropic glutamate receptor 5 in the striatum. J Comp Neurol 320:570–589

Boot BP, Orr CF, Ahlskog JE et al (2013) Risk factors for dementia with Lewy bodies: a case-control study. Neurology 81:833–840

Borota D, Murray E, Keceli G et al (2014) Post-study caffeine administration enhances memory consolidation in humans. Nat Neurosci 17:201–203

Bouchard TP, Malykhin N, Martin WR et al (2008) Age and dementia-associated atrophy predominates in the hippocampal head and amygdala in Parkinson's disease. Neurobiol Aging 29:1027–1039

Braak H, Del Tredici K, Rub U et al (2003) Staging of brain pathology related to sporadic Parkinson's disease. Neurobiol Aging 24:197–211

Braak H, Ghebremedhin E, Rub U et al (2004) Stages in the development of Parkinson's disease-related pathology. Cell Tissue Res 318:121–134

Braun N, Sévigny J, Mishra SK et al (2003) Expression of the ecto-ATPase NTPDase2 in the germinal zones of the developing and adult rat brain. Eur J Neurosci 17:1355–1364

Bruns RF, Katims JJ, Annau Z et al (1983) Adenosine receptor interactions and anxiolytics. Neuropharmacology 22:1523–1529

Bura SA, Nadal X, Ledent C et al (2008) A_{2A} adenosine receptor regulates glia proliferation and pain after peripheral nerve injury. Pain 140:95–103

Cameron IG, Watanabe M, Pari G et al (2010) Executive impairment in Parkinson's disease: response automaticity and task switching. Neuropsychologia 48:1948–1957

Canals M, Marcellino D, Fanelli F et al (2003) Adenosine A_{2A}-dopamine D2 receptor-receptor heteromerization: qualitative and quantitative assessment by fluorescence and bioluminescence energy transfer. J Biol Chem 278:46741–46749

Canas PM, Porciúncula LO, Cunha GM et al (2009) Adenosine A_{2A} receptor blockade prevents synaptotoxicity and memory dysfunction caused by beta-amyloid peptides via p38 mitogen-activated protein kinase pathway. J Neurosci 29:14741–14751

Cao C, Cirrito JR, Lin X et al (2009) Caffeine suppresses amyloid-beta levels in plasma and brain of Alzheimer's disease transgenic mice. J Alzheimers Dis 17:681–697

Cao C, Loewenstein DA, Lin X et al (2012) High Blood caffeine levels in MCI linked to lack of progression to dementia. J Alzheimers Dis 30:559–572

Charney DS, Heninger GR, Jatlow PI (1985) Increased anxiogenic effects of caffeine in panic disorders. Arch Gen Psychiatry 42:233–243

Chaudhuri KR, Healy DG, Schapira AH et al (2006) Non-motor symptoms of Parkinson's disease: diagnosis and management. Lancet Neurol 5:235–245

Chen JF, Xu K, Petzer JP et al (2001) Neuroprotection by caffeine and A_{2A} adenosine receptor inactivation in a model of Parkinson's disease. J Neurosci 21:RC143

Chesselet MF, Richter F, Zhu C et al (2012) A progressive mouse model of Parkinson's disease: the Thy1-aSyn ("Line 61") mice. Neurother 9:297–314

Childs E, Hohoff C, Deckert J et al (2008) Association between $ADORA_{2A}$ and DRD2 polymorphisms and caffeine-induced anxiety. Neuropsychopharmacology 33:2791–2800

Ciruela F, Casadó V, Rodrigues RJ et al (2006) Presynaptic control of striatal glutamatergic neurotransmission by adenosine A_1-A_{2A} receptor heteromers. J Neurosci 26:2080–2087

Coelho JE, Alves P, Canas PM et al (2014) Overexpression of adenosine A$_{2A}$ receptors in rats: effects on depression, locomotion, and anxiety. Front Psychiatry 5:67

Cognato GP, Agostinho PM, Hockemeyer J et al (2010) Caffeine and an adenosine A$_{2A}$ receptor antagonist prevent memory impairment and synaptotoxicity in adult rats triggered by a convulsive episode in early life. J Neurochem 112:453–462

Cools R, Barker RA, Sahakian BJ et al (2003) L-DOPA medication remediates cognitive inflexibility, but increases impulsivity in patients with Parkinson's disease. Neuropsychologia 41:1431–1441

Cools R, Miyakawa A, Sheridan M et al (2010) Enhanced frontal function in Parkinson's disease. Brain 133:225–233

Correa M, Font L (2008) Is there a major role for adenosine A$_{2A}$ receptors in anxiety? Front Biosci 13:4058–4070

Costa MS, Botton PH, Mioranzza S et al (2008) Caffeine prevents age-associated recognition memory decline and changes brain-derived neurotrophic factor and tirosine kinase receptor (TrkB) content in mice. Neuroscience 153:1071–1078

Costa J, Lunet N, Santos C et al (2010) Caffeine exposure and the risk of Parkinson's disease: a systematic review and meta-analysis of observational studies. J Alzheimers Dis 20: S221–S238

Costenla AR, Diogenes MJ, Canas PM et al (2011) Enhanced role of adenosine A$_{2A}$ receptors in the modulation of LTP in the rat hippocampus upon ageing. Eur J Neurosci 34:12–21

Cunha RA (2008) Different cellular sources and different roles of adenosine: A$_1$ receptor-mediated inhibition through astrocytic-driven volume transmission and synapse-restricted A$_{2A}$ receptor-mediated facilitation of plasticity. Neurochem Int 52:65–72

Cunha RA, Agostinho PM (2010) Chronic caffeine consumption prevents memory disturbance in different animal models of memory decline. J Alzheimers Dis 20:S95–S116

Cunha GM, Canas PM, Oliveira CR et al (2006) Increased density and synapto-protective effect of adenosine A$_{2A}$ receptors upon sub-chronic restraint stress. Neuroscience 141:1775–1781

Cunha RA, Ferre S, Vaugeois JM et al (2008a) Potential therapeutic interest of adenosine A$_{2A}$ receptors in psychiatric disorders. Curr Pharm Des 14:1512–1524

Cunha GM, Canas PM, Melo CS et al (2008b) Adenosine A$_{2A}$ receptor blockade prevents memory dysfunction caused by beta-amyloid peptides but not by scopolamine or MK-801. Exp Neurol 210:776–781

Cutler DL, Tendolkar A, Grachev ID (2012) Safety, tolerability and pharmacokinetics after single and multiple doses of preladenant (SCH420814) administered in healthy subjects. J Clin Pharm Ther 37:578–587

d'Alcantara P, Ledent C, Swillens S et al (2001) Inactivation of adenosine A$_{2A}$ receptor impairs long term potentiation in the accumbens nucleus without altering basal synaptic transmission. Neuroscience 107:455–464

da Cunha C, Angelucci ME, Canteras NS et al (2002) The lesion of the rat substantia nigra pars compacta dopaminergic neurons as a model for Parkinson's disease memory disabilities. Cell Mol Neurobiol 22:227–237

Dall'Igna OP, Fett P, Gomes MW et al (2007) Caffeine and adenosine A(2a) receptor antagonists prevent beta-amyloid (25–35)-induced cognitive deficits in mice. Exp Neurol 203:241–245

Dando R, Dvoryanchikov G, Pereira E et al (2012) Adenosine enhances sweet taste through A$_{2B}$ receptors in the taste bud. J Neurosci 32:322–330

Dauer W, Przedborski S (2003) Parkinson's disease: mechanisms and models. Neuron 39:889–909

Deckert J, Nothen MM, Franke P et al (1998) Systematic mutation screening and association study of the A$_1$ and A$_{2a}$ adenosine receptor genes in panic disorder suggest a contribution of the A$_{2a}$ gene to the development of disease. Mol Psychiatry 3:81–85

Derry CJ, Derry S, Moore RA (2012) Caffeine as an analgesic adjuvant for acute pain in adults. Cochrane Database Syst Rev 3:CD009281

Dixon AK, Gubitz AK, Sirinathsinghji DJ et al (1996) Tissue distribution of adenosine receptor mRNAs in the rat. Br J Pharmacol 118:1461–1468

Doengi M, Deitmer JW, Lohr C (2008) New evidence for purinergic signaling in the olfactory bulb: A_{2A} and P2Y1 receptors mediate intracellular calcium release in astrocytes. FASEB J 22:2368–2378

Doty RL, Deems DA, Stellar S (1988) Olfactory dysfunction in parkinsonism: a general deficit unrelated to neurologic signs, disease stage, or disease duration. Neurology 38:1237–1244

Drolet RE, Cannon JR, Montero L et al (2009) Chronic rotenone exposure reproduces Parkinson's disease gastrointestinal neuropathology. Neurobiol Dis 36:96–102

Duarte JM, Agostinho PM, Carvalho RA et al (2012) Caffeine consumption prevents diabetes-induced memory impairment and synaptotoxicity in the hippocampus of NONcZNO10/LTJ mice. PLoS One 7:e21899

Dunwiddie TV, Masino SA (2001) The role and regulation of adenosine in the central nervous system. Annu Rev Neurosci 24:31–55

Eagle DM, Baunez C (2010) Is there an inhibitory-response-control system in the rat? Evidence from anatomical and pharmacological studies of behavioral inhibition. Neurosci Biobehavior Rev 34:50–72

El Yacoubi M, Ledent C, Parmentier M et al (2000) The anxiogenic-like effect of caffeine in two experimental procedures measuring anxiety in the mouse is not shared by selective A(2A) adenosine receptor antagonists. Psychopharmacology 148:153–163

El Yacoubi M, Ledent C, Parmentier M et al (2001) Adenosine A_{2A} receptor antagonists are potential antidepressants: evidence based on pharmacology and A_{2A} receptor knockout mice. Br J Pharmacol 134:68–77

El Yacoubi M, Costentin J, Vaugeois JM (2003) Adenosine A_{2A} receptors and depression. Neurology 61: S82–S87

Eskelinen MH, Ngandu T, Tuomilehto J et al (2009) Midlife coffee and tea drinking and the risk of late-life dementia: a population-based CAIDE study. J Alzheimers Dis 16:85–89

Espinosa J, Rocha A, Nunes F et al (2013) Caffeine consumption prevents memory impairment, neuronal damage, and adenosine A_{2A} receptors upregulation in the hippocampus of a rat model of sporadic dementia. J Alzheimers Dis 34:509–518

Factor SA, Wolski K, Togasaki DM et al (2013) Long-term safety and efficacy of preladenant in subjects with fluctuating Parkinson's disease. Mov Disord 28:817–820

Ferré S (2008) An update on the mechanisms of the psychostimulant effects of caffeine. J Neurochem 105:1067–1079

Ferré S, Karcz-Kubicha M, Hope BT et al (2002) Synergistic interaction between adenosine A_{2A} and glutamate mGlu5 receptors: implications for striatal neuronal function. Proc Natl Acad Sci U S A 99:11940–11945

Ferré S, Ciruela F, Quiroz C et al (2007) Adenosine receptor heteromers and their integrative role in striatal function. ScientificWorldJournal 7:74–85

Ferreira SG, Gonçalves FQ, Marques JM et al (2015) Presynaptic A_{2A} adenosine receptors dampen CB_1 cannabinoid receptor-mediated inhibition of corticostriatal glutamatergic transmission. Br J Pharmacol 172:1074–1086

Fil A, Cano-de-la-Cuerda R, Munoz-Hellin E et al (2013) Pain in Parkinson disease: a review of the literature. Parkinsonism Relat Disord 19:285–294

Finger TE, Danilova V, Barrows J et al (2005) ATP signaling is crucial for communication from taste buds to gustatory nerves. Science 310:1495–1499

Fink JS, Weaver DR, Rivkees SA et al (1992) Molecular cloning of the rat A_2 adenosine receptor: selective co-expression with D2 dopamine receptors in rat striatum. Mol Brain Res 14:186–195

Flajolet M, Wang Z, Futter M et al (2008) FGF acts as a co-transmitter through adenosine A_{2A} receptor to regulate synaptic plasticity. Nature Neurosci 11:1402–1409

Florio C, Prezioso A, Papaioannou A et al (1998) Adenosine A_1 receptors modulate anxiety in CD1 mice. Psychopharmacology 136:311–319

Foltynie T, Brayne CE, Robbins TW et al (2004) The cognitive ability of an incident cohort of Parkinson's patients in the UK. The CamPaIGN study. Brain 127:550–560

Ford B (2010) Pain in Parkinson's disease. Mov Disord 25:S98–S103

Fredholm BB, Bättig K, Holmén J et al (1999) Actions of caffeine in the brain with special reference to factors that contribute to its widespread use. Pharmacol Rev 51:83–133

Fredholm BB, Chen JF, Cunha RA et al (2005) Adenosine and brain function. Int Rev Neurobiol 63:191–270

Freitag CM, Agelopoulos K, Huy E et al (2010) Adenosine A$_{2A}$ receptor gene (ADORA$_{2A}$) variants may increase autistic symptoms and anxiety in autism spectrum disorder. Eur Child Adolesc Psychiatry 19:67–74

Gelber RP, Petrovitch H, Masaki KH et al (2011) Coffee intake in midlife and risk of dementia and its neuropathologic correlates. J Alzheimers Dis 23:607–615

Gerevich Z, Wirkner K, Illes P (2002) Adenosine A$_{2A}$ receptors inhibit the N-methyl-D-aspartate component of excitatory synaptic currents in rat striatal neurons. Eur J Pharmacol 451:161–164

Gerfen CR, Engber TM, Mahan LC et al (1990) D1 and D2 dopamine receptor-regulated gene expression of striatonigral and striatopallidal neurons. Science 250:1429–1432

Gerlach M, Riederer P (1996) Animal models of Parkinson's disease: an empirical comparison with the phenomenology of the disease in man. J Neural Transm 103:987–1041

Gevaerd MS, Takahashi RN, Silveira R et al (2001) Caffeine reverses the memory disruption induced by intra-nigral MPTP-injection in rats. Brain Res Bull 55:101–106

Giménez-Llort L, Fernández-Teruel A, Escorihuela RM et al (2002) Mice lacking the adenosine A1 receptor are anxious and aggressive, but are normal learners with reduced muscle strength and survival rate. Eur J Neurosci 16:547–550

Gimenez-Llort L, Schiffmann SN, Shmidt T et al (2007) Working memory deficits in transgenic rats overexpressing human adenosine A$_{2A}$ receptors in the brain. Neurobiol Learn Mem 87:42–56

Gomes CV, Kaster MP, Tomé AR (2011) Adenosine receptors and brain diseases: neuroprotection and neurodegeneration. Biochim Biophys Acta 1808:1380–1399

Gotham AM, Brown RG, Marsden CD (1988) 'Frontal' cognitive function in patients with Parkinson's disease 'on' and 'off' levodopa. Brain 111:299–321

Granovsky Y, Schlesinger I, Fadel S et al (2013) Asymmetric pain processing in Parkinson's disease. Eur J Neurol 20:1375–1382

Greene JG, Noorian AR, Srinivasan S (2009) Delayed gastric emptying and enteric nervous system dysfunction in the rotenone model of Parkinson's disease. Exp Neurol 218:154–161

Hadfield MG (1997) Caffeine and the olfactory bulb. Mol Neurobiol 15:31–39

Hameleers PA, Van Boxtel MP, Hogervorst E et al (2000) Habitual caffeine consumption and its relation to memory, attention, planning capacity and psychomotor performance across multiple age groups. Hum Psychopharmacol 15:573–581

Hamilton SP, Slager SL, De Leon AB et al (2004) Evidence for genetic linkage between a polymorphism in the adenosine 2$_A$ receptor and panic disorder. Neuropsychopharmacology 29:558–565

Harris A, Ursin H, Murison R et al (2007) Coffee, stress and cortisol in nursing staff. Psychoneuroendocrinology 32:322–330

Haskell CF, Kennedy DO, Wesnes KA et al (2005) Cognitive and mood improvements of caffeine in habitual consumers and habitual non-consumers of caffeine. Psychopharmacology 179:813–825

Hauser RA (2011) Future treatments for Parkinson's disease: surfing the PD pipeline. Int J Neurosci 121:53–62

Hauser RA, Shulman LM, Trugman JM et al (2008) Study of istradefylline in patients with Parkinson's disease on levodopa with motor fluctuations. Mov Disord 23:2177–2185

Hauser RA, Cantillon M, Pourcher E et al (2011) Preladenant in patients with Parkinson's disease and motor fluctuations: a phase 2, double-blind, randomised trial. Lancet Neurol 10:221–229

Hegg CC, Greenwood D, Huang W et al (2003) Activation of purinergic receptor subtypes modulates odor sensitivity. J Neurosci 23:8291–8301

Hely MA, Reid WG, Adena MA et al (2008) The Sydney multicenter study of Parkinson's disease: the inevitability of dementia at 20 years. Mov Disord 23:837–844

Higginson CI, Wheelock VL, Carroll KE et al (2005) Recognition memory in Parkinson's disease with and without dementia: evidence inconsistent with the retrieval deficit hypothesis. J Clin Exp Neuropsychol 27:516–528

Higley MJ, Sabatini BL (2010) Competitive regulation of synaptic Ca^{2+} influx by D2 dopamine and A$_{2A}$ adenosine receptors. Nat Neurosci 13:958–966

Hillion J, Canals M, Torvinen M et al (2002) Coaggregation, cointernalization, and codesensitization of adenosine A_{2A} receptors and dopamine D2 receptors. J Biol Chem 277:18091–18097

Hilton D, Stephens M, Kirk L et al (2014) Accumulation of alpha-synuclein in the bowel of patients in the pre-clinical phase of Parkinson's disease. Acta Neuropathol 127:235–241

Hirsch E, Graybiel AM, Agid YA (1988) Melanized dopaminergic neurons are differentially susceptible to degeneration in Parkinson's disease. Nature 334:345–348

Hodgson RA, Bertorelli R, Varty GB et al (2009) Characterization of the potent and highly selective A_{2A} receptor antagonists preladenant and SCH 412348 [7-[2-[4-2,4-difluorophenyl]-1-piperazinyl]ethyl]-2-(2-furanyl)-7H-pyrazolo[4,3-e][1,2,4]triazolo[1,5-c]pyrimidin-5-amine] in rodent models of movement disorders and depression. J Pharmacol Exp Ther 330:294–303

Hohoff C, McDonald JM, Baune BT et al (2005) Interindividual variation in anxiety response to amphetamine: possible role for adenosine A_{2A} receptor gene variants. Am J Med Gen 139B:42–44

Hohoff C, Mullings EL, Heatherley SV et al (2010) Adenosine A_{2A} receptor gene: evidence for association of risk variants with panic disorder and anxious personality. J Psychiatr Res 44:930–937

Housley GD, Bringmann A, Reichenbach A (2009) Purinergic signaling in special senses. Trends Neurosci 32:128–141

Hussey MJ, Clarke GD, Ledent C et al (2007) Reduced response to the formalin test and lowered spinal NMDA glutamate receptor binding in adenosine A_{2A} receptor knockout mice. Pain 129:287–294

Ibarretxe-Bilbao N, Tolosa E, Junque C et al (2009) MRI and cognitive impairment in Parkinson's disease. Mov Disord 24:S748–S753

Jahfari S, Waldorp L, van den Wildenberg WP et al (2011) Effective connectivity reveals important roles for both the hyperdirect (fronto-subthalamic) and the indirect (fronto-striatal-pallidal) fronto-basal ganglia pathways during response inhibition. J Neurosci 31:6891–6899

Jia C, Sangsiri S, Belock B et al (2011) ATP mediates neuroprotective and neuroproliferative effects in mouse olfactory epithelium following exposure to satratoxin G in vitro and in vivo. Toxicol Sci 124:169–178

Johansson B, Halldner L, Dunwiddie TV et al (2001) Hyperalgesia, anxiety, and decreased hypoxic neuroprotection in mice lacking the adenosine A_1 receptor. Proc Natl Acad Sci U S A 98:9407–9412

Jost WH (2010) Gastrointestinal dysfunction in Parkinson's disease. J Neurol Sci 289:69–73

Kadowaki Horita T, Kobayashi M, Mori A et al (2013) Effects of the adenosine A_{2A} antagonist istradefylline on cognitive performance in rats with a 6-OHDA lesion in prefrontal cortex. Psychopharmacology 230:345–352

Kaelin-Lang A, Lauterburg T, Burgunder JM (1999) Expression of adenosine A_{2a} receptors gene in the olfactory bulb and spinal cord of rat and mouse. Neurosci Lett 261:189–191

Kale PP, Addepalli V (2014) Augmentation of antidepressant effects of duloxetine and bupropion by caffeine in mice. Pharmacol Biochem Behav 124:238–244

Kano O, Ikeda K, Cridebring D et al (2011) Neurobiology of depression and anxiety in Parkinson's disease. Parkinson's Dis 2011:143547

Kataoka S, Baquero A, Yang D et al (2012) $A_{2B}R$ adenosine receptor modulates sweet taste in circumvallate taste buds. PLoS One 7(1):e30032

Kawachi I, Willett WC, Colditz GA et al (1996) A prospective study of coffee drinking and suicide in women. Arch Intern Med 156:521–525

Kehagia AA, Barker RA, Robbins TW (2010) Neuropsychological and clinical heterogeneity of cognitive impairment and dementia in patients with Parkinson's disease. Lancet Neurol 9:1200–1213

Kéri S, Moustafa AA, Myers CE et al (2010) a-Synuclein gene duplication impairs reward learning. Proc Natl Acad Sci U S A 107:15992–15994

Kim JJ, Diamond DM (2002) The stressed hippocampus, synaptic plasticity and lost memories. Nature Rev Neurosci 3:453–662

Kinnamon SC, Finger TE (2013) A taste for ATP: neurotransmission in taste buds. Front Cell Neurosci 7:264

Klepac N, Trkulja V, Relja M et al (2008) Is quality of life in non-demented Parkinson's disease patients related to cognitive performance? A clinic-based cross-sectional study. Eur J Neurol 15:128–133

Knowlton BJ, Mangels JA, Squire LR (1996) A neostriatal habit learning system in humans. Science 273:1399–1402

Kulesskaya N, Voikar V, Peltola M et al (2013) CD73 is a major regulator of adenosinergic signalling in mouse brain. PLoS One 8:e66896

Kuo YM, Li ZS, Jiao Y et al (2010) Extensive enteric nervous system abnormalities in mice transgenic for artificial chromosomes containing Parkinson disease-associated alpha-synuclein gene mutations precede central nervous system changes. Hum Mol Genet 19:1633–1650

Lambert NA, Teyler TJ (1991) Adenosine depresses excitatory but not fast inhibitory synaptic transmission in area CA1 of the rat hippocampus. Neurosci Lett 122:50–52

Lange KW, Robbins TW, Marsden CD et al (1992) L-DOPA withdrawal in Parkinson's disease selectively impairs cognitive performance in tests sensitive to frontal lobe dysfunction. Psychopharmacology 107:394–404

Laurent C, Eddarkaoui S, Derisbourg M et al (2014a) Beneficial effects of caffeine in a transgenic model of Alzheimer's disease-like tau pathology. Neurobiol Aging 35:2079–2090

Laurent C, Burnouf S, Ferry B et al (2014b) A$_{2A}$ adenosine receptor deletion is protective in a mouse model of Tauopathy. Mol Psychiatry. doi:10.1038/mp.2014.151

Ledent C, Vaugeois JM, Schiffmann SN et al (1997) Aggressiveness, hypoalgesia and high blood pressure in mice lacking the adenosine A$_{2a}$ receptor. Nature 388:674–678

Leentjens AF, Dujardin K, Marsh L et al (2011) Anxiety rating scales in Parkinson's disease: a validation study of the Hamilton anxiety rating scale, the Beck anxiety inventory, and the hospital anxiety and depression scale. Mov Disord 26:407–415

Leite MR, Wilhelm EA, Jesse CR et al (2011) Protective effect of caffeine and a selective A$_{2A}$ receptor antagonist on impairment of memory and oxidative stress of aged rats. Exp Gerontol 46:309–315

Lemke MR (2008) Depressive symptoms in Parkinson's disease. Eur J Neurol 15:21–25

Lerner TN, Kreitzer AC (2012) RGS4 is required for dopaminergic control of striatal LTD and susceptibility to parkinsonian motor deficits. Neuron 73:347–359

Lewis SJ, Dove A, Robbins TW et al (2003) Cognitive impairments in early Parkinson's disease are accompanied by reductions in activity in frontostriatal neural circuitry. J Neurosci 23:6351–6356

Li P, Rial D, Canas PM et al (2015) Optogenetic activation of intracellular adenosine A$_{2A}$ receptor signaling in hippocampus is sufficient to trigger CREB phosphorylation and impair memory. Mol Psychiatry doi:10.1038/mp.2014.182

Lindsay J, Laurin D, Verreault R et al (2002) Risk factors for Alzheimer's disease: a prospective analysis from the Canadian Study of Health and Aging. Am J Epidemiol 156:445–453

Liu Y, Emeson RB, Samuel CE (1999) Serotonin-2C receptor pre-mRNA editing in rat brain and in vitro by splice site variants of the interferon-inducible double-stranded RNA-specific adenosine deaminase ADAR1. J Biol Chem 274:18351–18358

Lopes LV, Sebastião AM, Ribeiro JA (2011) Adenosine and related drugs in brain diseases: present and future in clinical trials. Curr Top Med Chem 11:1087–1101

Loram LC, Harrison JA, Sloane EM et al (2009) Enduring reversal of neuropathic pain by a single intrathecal injection of adenosine 2$_A$ receptor agonists: a novel therapy for neuropathic pain. J Neurosci 29:14015–14025

Lucas M, Mirzaei F, Pan A et al (2011) Coffee, caffeine, and risk of depression among women. Arch Intern Med 171:1571–1578

Lucas M, O'Reilly EJ, Pan A et al (2013) Coffee, caffeine, and risk of completed suicide: results from three prospective cohorts of American adults. World J Biol Psychiatry 15:377–386

Manzoni OJ, Manabe T, Nicoll RA (1994) Release of adenosine by activation of NMDA receptors in the hippocampus. Science 265:2098–2101

Marangos PJ, Boulenger JP (1985) Basic and clinical aspects of adenosinergic neuromodulation. Neurosci Biobehav Rev 9:421–430

Martire A, Tebano MT, Chiodi V et al (2011) Pre-synaptic adenosine A_{2A} receptors control cannabinoid CB1 receptor-mediated inhibition of striatal glutamatergic neurotransmission. J Neurochem 116:273–280

Matos M, Augusto E, Santos-Rodrigues AD et al (2012) Adenosine A_{2A} receptors modulate glutamate uptake in cultured astrocytes and gliosomes. Glia 60:702–716

Maximino C, Lima MG, Olivera KR et al (2011) Adenosine A1, but not A2, receptor blockade increases anxiety and arousal in Zebrafish. Basic Clin Pharmacol Toxicol 109:203–207

McEwen BS (2007) Physiology and neurobiology of stress and adaptation: central role of the brain. Physiol Rev 87:873–904

Minor TR, Winslow JL, Chang WC (1994) Stress and adenosine: II. Adenosine analogs mimic the effect of inescapable shock on shuttle-escape performance in rats. Behav Neurosci 108:265–276

Minor TR, Rowe M, Cullen PK et al (2008) Enhancing brain adenosine signaling with the nucleoside transport blocker NBTI (S-(4-nitrobenzyl)-6-theoinosine) mimics the effects of inescapable shock on later shuttle-escape performance in rats. Behav Neurosci 122:1236–1247

Mizuno Y, Hasegawa K, Kondo T et al (2010) Clinical efficacy of istradefylline (KW-6002) in Parkinson's disease: a randomized, controlled study. Mov Disord 25:1437–1443

Monchi O, Petrides M, Doyon J et al (2004) Neural bases of set-shifting deficits in Parkinson's disease. J Neurosci 24:702–710

Morley JF, Duda JE (2014) Use of hyposmia and other non-motor symptoms to distinguish between drug-induced parkinsonism and Parkinson's disease. J Parkinsons Dis 4:169–173

Mott AM, Nunes EJ, Collins LE et al (2009) The adenosine A_{2A} antagonist MSX-3 reverses the effects of the dopamine antagonist haloperidol on effort-related decision-making in a T-maze cost/benefit procedure. Psychopharmacology 204:103–112

Nam HW, Hinton DJ, Kang NY et al (2013) Adenosine transporter ENT1 regulates the acquisition of goal-directed behavior and ethanol drinking through A_{2A} receptor in the dorsomedial striatum. J Neurosci 33:4329–4338

Ongini E (2003) Adenosine A_{2A} receptors in nonlocomotor features of Parkinson's disease: introduction. Neurology 61:S72–S73

Orr AG, Orr AL, Li XJ et al (2009) Adenosine A_{2A} receptor mediates microglial process retraction. Nat Neurosci 12:872–878

Owen AM, Iddon JL, Hodges JR et al (1997) Spatial and non-spatial working memory at different stages of Parkinson's disease. Neuropsychologia 35:519–532

Palacios N, Gao X, McCullough ML et al (2012) Caffeine and risk of Parkinson's disease in a large cohort of men and women. Mov Disord 27:1276–1282

Pandolfo P, Machado NJ, Köfalvi A et al (2013) Caffeine regulates frontocorticostriatal dopamine transporter density and improves attention and cognitive deficits in an animal model of attention deficit hyperactivity disorder. Eur Neuropsychopharmacol 23:317–328

Pardo M, Lopez-Cruz L, Valverde O et al (2012) Adenosine A_{2A} receptor antagonism and genetic deletion attenuate the effects of dopamine D2 antagonism on effort-based decision making in mice. Neuropharmacology 62:2068–2077

Partiot A, Vérin M, Pillon B et al (1996) Delayed response tasks in basal ganglia lesions in man: further evidence for a striato-frontal cooperation in behavioural adaptation. Neuropsychologia 34:709–721

Pechlivanova DM, Tchekalarova JD, Alova LH et al (2012) Effect of long-term caffeine administration on depressive-like behavior in rats exposed to chronic unpredictable stress. Behav Pharmacol 23:339–347

Pereira GS, Rossato JI, Sarkis JJ et al (2005) Activation of adenosine receptors in the posterior cingulate cortex impairs memory retrieval in the rat. Neurobiol Learn Mem 83:217–223

Petersen KU (2014) Caffeine in analgesics—myth or medicine?. MMW Fortschr Med 156:60

Peterson DA, Elliott C, Song DD et al (2009) Probabilistic reversal learning is impaired in Parkinson's disease. Neuroscience 163:1092–1101

Pfeiffer RF (2003) Gastrointestinal dysfunction in Parkinson's disease. Lancet Neurol 2:107–116

Ponsen MM, Stoffers D, Booij J et al (2004) Idiopathic hyposmia as a preclinical sign of Parkinson's disease. Ann Neurol 56:173–181

Postuma RB, Lang AE, Munhoz RP et al (2012) Caffeine for treatment of Parkinson disease: a randomized controlled trial. Neurology 79:651–658

Pouclet H, Lebouvier T, Coron E et al (2012) Lewy pathology in gastric and duodenal biopsies in Parkinson's Disease. Mov Dis 27:708–708

Prediger RD (2010) Effects of caffeine in Parkinson's disease: from neuroprotection to the management of motor and non-motor symptoms. J Alzheimers Dis 20:S205–S220

Prediger RD, Takahashi RN (2003) Ethanol improves short-term memory in rats. Involvement of opioid and muscarinic receptors. Eur J Pharmacol 462:115–123

Prediger RD, Batista LC, Takahashi RN (2005a) Caffeine reverses age-related deficits in olfactory discrimination and social recognition memory in rats. Involvement of adenosine A$_1$ and A$_{2A}$ receptors. Neurobiol Aging 26:957–964

Prediger RD, Da Cunha C, Takahashi RN (2005b) Antagonistic interaction between adenosine A$_{2A}$ and dopamine D2 receptors modulates the social recognition memory in reserpine-treated rats. Behav Pharmacol 16:209–218

Prediger RD, Fernandes D, Takahashi RN (2005c) Blockade of adenosine A$_{2A}$ receptors reverses short-term social memory impairments in spontaneously hypertensive rats. Behav Brain Res 159:197–205

Prediger RD, da Silva GE, Batista LC et al (2006) Activation of adenosine A$_1$ receptors reduces anxiety-like behavior during acute ethanol withdrawal (hangover) in mice. Neuropsychopharmacology 31:2210–2220

Prediger RD, Rial D, Medeiros R et al (2009) Risk is in the air: an intranasal MPTP (1-methyl-4-phenyl-1,2,3,6-tetrahydropyridine) rat model of Parkinson's disease. Ann N Y Acad Sci 1170:629–636

Prediger RD, Aguiar AS Jr, Rojas-Mayorquin AE et al (2010) Single intranasal administration of 1-methyl-4-phenyl-1,2,3,6-tetrahydropyridine in C57BL/6 mice models early preclinical phase of Parkinson's disease. Neurotox Res 17:114–129

Prediger RD, Aguiar AS Jr, Moreira EL et al (2011) The intranasal administration of 1-methyl-4-phenyl-1,2,3,6-tetrahydropyridine (MPTP): a new rodent model to test palliative and neuroprotective agents for Parkinson's disease. Curr Pharm Des 17:489–507

Prediger RD, Aguiar AS Jr, Matheus FC et al (2012a) Intranasal administration of neurotoxicants in animals: support for the olfactory vector hypothesis of Parkinson's disease. Neurotox Res 21:90–116

Prediger RD, Matheus FC, Schwarzbold ML et al (2012b) Anxiety in Parkinson's disease: a critical review of experimental and clinical studies. Neuropharmacology 62:115–124

Rana AQ, Saeed U, Sufian Masroor M et al (2014) A cross-sectional study investigating clinical predictors and physical experiences of pain in Parkinson's disease. Funct Neurol:1–8

Rebola N, Pinheiro PC, Oliveira CR et al (2003) Subcellular localization of adenosine A$_1$ receptors in nerve terminals and synapses of the rat hippocampus. Brain Res 987:49–58

Rebola N, Canas PM, Oliveira CR et al (2005) Different synaptic and subsynaptic localization of adenosine A$_{2A}$ receptors in the hippocampus and striatum of the rat. Neuroscience 132:893–903

Rebola N, Lujan R, Cunha RA et al (2008) Adenosine A$_{2A}$ receptors are essential for long-term potentiation of NMDA-EPSCs at hippocampal mossy fiber synapses. Neuron 57:121–134

Riederer P, Wuketich S (1976) Time course of nigrostriatal degeneration in Parkinson's disease. A detailed study of influential factors in human brain amine analysis. J Neural Transm 38:277–301

Ritchie K, Carrière I, de Mendonça A et al (2007) The neuroprotective effects of caffeine: a prospective population study (the Three City Study). Neurology 69:536–545

Rodrigues RJ, Alfaro TM, Rebola N et al (2005) Co-localization and functional interaction between adenosine A$_{2A}$ and metabotropic group 5 receptors in glutamatergic nerve terminals of the rat striatum. J Neurochem 92:433–441

Rosin DL, Hettinger BD, Lee A et al (2003) Anatomy of adenosine A$_{2A}$ receptors in brain: morphological substrates for integration of striatal function. Neurology 61:S12–S18

Savica R, Carlin JM, Grossardt BR et al (2009) Medical records documentation of constipation preceding Parkinson disease: a case-control study. Neurology 73:1752–1758

Sawamoto N, Piccini P, Hotton G et al (2008) Cognitive deficits and striato-frontal dopamine release in Parkinson's disease. Brain 131:1294–1302

Scatton B, Javoy-Agid F, Rouquier L et al (1983) Reduction of cortical dopamine, noradrenaline, serotonin and their metabolites in Parkinson's disease. Brain Res 275:321–328

Schiffmann SN, Vanderhaeghen JJ (1993) Adenosine A_2 receptors regulate the gene expression of striatopallidal and striatonigral neurons. J Neurosci 13:1080–1087

Schiffmann SN, Fisone G, Moresco R et al (2007) Adenosine A_{2A} receptors and basal ganglia physiology. Progr Neurobiol 83:277–292

Schwarzschild MA, Chen JF, Ascherio A (2002) Caffeinated clues and the promise of adenosine A_{2A} antagonists in PD. Neurology 58:1154–1160

Schwarzschild MA, Agnati L, Fuxe K et al (2006) Targeting adenosine A_{2A} receptors in Parkinson's disease. Trends Neurosci 29:647–654

Senba E, Daddona PE, Nagy JI (1987) Adenosine deaminase-containing neurons in the olfactory system of the rat during development. Brain Res Bull 18:635–648

Serrano A, Haddjeri N, Lacaille JC et al (2006) GABAergic network activation of glial cells underlies hippocampal heterosynaptic depression. J Neurosci 26:5370–5382

Shen W, Flajolet M, Greengard P et al (2008) Dichotomous dopaminergic control of striatal synaptic plasticity. Science 321:848–851

Shohamy D, Myers CE, Grossman S et al (2004) Cortico-striatal contributions to feedback-based learning: converging data from neuroimaging and neuropsychology. Brain 127:851–859

Siderowf A, Jennings D, Connolly J et al (2007) Risk factors for Parkinson's disease and impaired olfaction in relatives of patients with Parkinson's disease. Mov Disord 22:2249–2255

Smith AP (2009) Caffeine, cognitive failures and health in a non-working community sample. Hum Psychopharmacol 24:29–34

Smith KM, Browne SE, Jayaraman S et al (2014) Effects of the selective adenosine A_{2A} receptor antagonist, SCH 412348, on the parkinsonian phenotype of MitoPark mice. Eur J Pharmacol 728:31–38

Snyder SH, Bruns RF, Daly JW et al (1981) Multiple neurotransmitter receptors in the brain: amines, adenosine, and cholecystokinin. Fed Proc 40:142–146

Spinetta MJ, Woodlee MT, Feinberg LM et al (2008) Alcohol-induced retrograde memory impairment in rats: prevention by caffeine. Psychopharmacology 201:361–371

Stacy M, Silver D, Mendis T et al (2008) A 12-week, placebo-controlled study (6002-US-006) of istradefylline in Parkinson disease. Neurology 70:2233–2240

Svenningsson P, Le Moine C, Fisone G et al (1999) Distribution, biochemistry and function of striatal adenosine A_{2A} receptors. Progr Neurobiol 59:355–396

Takahashi RN, Pamplona FA, Prediger RD (2008) Adenosine receptor antagonists for cognitive dysfunction: a review of animal studies. Front Biosci 13:2614–2632

Taylor AE, Saint-Cyr JA, Lang AE (1987) Parkinson's disease. Cognitive changes in relation to treatment response. Brain 110:35–51

Tebano MT, Martire A, Rebola N et al (2005) Adenosine A_{2A} receptors and metabotropic glutamate 5 receptors are co-localized and functionally interact in the hippocampus: a possible key mechanism in the modulation of N-methyl-D-aspartate effects. J Neurochem 95:1188–1200

Tetzlaff W, Schubert P, Kreutzberg GW (1987) Synaptic and extrasynaptic localization of adenosine binding sites in the rat hippocampus. Neuroscience 21:869–875

Thompson SM, Haas HL, Gähwiler BH (1992) Comparison of the actions of adenosine at pre- and postsynaptic receptors in the rat hippocampus in vitro. J Physiol 451:347–363

Todorova A, Jenner P, Ray Chaudhuri K (2014) Non-motor Parkinson's: integral to motor Parkinson's, yet often neglected. Pract Neurol 14:310–322

Uchida S, Tashiro T, Kawai-Uchida M et al (2014) The adenosine A_{2A}-receptor antagonist istradefylline enhances the motor response of L-DOPA without worsening dyskinesia in MPTP-treated common marmosets. J Pharmacol Sci 124:480–485

van Boxtel MP, Schmitt JA, Bosma H et al (2003) The effects of habitual caffeine use on cognitive change: a longitudinal perspective. Pharmacol Biochem Behav 75:921–927

van Gelder BM, Buijsse B, Tijhuis M et al (2007) Coffee consumption is inversely associated with cognitive decline in elderly European men: the FINE Study. Eur J Clin Nutr 61:226–232

Vandenbeuch A, Anderson CB, Parnes J et al (2013) Role of the ectonucleotidase NTPDase2 in taste bud function. Proc Natl Acad Sci U S A 110:14789–14794

Varty GB, Hodgson RA, Pond AJ et al (2008) The effects of adenosine A$_{2A}$ receptor antagonists on haloperidol-induced movement disorders in primates. Psychopharmacology 200:393–401

Vila-Luna S, Cabrera-Isidoro S, Vila-Luna L et al (2012) Chronic caffeine consumption prevents cognitive decline from young to middle age in rats, and is associated with increased length, branching, and spine density of basal dendrites in CA1 hippocampal neurons. Neuroscience 202:384–395

Wang JH, Ma YY, van den Buuse M (2006) Improved spatial recognition memory in mice lacking adenosine A$_{2A}$ receptors. Exp Neurol 199:438–445

Wang XP, Sun BM, Ding HL (2009) Changes of procedural learning in Chinese patients with non-demented Parkinson disease. Neurosci Lett 449: 161–163

Wasner G, Deuschl G (2012) Pains in Parkinson disease–many syndromes under one umbrella. Nat Rev Neurol 8:284–294

Wei CJ, Singer P, Coelho J et al (2011) Selective inactivation of adenosine A$_{2A}$ receptors in striatal neurons enhances working memory and reversal learning. Learn Mem 18:459–474

Wei CJ, Augusto E, Gomes CA et al (2014) Regulation of fear responses by striatal and extrastriatal adenosine A$_{2A}$ receptors in forebrain. Biol Psychiatry 75:855–863

Whittington CJ, Podd J, Stewart-Williams S (2006) Memory deficits in Parkinson's disease. J Clin Exp Neuropsychol 28:738–754

Williams-Gray CH, Evans JR, Goris A et al (2009) The distinct cognitive syndromes of Parkinson's disease: 5 year follow-up of the CamPaIGN cohort. Brain 132:2958–2969

Woodson JC, Minor TR, Job RF (1998) Inhibition of adenosine deaminase by erythro-9-(2-hydroxy-3-nonyl)adenine (EHNA) mimics the effect of inescapable shock on escape learning in rats. Behav Neurosci 112:399–409

Xiao Q, Chen S, Le W (2014) Hyposmia: a possible biomarker of Parkinson's disease. Neurosci Bull 30:134–140

Yamada K, Kobayashi M, Mori A et al (2013) Antidepressant-like activity of the adenosine A$_{2A}$ receptor antagonist, istradefylline (KW-6002), in the forced swim test and the tail suspension test in rodents. Pharmacol Biochem Behavior 114–115:23–30

Yamada K, Kobayashi M, Shiozaki S et al (2014) Antidepressant activity of the adenosine A$_{2A}$ receptor antagonist, istradefylline (KW-6002) on learned helplessness in rats. Psychopharmacology 231:2839–2849

Yang JN, Chen JF, Fredholm BB (2009) Physiological roles of A$_1$ and A$_{2A}$ adenosine receptors in regulating heart rate, body temperature, and locomotion as revealed using knockout mice and caffeine. Am J Physiol 296:H1141–H1149

Yoon KW, Rothman SM (1991) Adenosine inhibits excitatory but not inhibitory synaptic transmission in the hippocampus. J Neurosci 11:1375–1380

Yu C, Gupta J, Chen JF et al (2009) Genetic deletion of A$_{2A}$ adenosine receptors in the striatum selectively impairs habit formation. J Neurosci 29:15100–15103

Zhou SJ, Zhu ME, Shu D et al (2009) Preferential enhancement of working memory in mice lacking adenosine A$_{2A}$ receptors. Brain Res 1303:74–83

Chapter 11
Imaging Studies with A_{2A} Receptor Antagonists

Adriana Alexandre S. Tavares, Olivier Barret, John P. Seibyl
and Gilles D. Tamagnan

Abstract Single photon emission computed tomography (SPECT) and positron emission tomography (PET) are increasingly used to understand differential diagnosis and pathophysiological progression of a variety of neurodegenerative and neuropsychiatric disorders. These techniques have also been instrumental in the process of drug discovery and development. Over the last decades, the development of high affinity and subtype-selective adenosine 2A (A_{2A}) radiotracers has enable the non-invasive in vivo quantification of these receptors using SPECT and PET imaging. Data collected so far has confirmed the value of PET and SPECT techniques in assessing A_{2A} changes in brain. These findings can foster the rapid widespread use of PET and SPECT A_{2A} imaging, in particular now that suitable PET and SPECT probes with attractive in vivo properties are available for quantification of A_{2A} in brain. In particular, the recent report of radiotracers labelled with fluorine-18 or iodine-123 that displayed improved binding potentials in vivo compared with radiotracers previously developed, provides the opportunity to further expand the global use of in vivo pre-clinical and clinical A_{2A} imaging studies in neuroscience research. This book chapter provides a brief overview of the value of PET and SPECT in neuroscience, describes the key in vivo characteristics of PET and SPECT radiotracers developed to date for imaging A_{2A} in brain and offers examples of previous pre-clinical and clinical studies that used PET and SPECT with A_{2A} radiotracers to address a specific research question, with a particular focus on studies examining Parkinson's disease.

Keywords A_{2A} · PET · SPECT · Radiotracer · Parkinson's disease · [^{11}C]TMSX · [^{11}C]SCH442416 · [^{123}I]MNI-420 · [^{18}F]MNI-444

A. A. S. Tavares (✉) · O. Barret · J. P. Seibyl · G. D. Tamagnan
Molecular NeuroImaging, LLC, New Haven, CT, USA
e-mail: adriana_tavares@msn.com

O. Barret
e-mail: obarret@mnimaging.com

J. P. Seibyl
e-mail: jseibyl@mnimaging.com

G. D. Tamagnan
e-mail: gtamagnan@mnimaging.com

© Springer International Publishing Switzerland 2015
M. Morelli et al. (eds.), *The Adenosinergic System,* Current Topics in Neurotoxicity 10,
DOI 10.1007/978-3-319-20273-0_11

Introduction

Over the years, imaging techniques have often been the putative *eyes of science*, insofar as imaging provides non-invasive, in vivo quantification of multidimensional, multiparameter data. Molecular imaging techniques, such as single photon emission computed tomography (SPECT) and positron emission tomography (PET), are increasingly used to measure parameters such as concentration, tissue kinetic properties and receptor density changes. Neuroimaging using PET and SPECT has assumed a key role in improving the understanding of differential diagnosis and pathophysiological progression of a variety of neurodegenerative and neuropsychiatric disorders, by employing imaging biomarkers to complement clinical measures. These techniques have also been critical to the process of drug discovery and development.

The success of brain PET and SPECT dopaminergic imaging in Parkinson's disease (PD) (Loane and Politis 2011; Marek et al. 2001; Parkinsons Study Group 2002; Schwarz et al. 2004) has been important to the rapid growth of the use of these imaging modalities. More selective PET and SPECT brain radiotracers that label specific dopaminergic and non-dopaminergic targets in patients in vivo will improve our understanding of the pathophysiology of Parkinson's disease (PD), and facilitate more informed strategies for the use of current or novel therapies. The adenosine 2A receptor (A_{2A}), in particular, has attracted significant interest from the scientific community as a novel therapeutic target in late stage PD, mainly prompted by multiple studies demonstrating the co-expression of A_{2A} and dopamine D_2 receptors in basal ganglia neurons important in the control of movement (Ferré et al. 2007, 2008; Fuxe et al. 2003; Ikeda et al. 2002). Imaging A_{2A} in brain using selective PET and SPECT radiotracers can provide the opportunity to further understand the role of these receptors in PD and help drug discovery programmes developing improved adenosinergic strategies for treatment of PD.

This chapter provides a brief overview of the value of PET and SPECT in neuroscience. Subsequently, PET and SPECT radiotracers developed to date for imaging A_{2A} in brain will be presented, highlighting their main characteristics. Examples of previous pre-clinical and clinical studies that used PET and SPECT with A_{2A} radiotracers to address a specific research question will also be discussed here, with a particular focus on studies examining PD.

PET and SPECT Imaging in Neuroscience

Molecular imaging has been defined as the in vivo characterization and measurement of biological processes at the cellular and molecular level (Haberkorn and Eisenhurt 2005). Radionuclide imaging, namely PET and SPECT, is at the leading edge of molecular imaging as it enables the in vivo quantitative measurement of the distribution of a radiotracer, in order to provide information on a specific biological or biochemical process in the living body (Salvadori 2008). The use of a radiotracer allows for exceptional target specificity at the molecular level that cannot be accomplished with other imaging techniques (de Kemp et al. 2010). In addition, radiotracer imaging provides a large and ever-expanding number of brain targets for scintigraphic interrogation. This permits the design of more sophisticated studies of the interdependence and interaction of pathophysiologic changes involving multiple neurochemical systems in neuropsychiatric diseases. Imaging with radiotracers is based on the principle that the radiotracer does not alter or perturb the biological system under investigation. For this to be possible, the injected mass of a radiotracer should be as low as possible so that it occupies only a small percentage of the target, i.e. the microdosing principle (also denoted as tracer principle). For example, in brain receptor imaging, radiotracers should not occupy more than 1 % of the available receptors (Ruth 2009).

PET and SPECT imaging provide the means for examining regional cerebral blood flow, metabolism, and pharmacology in vivo. These imaging modalities have helped to establish the diagnosis of multiple neurodegenerative and neuropsychiatric disorders, where this is in doubt, and to provide potential biomarkers for following disease development and the effects of drugs on disease progression. Furthermore, PET and SPECT can advance our understanding of different diseases and determine the functional effects of therapy on neurotransmission and metabolism. These imaging modalities are also important tools in drug discovery and development programs that target central nervous system disorders (Fig. 11.1). For example, developing radiotracers analogues of novel drugs can help outlining the regional distribution in brain of the novel therapeutic agent and it can also establish proof of principle that the new drugs cross the blood-brain barrier to reach their targeted receptor, transporter or enzyme with high specificity, i.e. evaluation of drug-target engagement. This approach has an added advantage of allowing for multiple drug analogues to be tested with less strict regulatory requisites than those mandatory for novel medicinal drugs, owing to the extremely low mass dose injected. PET and

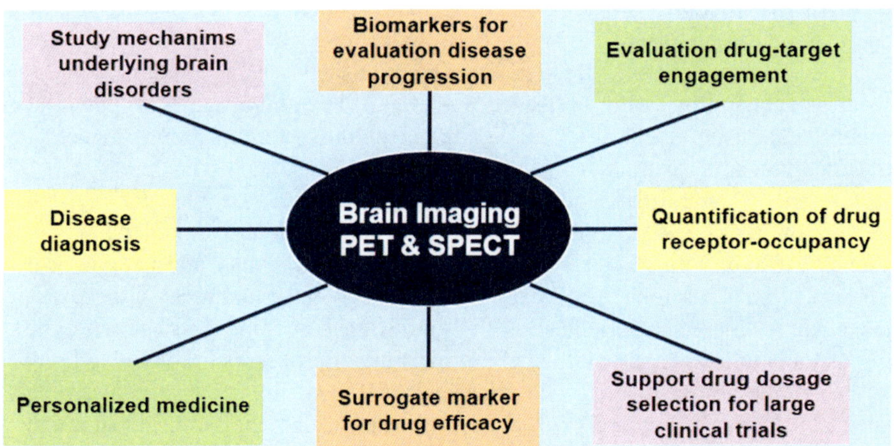

Fig. 11.1 Key applications of PET and SPECT imaging in neuroscience research

SPECT imaging studies can also provide receptor, transporter and enzyme active site dose-occupancy profiles, thus guiding dosage selection for Phase I and Phase II clinical trials (Brooks 2005). Ultimately, radionuclide imaging might provide a surrogate marker for drug efficacy and accelerate the implementation of the concept of personalized medicine, where receptor, transporter or enzyme binding profiles would help predict therapeutic outcome. This potential has not yet been completely realized, but there is a tremendous ongoing effort within the scientific community to implement the use of PET and SPECT imaging in patient's stratification for clinical trials and in more drug receptor-occupancy studies, as suggested by the growing number of publications in this area.

Imaging the Adenosinergic System in Brain

Adenosine is a neuromodulator produced by conversion of intra- and extracellular adenine nucleotides, which acts both in the central nervous system and in the periphery via four different G-protein coupled receptors: A_1, A_{2A}, A_{2B} and A_3 (Hirani et al. 2001; Mishina et al. 2007; Moresco et al. 2005; Müller and Jacobson 2011) Dysregulation of these receptors has been implicated in a variety of neurodegenerative and neuropsychiatric disorders. For example, the A_1 receptors have been found to play an important role in the regulation of alertness and sleep-wakefulness (Elmenhorst et al. 2007), as well as, in the pathophysiology of Alzheimer's disease (Angulo et al. 2003), stroke (Heurteaux et al. 1995), epilepsy and anxiety disorders (Gouder et al. 2003; Johansson et al. 2001; Plamondon et al. 1999). Previous studies have demonstrated that dysregulation of A_{2A} receptors is involved in the development of movement disorders, namely, PD (Hirani et al. 2001; Holschbach et al. 2006; Ikeda et al. 2002; Moresco et al. 2005), as well as, in Alzheimer's disease, Huntington's disease (HD), mood disorders, panic disorders, schizophrenia, attention deficit hyperactivity disorder, depression and addiction (Chen et al. 2013; El Yacoubi et al. 2001; Holschbach et al. 2006; Ledent et al. 1997; Müller and Jacobson 2011). The A_{2B} receptors role in neuroinflammatory processes has also been described (Rosi et al. 2003). Further, A_3 receptors have been investigated as potential therapeutic targets in cerebral ischemia (Chen et al 2006), chronic pain (Chen et al. 2012) and glaucoma treatment (Fishman et al. 2013).

As research establishes the relevance of adenosinergic receptor system dysregulation in multiple neurodegenerative and neuropsychiatric disorders, the development of tools to investigate A_1, A_{2A}, A_{2B} and A_3 in healthy and pathological conditions and to determine the efficacy of novel drugs in the treatment of these disorders becomes increasingly valuable. PET and SPECT imaging have proven to be useful as tools to investigate biological or biochemical processes in the living body with the unique ability to image neuroreceptors in vivo. Enhancing the available number of selective radiotracers will increase the multiplicity of biological sites and processes that can be imaged in vivo (Mariani et al. 2008; Ruth 2009). Consequently, the development of novel radiotracers is necessary to study and understand multiple pathophysiological processes and also to accelerate and aid drug discovery (Frank et al. 2007; Mariani et al. 2008; Ruth 2009; Salvadori 2008). In this context, selective radiotracers for imaging of the adenosinergic receptors have been developed.

Several radiotracers have been proposed for imaging A_1 in brain. Xanthine derivatives, such as [^{11}C]KF15372, [^{11}C]MPDX, [^{11}C]EPDX and [^{18}F]CPFPX, were synthesized as candidates for in vivo imaging of A_1 in brain (Bauer et al. 2003a,b; Fukumitsu et al. 2003; Ishiwata et al.1995, 2007). [^{11}C]FR194921, a highly selective non-xanthine derived antagonist, has also been proposed as a suitable radiotracer for in vivo imaging of A_1 in brain (Ishiwata et al. 2007; Matsuya et al. 2005). Another radiotracer labelled with the positron emitter ^{75}Se has been developed for A_1 imaging (5'-(methyl[^{75}Se]seleno)-N^6-cyclopentyladenosine), but no biological evaluation of this radiotracer has been reported (Blum et al. 2004). Within these PET radiotracers, only [^{18}F]CPFPX and [^{11}C]MPDX have been used in human clinical studies (Elmenhorst et al. 2012; Fukumitsu et al. 2003; Ishiwata et al. 2007; Meyer et al. 2006a, b).

Radiotracers for imaging A_{2A} receptors reported to date include, but are not limited to [^{11}C]TMSX, [^{11}C]SCH442416, [^{123}I]MNI-420 and [^{18}F]MNI-444. Detailed description of these and other A_{2A} radiotracers developed to date will be provided in a separate section on radiotracers for imaging A_{2A} receptors in brain, presented below. Recently, autoradiography studies using the tritiated radiotracer [^{3}H]MRS1754 have been used to successfully describe A_{2B} distribution in mouse, rabbit and dog brain (Auchampach et al. 2009). Although in vivo imaging of A_{2B} using PET and SPECT is still at an embryonic stage, with the continual development of highly selective A_{2B} antagonists, such as PSB-603 (for review on recent developments in adenosine receptor ligands see, for example Müller and Jacobson 2011), in vivo imaging of these receptors using PET or SPECT has potential for a rapid breakthrough. Two ^{18}F-labelled radiotracers have been recently introduced as potential PET probes for imaging A_3, [^{18}F]FE@SUPPY and [^{18}F]FE@SUPPY:2; and several nucleoside derivatives that contain ^{76}Br for PET imaging were recently reported, including MRS5147 (Kiesewetter et al. 2009; Mitterhauser et al. 2009).

Radiotracers for Imaging A_{2A} Receptors in Brain

The development of novel PET and SPECT radiotracers targeting A_{2A} has experienced significant growth as selective A_{2A} antagonists have become available. Several compounds have been developed as radiotracers for in vivo imaging of adenosine A_{2A} in brain (Fig. 11.2).

The xanthine derived A_{2A} antagonist radiotracer, [^{11}C]KF17837, that had shown promise in rodent studies, was unsuccessful in imaging A_{2A} in the monkey brain, where a limited brain extraction and high non-specific binding was observed (Noguchi et al. 1998; Stone-Elander et al. 1997). Further investigation of [^{11}C]KF17837 showed this radiotracer had poor selectivity for A_{2A}, as additional binding sites were subsequently identified. Other xanthine derived radiotracers that were prepared and

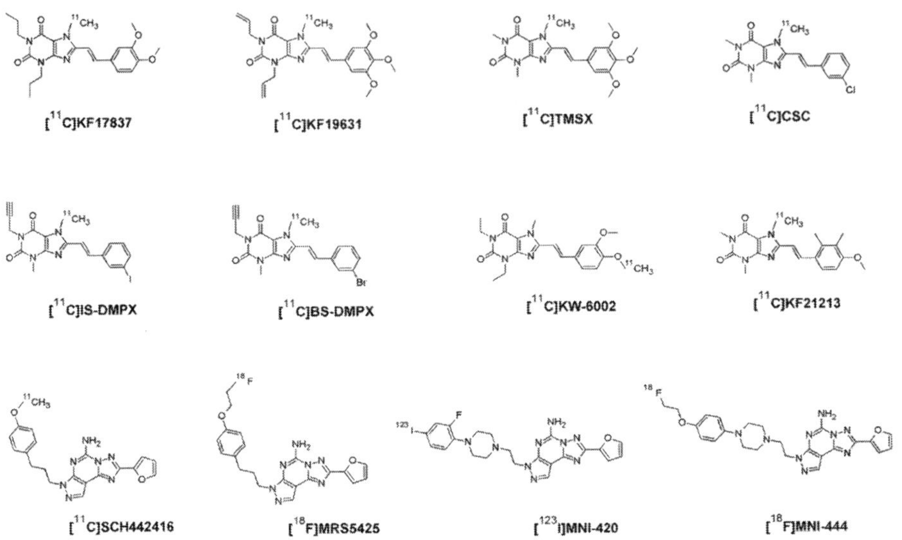

Fig. 11.2 Radiotracers for imaging A_{2A} in brain using PET and SPECT

evaluated in rodents included the [^{11}C]KF19631 and the [^{11}C]CSC (Ishiwata et al. 2000a, 2007; Márián et al. 1999). These radiotracers had similar properties to [^{11}C] KF17837 and thus, were deemed unsuitable for in vivo imaging of A_{2A} receptors in brain (Ishiwata et al. 2007). Later, [^{11}C]BS-DMPX and [^{11}C]IS-DMPX were developed and their biological properties were investigated by Ishiwata and co-workers. It was suggested that these radiotracers could be brominated and iodinated based on BS-DMPX and its chlorinated analogue CS-DMPX. The former could potentially be labelled with the positron emitter bromine-75 (half-life of 1.7 h) or bromine-76 (half-life of 16.1 h) and the latter with iodine-124 (half-life of 4.18 days) and iodine-123 (half-life of 13.3 h), allowing for the production of different PET and SPECT radiotracers (Ishiwata et al. 2007). However, despite the good in vitro affinity and selectivity for A_{2A}, [^{11}C]IS-DMPX and [^{11}C]BS-DMPX displayed high nonspecific binding and limited selectivity for the target in vivo, indicating that these radiotracers were not suitable for imaging of A_{2A} in brain (Ishiwata et al. 2000d). Another xanthine derived radiotracer, [^{11}C]KW-6002, was also developed, but its high concentration in extra-striatal regions meant [^{11}C]KW-6002 in vivo selectivity was questionable and limited its utility as a selective radiotracer for mapping A_{2A} in brain (Hirani et al. 2001). Remarkably, the non-labelled compound KW-6002 (also known as istradefylline) has been successfully used in several studies as an anti-parkinsonian agent (see for example, Bara-Jimenez et al. 2003; Hauser et al. 2003; Kase et al. 2001), having recently received the first global approval as a novel drug for treatment of PD in Japan (Dungo and Deeks 2013).

The continual search for improved xanthine derived radiotracers with higher A_{2A} selectivity, led to the development of [^{11}C]KF18446 (more frequently known as [^{11}C]TMSX) and [^{11}C]KF21213. In vivo evaluation of [^{11}C]TMSX in rodents and monkeys showed this radiotracer holds promise for imaging A_{2A} in brain (Ishiwata et al. 2005a). In mice the highest striatum:cerebellum uptake ratios of $[^{11}C]$TMSX were found to be 2.7 at 15 min post-injection; while in rats and monkeys the highest striatum:cerebellum uptake ratios reported were 2.67 (at 15 min post-injection) and 1.56 (at 60 min post-injection), respectively (Ishiwata et al. 2000a). In rodents, $[^{11}C]$KF21213 displayed a good striatal uptake ratio relative to cerebellum of 10.5 at 60 min post-injection and no specific uptake was observed in the cortex nor in the cerebellum (Wang et al. 2000). However, in nonhuman primate brain, $[^{11}C]$ KF21213 had a lower signal-to-noise ratio than $[^{11}C]$TMSX (Ishiwata et al. 2005a), suggesting that $[^{11}C]$TMSX is the most suitable radiotracer for mapping A_{2A} in brain among the xanthine derived radiotracers proposed to date.

All radiotracers presented above are xanthine derived radiotracers and it is known that the styryl group in xanthine derivatives is isomerized by exposure to visible light (Nonaka et al. 1993). This means that, in experimental pre-clinical and clinical studies, all procedures, from radiosynthesis to metabolite analysis of plasma samples from animals or humans required for quantitative evaluation of radiotracer binding, should be carefully performed under the exclusion of light. To overcome this issue, a non-xanthine derived A_{2A} antagonist radiotracer, $[^{11}C]$SCH442416, was developed and its potential as an A_{2A} imaging agent was investigated in vivo. Rodent studies showed that $[^{11}C]$SCH442416 had a good target:non-target ratio (in rats, a striatum:cerebellum ratio of around 4.6 was determined at 15 min post-injection) and a low amount of radioactive metabolites in brain and periphery (Moresco et al. 2005; Müller and Jacobson 2011). In addition, kinetic modelling showed that, in rats, $[^{11}C]$SCH442416 PET data could be modelled using both the 1- and 2-tissue compartmental models (1 T and 2 T) (details on nomenclature used for in vivo imaging of reversibly binding radiotracers can be found at Innis et al. 2007). The simplified reference tissue model (SRTM) (Lammerstma and Hume 1996) was able to estimate binding potentials (BP_{ND}) in the striatum, although the use of this method resulted in a small underestimation of the binding potential values by about 16% as compared with 1 T and 2 T models. The test-retest variability of BP_{ND} was lowest when using SRTM for data quantification, on average < 10%. $[^{11}C]$SCH442416 dosimetry estimates using the rodent model showed this radiotracer displayed a favourable dosimetry profile for imaging in humans, where an effective dose of 4.1 μSv/MBq was determined (Wells et al. 2013). Despite the promising rodent data, a relatively high non-specific binding and a striatal binding potential BP_{ND} of 0.74 was measured in monkey brain (Moresco et al. 2005; Müller and Jacobson 2011). Still, $[^{11}C]$SCH442416 has been used in several human studies investigating A_{2A} densities in specific brain disorders and in drug receptor-occupancy studies targeting A_{2A} in brain (Brooks et al. 2010; Mihara et al. 2008; Ramlackhansingh et al. 2011). In healthy human subjects, the binding potentials determined in the caudate, putamen and thalamus were around 0.53, 0.99 and 0.12, respectively (Ramlackhansingh et al. 2011).

Until recently all A_{2A} radiotracers developed (either xanthine or non-xanthine derived radiotracers) have been labelled with the positron emitter carbon-11. Radiotracers labelled with longer-lived radioisotopes, such as iodine-123 or fluorine-18, which could be dispensed from a central pharmacy rather than generated on-site, would be advantageous, in particular in the conventional clinical setting where no cyclotron facilities are typically available. They could also potentially allow for data to be acquired over a wider time window, without jeopardizing adequate data sampling with minimal noise, an issue more commonly encountered when using ^{11}C-labelled radiotracers. In 2011, Bhattacharjee and co-workers developed [^{18}F]MRS5425, a derivative of SCH442416, for in vivo imaging of A_{2A} in brain. Preliminary data in rodents demonstrated this radiotracer had a peak percentage injected dose in the striatum of 0.75%/g at 90 s post-injection, followed by a plateau at 3.5 min and a slow decline thereafter. Furthermore, they found that [^{18}F]MRS5425 striatal binding was blocked by pre-administration of SCH442416 (Bhattacharjee et al. 2011). Further evaluation of this radiotracer in nonhuman primates or humans has not been reported so far. In 2013, the Molecular NeuroImaging group publish data on the first successful SPECT radiotracer for imaging A_{2A} in brain, [^{123}I]MNI-420, a non-xanthine derived compound, analogue of preladenant. Shortly after, the same group reported the development of a MNI-420 analogue labelled with fluorine-18, [^{18}F]MNI-444, that has also shown to be promising as a PET radiotracer for imaging A_{2A} in brain. The striatum:cerebellum ratios determined for [^{123}I]MNI-420 and [^{18}F]MNI-444 in monkey brain were found to be~3.0–3.5 and~7.0–9.0, respectively (Alagille et al. 2013; Tavares et al. 2013a, b), indicating that [^{123}I]MNI-420 and [^{18}F]MNI-444 are the most suitable radiotracers for mapping A_{2A} in brain among the non-xanthine derived radiotracers proposed to date.

Other attempts to develop a selective A_{2A} radiotracer have been reported in the literature, including a study by Holschbach et al. in 2006 that examined the use of oxazolopyrimidines as potentially amenable compounds for A_{2A} imaging. Although the developed library had affinities for the A_{2A} in the low-nanomolar range, and some were quite selective over the A_1, a high level of non-specific binding obscuring specific binding in in vitro autoradiographic experiments deemed those oxazolopyrimidines unsuitable candidates for brain imaging of A_{2A} receptors by PET (Holschbach et al. 2006).

The sections below will describe in detail the pre-clinical and clinical validation data obtained so far for the most suitable and promising xanthine and non-xanthine derived radiotracers developed to date for imaging A_{2A} in brain using SPECT and PET.

[^{11}C]TMSX

In 2000, Ishiwata and co-workers reported the development of [^{11}C]TMSX. In mice, the striatal uptake of [^{11}C]TMSX gradually decreased over time, with the striatum:cerebellum ratio peaking at 2.71 at 15 min post-injection. [^{11}C]TMSX uptake was blocked by xanthine and non-xanthine derived compounds. However, blockade with cold TMSX was able to reduce the cortical uptake by about 50%, while the non-xanthine derived compound SCH 58261 only reduced the cortical uptake by~25% (Ishiwata et al. 2000a). Autoradiography experiments demonstrated that [^{11}C]TMSX had a different affinity for A$_{2A}$ in the striatum compared with the cerebral cortex, 9.8 nM and 16 nM, respectively (Ishiwata et al. 2000b). These findings were also observed by Fredholm and co-workers using tritiated A$_{2A}$ ligands and autoradiography, leading the authors to suggest the observed uptake in cortical areas was due to the presence of atypical A$_{2A}$ receptor subtypes that contrasted with the classical A$_{2A}$ receptors in the striatum (Cunha et al. 1996; Lindström et al. 1996). This hypothesis has also been suggested by Noguchi et al. in 1998 as the mechanism underlying the observed unknown binding sites with [^{11}C]KF17837, another xanthine-derived PET radiotracer (Noguchi et al. 1998). These differences in binding affinity determined for [^{11}C]TMSX do not seem to be related with this radiotracer binding to other receptors in brain, as prior in vitro studies showed [^{11}C]TMSX had negligible affinity for 13 neuroreceptors, including adrenergic, dopamine, acetylcholine and serotonin receptors (Ishiwata et al. 2000b).

Further evaluation of [^{11}C]TMSX in rats also demonstrated a striatum:cerebellum ratio similar to mice (about 2.67 at 15 min post-injection). Later, data from nonhuman primate studies showed that the striatal activity levels were retained high for the initial 20 min and then gradually decreased with time. In monkeys, the binding ratios of [^{11}C]TMSX were found to peak at around 1.56 at 60 min post-injection (Ishiwata et al. 2000a).

[^{11}C]TMSX metabolism in mice blood was slow and, when the metabolite analysis was carefully undertaken under dim light to prevent the radiotracer isomerization, over 80 % of the detected radioactivity in mice plasma at 30 min post-injection was parent unchanged compound. Furthermore, metabolism experiments conducted using mice brain demonstrated the radiolabelled metabolites present in this organ were negligible. In monkeys, peripheral degradation of this radiotracer was found to be faster than in mice, with parent compound in plasma at 30 min post-injection of about 40 % (Ishiwata et al. 2000a).

Prompted by the promising results in rodents and monkeys, human studies were subsequently undertaken to characterize [^{11}C]TMSX pharmacokinetics in brain, whole-body biodistribution and dosimetry. Kinetic modelling of [^{11}C]TMSX brain PET data demonstrated the 2 T model and the Logan graphical analysis (Innis et al. 2007; Logan 2000) were able to describe the obtained time-activity curves in humans. Furthermore, data published by Naganawa et al. in 2007 showed that A$_{2A}$ in the human brain could be visualized as a binding potential image using [^{11}C]TMSX and PET without arterial blood sampling (Naganawa et al. 2007).

Mishina et al. found about 30 % specific binding in human cerebellum, cerebral cortex and thalamus when imaging with [^{11}C]TMSX, indicating that in humans the cerebellum, cortex, or thalamus could not be used as reference region for quantification of [^{11}C]TMSX binding potentials (Mishina et al. 2007). Subsequently, the centrum semiovale was proposed as an alternative to the use of the cerebellum as reference region for quantification of [^{11}C]TMSX PET data, because this region had the lowest [^{11}C]TMSX binding of all investigated regions, and was considered to be devoid of specific binding due to the few neurons present there (Mishina et al. 2007; Naganawa et al. 2007). Additionally, it was found that the duration of 10–40 min after radiotracer administration was a practical choice for estimating the

[^{11}C]TMSX total distribution volumes accurately (Naganawa et al. 2007). When the centrum semiovale was used as reference region for data quantification, the [^{11}C]TMSX binding potential in human brain was highest in the anterior (1.25) and posterior putamen (1.20), followed by the head of the caudate nucleus (1.05) and thalamus (1.03), and it was low in the cerebral cortex, in particular in the frontal lobe (0.46). The highest binding potential in the striatum agrees with previous reports demonstrating enriched A$_{2A}$ expression in that brain region (Sihver et al. 2009; Svenningsson et al. 1997), however the binding of [^{11}C]TMSX was relatively larger in the thalamus when compared with other mammals and prior studies with human brain tissue. Mishina et al. suggested that atypical A$_{2A}$ receptors may be involved in the cerebral cortex, cerebellum and thalamus observed specific binding with [^{11}C]TMSX (Mishina et al. 2007). Overall, A$_{2A}$ imaging by [^{11}C]TMSX PET reflects the distribution of A$_{2A}$ in brain previously reported, although regional differences in the signals of specific binding were relatively smaller compared with those of other radioligands. For example, the binding potential of [^{11}C]TMSX in the putamen was only 1.2 times of that in the thalamus and 3.7 times of that in the frontal cortex; while post-mortem studies have revealed that the density of A$_{2A}$ in the putamen and caudate nucleus was 5 times that of the thalamus and 3–5 times of that in the cerebral cortices (Svenningsson et al. 1997). This discrepancy may be due to differences in the methodology used, such as, PET imaging versus autoradiography and differences in the radiotracers used.

In humans, [^{11}C]TMSX peripheral metabolism in blood was extremely slow with about 95 % parent present at 30 min post-injection, indicating the labelled metabolites may be negligible in the human [^{11}C]TMSX PET examination (Mishina et al. 2007). This contrasts with pre-clinical monkey data that found 40 % of parent compound in plasma at 30 min post-injection, but it is in line with mice studies data demonstrating that the percentage of unchanged [^{11}C]TMSX in plasma at 30 min post-injection was about 80 % (Márián et al. 1999).

Despite the discrepancies in [^{11}C]TMSX regional distribution in brain in comparison with expected mapping from autoradiography studies, that requires further investigation, [^{11}C]TMSX whole-body and biodistibution was investigated in three human subjects. These data were recently published and demonstrated that [^{11}C]TMSX main elimination route in humans was hepatobiliary. The whole-body effective dose was determined to be about 3.6±0.29 µSv/MBq, which is in line with other ^{11}C-labelled radiotracers previously developed and currently in human use (Sakata et al. 2013).

[^{123}I]MNI-420

In 2013, Tavares and co-workers reported the development of [^{123}I]MNI-420 as a SPECT radiotracer for imaging A$_{2A}$ in brain. A rapid brain penetration was observed following intravenous injection of [^{123}I]MNI-420 in two different species of nonhuman primates (cynomolgus monkeys and baboons). The regional brain accumulation of [^{123}I]MNI-420 was consistent with the known distribution of A$_{2A}$

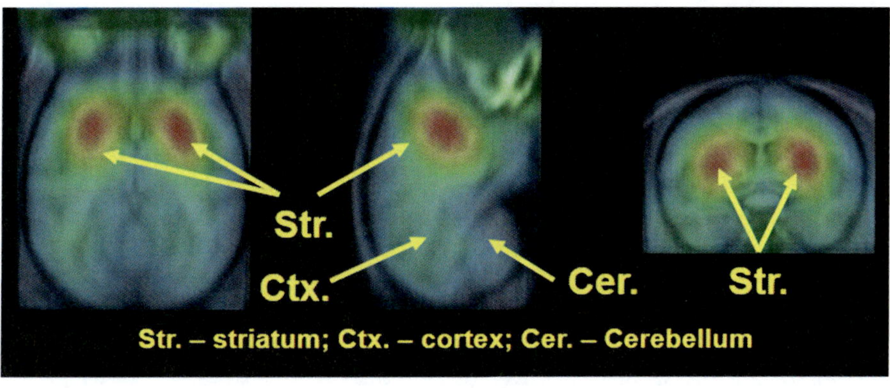

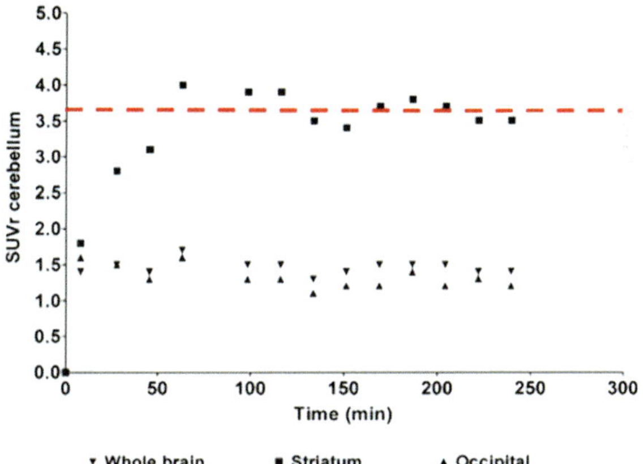

Fig. 11.3 Representative cynomolgus monkeys [^{123}I]MNI-420 brain SPECT images (*top row*) and standardized uptake value ratios (*SUVr*) in different brain regions (*bottom row*). SPECT images presented in the *top row* are co-registered with magnetic resonance images for better anatomical localization of different brain structures. Sum SPECT images from 0 to 240 min post-radiotracer injection. Note the high target:cerebellum ratios determined (*cerebellum as reference region*)

in brain. Striatum:cerebellum ratios and binding potentials of around 3.0–3.5 and 2.0–2.5, respectively, were measured in monkey and baboon brain (Tavares et al. 2013a) (Fig. 11.3). These results compared favourably with promising radiotracers previously developed for imaging of A_{2A} in brain. For example, [^{11}C]TMSX and [^{11}C]SCH442416, both already used in multiple human studies, had a maximum uptake ratio to cerebellum and binding potential of around 1.6 and 0.7 in monkey striatal region, respectively (Mishina et al. 2007; Moresco et al. 2005). The cerebellum was used as the reference region for estimation of tissue ratios and binding potentials of [^{123}I]MNI-420. Collected data in nonhuman primates showed the cerebellum had the lowest uptake of all brain regions and its binding did not appear to be reduced during pre-blocking experiments with high doses of a selective A_{2A} drug, preladenant (Tavares et al. 2013a). Unlike previously developed xanthine-derived radiotracers for imaging A_{2A} in brain, including [^{11}C]TMSX, that have shown evidence of undefined extra-striatal specific binding in the cerebellum in vivo, [^{123}I] MNI-420 SPECT showed no specific binding in that region. This agrees with high resolution autoradiography experiments showing that the cerebellum is a region with low to negligible A_{2A} density (Sihver et al. 2009), supporting the use of the cerebellum as a reference region. It is possible that the xanthine-derived radiotracers have a different in vivo behaviour compared to the non-xanthine ones. In fact, it was proposed that the reduction of the xanthine-based radiotracers extra-striatal binding (such as [^{11}C]TMSX) by xanthine but not by non-xanthine A_{2A} antagonists, was either due to additional binding of those radiotracers to "undefined" sites or that a slower in vivo association kinetics for non-xanthine antagonists was the reason for such unexpected observation (Hirani et al. 2001; Ishiwata et al. 2000; Noguchi et al. 1998; Wang et al. 2000). However, one can also hypothesize that the observed results with xanthine radiotracers previously developed could be simply due to the lack of in vivo selectivity, the high reported non-specific binding or a combination of those two factors.

Encouraged by the promising results in nonhuman primates, [^{123}I]MNI-420 validation in humans was subsequently undertaken. Similar to the collected monkey data, [^{123}I]MNI-420 accumulation in human brain in vivo was consistent with known A_{2A} distribution and at the optimal imaging time point (>90 min post-injection) stable striatum:cerebellum ratios of 1.6, 2.0 and 1.8 were measured in the caudate, putamen and striatum, respectively (Tavares et al. 2013b). These values compare favourably with the highest distribution volume ratios reported for [^{11}C]TMSX in human subjects (1.4 and 1.5 in the caudate and putamen, respectively; when the frontal lobe, temporal lobe and occipital lobe were used as reference region) (Mishina et al. 2011). Human studies using [^{11}C]TMSX have demonstrated radiotracer binding to the thalamus (Mishina et al. 2011), where the thalamic binding was translated into an average binding potential value of around 1.03 versus 1.25 in the putamen. This thalamic binding has been attributed to the presence of atypical A_{2A} receptors in that brain region. However, it is interesting to note that human SPECT data acquired using [^{123}I]MNI-420 did not display detectable binding to the thalamus (Tavares et al. 2013b). This further supports the hypothesis that xanthine derived radiotracers have a different in vivo behaviour compared with the non-xanthine ones. Alternatively, one could also hypothesise that the xanthine derived radiotracers may be less stable than non-xanthine derived radiotracers and their metabolite(s) could potentially bind to other non-target sites in brain.

Kinetic modelling of [^{123}I]MNI-420 human SPECT data demonstrated the 2 T model was able to describe this radiotracer kinetics in brain. Furthermore, it was found that non-invasive methods of quantification, namely, SRTM and non-invasive Logan graphical analysis, using the cerebellum as a reference region, were able to describe the SPECT data. However, the results obtained using these non-invasive methods were slightly underestimated by 14 %. Similarly the use of the stable striatum:cerebellum ratios at t>90 min post-injection allowed for the quantification of the data, but marginally underestimated the binding potentials by about 6 %. The [^{123}I]MNI-420 striatal binding potentials were found to range between 0.8 and 1.2 (Tavares et al. 2013b), that is in line with data reported for [^{11}C]SCH442416 (binding potentials ranging between 0.5 and 1.0) and [^{11}C]TMSX (binding potentials ranging between 1.0 and 1.3) (Naganawa et al. 2007; Ramlack-hansingh et al. 2011).

The test-retest variability in the striatum determined using [^{123}I]MNI-420 SPECT was found to be, on average, 4.8 % when using SRTM method, 3.5 % when using Logan reference and 6.5 % when using SUVr methods (Tavares et al. 2013b). These values compare favourably with previously reported test-retest variability determined in humans with other ^{123}I-labelled SPECT radiotracers developed for brain receptor imaging and support the use of [^{123}I]MNI-420 SPECT for mapping A$_{2A}$ in brain.

Peripheral metabolism of [^{123}I]MNI-420 in humans post intravenous injection was slow and at 30 min and 2 h after radiotracer injection, the parent fraction in arterial plasma was determined to be around 87 % and 63 %, respectively. Whole-body biodistribution and dosimetry studies were subsequently conducted in healthy male and female human volunteers. Collected data showed [^{123}I]MNI-420 main elimination was mainly hepatobiliary, being minimally excreted via the urinary system. The mean effective dose was determined to be around 0.036 mSv/MBq, suggesting acceptably low radiation exposure associated with [^{123}I]MNI-420 imaging in human subjects and allowing multiple scans to be performed in the same research subjects per year (Tavares et al. 2013b). The determined mean effective dose is within the typical range of doses for ^{123}I-labelled radiotracers.

[^{18}F]MNI-444

Subsequent to the development of the SPECT A$_{2A}$ radiotracer [^{123}I]MNI-420, a ^{18}F-labelled PET A$_{2A}$ radiotracer analogue of MNI-420 was developed and named [^{18}F]MNI-444. Following intravenous bolus injection, [^{18}F]MNI-444 rapidly entered the nonhuman primate brain and distributed in the tissue accordingly with the known densities of A$_{2A}$ in brain (Fig. 11.4) (Alagille et al. 2013).

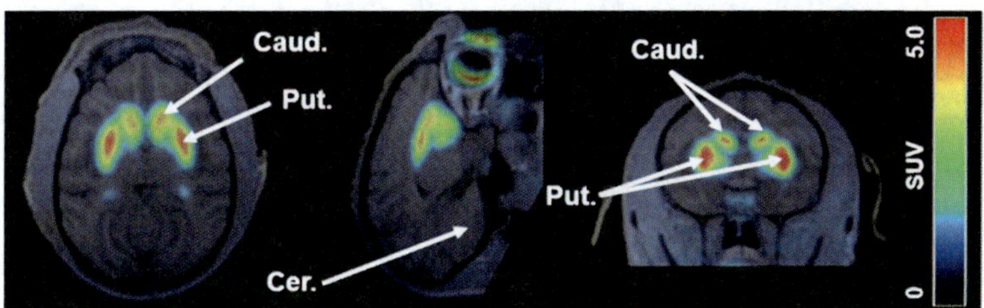

Fig. 11.4 Representative rhesus macaque [^{18}F]MNI-444 PET images fused to magnetic resonance images. Average [^{18}F]MNI-444 PET images over 180 min of acquisition. Legend: *Caud* Caudate, *Put* Putamen, *Cer* Cerebellum

Kinetic modelling of acquired monkey PET data with [^{18}F]MNI-444 showed this radiotracer had exceptionally high striatal binding potentials ranging between ~5.5 and 8.0 when using the invasive Logan graphical analysis. Non-invasive methods of analysis using the cerebellum as a reference region were able to quantify [^{18}F]MNI-444 PET data although a slight < 10 % bias was determined when using 180 min of data or < 5 % when using 120 min of data. The binding potentials measured for [^{18}F]MNI-444 were the highest reported thus far for an A$_{2A}$ radiotracer in nonhuman primate brain and can allow for superior in vivo inspection of these receptors, either in studies investigating mechanisms underlying brain disorders or the efficacy of novel drugs, than previously developed ligands. Test-retest data in monkey showed a good reproducibility (< 10 % variability) when binding potentials were determined using only 120 min of PET data [results pending publication]. Furthermore, [^{18}F]MNI-444 binding was blocked in a dose-dependent mode by the selective A$_{2A}$ antagonist, preladenant, (Alagille et al. 2013) and the occupancy estimates obtained using the plasma-based and reference-region-based methods were in good agreement, indicating that receptor-occupancy could be estimated without the need for arterial sampling.

Whole-body PET imaging following intravenous bolus injection of [^{18}F]MNI-444 in adult rhesus monkeys showed that the radiotracer was eliminated via the hepatobiliary and the urinary systems. The mean effective dose determined using the nonhuman primate model was found to be around 0.02–0.03 mSv/MBq, suggesting only modest radiation exposure associated with [^{18}F]MNI-444 imaging [unpublished data]. The whole body effective dose of [^{18}F]MNI-444 determined was similar to that of other ^{18}F-labeled radiotracers currently used in neuroreceptor human studies and would potentially allow multiple scans to be performed in the same research subject per year.

Preliminary data from ongoing humans studies demonstrated that [^{18}F]MNI-444 had a good brain penetration, distribution consistent with known A$_{2A}$ densities and excellent binding potentials of about 4–5 in target regions. A test-retest reproducibility of < 10 % was determined for the first group of human subjects imaged, indicating this radiotracer has potential for imaging A$_{2A}$ in human brain.

Imaging Studies with Selective A$_{2A}$ PET and SPECT Radiotracers in Neuroscience Research

As discussed above, over the years, multiple xanthine and non-xanthine derived radiotracers have been developed and evaluated for in vivo imaging of A$_{2A}$. The availability of these radiotracers allowed for in vivo inspection of changes of A$_{2A}$ expression in brain in different research projects, as well as, the effects of drugs on those receptors. Below are presented examples of pre-clinical and clinical imaging experiments with PET and SPECT radiotracers used to estimate drug receptor-occupancy or to quantify changes in A$_{2A}$ expression in the living brain, with a particular emphasis on PD research.

Effects of Caffeine and Theophylline on A$_{2A}$ Receptors in Brain

The natural stimulants caffeine and theophylline are the prototypical A$_{2A}$ antagonists. Although these alkylxanthine derivates are non-selective A$_{2A}$ antagonists, studies have shown evidence of their motor stimulant and neuroprotection effects in PD (Ferré et al. 1992; Schwarzschild et al. 2002; Xu et al. 2010), albeit the usefulness of theophylline has been less consensual than the usefulness of caffeine. In 2002, Kulisevsky et al. published a report showing that theophylline consistently failed to potentiate the anti-PD action of L-DOPA (Kulisevsky et al. 2002). Conversely, caffeine has dopamine agonist-like effects (Garrett and Griffith 1997) and population studies have shown that caffeine consumption may reduce the risk of developing PD, where the incidence of PD decreases as a function of caffeine intake. Data from those studies have shown that the PD incidence drops from 9.2/1000 persons for non-drinkers to 3.1/1000 persons for individuals whose caffeine levels per day are around 106.8–705.3 mg (Ascherio et al. 2001; Ross et al. 2000).

The effects of caffeine in the brain occur mostly by inhibition of A$_{2A}$ receptors (Ross et al. 2000). In receptor-occupancy studies using SPECT with [^{123}I]MNI-420, intravenous injection of caffeine reduced the radiotracer striatal binding in a dose-dependent mode, where the drug dose able to induce 50 % receptor-occupancy (ED$_{50}$) was found to be 3.8 mg/kg. Furthermore, at an acute dose of 20 mg/kg (around 300 mg per study), the A$_{2A}$ occupancy by caffeine was found to be around 98 % (Fig. 11.5) and ~54 % receptor-occupancy was achieved when injecting 5 mg/kg of caffeine intravenously (about 75 mg per study) (Tavares et al. 2013a). Later, during the human validation of [^{123}I]MNI-420, one of the human subjects enrolled in the test-retest studies inadvertently ingested a caffeinated beverage shortly before the retest scanning session. This resulted in about 60 % reduction of the binding potentials in the striatum compared with the test values (Fig. 11.6) (Tavares et al. 2013b). The observed decrease in that human subject retest binding potential values in comparison with the test data is not surprising and agrees with the pre-clinical data collected in baboons (Tavares et al. 2013b). Both the animal and human studies with SPECT and [^{123}I]MNI-420 were able to clearly demonstrate and directly quantify the significant effects of caffeine on A$_{2A}$ in brain.

In mice, intraperitoneal pre-treatment with 10 mg/kg theophylline 15 min before [^{11}C]TMSX administration induced a 27 % reduction in the striatal uptake (Ishiwata et al. 2000b). When the effects of theophylline on [^{11}C]TMSX binding were evaluated in mice heart, about 34 % of radiotracer uptake in that organ was blocked at doses of 100 mg/kg (Ishiwata et al. 2003). Later in 2005, Ishiwata and co-workers reported that a theophylline-infusion in human subjects was only able to slightly reduce [^{11}C]TMSX binding in the caudate nucleus (8 % reduction) and in the putamen (4.5 % reduction). This small reduction in [^{11}C]TMSX binding induced by theophylline-infusion was attributed to the low dose of theophylline used (~200 mg, the recommended limit for clinical use of theophylline, which corresponds to about 3.5 mg/kg) and the relatively weak affinity of theophylline in comparison with the radiotracer affinity: the affinity values (K$_i$) for the A$_1$ and A$_{2A}$ are 1600 nM and 5.9 nM for TMSX, and 23,000 nM and 16,000 nM for theophylline, respectively

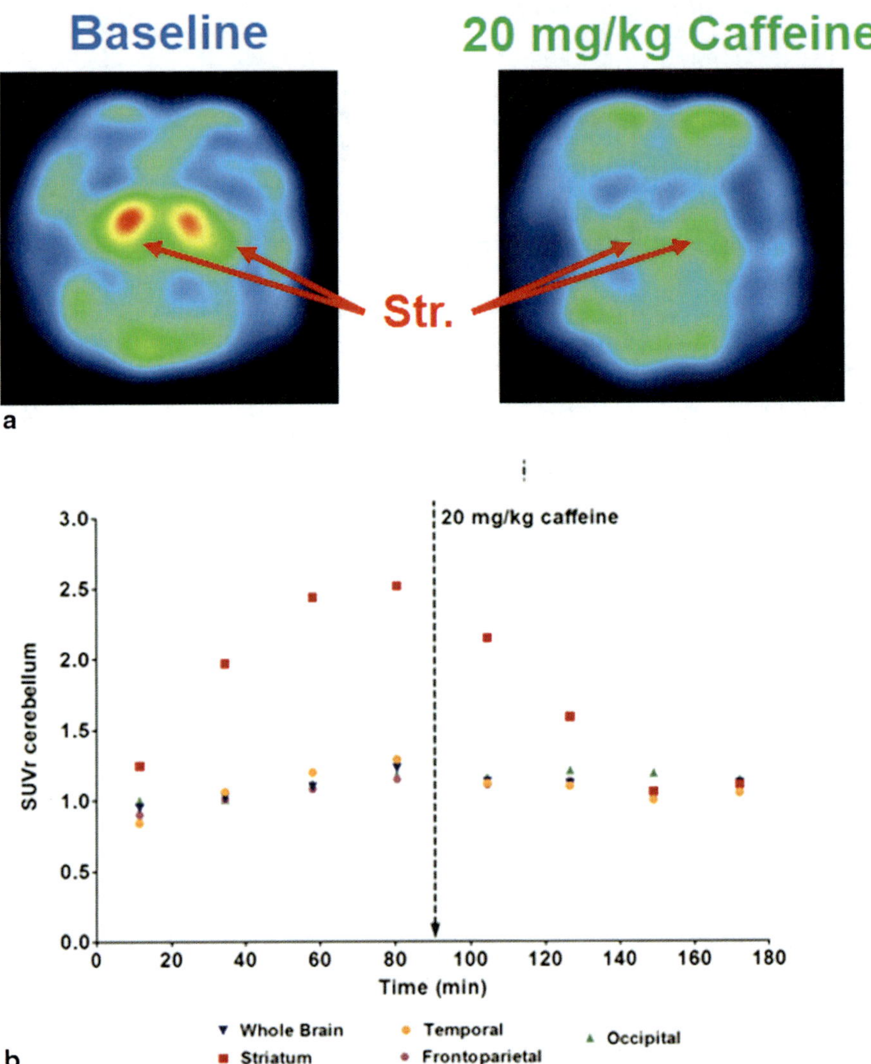

Fig. 11.5 **a** Representative SPECT image of [^{123}I]MNI-420 distribution in baboon brain at baseline, i.e. prior to displacement (*left figure*), and post caffeine administration (*right image*). **b** [^{123}I] MNI-420 standardized uptake value ratios (SUVr) curves obtained in baboons at displacement conditions. Sum SPECT images before (0–90 min) and following displacement (90–180 min). Note significant reduction in SUVr values in striatum following intravenous injection of 20 mg/kg caffeine (*cerebellum used as reference region*)

(Ishiwata et al. 2005b). The reduction of [^{11}C]TMSX binding determined in mice brain and heart post-administration of theophylline contrasts with the human data, where a marginal reduction of the striatal uptake was noted. This difference is most likely due to the ~28 times higher dose used in the mice studies compared with the human study.

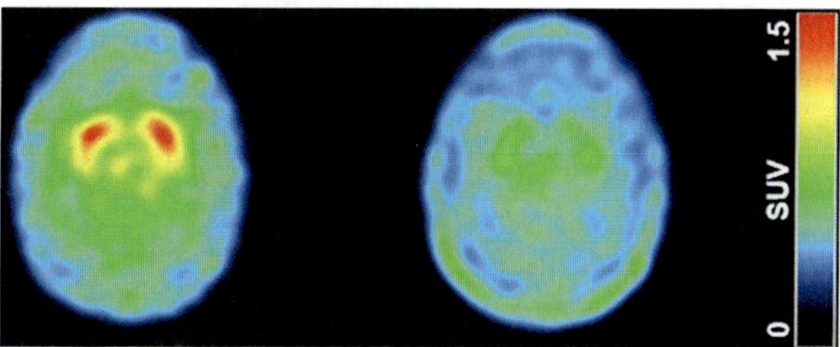

Fig. 11.6 Effects of caffeine on [^{123}I]MNI-420 binding in human brain. SPECT standardized uptake value (SUV) sum images (0–240 min) of [^{123}I]MNI-420 distribution in a human subject at baseline (*left side image*) and post ingestion of a caffeinated beverage (*right side image*). Note the significant reduction in radiotracer uptake in the SPECT images post-caffeinated beverage ingestion in comparison with baseline SPECT images

Although the human data collected using SPECT with [^{123}I]MNI-420 or PET with [^{11}C]TMSX cannot be directly compared, as a result of differences in the radiotracer and imaging modality used, it is interesting to note that ∼100 mg of caffeine seem to be able to induce considerably higher A$_{2A}$ occupancy than ∼200 mg of theophylline (50 % versus 5 %). This may explain the inconsistency and controversy surrounding the usefulness of theophylline in the treatment of PD motor signs, in contrast with the putative value of caffeine as motor stimulant and neuroprotecting agent, given that very high doses of theophylline seem to be needed to yield results that would approach those obtained with caffeine. But the human imaging data collected thus far is limited, and more refined imaging studies are needed to better determine the value of theophylline and caffeine in PD.

Drug Receptor-Occupancy Studies Targeting A$_{2A}$ in Brain

Caffeine and theophylline are known to be non-selective A$_{2A}$ xanthine derived antagonists. And over the years, significant efforts have been done to develop A$_{2A}$ antagonists with improved affinity and selectivity.

In 2007, Mihara et al. reported the pharmacological characterization of a novel and potent A$_1$ and A$_{2A}$ dual antagonist drug, ASP5854, for treatment of PD. This group rational for developing a dual antagonist rather than a selective A$_{2A}$ agent was based on the known roles of A$_1$ in memory and cognition (Costenla et al. 1999; Maemoto et al. 2004; Normile and Barraco 1991) and A$_{2A}$ in motor control (Fredholm et al. 1999; Kanda et al. 1998; Koga et al. 2000; Ikeda et al. 2002). They proposed that the blockade of both A$_1$ and A$_{2A}$ might have therapeutic implications for different neurodegenerative diseases, in particular PD, which presents both the motor disability and the cognitive impairment. ASP5854 affinities for human adenosine receptors were determined to be as follows: K$_i$ A$_{2A}$ of 1.7 nM, K$_i$ A$_1$ of 9.0 nM and K$_i$ A$_3$ higher than 557 nM (Mihara et al. 2007). ASP5854 receptor-occupancy studies were acquired in nonhuman primates using PET with [^{11}C]SCH442416. Collected data showed that intravenous administration of ASP5854 1 h prior to radiotracer injection blocked [^{11}C]SCH442416 in a dose-dependent mode and induced long-lasting occupancy. Doses of 0.1 mg/kg resulted in plasma exposures of about 98,000 ng/mL and corresponded to 85–95 % A_{2A} occupancy. The authors concluded that the novel A_1 and A_{2A} antagonist, ASP5854, was an active and brain-penetrable drug in nonhuman primates and might represent a novel treatment for PD. Furthermore, the obtained findings supported the use of PET imaging with selective A_{2A} radiotracers for estimating the effective doses of A_{2A} antagonists in humans (Mihara et al. 2008).

Vipadenant (previously known as BIIB014), a non-xanthine A_{2A} antagonist, has been proposed for treatment of early and late PD (K_i A_{2A} of 1.3 nM, K_i A_1 of 68 nM, K_i A_{2B} of 63 nM and K_i A_3 of 1005 nM in human brain) (Müller and Jacobson 2011). An open-label human PET study with [^{11}C]SCH442416 was used to determine the relationship among dose, steady-state plasma levels and A_{2A} occupancy in brain. Fifteen healthy human subjects underwent a baseline PET scan prior to drug administration and a post-blockade PET scan after daily oral vipadenant (2.5–100 mg/day for 10 or 11 days). In brain target regions, vipadenant induced a receptor occupancy ranging between 74 % and 94 % at the lowest investigated dose (2.5 mg). Receptor saturation (> 90 % occupancy) was achieved at doses of 100 mg. The minimal daily dose required for receptor saturation was found to be around 10.2 mg, which was equivalent to 0.097 µg/ml in plasma (steady state minimum concentration). The authors concluded that the imaging results, together with the previously acquired efficacy results in animals, support the continued development of vipadenant as a potential treatment for PD (Brooks et al. 2010).

Peladenant, a highly selective and potent non-xanthine derived A_{2A} antagonist (K_i A_{2A} of 0.9 nM and K_i A_1, A_{2B} and A_3 > 1000 nM in human brain), was recently developed for treatment of PD (Müller and Ferré 2007; Müller and Jacobson 2011). In cynomolgus monkeys, a SPECT study with [^{123}I]MNI-420 demonstrated that intravenous administration of preladenant 15 min prior to radiotracer injection reduced the [^{123}I]MNI-420 uptake in the striatum in a dose dependent mode (Fig. 11.7). At doses of 1.2 mg/kg of preladenant around 98 % of the A_{2A} receptor were occupied by the drug. Preladenant ED_{50} in vivo was determined to be around 0.06 mg/kg (Tavares et al. 2013a).

Animal Models and In Vivo Imaging of A_{2A} in Brain

PD can be modelled in laboratory animals by unilateral injection of the selective monoaminergic toxin, 6-hydroxydopamine (6-OHDA), into the substantia nigra or medial forebrain bundle, which causes neuronal death. Recently, A_{2A} receptor changes in the 6-OHDA rat model for PD were investigated using [^{18}F]MRS5425 and PET. Collected data showed that in 6-OHDA-lesioned rats, the measured %ID/g was significantly higher in the lesioned side compared to the intact side (increase of 9–12 % in the %ID/g) (Bhattacharjee et al. 2011). This study results further supports the important role of imaging A_{2A} in PD research.

The quinolinic acid-lesioned striatum HD rat model has been developed by inducing degeneration of striatopallidal γ-aminobutyric acid-ergic-enkephalin

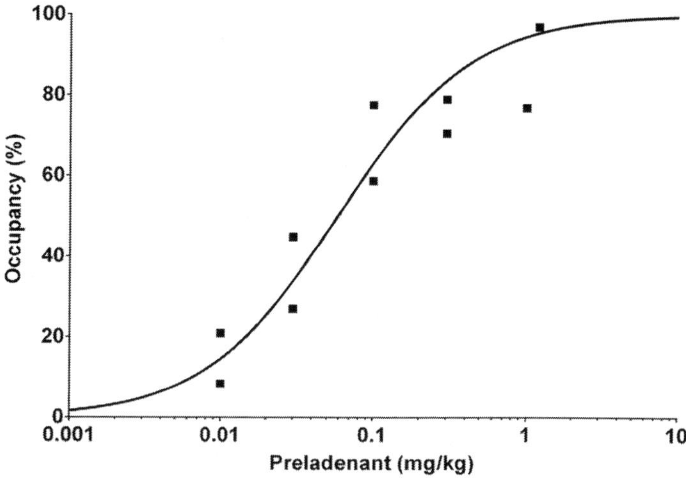

Fig. 11.7 Preladenant dose-dependent occupancy curve measured using SPECT with [^{123}I]MNI-420 in monkeys

neurons via intrastriatal injection of quinolinic acid. Prior studies have used [^{11}C] TMSX PET to investigate changes in the striatal A$_{2A}$ receptors in that model. Acquired PET data demonstrated that the radiotracer binding was significantly reduced in the quinolinic acid-lesioned striatum (binding potential of 0.6 versus 0.4 for intact and lesioned side, respectively). This decrease in signal was comparable to the decrease in binding potentials measured with the dopamine D$_2$ radiotracer [^{11}C]raclopride and was further confirmed by ex vivo A$_{2A}$ autoradiography (Ishiwata et al. 2002). On a different study with the same animal model, ex vivo autoradiography with [^{11}C]TMSX showed this radiotracer, but not [^{11}C]raclopride, was incorporated into the rat globus pallidus with a striatum:globus pallidus ratio of~0.6 (Ishiwata et al. 2000c), indicating that imaging A$_{2A}$ with PET or SPECT and selective radiotracers can be useful as a marker of the terminals projecting from the striatum to the globus pallidus.

Human Clinical Studies Evaluating A$_{2A}$ in Brain

Ramlackhansingh and co-workers in 2011 reported human clinical studies investigating the striatal A$_{2A}$ receptor availability in PD patients with and without L-DOPA induced dyskinesias (LIDs). Six patients with PD with and 6 without LIDs were enrolled in this PET study. [^{11}C]SCH442416 was used as a biomarker for quantification of brain A$_{2A}$ in vivo and collected data demonstrated A$_{2A}$ binding potentials in the caudate and putamen of PD patients with LIDs was significantly higher than that of subjects with PD without LIDs, which lay within the control range. Furthermore, the authors also reported that thalamic A$_{2A}$ binding was similar across all three groups. The authors concluded that the increased A$_{2A}$ receptor availability in the striatum of PD patients with LIDs was compatible with altered adenosine transmission playing a role in LIDs, which further provides rationale for trials of A_{2A} receptor antagonist in the treatment of PD motor complications (Ramlackhansingh et al. 2011).

On another study reported by Mishina et al. in 2011, A_{2A} receptor changes in the striata of PD patients were investigated. In that study, [^{11}C]TMSX and PET were used to measure brain A_{2A} binding in vivo in nine drug-naïve PD patients, seven PD patients with mild dyskinesia and six elderly control subjects. Seven of the drug-naïve PD patients underwent a second series of PET scans following antiparkinsonian therapy. The L-DOPA equivalent dose for these PD patients ranged from 75.0 to 825.5 mg at the time of post-therapeutic PET scanning. None of these patients developed dyskinesia during the study period. Results from this study showed that the distribution volume ratios in the putamen were significantly larger in the dyskinesic patients than in the control subjects. Furthermore, it was found that in the drug-naïve PD patients, the radiotracer binding in the putamen, but not in the head of the caudate nucleus, was significantly lower on the more affected side than on the less affected side. In this same group, the A_{2A} receptors were significantly increased after antiparkinsonian therapy in both putamen, but not in the head of the caudate nucleus (Mishina et al. 2011). The authors of this study concluded that A_{2A} receptors were increased in PD patients with dyskinesia, which agrees with the data published by Ramlackhansingh and co-workers in the same year (Ramlackhansingh et al. 2011). Furthermore, the Mishina et al. study results suggest that A_{2A} receptors in the putamen may compensate for the asymmetrical decrease of dopamine in drug-naïve PD patients. This is in agreement with prior research demonstrating a co-expression of A_{2A} and dopaminergic D_2 receptors in basal ganglia neurons. Previous work has shown that agonists at A_{2A} inhibit D_2 receptor-mediated neurotransmission, while blockade of A_{2A} potentiates D_2 receptor stimulation (Hirani et al. 2001). The reciprocal actions of these receptors are important in the control of movement (Holschbach et al. 2006; Moresco et al. 2005). The imaging data collected by Mishina et al. warrants for further imaging studies investigating this mechanism in larger cohorts. Finally, the Mishina et al. study also showed that antiparkinsonian therapy increases the A_{2A} in the putamen.

Taken together, the data presented by Mishina et al. and Ramlackhansingh et al. demonstrated, by means of in vivo imaging, that A_{2A} plays an important role in regulation of parkinsonism in PD.

The Molecular NeuroImaging group is currently investigating the changes in A_{2A} binding potentials in HD and PD patients compared with healthy controls using [^{123}I]MNI-420 and SPECT. This clinical trial is ongoing, but preliminary data from seven healthy controls and five HD patients demonstrated on average a $\sim$50 % reduction in striatal binding in HD patients compared with healthy controls. These data agree with the pre-clinical studies in a HD rat model that also found a reduction in radiotracer binding in the lesioned side compared with the intact side (Ishiwata et al. 2002).

Table 11.1 Summary of changes in A$_{2A}$ binding in human brain in healthy controls, PD and HD patients. Data from three different studies

Study group	Binding potentials and distribution volumes in different brain regions				
	Ramlackhansingh study binding potentials		Mishina study distribution volumes		Molecular NeuroImaging study binding potentials[1]
	Caudate	Putamen	Caudate	Putamen	Striatum
Healthy controls	0.53±0.24	0.99±0.21	1.38±0.08	1.47±0.11	0.94±0.27
Drug-naïve PD	---	---	1.37±0.09	1.48±0.33	---
PD without LID	0.40±0.24	0.97±0.21	---	---	---
PD with LID	0.96±0.46*	1.67±0.62*	1.44±0.15	1.58±0.15*	---
HD	---	---	---	---	0.48±0.07*

PD, Parkinson's disease patients; *LID*, L-DOPA-induced dyskinesia; *HD*, Huntington's disease patients [1]preliminary data from ongoing clinical study *$p < 0.05$ versus healthy controls

A summary with key findings from these three human imaging studies is presented in Table 11.1.

Concluding Remarks

The development of high affinity and subtype-selective A$_{2A}$ radiotracers over the last decades has enabled the non-invasive in vivo quantitative measurement of these receptors in monkey and human brain, by means of PET and SPECT imaging. Imaging data collected so far has confirmed the value of PET and SPECT techniques in assessing A$_{2A}$ changes in brain. These findings can foster the rapid widespread use of PET and SPECT A$_{2A}$ imaging not only in PD research, but in other neurodegenerative and neuropsychiatric disorders research, including stroke, traumatic brain injury, Alzheimer's disease, HD, depression, schizophrenia, attention deficit hyperactivity disorder and addiction. Exciting times are ahead for brain imaging research with selective A$_{2A}$ radiotracers, now that suitable PET and SPECT probes with exquisite in vivo properties are available for quantification of A$_{2A}$ in brain. In particular, the recent report of [^{123}I]MNI-420 and [^{18}F]MNI-444, two radiotracers with improved binding potentials in vivo compared with radiotracers previously developed, provides the opportunity to expand the use of in vivo pre-clinical and clinical A$_{2A}$ imaging studies in neuroscience research, as global dissemination of radiotracers labelled with fluorine-18 or iodine-123 is feasible. Other adenosinergic imaging targets will likely become amenable for imaging by PET and SPECT in the near future, as a result of the medicinal chemistry continued efforts to obtain high-affinity and selective radiotracers for those targets.

References

Alagille D, Barret O, Vala C et al (2013) Preclinical evaluation of [F-18]MNI-444. A promising PET imaging agent of adenosine $A_{(2a)}$ receptors. J Labelled Cpd Radiopharm 56:S308

Angulo E, Casadó V, Mallol J et al (2003) A1 adenosine receptors accumulate in neurodegenerative structures in Alzheimer disease and mediate both amyloid precursor protein processing and tau phosphorylation and translocation. Brain Pathol 13:440–451

Ascherio A, Zhang SM, Hernán MA et al (2001) Prospective study of caffeine consumption and risk of Parkinson's disease in men and women. Ann Neurol 50:56–63

Auchampach JA, Kreckler LM, Wan TC et al (2009) Characterization of the A2B Adenosine Receptor from Mouse, Rabbit, and Dog. J Pharmacol Exp Ther 329:2–13

Bara-Jimenez W, Sherzai A, Dimitrova T et al (2003) Adenosine A_{2A} receptor antagonist treatment of Parkinson's disease. Neurology 61:293–296

Bauer A, Holschbach MH, Cremer M et al (2003a) Evaluation of 18F-CPFPX, a novel adenosine A1 receptor ligand: in vitro autoradiography and high-resolution small animal PET. J Nucl Med 44:1682–1689

Bauer A, Holschbach MH, Meyer PT et al (2003b) In vivo imaging of adenosine A1 receptors in the human brain with [18F]CPFPX and positron emission tomography. Neuroimage 19:1760–1769

Bhattacharjee AK, Lang L, Jacobson O et al (2011) Striatal adenosine A_{2A} receptor mediated PET Imaging in 6-hydroxydopamine lesioned rats using [18F]-MRS5425. Nucl Med Biol 38:897–906

Blum T, Ermert J, Wutz W et al (2004) First no-carrier-added radioselenation of an adenosine-A1 receptor ligand. J Labelled Cpd Radiopharm 47:415–427

Brooks D (2005) Positron emission tomography and single-photon emission computed tomography in central nervous system drug development. NeuroRx 2:226–236

Brooks DJ, Papapetropoulos S, Vandenhende F et al (2010) An open-label, positron emission tomography study to assess adenosine A_{2A} brain receptor occupancy of vipadenant (BIIB014) at steady-state levels in healthy male volunteers. Clin Neuropharmacol 33:55–60

Chen GJ, Harvey BK, Shen H et al (2006) Activation of adenosine A3 receptors reduces ischemic brain injury in rodents. J Neurosci Res 84:1848–1855

Chen Z, Janes K, Chen C et al (2012) Controlling murine and rat chronic pain through A3 adenosine receptor activation. FASEB J 26:1855–1865

Chen JF, Eltzsching HK, Fredholm BB (2013) Adenosine receptors as drug targets—what are the challenges? Nat Rev Drug Discov 12:265–286

Costenla AR, De Mendonça A, Ribeiro JA (1999) Adenosine modulates synaptic plasticity in hippocampal slices from aged rats. Brain Res 851:228–234

Cunha RA, Johansson B, Constantino MD et al (1996) Evidence for high-affinity binding sites for the adenosine A_{2A} receptor agonist [3H]CGS 21680 in the rat hippocampus and cerebral cortex that are different from striatal A_{2A} receptors. Naunyn-Schmiedebergs Arch Pharmacol 353:261–271

de Kemp RA, Epstein FH, Catana C et al (2010) Small-animal molecular imaging methods. J Nucl Med 51:18 S–32 S

Dungo R, Deeks E (2013) Istradefylline: first global approval. Drugs 73:875–882

Elmenhorst D, Meyer PT, Winz OH et al (2007) Sleep deprivation increases A1 adenosine receptor binding in the human brain: a positron emission tomography study. J Neurosci 27:2410–2415

Elmenhorst D, Meyer PT, Matusch A et al (2012) Caffeine occupancy of human cerebral A1 adenosine receptors: in vivo quantification with 18F-CPFPX and PET. J Nucl Med 53:1723–1729

El Yacoubi M E, Ledent C, Parmentier M et al (2001) Adenosine A_{2A} receptor antagonists are potential antidepressants: evidence based on pharmacology and A_{2A} receptor knockout mice. Br J Pharmacol 134:68–77

Ferré S, Fuxe K, von Euler G et al (1992) Adenosine-dopamine interactions in the brain. Neuroscience 51:501–512

Ferré S, Ciruela F, Quiroz C et al (2007) Adenosine receptor heteromers and their integrative role in striatal function. ScientificWorldJournal 7:74–85

Ferré S, Quiroz C, Woods AS et al (2008) An update on adenosine A$_{2A}$-dopamine D2 receptor interactions: implications for the function of g protein-coupled receptors. Curr Pharm Des 14:1468–1474

Fishman P, Cohen S, Bar-Yehuda S (2013) Targeting the A3 adenosine receptor for glaucoma treatment. Mol Med Rep 7:1723–1725

Frank RA, Langstrom B, Antoni G et al (2007) The imaging continuum: bench to biomarkers to diagnostics. J Labelled Cpd Radiopharm 50:746–769

Fredholm BB, Bättig K, Holmén J et al (1999) Actions of caffeine in the brain with special reference to factors that contribute to its widespread use. Pharmacol Rev 51:83–125

Fukumitsu N, Ishii K, Kimura Y et al (2003) Imaging of adenosine A1 receptors in the human brain by positron emission tomography with [11C]MPDX. Ann Nucl Med 17:511–515

Fuxe K, Agnati LF, Jacobsen K et al (2003) Receptor heteromerization in adenosine A$_{2A}$ receptor signaling. Relevance for striatal function and Parkinson's disease. Neurology 61:S19–S23

Garrett B, Griffiths RR (1997) The role of dopamine in the behavioral effects of caffeine in animals and humans. Pharmacol Biochem Behav 57:533–541

Gouder N, Fritschy J, Boison D (2003) Seizure suppression by adenosine A1 receptor activation in a mouse model of pharmacoresistant epilepsy. Epilepsia 44:877–885

Haberkorn U, Eisenhurt M (2005) Molecular imaging and therapy–a programme based on the development of new biomolecules. Eur J Nucl Med Mol Imaging 32:1354–1359

Hauser RA, Hubble JP, Truong DD et al (2003) Randomized trial of the adenosine A$_{2A}$ receptor antagonist istradefylline in advanced PD. Neurology 61:297–303

Heurteaux C, Lauritzen I, Widmann C et al (1995) Essential role of adenosine, adenosine A1 receptors, and ATP-sensitive K + channels in cerebral ischemic preconditioning. Proc Natl Acad Sci U S A 92:4666–4670

Hirani E, Gillies J, Karasawa A et al (2001) Evaluation of [4-O-methyl-11C]KW-6002 as a potential PET ligand for mapping central adenosine A$_{2A}$ receptors in rats. Synapse 42:164–176

Holschbach MH, Bier D, Stüsgen S et al (2006) Synthesis and evaluation of 7-amino-2-(2(3)-furyl)-5-phenylethylamino-oxazolo[5,4-d]pyrimidines as potential A$_{2A}$ adenosine receptor antagonists for positron emission tomography (PET). Eur J Med Chem 41:7–15

Ikeda K, Kurokawa M, Aoyama S et al (2002) Neuroprotection by adenosine A$_{2A}$ receptor blockade in experimental models of Parkinson's disease. J Neurochem 80:262–270

Innis RB, Cunningham VJ, Delforge J et al (2007) Consensus nomenclature for in vivo imaging of reversibly binding radioligands. J Cereb Blood Flow Metab 27:1533–1539

Ishiwata K, Furuta R, Shimada J et al (1995) Synthesis and preliminary evaluation of [11C] KF15372, a selective adenosine A1 antagonist. Appl Radiat Isot 46:1009–1013

Ishiwata K, Noguchi J, Wakabayashi S et al (2000) 11 C-Labeled KF18446: a potential central nervous system adenosine A2a receptor ligand. J Nucl Med 41:345–35

Ishiwata K, Noguchi J, Wakabayashi S et al (2000a) 11 C-Labeled KF18446: a potential central nervous system adenosine A$_{2a}$ receptor ligand. J Nucl Med 2000 41:345–335

Ishiwata K, Ogi N, Shimada J et al (2000b) Further characterization of a CNS adenosine A$_{2a}$ receptor ligand [11C] KF18446 with in vitro autoradiography and in vivo tissue uptake. Ann Nucl Med 14:81–89

Ishiwata K, Ogi N, Shimada J et al (2000c) Search for PET probes for imaging the globus pallidus studied with rat brain ex vivo autoradiography. Ann Nucl Med 14:461–466

Ishiwata K, Shimada J, Wang WF et al (2000d) Evaluation of iodinated and brominated [11C] styrylxanthine derivatives as in vivo radioligands mapping adenosine A$_{2A}$ receptor in the central nervous system. Ann Nucl Med 14:247–253

Ishiwata K, Ogi N, Hayakawa N et al (2002) Adenosine A$_{2A}$ receptor imaging with [11C]KF18446 PET in the rat brain after quinolinic acid lesion: Comparison with the dopamine receptor imaging. Ann Nucl Med 16:467–475

Ishiwata K, Kawamura K, Kimura Y et al (2003) Potential of an adenosine A_{2A} receptor antagonist [11C]TMSX for myocardial imaging by positron emission tomography: a first human study. Ann Nucl Med 17:457–462

Ishiwata K, Mishina M, Kimura Y et al (2005a) First visualization of adenosine A_{2A} receptors in the human brain by positron emission tomography with [11C]TMSX. Synapse 55:133–136

Ishiwata K, Tsukada H, Kimura Y et al (2005b) In vivo evaluation of [11C]TMSX and [11C] KF21213 for mapping adenosine A_{2A} receptors: brain kinetics in the conscious monkey and P-glycoprotein modulation in the mouse brain. J Cereb Blood Flow Metab 25:S658

Ishiwata K, Kimura Y, de Vries EFJ et al (2007) PET tracers for mapping adenosine receptors as probes for diagnosis of CNS disorders. CNS Agents Med Chem 7:57–77

Johansson B, Halldner L, Dunwiddie TV et al (2001) Hyperalgesia, anxiety, and decreased hypoxic neuroprotection in mice lacking the adenosine A1 receptor. Proc Natl Acad Sci U S A 98:9407–9412

Kanda T, Tashiro T, Kuwana Y et al (1998) Adenosine A_{2A} receptors modify motor function in MPTP-treated common marmosets. Neuroreport 24:2857–2860

Kase H (2001) New aspects of physiological and pathophysiological functions of adenosine A_{2A} receptor in basal ganglia. Biosci Biotechnol Biochem 65:1447–1457

Kiesewetter DO, Lang L, Ma Y et al (2009) Synthesis and characterization of [76Br]-labeled high affinity A3 adenosine receptor ligands for positron emission tomography. Nucl Med Biol 36:3–10

Koga K, Kurokawa M, Ochi M et al (2000) Adenosine A(2 A) receptor antagonists KF17837 and KW-6002 potentiate rotation induced by dopaminergic drugs in hemi-Parkinsonian rats. Eur J Pharmacol 408:249–255

Kulisevsky J, Barbanoj M, Gironell A et al (2002) A double-blind crossover, placebo-controlled study of the adenosine A_{2A} antagonist theophylline in Parkinson's disease. Clin Neuropharmacol 25:25–31

Lammerstma A, Hume S (1996) Simplified reference tissue model for PET receptor studies. Neuroimage 4:153–158

Ledent C, Vaugeois JM, Schiffmann SN et al (1997) Aggressiveness, hypoalgesia and high blood pressure in mice lacking the adenosine A_{2a} receptor. Nature 338:674–678

Lindström K, Ongini E, Fredholm BB (1996) The selective adenosine A_{2A} receptor antagonist SCH 58261 discriminates between two different binding sites for [3H]-CGS 21680 in the rat brain. Naunyn-Schmiedebergs Arch Pharmacol 354:539–541

Loane C, Politis M (2011) Positron emission tomography neuroimaging in Parkinson's disease. Am J Transl Res 3:323–341

Logan J (2000) Graphical analysis of PET data applied to reversible and irreversible tracers. Nucl Med Biol 27:661–670

Maemoto T, Tada M, Mihara T et al (2004) Pharmacological characterization of FR194921, a new potent, selective, and orally active antagonist for central adenosine A1 receptors. J Pharmacol Sci 96:42–52

Marek K, Innis R, van Dyck C et al (2001) [123I]beta-CIT SPECT imaging assessment of the rate of Parkinson's disease progression. Neurology 57:2089–2094

Márián T, Boros I, Lengyel Z et al (1999) Preparation and primary evaluation of [11C]CSC as a possible tracer for mapping adenosine A_{2A}^{*} receptors by PET. Appl Radiat Isot 50:887–893

Mariani G, Bruselli L, Duatti A (2008) Is PET always an advantage versus planar and SPECT imaging? Eur J Nucl Med Mol Imaging 35:1560–1565

Matsuya T, Takamatsu H, Murakami Y et al (2005) Synthesis and evaluation of [11C]FR194921 as a nonxanthine-type PET tracer for adenosine A1 receptors in the brain. Nucl Med Biol 32:837–844

Meyer PT, Elmenhorst D, Matusch A et al (2006a) 18F-CPFPX PET: on the generation of parametric images and the effect of scan duration. J Nucl Med 47:200–207

Meyer PT, Elmenhorst D, Matusch A et al (2006b) A1 adenosine receptor PET using [18F]CPFPX: displacement studies in humans. Neuroimage 32:1100–1105

Mihara T, Mihara K, Yarimizu J et al (2007) Pharmacological characterization of a novel, potent adenosine A1 and A$_{2A}$ receptor dual antagonist, 5-[5-Amino-3-(4-fluorophenyl)pyrazin-2-yl]–1-isopropylpyridine-2(1 H)-one (ASP5854), in models of Parkinson's disease and cognition. J Pharmacol Exp Ther 323:708–719

Mihara T, Noda A, Arai H et al (2008) Brain adenosine A$_{2A}$ receptor occupancy by a novel A1/A$_{2A}$ receptor antagonist, ASP5854, in Rhesus monkeys: relationship to anticataleptic effect. J Nucl Med 49:1183–1188

Mishina M, Ishiwata K, Kimura Y et al (2007) Evaluation of distribution of adenosine A$_{2A}$ receptors in normal human brain measured with [11C]TMSX PET. Synapse 61:778–784

Mishina M, Ishiwata K, Naganawa M et al (2011) Adenosine A$_{2A}$ receptors measured with [11C] TMSX PET in the striata of Parkinson's disease patients. PLoS One 6:e17338

Mitterhauser M, Haeusler D, Mien LK et al (2009) Automatisation and first evaluation of [18F] FE@SUPPY:2, an alternative PET-tracer for the adenosine A3 receptor: a comparison with [18F]FE@SUPPY. Open Nucl Med Journal 1:15–23

Moresco RM, Todde S, Belloli S et al (2005) In vivo imaging of adenosine A$_{2A}$ receptors in rat and primate brain using [11C]SCH442416. Eur J Nucl Med Mol Imaging 32:405–413

Müller CE, Ferré S (2007) Blocking striatal adenosine A$_{2A}$ receptors: a new strategy for basal ganglia disorders. Recent Pat CNS Drug Discov 2:1–21

Müller CE, Jacobson KA (2011) Recent developments in adenosine receptor ligands and their potential as novel drugs. Biochim Biophys Acta 1808:1290–1308

Naganawa M, Kimura Y, Mishina M et al (2007) Quantification of adenosine A$_{2A}$ receptors in the human brain using [11C]TMSX and positron emission tomography. Eur J Nucl Med Mol Imaging 34:679–687

Noguchi J, Ishiwata K, Wakabayashi S et al (1998) Evaluation of carbon-11-labeled KF17837: a potential CNS adenosine A$_{2a}$ receptor ligand. J Nucl Med 39:498–503

Nonaka Y, Shimada J, Nonaka H et al (1993) Photoisomerization of a potent and selective adenosine A2 antagonist, (E)-1,3-dipropyl-8-(3,4-dimethoxystyryl)-7-methylxanthine. J Med Chem 36:3731–3733

Normile HJ, Barraco RA (1991) N6-cyclopentyladenosine impairs passive avoidance retention by selective action at A1 receptors. Brain Res Bull 27:101–104

Parkinson study group (2002) Dopamine transporter brain imaging to assess the effects of pramipexole vs levodopa on Parkinson's disease progression. JAMA 287:1653–1661

Plamondon H, Blondeau N, Heurteaux C et al (1999) Mutually Protective actions of kainic acid epileptic preconditioning and sublethal global ischemia on hippocampal neuronal death: involvement of adenosine A1 receptors and KATP channels. J Cereb Blood Flow Metab 19:1296–1308

Ramlackhansingh AF, Bose SK, Ahmed I et al (2011) Adenosine 2 A receptor availability in dyskinetic and nondyskinetic patients with Parkinson disease. Neurology 76:1811–1816

Rosi S, McGann K, Hauss-Wegrzyniak B et al (2003) The influence of brain inflammation upon neuronal adenosine A2B receptors. J Neurochem 86:220–227

Ross GW, Abbott RD, Petrovitch H et al (2000) Association of coffee and caffeine intake with the risk of Parkinson disease. JAMA 283:2674–2679

Ruth T (2009) The uses of radiotracers in the life sciences. Rep Prog Phys 72:1–23

Sakata M, Oda K, Toyohara et al (2013) Direct comparison of radiation dosimetry of six PET tracers using human whole-body imaging and murine biodistribution studies. Ann Nucl Med 27:285–296

Salvadori P (2008) Radiopharmaceuticals, drugs development and pharmaceutical regulations in Europe. Curr Radiopharm 1:7–11

Schwarz J, Storch A, Koch W et al (2004) Loss of dopamine transporter binding in Parkinson's disease follows a single exponential rather than linear decline. J Nucl Med 45:1694–1697

Schwarzschild MA, Chen JF, Ascherio A (2002) Caffeinated clues and the promise of adenosine A$_{2A}$ antagonists in PD. Neurology 58:1154–1160

Sihver W, Schulze A, Wutz W et al (2009) Autoradiographic comparison of in vitro binding characteristics of various tritiated adenosine A$_{2A}$ receptor ligands in rat, mouse and pig brain and first ex vivo results. Eur J Pharmacol 616:107–114

Stone-Elander S, Thorell JO, Eriksson L et al (1997) In vivo biodistribution of [N-11 C-methyl] KF 17837 using 3-D-PET: evaluation as a ligand for the study of adenosine A_{2A} receptors. Nucl Med Biol 24:187–191

Svenningsson P, Hall H, Sedvall G et al (1997) Distribution of adenosine receptors in the postmortem human brain: an extended autoradiographic study. Synapse 27:322–335

Tavares AAS, Batis J, Barret O et al (2013a) In vivo evaluation of [123I]MNI-420: A novel single photon emission computed tomography radiotracer for imaging of adenosine 2 A receptors in brain. Nucl Med Biol 40:403–409

Tavares AAS, Batis JC, Papin C et al (2013b) Kinetic modeling, test–retest, and dosimetry of 123I-MNI-420 in humans. J Nucl Med 54:1760–1767

Wang WF, Ishiwata K, Nonaka H et al (2000) Carbon-11-labeled KF21213: a highly selective ligand for mapping CNS adenosine A_{2A} receptors with positron emission tomography. Nucl Med Biol 27:541–546

Wells L, Salinas C, Tang SP et al (2013) Evaluation of the adenosine A_{2A} receptor ligand [11C] SCH442416 in the rodent brain. J Nucl Med 54:382

Xu K, Xu YH, Chen JF et al (2010) Neuroprotection by caffeine: time course and role of its metabolites in the MPTP model of Parkinson's disease. Neuroscience 167:475–481

Chapter 12
Caffeine and Neuroprotection
in Parkinson's Disease

Filipe B. Rodrigues, Daniel Caldeira, Joaquim J. Ferreira and João Costa

Abstract Parkinson's disease (PD)—the second most common neurodegenerative condition worldwide—has no proven neuroprotective intervention. However PD belongs to the ever-growing group of diseases that occur less frequently in coffee-drinkers. Coffee is the major dietary source of caffeine—an adenosine A_{2A} receptor antagonist. This is presumed to be the main mechanism responsible for the decreased risk of developing PD among coffee drinkers. Furthermore, in view of other biochemical and cellular actions attributed to caffeine, it has been proposed based on basic science results that caffeine may have a neuroprotective role in PD. Animal data is supportive of this hypothesis by showing that caffeine is able to prevent neurodegeneration in PD animal models. Still, human data is lacking precluding the establishment of firm conclusions on the role of caffeine as a disease-modifying agent in PD.

Keywords Parkinson disease · Caffeine · Coffee · Risk factors · Systematic review · Meta-analysis

Introduction

Parkinson's disease (PD) is a frequent disease—the second most common neurodegenerative condition worldwide (de Lau and Breteler 2006). Its annual incidence varies across studies and countries from 5 to 20 affected individuals per

J. Costa (✉) · F. B. Rodrigues · D. Caldeira · J. J. Ferreira
Laboratory of Clinical Pharmacology and Therapeutics, Faculty of Medicine,
University of Lisbon, Lisbon, Portugal
e-mail: jncosta@medicina.ulisboa.pt

J. Costa
Clinical Pharmacology Unit, Instituto de Medicina Molecular, Lisbon, Portugal

Evidence Based Medicine Centre, Faculty of Medicine, University of Lisbon,
Lisbon, Portugal

Portuguese Collaborating Centre of the Cochrane Iberoamerican Network,
Faculty of Medicine, University of Lisbon, Lisbon, Portugal

© Springer International Publishing Switzerland 2015
M. Morelli et al. (eds.), *The Adenosinergic System,* Current Topics in Neurotoxicity 10,
DOI 10.1007/978-3-319-20273-0_12

100,000 individuals (Tanner and Goldman 1996). PD prevalence also varies from 44 to 347 per 100,000 individuals (Tanner and Goldman 1996), increasing steadily with age (de Rijk et al. 1997). The risk of acquiring PD when age-adjusted is higher for males comparing to females (de Lau and Breteler 2006; Van den Eeden et al. 2003).

PD is characterized pathologically by dopaminergic neuronal depletion at *substantia nigra pars compacta* (SNc) and by the presence of Lewy bodies and Lewy neuritis—both resulting from alpha-synuclein accumulation—elsewhere in the brain (Pollanen et al. 1993; Spillantini et al. 1997). Clinically, PD is characterized by parkinsonism—resting tremor, bradykinesia, rigidity and postural imbalance—usually with an asymmetric onset, a good initial response of the symptoms to L-DOPA treatment, and a progressive course (Hughes et al. 1992). Nevertheless, PD is no longer considered to be a "pure" disorder of motor control, and many non-motor clinical manifestations—such as REM-sleep behavioural disorder, depression and dementia—are now recognised as belonging to the natural history of the disease. These manifestations are linked to the widespread distribution of the abnormalities with alpha-synuclein accumulation not only in the brain but also in the autonomic and peripheral nervous systems and multiple organs—PD is a progressive multi-organ proteinopathy (Obeso et al. 2014).

The cause of PD remains partially unknown, and PD is thought to result from a complex interaction between genes and environment (de Lau and Breteler 2006).

Currently, there is no effective intervention to change disease progression (Ferreira et al. 2012; Suchowersky et al. 2006) and further basic and clinical research is needed to elucidate disease pathways and biomarkers, as equally as new therapeutic targets.

Surprisingly, PD belongs to the ever-growing group of diseases that occur less frequently in coffee-drinkers. Nefzger and his colleagues first appreciated this epidemiological relation—a negative association—in 1968 (Nefzger et al. 1968). Since then, several explanations have emerged to justify this relation, although none fully satisfactory. Here we hypothesize that substances present in caffeinated beverages—such as caffeine—have an action on the central nervous system (CNS) with a potential neuroprotective role, as far as dopaminergic degeneration is concerned.

Caffeine is Not Equivalent to Coffee

The next section discusses figures obtained from studies that took place in the U.S.A. (National Coffee Association of U.S.A. 1993). Nonetheless, more recent studies from the United Kingdom, Denmark, Canada and Australia substantiate the former, with the exception that the average daily caffeine consumption is higher in the latter (Barone and Roberts 1995).

Coffee Consumption

Coffee is probably the most widely consumed behaviourally active substance worldwide (Fredholm et al. 1993, 1999)—more than half of the population aged 10 years or older drink coffee (National Coffee Association of U.S.A. 1993); and on average each adult drinks 4 cups of coffee per day, which corresponds to a daily caffeine consumption of 244 mg (National Coffee Association of U.S.A. 1993).

These figures support and continue to uphold the major concern of governmental and regulatory agencies, as much as the world scientific community, on the overall effects of coffee on human health.

Coffee Composition and Caffeine

The history of caffeine consumption can be traced from the Palaeolithic period. There are more than 50 species of plants know to have caffeine in their composition and some are part of our diet since the human primordials.

Several beverages and foods contain caffeine including: coffee, either instant, brewed or decaffeinated coffee; teas; some soft drinks such as colas; and cocoa and chocolate derived products, as hot chocolate, chocolate milk and chocolate candies (Barone and Roberts 1995).

Having stated that caffeine is present in many sources, coffee is the major dietary source of caffeine (Pao 1982): the average adult consumes 3 mg of caffeine per kilogram of body weight daily and about 2 mg come from coffee (Barone and Roberts 1995).

On average, a 150 mL cup of brewed coffee has a caffeine content of 85 mg and a 150 mL cup of instant coffee has a caffeine content of 60 mg (Burg 1975). Even the non-caffeinated decaffeinated coffee has an approximate content of 3 mg of caffeine per 150 mL of coffee (Burg 1975).

Coffee has a complex formulation, with thousands of chemical components.

Caffeine (Fig. 12.1a)—1,3,7-trimethylpurine-2,6-dione—was first isolated in 1820. It is the most pharmacologically active substance in coffee but, many other coffee products could have a potential role on human health such as cafestol, kahweol, clorogenic acid, magnesium, potassium, niacin and vitamin E.

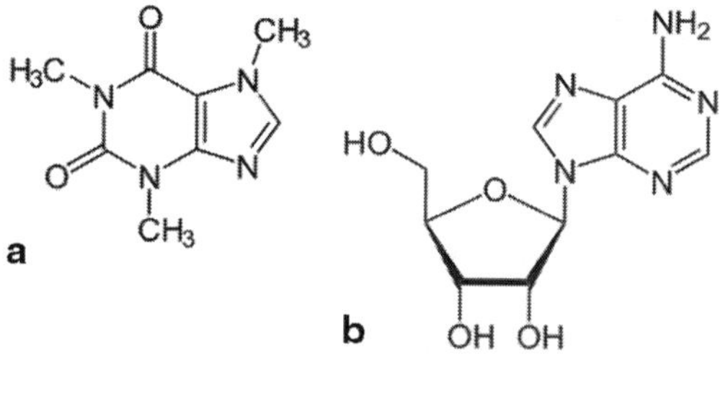

Fig. 12.1 Chemical structure, **a** Caffeine, **b** Adenosine

Within the human body, caffeine is metabolized to paraxanthine—its major degradation product—theobromine, theophylline and other minor metabolites (Lelo et al. 1986).

Caffeine is a natural occurring adenosine (Fig. 12.1b) receptor antagonist (Biaggioni et al. 1991; Li et al. 2001). It antagonizes both the A_1 and A_2 adenosine receptors in a competitive fashion (Chen et al. 2008; Daly et al. 1983). The A_1 receptor inhibits the enzyme adenylate cyclase and the A_2 receptor stimulates the enzyme adenylate cyclase. The A_{2A} adenosine receptor, a subtype of A_2 adenosine receptors, assumes paramount relevance within the brain.

Caffeine and Neuroprotection

Different reasons justify the huge investment on research made on this particular subset of receptors over the past years: adenosine A_{2A} antagonism might have the power to ameliorate PD symptoms and signs, plus—and more excitingly—could putatively prevent neurodegeneration.

The adenosine A_{2A} receptors are abundant in the striatum of several species (Parkinson and Fredholm 1990; Prémont et al. 1979) including humans (Martinez-Mir et al. 1991). This structure is an integrant part of the basal ganglia—a cluster of several deeply located groups of neuronal cell bodies within the brain matter—through which a functional connection is established with the brain motor areas. There is an important relation of the striatum with movement control and this relation was first interpreted using the classic physiopathological model of basal ganglia circuitry involving a direct and an indirect pathway (Alexander and Crutcher 1990) (Fig. 12.2a and b).

A comprehensive review of the basal ganglia circuit models and of the cortico-thalamic-basal ganglia interactions, in both physiological and pathological states, is beyond the scope of this chapter. Briefly, in PD there is an imbalance in the activity of the above-mentioned neuro-functional-anatomical pathways of the classic model due to the loss of SNc dopaminergic drive to striatal neurons. This results in a reduced activity of the direct pathway—due to loss of dopamine receptors type 1 (D_1) stimulation—and in an increased activity of the indirect pathway—due to release of dopamine receptors type 2 (D_2) inhibition. The net consequence is an increased inhibition of the ventrolateral thalamus, which in turn inhibits movement (Fig. 12.2c and d). Although greatly simplified, this concept of overactivity of the indirect pathway in PD is useful for purposes of understanding the rationale of a putative neuroprotective role of caffeine in PD (Ellens and Leventhal 2013; Wichmann and Dostrovsky 2011).

The GABAergic neurons of the indirect striato-pallido-nigral pathway are of particular interest for the modulation of this pathway by caffeine (Fink et al. 1992; Schiffmann and Vanderhaeghen 1993). In this pathway, the adenosine A_{2A} receptors strongly interact with the D_2 receptors of striato-pallidal GABAergic neurons. Adenosine A_{2A} receptors mediate an adenylate cyclase-independent (Yang et al. 1995) dopaminergic inhibition (Fuxe et al. 1993). Therefore, the overall net result of adenosine activation is the activation of the indirect pathway.

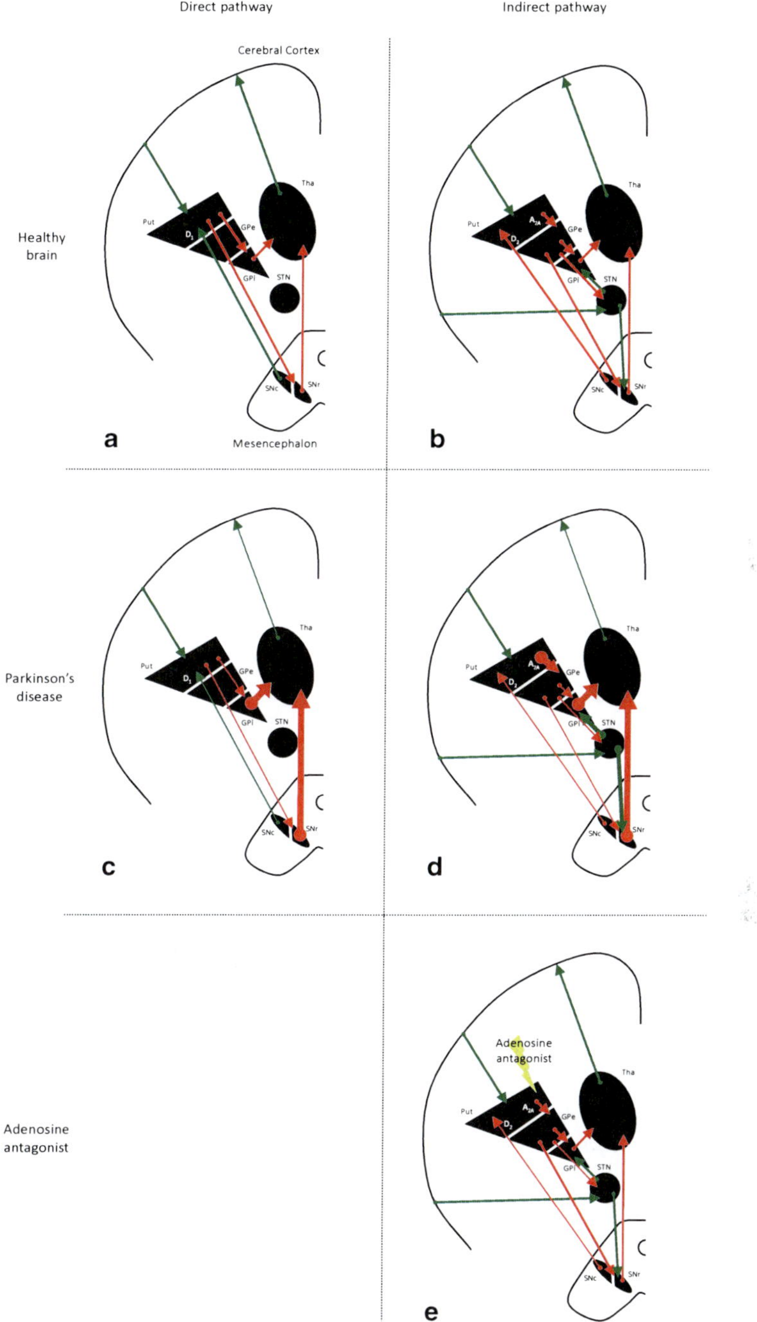

Fig. 12.2 Simplified motor circuitry, **a** Healthy direct pathway, **b** Healthy indirect pathway, **c** Direct pathway in PD, **d** Indirect pathway in PD, **e** Proposed effect of an adenosine antagonist in the indirect pathway. *Green arrows* are excitatory neurons and *red arrows* are inhibitory neurons. A_{2A} adenosine receptor type 2, subtype A; D_1 dopamine receptor type 1, D_2 dopamine receptor type 2, *GPe* globus pallidus externus, *GPi* globus pallidus internus, *Put* putamen, *SNc* substatia nigra pars compacta, *SNr* substatia nigra pars reticulata, *STN* nucleus subthalamicus, *Tha* thalamus

On the other hand, this means that blockade of the adenosine receptors (with adenosine antagonists) produces an enhancement of dopaminergic transmission—such as the one seen when applying dopamine D_2 receptor agonists—thus counteracting motor inhibition (Fig. 12.2e). This effect was first demonstrated in animal models of unilateral nigrostriatal lesion (Fredholm et al. 1976).

Apart from caffeine, several other xantine and non-xantine A_{2A} receptor antagonists—as KF17837 and SCH 58261, respectively—have shown to induce the same phenomenon (Fenu et al. 1997; Kanda et al. 1994).

Importantly the direct basal ganglia pathway also suffers the influence of adenosine A_{2A} receptors through dopaminergic D_1 striatonigral neurons (Morelli et al. 1994; Pinna et al. 1996).

On the other hand, PD is a multifactorial disease resulting from the net effects on the survival of *substantia nigra* (SN) dopaminergic neurons mediated by putative neurotoxic and neuroprotective factors and by their complex interactions with pre-existing genetic factors.

Adenosine also takes part in neuronal cell death control: animal models of cerebral ischemia demonstrated a substantial increment of excitatory substances, such as glutamate, during cell suffering (Fredholm et al. 1993). These amino acids contribute to nerve cell death (Coyle et al. 1981; Xu et al. 2006) through neuronal swelling (Rothman 1985; Xu et al. 2010) and excessive Ca^{2+} influx (Berdichevsky et al. 1983; Ungerstedt 1968). It has been shown that the release of these substances is controlled at least in part by adenosine receptor activation (Joghataie et al. 2004; O'Regan et al. 1992). This mechanism is called excitotoxic cell death (Aguiar et al. 2006; Olney 1986).

Having this in mind, a neuroprotective activity could be linked to adenosine receptors antagonists, which could reduce the presynaptic release of neurotoxic amino acids (Simpson et al. 1992). In fact, there is evidence that adenosine A_{2A} receptor antagonists can reduce excitotoxicity in models of cerebral ischemia (Lubitz et al. 1995; Thiruchelvam et al. 2000) and neurodegeneration (Kachroo et al. 2010; Popoli et al. 2002).

Neuroprotective mechanisms related to adenosine antagonists can also be traced from their interaction with glial cells. Adenosine is known to mediate glial cell glutamate efflux through adenosine A_{2A} receptors increasing the interstitial concentration of this amino acid (Li et al. 2001; Yazdani et al. 2006). Blockade with antagonists could be a target for neuroprotection.

Other proposed and less studied mechanisms of caffeine neuroprotection are: as blood-brain barrier stabilizing agent (Chen et al. 2008); as activator of intracellular survival pathways (Nakaso et al. 2008); and as membrane potential stabilizer (Mao et al. 2007).

More recently the adenosine A_{2A} receptor-mediated neurotoxicity theory was put to the proof by Kachroo and Schwarzschild (2012). Using mice with both adenosine A_{2A} receptor knockout and an α-synuclein mutation known to promote dopaminergic degeneration (Richfield et al. 2002), they have shown that the A_{2A} receptor is necessary to promote nigrostriatal denervation in this model (Kachroo and Schwarzschild 2012). This argument further supports the putative neuroprotective role of caffeine in PD.

At last, the neuroprotective effect of caffeine and other adenosine A_{2A} receptor antagonists extends far beyond PD. There is a growing body of animal evidence that supports its protective effect on other conditions characterized by neural tissue loss such as ischemic brain injury, Alzheimer's diseases (Dall'Igna et al. 2007) and Huntington's disease (Fink et al. 2003).

Taken together, all these data provide a strong biochemical, anatomical and physiological rationale for a putative neuroprotective role of caffeine in PD. In the following sections we present the results of a systematic literature review that aimed to evaluate animal and human data that further support this hypothesis. The authors looked with particular emphasis at human studies evaluating the effects of caffeine exposure on the incidence of PD, its natural history, and symptomatic relieve.

Finding the Evidence

A search strategy using "Parkinson's disease" AND "coffee" OR "caffeine" was applied to the following electronic databases: MEDLINE, Web of Science, EMBASE, LILACS and CENTRAL (Cochrane Library). There was no time or language restriction. The last search was conducted on April 2014. Additionally, reference lists from selected studies were crosschecked and experts were consulted. Data collection and analysis were performed independently by two of the authors.

The inclusion criteria used were: observational or experimental studies in animals and humans evaluating the relation between caffeine or coffee consumption and the risk, natural history, and symptomatic relief in PD.

Evidence from Animal Studies

There has been a plenty of animal model-based bench-research supporting the role of adenosine receptor antagonists, such as caffeine, on PD. A first group of studies revealed physiological data supporting the rationale for using this group of molecules. Some of that data were previously clarified in this chapter and a detailed explanation of these studies is beyond the scope of this chapter. A second group of studies was conducted as proof-of-concept to clinical studies and focus on the protective effect of caffeine in different animal models of PD. The results of these latter studies are discussed below.

MPTP Model of PD

MPTP (1-methyl-4-phenyl-1,2,3,6-tetrahydropyridine) is a neurotoxin capable of inducing striatal dopaminergic depletion (Jenner et al. 1984). When administered to laboratory animals, MPTP produces the most accurate model of PD (Gerlach and Riederer 1996).

Chen et al. (2001) were the first to demonstrate that caffeine in doses comparable to human consumption was able to attenuate MPTP-dependent striatal dopaminergic degeneration (Chen et al. 2001). They also showed similar results using specific adenosine A_{2A} receptor antagonists and utilizing adenosine A_{2A} receptor knock-out mice. A year later, Xu et al. (2002) proved that caffeine's neuroprotective effect did not undergo tolerance (Xu et al. 2002). On 2006, new clinical epidemiological data was released and appeared to show that estrogens could attenuate the neuroprotective effect of caffeine in women (further discussion on this matter is provided in the next section). In fact, the same was shown in this animal model (Xu et al. 2006).

Further conclusions were traced by Xu et al. 2010: the neuroprotective effect of caffeine is equally effective independently of the time relation with the insult; and the neuroprotective effect extends to caffeine metabolites, theophylline and paraxanthine (Xu et al. 2010).

6-OHDA *Model of PD*

6-OHDA (6-hydroxydopamine) is a neurotoxin with the particularity of being unable to trespass the blood-brain barrier. However, when injected directly in the brain parenchyma or within the brain ventricles 6-OHDA leads to a selective destruction of catecholaminergic neurons (Ungerstedt 1968). Therefore the injected region will suffer from a noradrenaline-, adrenaline- and dopamine-depletion while other neurotransmitters remain unchanged. Joghataie et al. (2004) demonstrated that caffeine could attenuate the neurotoxic effect of intraventricular 6-OHDA injection in rats (Joghataie et al. 2004) and Aguiar et al. (2006) confirmed the results by using striatal injections (Aguiar et al. 2006).

Paraquat plus Maneb Model of PD

This model uses two different pesticides and neurotoxins: paraquat is a herbicidal that has the ability to lower NADH-dehydrogenase activity, leading to the creation of superoxide free radicals (Turrens and Boveris 1980), maneb—manganese ethylenebisdithiocarbamate—is a fungicide. The administration of intraperitoneal injections of a mixture of these compounds in mice induces degeneration of the nigrostriatal dopamine systems and produces a model of PD (Thiruchelvam et al. 2000). Kachroo et al. (2010) shown that caffeine prevents neurodegeneration in this model (Kachroo et al. 2010).

MPP+ Model of PD

In this last model, MPP+ (methyl-4-phenylpyridinium) is used as neurotoxin. It is continuously infused into one of the lateral ventricles over a time course of 28 days.

The procedure causes an ipsilateral nigral dopaminergic neurons loss (Yazdani et al. 2006). Sonsalla et al. (2012) have shown that in this chronic model of dopaminergic lesion, caffeine, when consumed before the insult, can prevent neuronal damage and, when consumed in the drinking water after the insult, may arrest or delay neurodegeneration (Sonsalla et al. 2012).

Evidence from Human Studies

Caffeine Intake and the Risk of Parkinson's Disease

There is plenty of observational data, from cross-sectional, case-control and cohort studies, attesting the relation between caffeine consumption and the incidence of PD. However the results are heterogeneous—opening a window for further investigation.

In 2002, Hernán and colleagues summarized the available human epidemiological evidence and shown that when compared with non-coffee drinkers, regular coffee drinkers have a 31% reduction in the risk of developing PD (relative risk—RR—of 0.69; 95% CI: 0.59 to 0.80) (Hernán et al. 2002). Even when adjusting this result for smoking status—a well-established PD protective factor (Nefzger et al. 1968)—this inverse relation stood firmly: RR of 0.70 (95% CI: 0.59 to 0.84) (Hernán et al. 2002). In addition, a stronger inverse relation was found when comparing heavy coffee drinkers with light-coffee drinkers plus non-coffee drinkers, but not without the expense of lower levels of statistical confidence.

Still, this pivotal systematic review and meta-analysis was built from 8 case-control studies and 5 cohort studies, accepted a high degree of heterogeneity between studies, and did not address gender-associated relations. In fact, at least one important observational study—the Nurses' Health Study—suggested that the inverse relation between coffee consumption and PD incidence could not be observed in women (Ascherio et al. 2001).

This issue begged for clarification, and in 2010 we underlined the inverse relation between caffeine intake and the risk of PD by conducting another meta-analysis of systematically reviewed observational studies. We found a 25% risk reduction of developing PD among subjects with caffeine/coffee consumption (RR 0.75; 95% CI: 0.69 to 0.82), with low to moderate heterogeneity ($I^2 = 28.8\%$) among studies' results. When considering the female population, this inverse relation was weaker and failed to reach statistical significance (RR 0.86; 95% CI: 0.73 to 1.02). We also found an inverse linear relation for the level of coffee consumption: a RR of 0.76 (95% CI: 0.72 to 0.80) per 300 mg increase in caffeine intake (Costa et al. 2010).

This study included 27 reports—7 cohort studies, 2 nested case-control studies, 16 case-control studies and 1 cross sectional study—which produced a more homogeneous comparison of studies. Nevertheless, the possibility of publication bias was noticed. Further data was published in the last recent years—including another systematic review and meta-analysis that included 19 studies (Noyce et al.

2012), and the results reported strengthen ours. In this work, the gender issue was not definitively established.

Systematic Review of Observational Studies

For the purpose of this chapter, we have updated our previous systematic review to include the more recently available epidemiological human studies that have evaluated the effect of caffeine exposure on the incidence of PD.

Methods

We have followed the same methodology and assumptions previously published in our first systematic review and meta-analysis (Costa et al. 2010). We followed STROBE (Elm et al. 2007), MOOSE (Stroup et al. 2000) and PRISMA (Liberati et al. 2009) statements as guidelines. Reporting of statistical data followed SAMPL guidelines (Lang and Altman 2013).

Briefly, the search strategy was updated until April 2014. Cohort, case-control, or cross-sectional studies evaluating the relation between exposure to coffee/caffeine and the risk of PD (all diagnostic criteria were considered) or PD mortality were eligible for inclusion. Studies addressing the effects of short-term exposure to coffee or caffeine and those that evaluated associations other than the risk of PD, such as the rate of progression, were excluded. No studies were excluded a priori for weakness of design or data quality. Data extractions were obtained independently and cross-checked for accuracy. When different risk estimates were available in the same publication, we opted for those that reflected the greatest degree of control for potential confounders, to the largest number of categories of exposure among caffeine consumers, or to the most comprehensive assessment of caffeine intake, applying these criteria consecutively. If results were provided separately for different caffeine-containing beverages or food items, we opted for those referring to coffee consumption.

Quantitative data synthesis was accomplished through random effects meta-analysis (DerSimonian and Laird method). Relative risks (cumulative incidence ratios or incidence density ratios) and odd ratios (ORs) were treated the same and are referred to as RR. A cumulative random effects meta-analysis was conducted to allow a better understanding of the time trends in the understanding of the relation between caffeine intake and PD. Since more than one RR estimate was available from several studies, only the most precise measures of association were used from each report. This criterion was followed for selection of a single estimate per study when RRs were provided for different categories of exposure. If the precision of RR estimates was the same for more than one category, we conservatively chose the one corresponding to the RR closest to 1.

Heterogeneity was quantified using the I^2 statistic (Higgins and Thompson 2002). Publication and publication-related biases were examined through visual inspection of the funnel plot and with Begg adjusted rank correlation (Begg and Mazumdar 1994) and the Egger's regression asymmetry test (Egger et al. 1997).

Results

A total of 37 epidemiological studies met criteria for inclusion in this updated systematic review, including 11 cohort (Ascherio et al. 2001, 2003, 2004; Fink et al. 2001; Hu et al. 2007; Kyrozis et al. 2013; Liu et al. 2012; Palacios et al. 2012; Ross et al. 2000; Saaksjarvi et al. 2008; Tan et al. 2007b), 3 nested case-control (Paganini-Hill 2001; Skeie et al. 2010; Wirdefeldt et al. 2005), 22 case-control (Benedetti et al. 2000; Checkoway et al. 2002; Evans et al. 2006; Facheris et al. 2008; Fall et al. 1999; Haack et al. 1981; Hancock et al. 2007; Hellenbrand et al. 1996; Hosseini Tabatabaei et al. 2013; Jiméanez Jiméanez et al. 1992; Macleod and Counsell 2013; Morano et al. 1994; Nefzger et al. 1968; Nicoletti et al. 2010; Pereira and Garrett 2010; Powers et al. 2008; Preux et al. 2000; Ragonese et al. 2003; Sipetic et al. 2011; Tan et al. 2003, 2007a; Tanaka et al. 2011) and 1 cross-sectional study (Louis et al. 2003). The main characteristics of the studies are summarized in Table 12.1 and Fig. 12.3. The forest plot corresponding to Fig. 12.3 represents the RR estimates provided in each study for the association between caffeine intake and PD. Several estimates from the same study may be provided, referring to different exposure levels or to stratum-specific analyses.

The publication year ranged from 1968 to 2013. The studies were conducted mainly in the U.S.A. (15 out of 37, one of which in an Asian population); in Europe (2 in Spain, 2 in Sweden, 2 in Finland, 2 in Italy, 2 in the United Kingdom, 1 in Germany, 1 in France, 1 in Norway, 1 in Greece, 1 in Serbia, 1 in Portugal); 3 in Singapore, 1 in Japan, and 1 in Iran.

In cohort designs, the estimated mean age of the participants at the time of baseline evaluation ranged from 42 to 77 years.

Different sources of caffeine were accounted for in the reports reviewed, and the results used for meta-analysis refer to coffee consumption in most studies ($n=23$), to coffee and tea consumption in 5 studies, and 8 studies extended exposure assessment to all caffeinated beverages, or caffeinated beverages and products containing chocolate. Twenty seven out of the included 37 studies provided RR estimates for different categories of exposure, with an estimated daily exposure to caffeine ranging from 7.8 to 1507 mg, and the reference categories including different proportions of non-caffeine consumers and consumers of different amounts of caffeine. The clinical diagnosis of PD, based on a set of predefined clinical criteria, was the outcome in most studies. Information obtained from medical records and national medication or inpatient databases was occasionally considered as a complementary source in 10 studies, and the same was for information from death certificates in 4 studies. In 2 studies some patients had PD defined by self-report and not confirmed by a clinical diagnosis, death certificates or medical records. One study assessed PD mortality as the sole outcome.

Regarding potential confounding factors, smoking was taken into account in 17 studies. Exposure to heavy metals and use of pesticides or herbicides was accounted

Table 12.1 Main characteristics of the studies included in the systematic review

Reference Country Publication year	Type of study Sample characteristics Follow-up (Cohort studies)	Outcome assessment Definition of Parkinson Disease/ Parkinsonism	Evaluation of exposure Timing of exposure Validation of the method Items evaluated about caffeine exposure	Control of confounding
COHORT STUDIES				
Ross et al. *USA* 2000	Cohort Study Japanese-American men (Honolulu-Asia Aging Study) Age: 45–68 y M/F: 8006 (all M) Follow-up: Duration: 27 y (median); Completeness: NS	Review of hospital records, local neurologists records and death certificates (previous to 1991) After 1991, 3-step process: 1. Self declared diagnosis of PD (structured interview), PD medication 2. Evaluation by trained technician; recognition of PD clinical signs (tremor, gait disturbances, bradykinesia)' 3. Referral to study neurologist—criteria for PD diagnosis (consensus from 2 neurologists): Parkinsonism Progressive disorder Any two of: marked response to L-DOPA asymmetry at onset, or initial onset tremor Absence of other possible cause	24-h recall methods and food frequency questionnaires 1 week before Validated Coffee, tea (green and black), other caffeinated beverages, and caffeine from other sources	Age and smoking
Ascherio et al. *USA* 2001	Cohort Study Male health professionals (Health Professionals' Follow-Up Study) Age: 40–75 47,351: all M Follow-up: Duration: 9.2 y (mean); Completeness: >97%	Questionnaire (mail) Confirmation of diagnosis with the treating neurologist or by review of the medical records At least two: Tremor, rigidity, bradykinesia Response to L-DOPA	SFFQ (mail) 1 year before Validated Coffee, tea, cola and chocolate	Age, smoking, alcohol, BMI, physical activity

Table 12.1 (continued)

Reference Country Publication year	Type of study Sample characteristics Follow-up (Cohort studies)	Outcome assessment Definition of Parkinson Disease/ Parkinsonism	Evaluation of exposure Timing of exposure Validation of the method Items evaluated about caffeine exposure	Control of confounding
Fink et al. *USA* 2001	Cohort Study Participants in the Original Framingham Study who attended the 12th, 17th or 22nd biennial examination Age: 69 y (mean) M/F: 2382/3746 Follow-up: Duration: 10 y (for each index examination); Completeness: NS	Physical examinations Tremor, rigidity, bradykinesia Absence of other possible cause	Structured questionnaire NS NS Coffee, tea	Age, gender, smoking
Ascherio et al. *USA* 2003	Cohort Study Female registered nurses in 11 states (Nurses' Health Study) Age: 30–55 77,713 (1,039,434 person-years): all F Follow-up: Duration: 18 y (mean); Completeness: >98%	Questionnaire (mail) Confirmation of diagnosis with the treating neurologist or by review of the medical records At least two: Tremor, rigidity, bradykinesia Response to L-DOPA	SFFQ (mail) 1 year before Validated Coffee, tea, cola and chocolate	Age, smoking, alcohol, age at menopause, type of menopause, parity, use of oral contraceptives, and hormone use or duration of use
Ascherio et al. *USA* 2004	Cohort Study American Cancer Society volunteers (Cancer Prevention Study II) Age: ≥30 y (median: 57 for M; 56 for F) M/F: 301,164/238,058 Follow-up: Duration: 1989–1998; Completeness: 100%	PD as underlying or contributing cause of death National Death Index Idiopathic Parkinson's disease (ICD 9th revision)	Structured questionnaire NS NS Coffee, tea and sodas	Age, smoking, alcohol

Table 12.1 (continued)

Reference Country Publication year	Type of study Sample characteristics Follow-up (Cohort studies)	Outcome assessment Definition of Parkinson Disease/ Parkinsonism	Evaluation of exposure Timing of exposure Validation of the method Items evaluated about caffeine exposure	Control of confounding
Hu et al. *Finland* 2007	Cohort Study 4 cross-sectional population surveys in 1982, 1987, 1992 and 1997 Age: 25–74 y M/F: 14,293/15,042 Follow-up: Duration: 12.9 y; Completeness: 74–88%	National Social Insurance Institution's Register data Consultant (usually specialist in neurology) Medical history, clinical examination (tremor, bradykinesia, stiffness, etc) and other relevant diagnostic methods	Self-administered questionnaire at home NS NS Coffee and tea	Age, BMI, systolic blood pressure, total cholesterol, education, leisure-time physical activity, smoking, alcohol and tea consumption, and history of diabetes
Tan et al. *Singapore* 2007b	Cohort Study Ethnic Chinese, belonging to the two major dialect groups, and residing in government-built housing estates (Singapore Chinese Health Study) Age: 45–74 y (mean: 57 y) M/F: 27,956/35,262 Follow-up: Duration: 1993–2005; Completeness: 90%	Follow-up interviews, nationwide hospital discharge database and two hospital-specific Parkinson's disease registries 88% of the cases were evaluated by a movement disorder specialists/neurologist Diagnosis criteria were those from the Advisory Council of the US National Institute of Neurological Disorders and Stroke	SFFQ (in-person interview at home made by a trained interviewer) NS Validated Coffee, black and green tea and sodas	Age, year of interview, gender, dialect, smoking and level of education
Saaksjarvi et al. *Finland* 2008	Cohort Study Participants in a Finnish cohort study (Finnish Mobile Clinic Health Examination Survey) Age: 50–79 y (mean: 63 y) 6710 M/F: 3033/3677 Follow-up: Duration: 1973–1994; Completeness: NS	PD cases were identified through a nationwide registry of patients receiving medication reimbursement, which is based on certificates from neurologist Clinical diagnostic criteria (resting tremor, bradykinesia and/or muscle rigidity)	Self-administered health questionnaire Lifetime Validated Coffee (cups/day)	Age, gender, marital status, education, community density, alcohol consumption, leisure-time physical activity, smoking, body mass index, hypertension and serum cholesterol

Table 12.1 (continued)

Reference Country Publication year	Type of study Sample characteristics Follow-up (Cohort studies)	Outcome assessment Definition of Parkinson Disease/Parkinsonism	Evaluation of exposure Timing of exposure Validation of the method Items evaluated about caffeine exposure	Control of confounding
Liu et al. *USA* 2012	Cohort Study Participants in the National Institutes of Health-AARP Diet and Health Study Age: NS 318,260 M/F: 187,499/130,761 Follow up: Duration: 1995–2006; Completeness: NS	Screening questionnaire for participants. Possible PD cases' treating physician validation and/or patients medical records review by a movement disorder specialist Diagnosis (only 1): 1-Diagnosis confirmed by the treating physician 2-Medical record with a final diagnosis of PD 3-Evidence of two or more cardinal sign, with one being rest tremor or bradykinesia, progressive course, responsiveness to dopaminergic treatment	Grid-based diet history questionnaire 1 year Validated Caffeine intake (mg/day)	Age, gender, race, physical activity and smoking status
Palacios et al. *USA* 2012	Cohort Study Participants in a North American cohort the Cancer Prevention Study II Nutrition Cohort Age: men 75 y; women 74 y 112,122 M/F: 48,532/63,590 Follow-up: Duration: 1999–2007; Completeness: NS	Medical record review and treating physicians At least two out of four cardinal signs of PD (rigidity, postural instability, bradykinesia, rest tremor), progressive course and response to L-DOPA	Food Frequency Questionnaire NS NS Coffee, decaffeinated coffee, chocolate, caffeinated colas and tea	Age, gender, smoking and alcohol intake, total caloric intake, pesticide exposure, education, dairy intake, physical activity, use of ibuprofen and baseline BMI

Table 12.1 (continued)

Reference Country Publication year	Type of study Sample characteristics Follow-up (Cohort studies)	Outcome assessment Definition of Parkinson Disease/ Parkinsonism	Evaluation of exposure Timing of exposure Validation of the method Items evaluated about caffeine exposure	Control of confounding
Kyrozis et al. *Greece* 2013	Cohort Study Participants in the EPIC-Greece cohort Age: 20–86 y 25,407 (214,505 person-years) M/F: 10,344/16,063 Follow-up: Duration: 8.45 y (mean); Completeness: 91.6%	Review of EPIC-Greece routine follow-up evaluations until June 2009: possible PD cases were screened using a auto-matic, computed-assisted method first for self-reported diagnosis of PD, use of anti-parkinsonian drugs and then for negative censors (such as stroke, brain tumour, neuroleptic) Validation of cases was conducted using a focused three-item telephone questionnaire: 1-Were you given a diagnosis of PD by a neurologist? 2-Does the disease has a progressive course requiring medication? 3-Has there been a good response to the medication, at least during the first 3 years?	Semi-quantitative question-naire administered by trained interview 1 year Validated Caffeinated coffee and tea consumption (ml/day)	Gender, age, marital status, education, farming, smoking, BMI, physical activ-ity and energy intake

Table 12.1 (continued)

Reference Country Publication year	Type of study Sample characteristics Follow-up (Cohort studies)	Outcome assessment Definition of Parkinson Disease/ Parkinsonism	Evaluation of exposure Timing of exposure Validation of the method Items evaluated about caffeine exposure	Control of confounding
NESTED CASE-CONTROL STUDIES				
Paganini-Hill *USA* 2001	Nested Case-Control Study within a prospective cohort study of 13,979 residents of Leisure World Laguna Hills (Leisure World Study, California). Cases and controls were selected from all residents that answered the health survey questionnaire sent by mail. Controls from the large cohort study were matched for age, gender, vital status and death date Cases/Controls: 395 (M/F: NS)/2320 Age: 75 y	Cases identified through review of hospital discharge diagnoses of cohort members for PD, review of death certificates and report of physician diagnosis of PD	Health survey questionnaire NS NS Coffee and tea	Age, gender, smoking, alcohol, blood pressure medication, vitamin A and C
Wirdefeldt *et al.* *Sweden* 2005	Nested Case-Control Study Cases were 476 (M/F: 230/246) twins (mean age of 75 y) identified through the Swedish Inpatient Discharge Register and the Cause of Death Register Two control groups: (1) randomly selected twins unrelated to the cases matched for birth year, gender and questionnaire source of the exposure data (external control subjects; $n=2380$); and (2) co-twins of the cases (co-twin control subjects; $n=415$ same-sex twin pairs) Age: 75.3 y (mean)	Swedish Inpatient Discharge Register and the Cause of Death Register Idiopathic Parkinson's disease (ICD criteria)	Questionnaire sent to the twins in 1967 or in 1973 NS NS Coffee and tea	Matched for age, sex, genetic and familial environmental factors (twins) Smoking, alcohol and educational level

Table 12.1 (continued)

Reference Country Publication year	Type of study Sample characteristics Follow-up (Cohort studies)	Outcome assessment Definition of Parkinson Disease/ Parkinsonism	Evaluation of exposure Timing of exposure Validation of the method Items evaluated about caffeine exposure	Control of confounding
Skeie et al. *Norway* 2010	Nested Case-Control Study Cases (M/F: 126/122) were recruited from five neurological wards in western Norway from November 2004 to August 2006 Controls were recruited from a population of patient's spouses, friends and others, and matched for age and gender Cases/controls: 212/175 Age: NS	Screening visit and neurological examination At least two out of four cardinal motor signs, typical disease history with evidence of progressive parkinsonism, no dementia at onset of parkinsonism, and no severe atypical signs	Semi-structured interview NS NS Coffee and tea	Alcohol and smoking habits
CASE-CONTROL STUDIES				
Nefzger et al. *USA* 1968	Case-Control Study Cases were successive PD inpatients recruited from every neurologist (up to a maximum of ten patients) in Veteran's hospitals throughout the country Controls were the first patient registered after the PD case without psychiatric or extrapyramidal disease (matched for age) Cases/Controls: 198 (all M)/198 Age: NS (>50% had ≥65 y)	Neurological examination Clinical diagnosis (criteria NS)	In-person interview NS NS Coffee	Matched for age

Table 12.1 (continued)

Reference Country Publication year	Type of study Sample characteristics Follow-up (Cohort studies)	Outcome assessment Definition of Parkinson Disease/ Parkinsonism	Evaluation of exposure Timing of exposure Validation of the method Items evaluated about caffeine exposure	Control of confounding
Haack et al. *USA* 1981	Case-Control Study Cases were recruited from medical records of PD cases seen by a neurologist in central Kentucky Controls were identified trough door-to-door in neighbourhood (matched for age, gender and race) Cases/Controls: 237 (M/F: 127/110)/474 Age: 65 y (range, 25–89)	Neurological examination At least two: bradykinesia, resting tremor, rigidity	In-person interview (proxy if physically unable to answer questions) NS NS Coffee and tea	Matched for age, gender and race
Jiméanez *Jiméanez et al.* *Spain* 1992	Case-Control Study Cases were unselected PD patients recruited from an outpatient movement disorder clinic (Madrid) Controls were patients presenting in the emergency room at the same hospital complaining of minor non-neurologic ailments (matched for age and gender) Cases/Controls: 128 (M/F: 68/60)/256 Age: 65 y	Neurological examination Clinical diagnosis (criteria NS)	Personal interview assessing coffee drinking habits 5 years before NS Coffee	Matched for age and gender Smoking and alcohol
Morano et al. *Spain* 1994	Case-Control Study Cases were unselected outpatients making the first visit to one of two neurology clinics (Cáceres) Controls were patients presenting in the emergency room at the same hospital complaining of minor nonneurologic ailments (matched for age and gender) Cases/Controls: 74 (M/F: 33/41)/148 Age: 68 y	Neurological examination Clinical diagnosis (criteria NS)	Questionnaire NS NS Coffee and tea	Matched for age and gender

Table 12.1 (continued)

Reference Country Publication year	Type of study Sample characteristics Follow-up (Cohort studies)	Outcome assessment Definition of Parkinson Disease/ Parkinsonism	Evaluation of exposure Timing of exposure Validation of the method Items evaluated about caffeine exposure	Control of confounding
Hellenbrand et al. *Germany* 1996	Case-Control Study Cases were all PD patients recruited from nine German neurologic clinics (aged $\leq$65 y—older patients were not recruited in order to minimize memory deficits) Controls were randomly selected from the same neighbourhood or region Cases/Controls: 342 (M/F: 224/118)/342 Age: 56 y	Neurological examination UK Brain Bank criteria	SFFQ (in-person interview made by a trained interviewer) Before diagnosis of PD (cases) or 1 year before (controls) Validated Coffee and tea	Matched for age and gender Smoking, education and total energy intake
Fall et al. *Sweden* 1999	Case-Control Study Cases were selected from prescription records and medical reports Controls were randomly selected from population registry in the same area (Central Health Care District in Ostergotland County) Cases/Controls: 113 (M/F: NS)/263 Age: 40–75 y (mean age: 63 y for cases and 57 for controls)	Neurological examination At least one: hypokinesia, tremor, rigidity No earlier treatment with neuroleptic drugs Response to L-DOPA Progressive course Absence of atypical features	SFFQ (mail—proxy information when patients seemed unable to answer properly) 15 years before NS Coffee and tea	Age, gender, smoking, alcohol, occupation/exposure, food factors
Benedetti et al. *USA* 2000	Case-Control Study Cases were recruited from records linkage system of the Rochester Epidemiology Project Controls were randomly recruited from the community (matched for gender and age) Cases/Controls: 196 (M/F: 121/75)/196 Age: 71 y (range, 41–97)	Neurological revision of medical records (ICD codes) and full neurological examination in 27% of the cases At least two: bradykinesia, resting tremor, rigidity, postural instability Absence of other possible cause or atypical features Response to L-DOPA	Neurological revision of medical records NS Validated by telephone interview to a subsample of participants (direct and proxy interview) Coffee	Matched for age and gender Smoking, Alcohol

Table 12.1 (continued)

Reference Country Publication year	Type of study Sample characteristics Follow-up (Cohort studies)	Outcome assessment Definition of Parkinson Disease/ Parkinsonism	Evaluation of exposure Timing of exposure Validation of the method Items evaluated about caffeine exposure	Control of confounding
Preux et al. *France* 2000	Case-Control Study Cases were PD inpatients and outpatients of the Limoges University Hospital Controls were inpatients and outpatients from other hospital departments (matched for age and gender) Cases and controls had to live in the region of Limousin for at least 20 years Cases/Controls: 140 (M/F: 74/66)/280 Age: 71 y	Neurological examination UK Brain Bank criteria	Standard questionnaire (physician personal interview) Lifetime NS Coffee and tea	Matched for age and gender Age, smoking, PD familial history, urban area, toxic products
Checkoway et al. † *USA* 2002	Case-Control Study Cases selected from diagnosis logs at neurology and general clinics and from pharmacy database (Washington); MMSE to establish cognitive performance Controls were from health cooperative enrolees (matched for gender, age, geographic location and year of enrolment) Cases/Controls: 210 (M/F: 131/79)/347 Age: 71 y (range, 37–88)	Neurological examination or neurological panel review of charts At least two: bradykinesia, resting tremor, rigidity, postural instability Absence of other possible cause or atypical features	Structured in-person questionnaire by a nurse practitioner at subjects home During most of adult life NS Coffee, tea, cocoa, cola drinks and chocolate	Age, gender, ethnicity and education
Ragonese et al. *Italy* 2003	Case-Control Study Cases were consecutive outpatients at neurological clinics Controls randomly selected from population records of the municipality (matched for gender, age and place of residence) Cases/Controls: 150 (M/F: 68/82)/150 Age: 60 y (range, 31–81)	Neurological examination At least two: bradykinesia, resting tremor, rigidity, postural instability Unilateral onset or asymmetry Response to L-DOPA Progressive course	Structured questionnaire Years of coffee consumption (0; 1–40; >40) NS Coffee	Education, smoking and alcohol

Table 12.1 (continued)

Reference Country Publication year	Type of study Sample characteristics Follow-up (Cohort studies)	Outcome assessment Definition of Parkinson Disease/ Parkinsonism	Evaluation of exposure Timing of exposure Validation of the method Items evaluated about caf- feine exposure	Control of confounding
Tan et al. *Singapore* 2003	Case-Control Study Cases randomly selected from movement disorders database Controls were participants in community health screening programme (matched for gender, age and race) Cases/Controls: 200 (M/F: 115/85)/200 Age: 65 y (range, 43–88)	Neurological examination UK Brain Bank criteria	Structured questionnaire (interview) NS Validated (information from caregivers and family members) Coffee and tea	Matched for age, gender and race Tea, alcohol, smoking , head injury, stroke, hypertension, presence of heart conditions, toxin exposure and farm dwelling
Evans et al. *UK* 2006	Case-Control Study Cases were consecutive outpatients of Caucasian descent fulfilling Queen square brain bank criteria; MMSE to establish cognitive performance Controls were friends of participants, outpatients without PD and randomly recruited from a volunteer panel (matched for age and gender) Cases/Controls: 106 (M/F: 65/41)/106 Age: 65 y (range, 38–81)	Neurological examination Queen Square Brain Bank criteria	SFFQ (mail) 1 month before Validated Coffee, tea, chocolate milk, caffeinated soft drinks and chocolate	Matched for age and gender Sensation seeking score
Hancock et al. *USA* 2007	Family-based Case-Control Study Cases recruited through physician- and self-referrals to an academic medical center clinic (Miami) Controls were siblings, spouses, parents of subjects, other branches of family Cases/Controls: 356 (M/F: 235/121)/317 Age: 66 y (mean, PD cases)	Neurological examination At least two: bradykinesia, resting tremor, rigidity Absence of atypical features	Structured questionnaire (telephone) At reference age, 10 and 20 years before the reference age NS Coffee, tea and soft drinks	Age, gender, smoking and NSAIDs

Table 12.1 (continued)

Reference Country Publication year	Type of study Sample characteristics Follow-up (Cohort studies)	Outcome assessment Definition of Parkinson Disease/Parkinsonism	Evaluation of exposure Timing of exposure Validation of the method Items evaluated about caffeine exposure	Control of confounding
Tan et al. *Singapore* 2007b	Case-Control Study Cases were consecutive patients diagnosed with PD by a neurologist. Controls were volunteers from similar geographical regions (matched for age, gender and race). Cases/Controls: 418 (M/F: 243/175)/468 Age: 70 y (mean, PD cases)	Neurological examination UK Brain Bank criteria	Standard questionnaire with a semiquantitative food frequency section NS Validated Coffee and tea	Age, gender and smoking
Facheris et al. *USA* 2008	Family-based Case-Control Study Cases were 604 (M/F: 336/238) patients (mean age of 65 y, range 32–92) referred sequentially to the department of Neurology of the Mayo Clinic (Rochester) Two control groups: (1) siblings ($n=446$); and (2) unrelated controls ($n=158$) from the same geographic region matched for age, sex and ethnicity, selected randomly from Medicare and Medicaid services (if older than 65 y) or using random digit dialling (if younger than 65 y)	Neurological examination Detailed protocol (not specified)	Structured questionnaire (telephone—self or proxy for deceased or incapacitated subjects) From birth to the age at onset Validated Coffee, tea and caffeinated sodas	Age, gender, smoking and education

Table 12.1 (continued)

Reference Country Publication year	Type of study Sample characteristics Follow-up (Cohort studies)	Outcome assessment Definition of Parkinson Disease/ Parkinsonism	Evaluation of exposure Timing of exposure Validation of the method Items evaluated about caffeine exposure	Control of confounding
Powers et al. *USA* 2008	Case-Control Study Cases recruited sequentially through movement disorder clinics of NeuroGenetics Research Consortium (New York, Oregon, Washington and Georgia) Controls were spouses and blood relatives of patients, and community volunteers Cases/Controls: 1186 (M/F: 790/396)/928 Age: 69 y (range, 25–97)	Neurological examination UK Brain Bank criteria	Standardized, self-administered questionnaire. Lifetime NS Coffee	Age, gender, ethnicity, smoking, NSAIDs and state
Pereira and *Garrett* *Portugal* 2010	Case-control study Cases were recruited from a Portuguese hospital-based movement disorders outpatient clinic from December 2005 to October 2007 Controls were recruited from a physical therapy outpatient clinic and were matched for age (± 5 y), gender and place of residence Cases/controls: 88 (M/F 58/30)/176 Age (cases): 64.9 ± 9.5 y	Neurological examination UK Brain Bank criteria	Self-administered questionnaire NS Non-validated Coffee and tea	Age, gender and education level

Table 12.1 (continued)

Reference Country Publication year	Type of study Sample characteristics Follow-up (Cohort studies)	Outcome assessment Definition of Parkinson Disease/ Parkinsonism	Evaluation of exposure Timing of exposure Validation of the method Items evaluated about caf- feine exposure	Control of confounding
Nicoletti et al. *Italy* 2010	Case-control study Cases were successive PD patients recruited in five movement disorders centres in central and southern Italy from January 2005 to December 2005 Controls were PD patients' spouses and unrelated healthy individuals recruited from the population of subjects accompanying non-parkinsonian patients for hospital check-ups. The latter were matched for age ($\pm$5 y) and place of residence Cases/controls: 492 (M/F: 292/200)/459 Age (cases): 66$\pm$9.8 y (mean)	Neurological examination Gelb et al. 1999 diagnostic criteria	Face-to-face interview using a standardized structured questionnaire NS NS Coffee (cups/day)	Family history, gender, age, place of residence, cigarette smoking and wine consumption
Sipetic et al. *Serbia* 2011	Case-control study Cases were consecutive newly diagnosed PD patients recruited at the Institute of Neurology, Faculty of Medicine, Belgrade University, Serbia, from January 2001 to November 2005 Controls were patients with degenerative joint disease or digestive tract disease recruited at the University Medical Centre, Belgrade, Serbia, matched by sex, age ($\pm$2 y)) and place of residence (urban/rural) Cases/controls: 110 (M/F: 63/47)/220 Age (cases): 60.75 ($\pm$8.6) y	Neurological examination At least two cardinal signs (tremor, akinesia, rigidity) plus unequivocal response to L-DOPA	Interview using a standardized structured questionnaire NS NS Coffee	Family history, cigarette smoking and alcohol consumption

Table 12.1 (continued)

Reference Country Publication year	Type of study Sample characteristics Follow-up (Cohort studies)	Outcome assessment Definition of Parkinson Disease/ Parkinsonism	Evaluation of exposure Timing of exposure Validation of the method Items evaluated about caffeine exposure	Control of confounding
Tanaka et al. *Japan* 2011	Case-control study Cases were PD inpatients diagnosed by a collaborating neurologist from three university hospitals and one national hospital in Fukuoka Prefecture, and three university hospitals, three national hospitals and one municipal hospital in Osaka, Kyoto and Wakayama Prefectures between 1 April 2006 and 31 March 2008 Controls were non-matched out and inpatients recruited from other hospital departments without history of neurodegenerative disease from one university hospital in Fukuoka Prefecture and one university hospital and one national hospital in Osake, Kyoto and Wakayama Prefectures between 1 April 2006 and 31 March 2008 Cases/controls: 249 (M/F: 93/156)/368 Age (cases): 68.5±86 y (mean)	Neurological examination UK Brain Bank criteria	Self-administered diet history questionnaire at home Preceding month Validated Coffee, black tea, Japanese and Chinese teas	Gender, age, region of residence, educational level, pack-years of smoking, body mass index, the dietary glycemic index, and intake of cholesterol, vitamin E, b-carotene, vitamin B6, alcohol, and iron

Table 12.1 (continued)

Reference Country Publication year	Type of study Sample characteristics Follow-up (Cohort studies)	Outcome assessment Definition of Parkinson Disease/ Parkinsonism	Evaluation of exposure Timing of exposure Validation of the method Items evaluated about caffeine exposure	Control of confounding
Hosseini-Tabatabaei et al. *Iran* 2013	Case-control study Cases were PD inpatients recruited from the Shariati Hospital (Tehran) between May 2007 and May 2008 Controls were age (± 2 y) and gender matched inpatients recruited from other hospital departments without history of CNS or fertility diseases between May 2007 and May 2008 Cases/controls: 75 (M/F: 51/24)/75 Age (cases): 67.5 ± 2.6 y (mean)	Neurological examination At least two: rigidity, bradykinesia, postural instability and tremor OR Presence of only one if these signs together with the improvement of other signs with the use of anti-parkinson medication. Absence of: symmetric signs at the onset of the disease, dementia within the first year of onset, overt dementia, or a Mini-Mental State score of less than 24, bulbar signs and symptoms, early gait disorder, falling within the first year of onset of the disease, wheelchair dependence within 5 years, early autonomic failure, sleep apnea, gasping respirations, apraxia, alien limb, cortical sensory loss, abnormal metabolic tests, abnormal Wilson's disease test results in patients younger than 50 years, history of antipsychotic drug use, abnormal neuroimaging studies	In-person interview using questionnaires Lifetime NS Coffee and tea	Age and gender

Table 12.1 (continued)

Reference Country Publication year	Type of study Sample characteristics Follow-up (Cohort studies)	Outcome assessment Definition of Parkinson Disease/ Parkinsonism	Evaluation of exposure Timing of exposure Validation of the method Items evaluated about caf- feine exposure	Control of confounding
Macleod and *Counsell* *UK* 2013	Case-control study Cases were recruited from a non-specified community-based cohort of PD patients Controls were recruited from a non- specified population and matched for non-specified variables Cases/controls: 201 (M/F: NS)/249 Age: NS	NS NS	Face-to-face questionnaires NS NS NS (moderate and high caf- feine consumption)	Matched controls (variables not stated)
CROSS-SECTIONAL STUDIES				
Louis et al. *USA* 2003	Cross-sectional evaluation of coffee, smoke and parkinsonism Participants in the Washington Heights– Inwood Columbia Aging Project cohort (random selection from healthy Medicare beneficiaries) M/F: 655/1471 Mean age: 77 y	Neurologic and neuropsychological examinations At least two: bradykinesia, resting tremor, rigidity and postural instability	Food frequency questionnaire 1 year before Validated Coffee	Age, gender, ethnic- ity, smoking, years of education and dementia

AARP American Association of Retired Persons, *BMI* body mass index, *EPIC* European prospective investigation into cancer and nutrition, *ICD* international classification of diseases, *NS* not specified, *NSAIDs* non-steroidal antiinflammatory drugs, *PD* Parkinson's disease, *SFFQ* semiquantitative food-frequency questionnaire, *UPDRS* unified Parkinson's disease rating scale

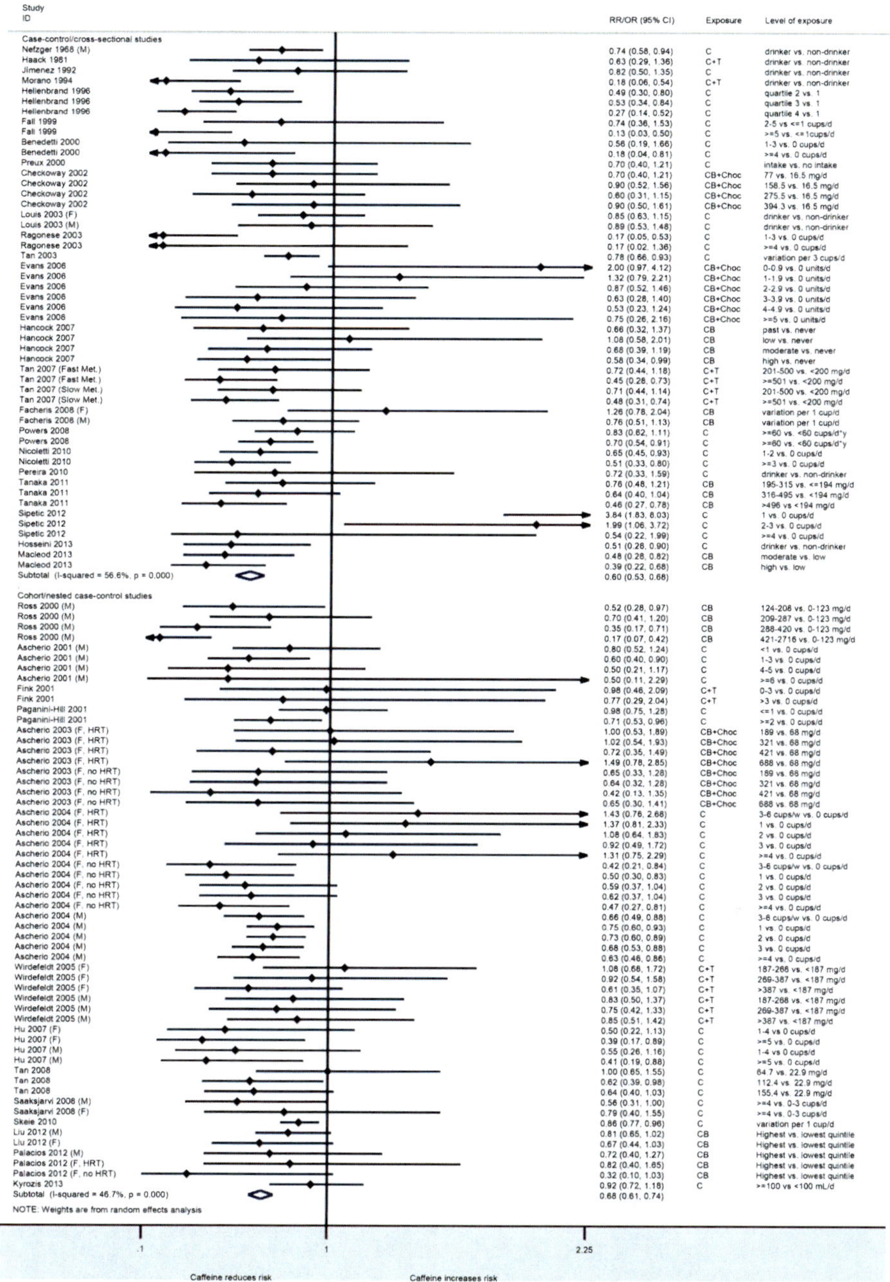

Fig. 12.3 Relative Risk estimates for the association between caffeine and Parkinson's disease, according to sources of caffeine intake and levels of exposure. Legend: *ID* identification, *OR/RR* odds ratio/relative risk, *M* male, *F* female, *C* coffee, *T* tea, *C + T* coffee and tea, *CB* caffeinated beverages, *CB + Choc* caffeinated beverages and chocolate, *Fast Met* fast metabolizers, *Slow Met* slow metabolizers, *HRT* hormonal replacement therapy, *d* day, *w* week

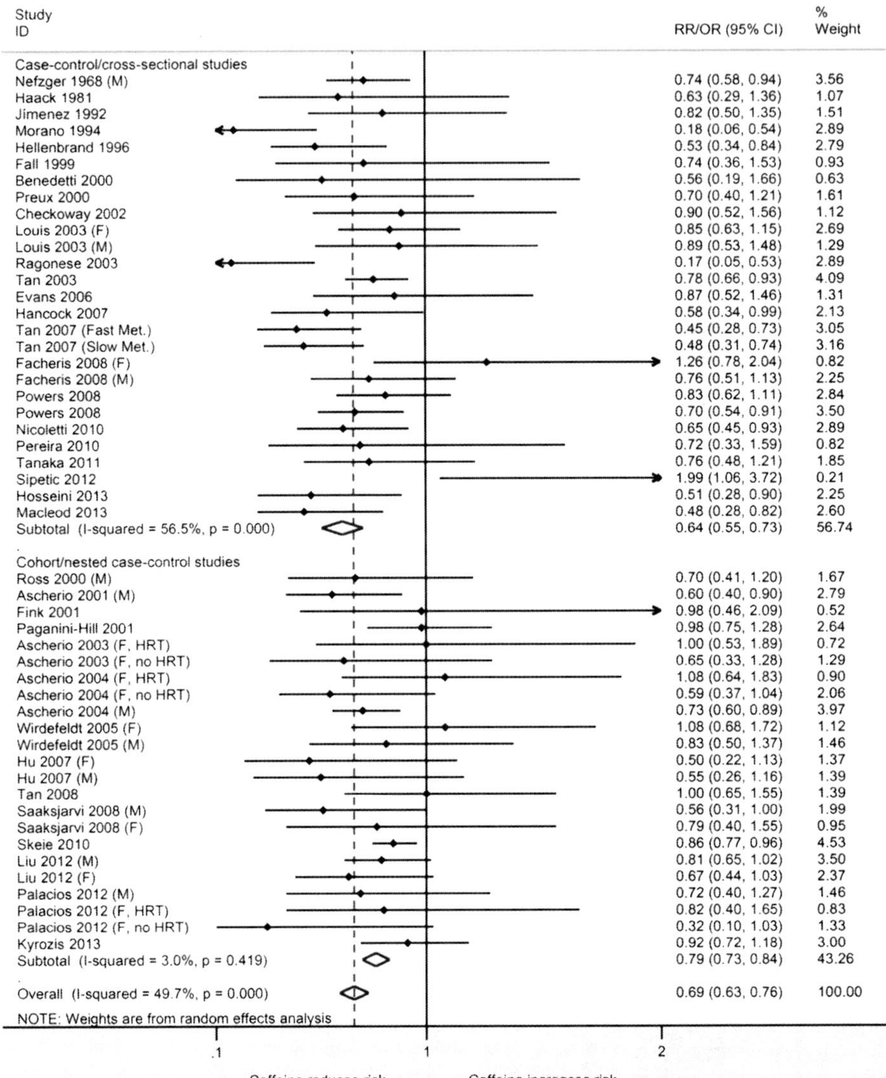

Fig. 12.4 Meta-analysis for the association between caffeine and Parkinson's disease, including the most precise RR estimates from each individual study. Legend: *ID* identification, *OR/RR* odds ratio/relative risk, *M* male, *F* female, *Fast Met* fast metabolizers, *Slow Met* slow metabolizers, *HRT* hormonal replacement therapy

for by three authors. Age and gender were controlled for in all studies except one, by stratified analysis, matching, or multiple regression.

Pooled results from both cohort and case-control studies (Fig. 12.4) showed that caffeine exposure was associated with a 31% reduction in the risk of developing PD (summary RR: 0.69; 95%CI: 0.63 to 0.76.), with moderate heterogeneity ($I^2 = 49.7\%$). No heterogeneity existed among results from cohort/nested case-control studies ($I^2=3.0\%$). Pooled analysis of these types of studies showed a 21 % reduction in the risk of developing PD (summary RR: 0.79; 95 %CI: 0.73 to 0.84) in comparison to a 38 % risk reduction found in pooled results from case-control studies (summary RR: 0.64; 95 %CI: 0.55 to 0.73), although in the later significant heterogeneity was documented ($I^2=56.5\%$).

The negative association was similar when only women were considered for analysis (summary RR=0.76, 95 %CI: 0.63 to 0.90, 13 estimates from 9 studies, $I^2=14.9\%$) and when only men were considered for analysis (summary RR=0.73, 95 %CI: 0.65 to 0.80, 11 estimates from 11 studies, $I^2=0.0\%$). A negative association was not observed in pooled results from studies reporting data on women under hormonal replacement therapy (summary RR = 0.97, 95 % CI: 0.60 to 1.33, 3 estimates from 3 studies, $I^2 = 0.0$ %).

In the last decade, the number of studies evaluating caffeine exposure and PD risk have more than duplicated (Fig. 12.5). However, the summary RR does not considerably vary since 2001, when it was 0.72 (95 %CI: 0.61 to 0.84). Nevertheless, the levels of heterogeneity have increased (I2: 26.6 % in 2001 vs.49.7 % in 2014).

We have previously documented (Costa et al. 2010) the existence of a linear relation between levels of exposure to caffeine and PD risk, with about 25 % risk reduction per 300 mg increase in caffeine intake.

Visual inspection of the funnel plot (Fig. 12.6) does not suggest the presence of publication bias. One possible outlier was a case-control study (Sipetic et al. 2011) showing an increased risk of PD among coffee drinkers. Egger's regression asymmetry test ($p=0.138$) and the Begg adjusted rank correlation test ($p=0.157$) also do not suggest publication bias.

Caffeine Intake and Parkinson's Disease Natural History

To date, several groups have tried to unravel whether caffeine has an effect on the natural history of PD. Age of onset and rate of clinical progression are surrogate markers of the natural history of the disease that are used to study the asymptomatic and the symptomatic phases of PD, respectively.

An early 2000 study signed by Benedetti et al. firstly verified a clinically and statistical significant difference in age of onset of motor symptoms between PD patients who drank coffee and those who never did—a median age of incidence of 72 years for the former and 64 years for the latter (Benedetti et al. 2000).

Conflicting evidence was published in 2009 by Kandinov et al., who showed that a more than 3 cup per day consumption of coffee could anticipate the age of PD onset by approximately 5 years (Kandinov et al. 2009).

Ever since, doubt was left above the responsibility of caffeine on PD age of onset until today.

In 2003, Schwarzschild et al. analysed data from the CALM-PD study—a randomised trial aiming to evaluate the rate of dopamine neuron degeneration by means of neuroimaging—and found no association between caffeine intake and rate of PD progression (Schwarzschild et al. 2003).

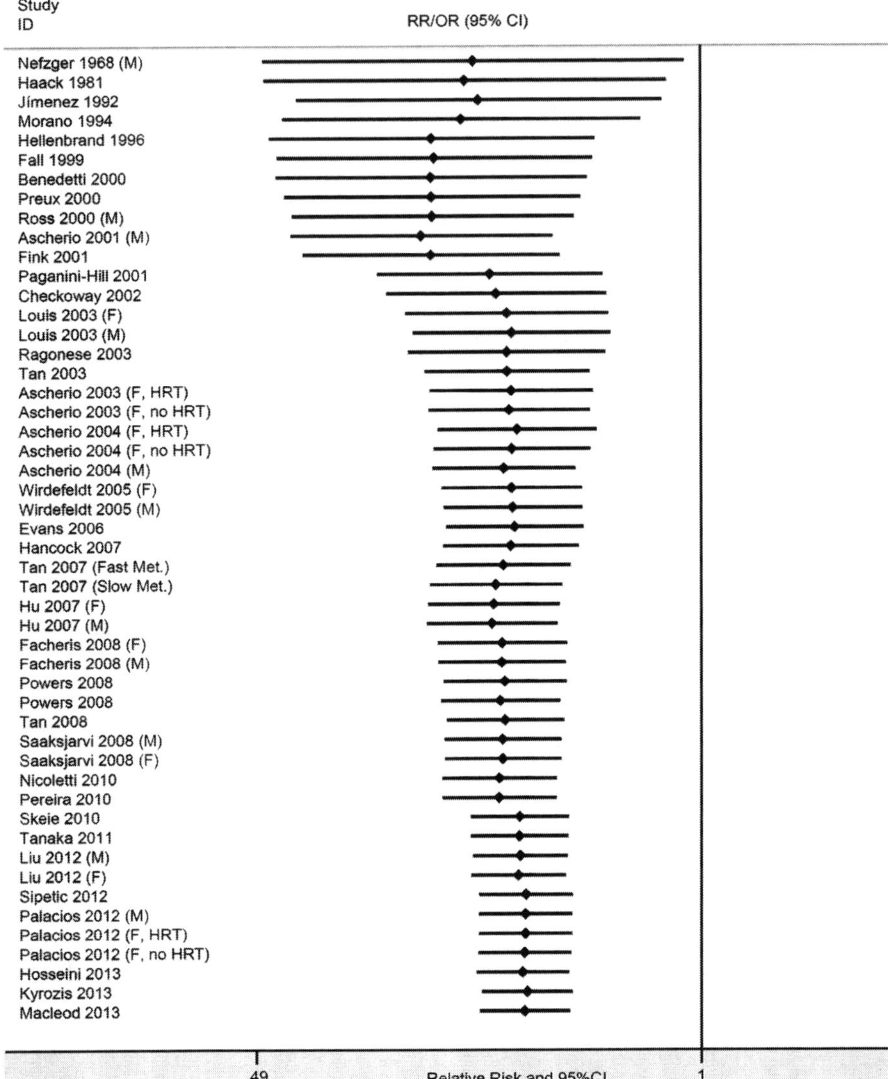

Fig. 12.5 Cumulative meta-analysis for the association between caffeine and Parkinson's disease, including the most precise Relative Risk estimates from each individual study. Legend: *M* male, *F* female, *Fast Met* fast metabolizers, *Slow Met* slow metabolizers, *HRT* hormonal replacement therapy

More recently, Simon et al. revisited this classic model of study by means of data from two PD futility trials. They reassessed the association between caffeine consumption and the rate of disease progression using 412 young non-medicated PD patients included in 1-year-long clinical trial. Statistics demonstrated no significant difference between the lowest and the highest quartiles for caffeine exposure as far as rate of progression is concerned (Simon et al. 2008).

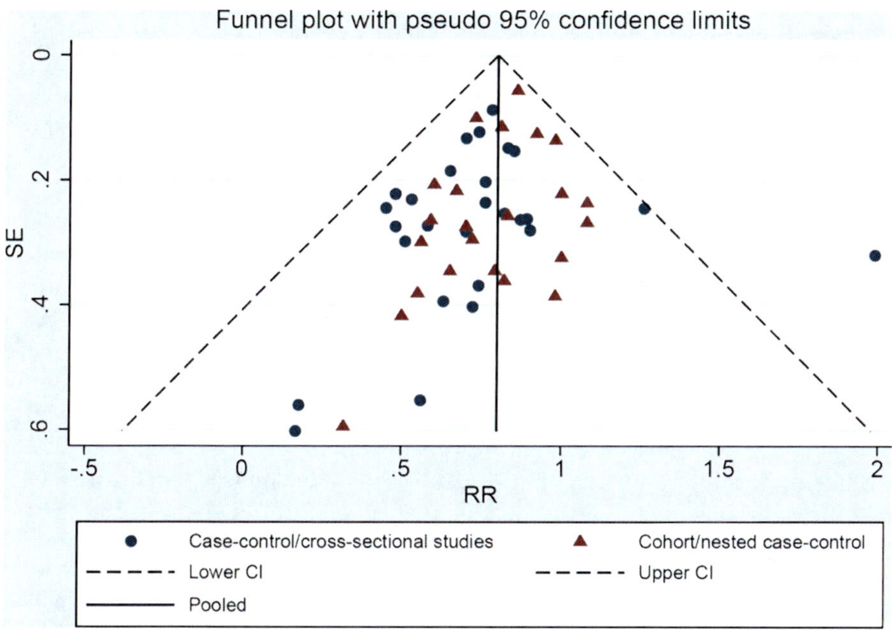

Fig. 12.6 Meta-analysis funnel plot, including the most precise Relative Risk estimates from each individual study

In addition, Kandinov et al. used retrospectively acquired data from 278 PD patients to study disease progression. Their results were compatible with the above-mentioned studies (Kandinov et al. 2007).

Altogether, these results may be view as supportive to the thesis that caffeine could prevent PD, but does not act as a disease-modifying agent. Still, there is plenty of space for further investigation.

Caffeine Intake and Parkinson's Disease Symptomatic Relief

In the late 70's, Shoulson (Shoulson and Chase 1975) and Kartzinel first tried to prove the role of caffeine as a dopamine receptor agonist-adjuvant in the treatment of PD motor symptoms and signs, but without success (Shoulson and Chase 1975; Kartzinel et al. 1976). However, these two small-sample controlled trials were limited by methodological issues.

In 2007, Kitagawa's Japanese group studied the efficacy of 100 mg of caffeine on freezing of gait in PD. Interestingly, caffeine was only beneficial in the akinetic-type gait freezing and the effect lasted until development of tolerance, which resumed after a 2-week period of withdrawal (Kitagawa et al. 2007). These results have limited clinical value due to lack of a control group, randomization, and reduced sample size.

A Phase II open-label dose-response clinical trial conducted from 2010 to 2011 in Canada by Altman et al. aiming to evaluate tolerability of caffeine by PD patients and its efficacy on alleviating motor and non-motor features of the disease found that 400 mg of caffeine daily could improve motor symptoms and sleep quality, and diminish daytime somnolence. The authors also found higher daily doses of caffeine to be difficult to tolerate by PD patients (Altman et al. 2011).

Nevertheless, the latter study also lacked a control group and warranted further investigation. In 2012, Postuma et al. also from Canada, conducted a 6-week randomized controlled clinical trial evaluating 100–200 mg of caffeine twice daily comparing to placebo. This 60-patient trial established no role for the alkaloid on excessive daytime somnolence. However, a statistic significant improvement on motor scales was observed after 6 weeks of caffeine intake (Postuma et al. 2012).

According to the clinicaltrials.gov website last visited on July 2014, there are no ongoing clinical trials investigating the effects of caffeine on PD, though there is a promising protocol for a future Phase III double-blind randomized parallel assignment efficacy trial aiming to investigate in a large cohort of PD patients the short term motor benefits of caffeine and the long term effect over disease progression. Until completion of such a study, no strong conclusions can be made on the effect of caffeine on PD symptoms relieve.

Conclusion and Next Steps

As a relentless neurodegenerative condition, PD is a growing cause of disability and mortality. There are several efficacious drugs on the market that alleviate motor symptoms, however there is no disease-modifying drug for PD. The need of identifying new biomarkers and potential targets for this disease is real, and a growing body of evidence from bench and epidemiological research points towards caffeine and other more selective adenosine A_{2A} receptor antagonists as potential targets.

For several reasons, such as well-known safety, widespread availability and inexpensiveness, caffeine would be a perfect molecule.

Still there is an evidence gap between basic and clinical sciences, calling for a role for translational research. Epidemiological data from human observational studies in the two last decades suggest a clinically relevant inverse association between exposure to caffeine and development of PD, providing empirical evidence for a neuroprotective role of this adenosine antagonist in PD.

However clinical studies are lacking and their results, although encouraging for further ones, are still disappointing for patients waiting for better days to come.

We expect more clinical studies, including randomized placebo-controlled clinical trials, in the future. In fact, recent findings on the mechanism of action of adenosine A_{2A} receptor antagonists have opened the perspective for further studies.

Copying Barone and Roberts (1995): "caffeine has been the subject of extensive research for two reasons—its wide occurrence in nature and its long history of use" (Barone and Roberts 1995). At present, we may add a third reason: its potential to prevent, alleviate and cure.

References

Aguiar LMV, Nobre HVJ, Macedo DS et al (2006) Neuroprotective effects of caffeine in the model of 6-hydroxydopamine lesion in rats. Pharmacol Biochem Behav 84:415–419

Alexander GE, Crutcher MD (1990) Functional architecture of basal ganglia circuits: neural substrates of parallel processing. Trends Neurosci 13:266–271

Altman RD, Lang AE, Postuma RB (2011) Caffeine in Parkinson's disease: a pilot open-label, dose-escalation study. Mov Disord 26:2427–2431

Ascherio A, Zhang SM, Hernan MA et al (2001) Prospective study of caffeine consumption and risk of Parkinson's disease in men and women. Ann Neurol 50:56–63

Ascherio A, Chen H, Schwarzschild MA et al (2003) Caffeine, postmenopausal estrogen, and risk of Parkinson's disease. Neurology 60:790–795

Ascherio A, Weisskopf MG, O'Reilly EJ et al (2004) Coffee consumption, gender, and Parkinson's disease mortality in the cancer prevention study II cohort: the modifying effects of estrogen. Am J Epidemiol 160:977–984

Barone JJ, Roberts HR (1995) Caffeine consumption. Food Chem Toxicol 34:119–129

Begg CB, Mazumdar M (1994) Operating characteristics of a rank correlation test for publication bias. Biometrics 50:1088–1101

Benedetti MD, Bower JH, Maraganore DM et al (2000) Smoking, alcohol, and coffee consumption preceding Parkinson's disease: a case-control study. Neurology 55:1350–1358

Berdichevsky E, Riveros N, Sánchez-Armáss S et al (1983) Kainate, N-methylaspartate and other excitatory amino acids increase calcium influx into rat brain cortex cells in vitro. Neurosci Lett 36:75–80

Biaggioni I, Paul S, Puckett A et al (1991) Caffeine and theophylline as adenosine receptor antagonists in humans. J Pharmacol Exp Ther 258:588–593

Burg AW (1975) Effects of caffeine in the human system. Tea Coffee Trade J 147:40–42

Checkoway H, Powers K, Smith-Weller T et al (2002) Parkinson's disease risks associated with cigarette smoking, alcohol consumption, and caffeine intake. Am J Epidemiol 155:732–738

Chen J-F, Xu K, Petzer JP et al (2001) Neuroprotection by caffeine and A(2A) adenosine receptor inactivation in a model of Parkinson's disease. J Neurosci 21:RC143

Chen X, Lan X, Roche I et al (2008) Caffeine protects against MPTP-induced blood-brain barrier dysfunction in mouse striatum. J Neurochem 107:1147–1157

Costa J, Lunet N, Santos C et al (2010) Caffeine exposure and the risk of Parkinson's disease: a systematic review and meta-analysis of observational studies. J Alzheimers Dis 20:S221–S238

Coyle JT, Bird SJ, Evans RH et al (1981) Excitatory amino acid neurotoxins: selectivity, specificity, and mechanisms of action. Based on an NRP one-day conference held June 30, 1980. Neurosci Res Program Bull 19:1–427

Dall'Igna OP, Fett P, Gomes MW et al (2007) Caffeine and adenosine A(2a) receptor antagonists prevent b-amyloid (25–35)-induced cognitive deficits in mice. Exp Neurol 203:241–245

Daly JW, Butts-Lamb P, Padgett W (1983) Subclasses of adenosine receptors in the central nervous system: interaction with caffeine and related methylxanthines. Cell Mol Neurobiol 3:69–80

de Lau LML, Breteler MMB (2006) Epidemiology of Parkinson's disease. Lancet Neurol 5:525–535

de Rijk MC, Tzourio C, Breteler MM et al (1997) Prevalence of parkinsonism and Parkinson's disease in Europe: the EUROPARKINSON collaborative study. European community concerted action on the epidemiology of Parkinson's disease. J Neurol Neurosurg Psychiatry 62:10–15

Egger M, Smith GD, Schneider M et al (1997) Bias in meta-analysis detected by a simple, graphical test. Br Med J 315:629–634

Ellens DJ, Leventhal DK (2013) Review: electrophysiology of basal ganglia and cortex in models of Parkinson disease. J Parkinsons Dis 3:241–254

Elm von E, Altman DG, Egger M et al (2007) The Strengthening the Reporting of Observational Studies in Epidemiology (STROBE) statement: guidelines for reporting observational studies. Epidemiology 18:800–804

Evans AH, Lawrence AD, Potts J et al (2006) Relationship between impulsive sensation seeking traits, smoking, alcohol and caffeine intake, and Parkinson's disease. J Neurol Neurosurg Psychiatry 77:317–321

Facheris MF, Schneider NK, Lesnick TG et al (2008) Coffee, caffeine-related genes, and Parkinson's disease: a case-control study. Mov Disord 23:2033–2040

Fall PA, Fredrikson M, Axelson O et al (1999) Nutritional and occupational factors influencing the risk of Parkinson's disease: a case-control study in southeastern Sweden. Mov Disord 14:28–37

Fenu S, Pinna A, Ongini E et al (1997) Adenosine A(2A) receptor antagonism potentiates L-DOPA-induced turning behaviour and c-fos expression in 6-hydroxydopamine-lesioned rats. Eur J Pharmacol 321:143–147

Ferreira JJ, Katzenschlager R, Bloem BR et al (2012) Summary of the recommendations of the EFNS/MDS-ES review on therapeutic management of Parkinson's disease. Eur J Neurol 20:5–15

Fink JS, Weaver DR, Rivkees SA et al (1992) Molecular cloning of the rat A2 adenosine receptor: selective co-expression with D2 dopamine receptors in rat striatum. Mol Brain Res 14:186–195

Fink JS, Bains LA, Beiser A et al (2001) Caffeine intake and the risk of incident Parkinson's disease: the Framingham study. Mov Disord 16:984

Fink JS, Kalda A, Ryu H et al (2003) Genetic and pharmacological inactivation of the adenosine A2A receptor attenuates 3-nitropropionic acid-induced striatal damage. J Neurochem 88:538–544

Fredholm BB, Fuxe K, Agnati L (1976) Effect of some phosphodiesterase inhibitors on central dopamine mechanisms. Eur J Pharmacol 38:31–38

Fredholm BB, Johansson B, van der Ploeg I et al (1993) Neuromodulatory roles of purines. Drug Dev Res 28:349–353

Fredholm BB, Bättig K, Holmén J et al (1999) Actions of caffeine in the brain with special reference to factors that contribute to its widespread use. Pharmacol Rev 51:83–133

Fuxe K, Ferré S, Snaprud P et al (1993) Antagonistic A2A/D2 receptor interactions in the striatum as a basis for adenosine/dopamine interactions in the central nervous system. Drug Dev Res 28:374–380

Gerlach M, Riederer P (1996) Animal models of Parkinson's disease: an empirical comparison with the phenomenology of the disease in man. J Neural Transm 103:987–1041

Haack DG, Baumann RJ, McKean HE et al (1981) Nicotine exposure and Parkinson disease. Am J Epidemiol 114:191–200

Hancock DB, Martin ER, Stajich JM et al (2007) Smoking, caffeine, and nonsteroidal anti-inflammatory drugs in families with Parkinson disease. Arch Neurol 64:576–580

Hellenbrand W, Boeing H, Robra BP et al (1996) Diet and Parkinson's disease. II: a possible role for the past intake of specific nutrients. Results from a self-administered food-frequency questionnaire in a case-control study. Neurology 47:644–650

Hernán MA, Takkouche B, Caamaño-Isorna F et al (2002) A meta-analysis of coffee drinking, cigarette smoking, and the risk of Parkinson's disease. Ann Neurol 52:276–284

Higgins JPT, Thompson SG (2002) Quantifying heterogeneity in a meta-analysis. Statist Med 21:1539–1558

Hosseini Tabatabaei N, Babakhani B, Hosseini-Tabatabaei A et al (2013) Non-genetic factors associated with the risk of Parkinson's disease in Iranian patients. Funct Neurol 28:107

Hu G, Bidel S, Jousilahti P et al (2007) Coffee and tea consumption and the risk of Parkinson's disease. Mov Disord 22:2242–2248

Hughes AJ, Daniel SE, Kilford L et al (1992) Accuracy of clinical diagnosis of idiopathic Parkinson's disease: a clinico-pathological study of 100 cases. J Neurol Neurosurg Psychiatry 55:181–184

Jenner P, Rupniak NM, Rose S et al (1984) 1-Methyl-4-phenyl-1,2,3,6-tetrahydropyridine-induced parkinsonism in the common marmoset. Neurosci Lett 50:85–90

Jiméanez Jiméanez FJ, Mateo D, Giméanez Roldan S (1992) Premorbid smoking, alcohol consumption, and coffee drinking habits in Parkinson's disease: a case-control study. Mov Disord 7:339–344

Joghataie MT, Roghani M, Negahdar F et al (2004) Protective effect of caffeine against neurodegeneration in a model of Parkinson's disease in rat: behavioral and histochemical evidence. Parkinsonism Relat Disord 10:465–468

Kachroo A, Schwarzschild MA (2012) Adenosine A2A receptor gene disruption protects in an alpha-synuclein model of Parkinson's disease. Ann Neurol 71:278–282

Kachroo A, Irizarry MC, Schwarzschild MA (2010) Caffeine protects against combined paraquat and maneb-induced dopaminergic neuron degeneration. Exp Neurol 223:657–661

Kanda T, Shiozaki S, Shimada J et al (1994) KF17837: a novel selective adenosine A 2A receptor antagonist with anticataleptic activity. Eur J Pharmacol 256:263–268

Kandinov B, Giladi N, Korczyn AD (2007) The effect of cigarette smoking, tea, and coffee consumption on the progression of Parkinson's disease. Parkinsonism Relat Disord 13:243–245

Kandinov B, Giladi N, Korczyn AD (2009) Smoking and tea consumption delay onset of Parkinson's disease. Parkinsonism Relat Disord 15:41–46

Kartzinel R, Shoulson I, Calne DB (1976) Studies with bromocriptine: III. concomitant administration of caffeine to patients with idiopathic parkinsonism. Neurology 26:741–743

Kitagawa M, Houzen H, Tashiro K (2007) Effects of caffeine on the freezing of gait in Parkinson's disease. Mov Disord 22:710–712

Kyrozis A, Ghika A, Stathopoulos P et al (2013) Dietary and lifestyle variables in relation to incidence of Parkinson's disease in Greece. Eur J Epidemiol 28:67–77

Lang TA, Altman DG (2013) Basic statistical reporting for articles published in Biomedical Journals: the "statistical analyses and methods in the published literature" or the "SAMPL guidelines." science editors' handbook, European Association of Science Editors

Lelo A, Miners JO, Robson RA et al (1986) Quantitative assessment of caffeine partial clearances in man. Br J Clin Pharmacol 22:183–186

Li XX, Nomura T, Aihara H et al (2001) Adenosine enhances glial glutamate efflux via A2a adenosine receptors. Life Sci 68:1343–1350

Liberati A, Altman DG, Tetzlaff J et al (2009) The PRISMA statement for reporting systematic reviews and meta-analyses of studies that evaluate health care interventions: explanation and elaboration. PLoS Med 6:e1000100

Liu R, Guo X, Park Y et al (2012) Caffeine intake, smoking, and risk of Parkinson disease in men and women. Am J Epidemiol 175:1200–1207

Louis ED, Luchsinger JA, Tang MX et al (2003) Parkinsonian signs in older people: prevalence and associations with smoking and coffee. Neurology 61:24–28

Macleod AD, Counsell CE (2013) cigarette smoking, alcohol consumption and caffeine intake in Pd and Pd subtypes: a community-based, incident cohort with matched controls. J Neurol Neurosurg Psychiatry 84:e2

Mao X, Chai Y, Lin YF (2007) Dual regulation of the ATP-sensitive potassium channel by caffeine. Am J Physiol Cell Physiol 292:C2239–C2258

Martinez-Mir MI, Probst A, Palacios JM (1991) Adenosine A2 receptors: selective localization in the human basal ganglia and alterations with disease. Neuroscience 42:697–706

Morano A, Jimenez-Jimenez FJ, Molina JA et al (1994) Risk-factors for Parkinson's disease: case-control study in the province of Cáceres, Spain. Acta Neurol Scand 89:164–170

Morelli M, Fenu S, Pinna A et al (1994) Adenosine A2 receptors interact negatively with dopamine D1 and D2 receptors in unilaterally 6-hydroxydopamine-lesioned rats. Eur J Pharmacol 251:21–25

Nakaso K, Ito S, Nakashima K (2008) Caffeine activates the PI3K/Akt pathway and prevents apoptotic cell death in a Parkinson's disease model of SH-SY5Y cells. Neurosci Lett 432:146–150

National Coffee Association of U.S.A (1993) United States of America Coffee Drinking Study, Winter 1993

Nefzger MD, Quadfasel FA, Karl VC (1968) A retrospective study of smoking in Parkinson's disease. Am J Epidemiol 88:149–158

Nicoletti A, Pugliese P, Nicoletti G et al (2010) Voluptuary habits and clinical subtypes of Parkinson's disease: the FRAGAMP case-control study. Mov Disord 25:2387–2394

Noyce AJ, Bestwick JP, Silveira-Moriyama L et al (2012) Meta-analysis of early nonmotor features and risk factors for Parkinson disease. Ann Neurol 72:893–901

Obeso JA, Rodriguez-Oroz MC, Stamelou M et al (2014) The expanding universe of disorders of the basal ganglia. Lancet 384:523–531

Olney JW (1986) Inciting excitotoxic cytocide among central neurons. Adv Exp Med Biol 203:631–645

O'Regan MH, Simpson RE, Perkins LM et al (1992) The selective A2 adenosine receptor agonist CGS 21680 enhances excitatory transmitter amino acid release from the ischemic rat cerebral cortex. Neurosci Lett 138:169–172

Paganini-Hill A (2001) Risk factors for parkinson's disease: the leisure world cohort study. Neuroepidemiology 20:118–124

Palacios N, Gao X, McCullough ML et al (2012) Caffeine and risk of Parkinson's disease in a large cohort of men and women. Mov Disord 27:1276–1282

Pao EM, Fleming KH, Guenther PM et al (1982) Foods commonly eaten by individuals: amount per day and per eating occasion. Consumer Nutrition Center, Human Nutrition Information Service. United States Department of Agriculture, Washington, D.C.

Parkinson FE, Fredholm BB (1990) Autoradiographic evidence for G-protein coupled A2-receptors in rat neostriatum using [3H]-CGS 21680 as a ligand. Naunyn Schmiedebergs Arch Pharmacol 342:85–89

Pereira D, Garrett C (2010) Factores de risco da doença de Parkinson: um estudo epidemiológico. Acta Med Port 23:15–24

Pinna A, Di Chiara G, Wardas J et al (1996) Blockade of A2a adenosine receptors positively modulates turning behaviour and c-Fos expression induced by D1 agonists in dopamine-denervated rats. Eur J Neurosci 8:1176–1181

Pollanen MS, Dickson DW, Bergeron C (1993) Pathology and biology of the Lewy body. J Neuropathol Exp Neurol 52:183–191

Popoli P, Pintor A, Domenici MR et al (2002) Blockade of striatal adenosine A2A receptor reduces, through a presynaptic mechanism, quinolinic acid-induced excitotoxicity: possible relevance to neuroprotective interventions in neurodegenerative diseases of the striatum. J Neurosci 22:1967–1975

Postuma RB, Lang AE, Munhoz RP et al (2012) Caffeine for treatment of Parkinson disease: a randomized controlled trial. Neurology 79:651–658

Powers KM, Kay DM, Factor SA et al (2008) Combined effects of smoking, coffee, and NSAIDs on Parkinson's disease risk. Mov Disord 23:88–95

Prémont J, Perez M, Blanc G et al (1979) Adenosine-sensitive adenylate cyclase in rat brain homogenates: kinetic characteristics, specificity, topographical, subcellular and cellular distribution. Mol Pharmacol 16:790–804

Preux PM, Condet A, Anglade C et al (2000) Parkinson's disease and environmental factors. Matched case-control study in the Limousin region, France. Neuroepidemiology 19:333–337

Ragonese P, Salemi G, Morgante L et al (2003) A case-control study on cigarette, alcohol, and coffee consumption preceding Parkinson's disease. Neuroepidemiology 22:297–304

Richfield EK, Thiruchelvam MJ, Cory-Slechta DA et al (2002) Behavioral and neurochemical effects of wild-type and mutated human alpha-synuclein in transgenic mice. Exp Neurol 175:35–48

Ross GW, Abbott RD, Petrovitch H et al (2000) Association of coffee and caffeine intake with the risk of Parkinson disease. JAMA 283:2674–2679

Rothman SM (1985) The neurotoxicity of excitatory amino acids is produced by passive chloride influx. J Neurosci 5:1483–1489

Saaksjarvi K, Knekt P, Rissanen H et al (2008) Prospective study of coffee consumption and risk of Parkinson's disease. Eur J Clin Nutr 62:908–915

Schiffmann SN, Vanderhaeghen JJ (1993) Adenosine A2 receptors regulate the gene expression of striatopallidal and striatonigral neurons. J Neurosci 13:1080–1087

Schwarzschild MA, Chen J-F, Tennis M et al (2003) Relating caffeine consumption to Parkinson's disease progression and dyskinesias development. Mov Disord 18:1082–1083

Shoulson I, Chase T (1975) Caffeine and the antiparkinsonian response to levodopa or piribedil. Neurology 25:722–724

Simon DK, Swearingen CJ, Hauser RA et al (2008) Caffeine and progression of Parkinson disease. Clin Neuropharmacol 31:189–196

Simpson RE, O'Regan MH, Perkins LM et al (1992) Excitatory transmitter amino acid release from the ischemic rat cerebral cortex: effects of adenosine receptor agonists and antagonists. J Neurochem 58:1683–1690

Sipetic SB, Vlajinac HD, Maksimovic JM et al (2011) Cigarette smoking, coffee intake and alcohol consumption preceding Parkinson's disease: a case-control study. Acta Neuropsychiatr 24:109–114

Skeie GO, Muller B, Haugarvoll K et al (2010) Differential effect of environmental risk factors on postural instability gait difficulties and tremor dominant Parkinson's disease. Mov Disord 25:1847–1852

Sonsalla PK, Wong L-Y, Harris SL et al (2012) Delayed caffeine treatment prevents nigral dopamine neuron loss in a progressive rat model of Parkinson's disease. Exp Neurol 234:482–487

Spillantini MG, Schmidt ML, Lee VM et al (1997) Alpha-synuclein in Lewy bodies. Nature 388:839–840

Stroup DF, Berlin JA, Morton SC et al (2000) Meta-analysis of observational studies in epidemiology. JAMA 283:2008–2012

Suchowersky O, Gronseth G, Perlmutter J et al (2006) Practice parameter: neuroprotective strategies and alternative therapies for Parkinson disease (an evidence-based review): report of the quality standards subcommittee of the American Academy of Neurology. Neurology 66:976–982

Tan E-K, Tan C, Fook-Chong SMC et al (2003) Dose-dependent protective effect of coffee, tea, and smoking in Parkinson's disease: a study in ethnic Chinese. J Neurol Sci 216:163–167

Tan E-K, Chua E, Fook-Chong SM et al (2007a) Association between caffeine intake and risk of Parkinson's disease among fast and slow metabolizers. Pharmacogenet Genomics 17:1001–1005

Tan LC, Koh W-P, Yuan J-M et al (2007b) Differential effects of black versus green tea on risk of Parkinson's disease in the Singapore Chinese Health Study. Am J Epidemiol 167:553–560

Tanaka K, Miyake Y, Fukushima W et al (2011) Intake of Japanese and Chinese teas reduces risk of Parkinson's disease. Parkinsonism Relat Disord 17:446–450

Tanner CM, Goldman SM (1996) Epidemiology of Parkinson's disease. Neurol Clin 14:317–335

Thiruchelvam M, Brockel BJ, Richfield EK et al (2000) Potentiated and preferential effects of combined paraquat and maneb on nigrostriatal dopamine systems: environmental risk factors for Parkinson's disease? Brain Res 873:225–234

Turrens JF, Boveris A (1980) Generation of superoxide anion by the NADH dehydrogenase of bovine heart mitochondria. Biochem J 191:421–427

Ungerstedt U (1968) 6-Hydroxy-dopamine induced degeneration of central monoamine neurons. Eur J Pharmacol 5:107–110

Van den Eeden SK, Tanner CM, Bernstein AL et al (2003) Incidence of Parkinson's disease: variation by age, gender, and race/ethnicity. Am J Epidemiol 157:1015–1022

von Lubitz DK, Lin RC, Jacobson KA (1995) Cerebral ischemia in gerbils: effects of acute and chronic treatment with adenosine A2A receptor agonist and antagonist. Eur J Pharmacol 287:295–302

Wichmann T, Dostrovsky JO (2011) Pathological basal ganglia activity in movement disorders. Neuroscience 198:232–244

Wirdefeldt K, Gatz M, Pawitan Y et al (2005) Risk and protective factors for Parkinson's disease: a study in Swedish twins. Ann Neurol 57:27–33

Xu K, Xu Y-H, Chen J-F et al (2002) Caffeine's neuroprotection against 1-methyl-4-phenyl-1,2,3,6-tetrahydropyridine toxicity shows no tolerance to chronic caffeine administration in mice. Neurosci Lett 322:13–16

Xu K, Xu Y-H, Brown-Jermyn D et al (2006) Estrogen prevents neuroprotection by caffeine in the mouse 1-methyl-4-phenyl-1,2,3,6-tetrahydropyridine model of Parkinson's disease. J Neurosci 26:535–541

Xu K, Xu Y-H, Chen J-F et al (2010) Neuroprotection by caffeine: time course and role of its metabolites in the MPTP model of Parkinson's disease. Neuroscience 167:475–481
Yang SN, Dasgupta S, Lledo PM et al (1995) Reduction of dopamine D2 receptor transduction by activation of adenosine A2a receptors in stably A2a/D2 (long-form) receptor co-transfected mouse fibroblast cell lines: studies on intracellular calcium levels. Neuroscience 68:729–736
Yazdani U, German DC, Liang CL et al (2006) Rat model of Parkinson's disease: chronic central delivery of 1-methyl-4-phenylpyridinium (MPP+). Exp Neurol 200:172–183

Chapter 13
The Story of Istradefylline—The First Approved A_{2A} Antagonist for the Treatment of Parkinson's Disease

Akihisa Mori, Peter LeWitt and Peter Jenner

Abstract Istradefylline is the first selective adenosine A_{2A} receptor antagonist which has recently been approved in Japan for Parkinson's disease therapy. Its launch followed a journey through drug development over a period of more than 20 years. This chapter details the progression of istradefylline from identification of the receptor target for Parkinson's disease therapy, to characterisation as a development candidate, to elucidation of its mechanism of action, and finally, to progression through clinical evaluation and eventual registration. Initially, istradefylline was shown to be a highly selective antagonist for adenosine A_{2A} receptors and to have a highly localised site of action linked to the indirect output pathway from the striatum. Subsequently, it was found to be effective at reversing motor impairments in rodent and primate models of Parkinson's disease without provoking dyskinesia in primates. In clinical trials, istradefylline (as an adjunct to L-DOPA therapy) decreased 'OFF' time without increasing troublesome dyskinesia. The latter findings were the basis for its registration as a treatment for '*wearing off*' in Parkinson's disease. However, this sequence of apparently logical events was interrupted by many challenges that had to be overcome—the topic of a still unfolding story. At this time, the development of istradefylline in Parkinson's disease is still incomplete and under further clinical investigation. Recently, the drug has shown effectiveness in experimental models of non-motor features of Parkinson's disease. The latter findings and further experience from the clinical use of istradefylline in Parkinson's disease will provide future scope for the development of A_{2A} antagonists in treating human disorders.

Keywords Istradefylline · Adenosine A_{2A} antagonist · Parkinson's disease · Striatal output · MPTP-treated primate · Clinical trials

A. Mori (✉)
Medical Affairs Department, Kyowa Hakko Kirin Co., Ltd., Tokyo, Japan
e-mail: akihisa.mori@kyowa-kirin.co.jp

P. LeWitt
Departments of Neurology, Henry Ford Hospital and Wayne State University School of Medicine, West Bloomfield, Michigan, USA

P. Jenner
Neurodegenerative Diseases Research Group, Institute of Pharmaceutical Sciences, Faculty of Life Sciences and Medicine, King's College, London, UK

© Springer International Publishing Switzerland 2015
M. Morelli et al. (eds.), *The Adenosinergic System*, Current Topics in Neurotoxicity 10,
DOI 10.1007/978-3-319-20273-0_13

Introduction

Research into the treatment of Parkinson's disease (PD) has been dominated by dopamine replacement therapy with L-DOPA (plus decarboxylase inhibitors) and by dopamine agonist drugs. However, with disease progression and increasing duration of drug treatment, dopaminergic medication often fails to control both motor and non-motor features of PD. Problems that can emerge include increasingly long periods of immobility (*'wearing off'*; 'ON-OFF') and motor complications (such as dyskinesia and freezing of gait) that are inadequately treated even with optimised dopaminergic therapy (Fahn 2008; Stocchi et al. 2008). In addition, a wide range of non-motor symptoms in autonomic, sensory, sleep, and neuropsychiatric realms can occur before, concomitant with, or after the development of PD motor symptoms; and most are not alleviated by dopaminergic medication (Chaudhuri and Schapira 2009; Chaudhuri et al. 2011; Martinez-Martin et al. 2011; Rektorova et al. 2011). As a consequence, there is a continuing need for new treatment approaches for the control of both motor and non-motor features of PD.

Non-dopaminergic approaches to the treatment of PD appear promising for two reasons. First, the pathology of PD is widespread, affecting a range of cell groups from the brain stem to the forebrain (including dorsal motor nucleus of the vagus, locus coeruleus, raphe nuclei, pedunculopontine nucleus, and nucleus basalis of Meynert). None of these sites are dopaminergic in nature, but instead, use other neurotransmitters—including acetylcholine, glutamate, serotonin, and noradrenaline (Braak et al. 2004; Javoy-Agid et al. 1984). Neuronal loss in these areas may contribute not only to motor problems in PD but also its non-motor signs and symptoms. Second, loss of the dopaminergic nigro-striatal pathway in PD leads to alterations in the function of the neuronal networks making up the cortical-striatal-thalamic loop (the basal ganglia thalamo-cortical circuit) that controls voluntary movement (Obeso et al. 2008). Specifically, there are alterations in the activity of the direct strio-internal globus pallidus (striatonigral) and the indirect strio-external globus pallidus (striatopallidal) pathway, both of which are constituted by medium spiny neurons (MSNs), that are integral to controlling motor function and in the expression of dyskinesia. Again, these pathways are not dopaminergic in nature and largely rely on γ-Aminobutyric acid (GABA), but are also affected by dopamine, acetylcholine, and glutamate as neurotransmitters in the striatum. They also have a wide range of other neurotransmitter receptors located on them—including serotonin, noradrenaline, histamine, opiate, cannabinoid, and adenosine. All of these pathways represent potential targets for the manipulation of motor function in PD using non-dopaminergic approaches.

Until recently, there have been few non-dopaminergic treatments for PD motor symptoms. Anticholinergic drugs have been used to lessen resting tremor but can be associated with a wide range of side-effects undesirable in an elderly PD population—cognitive impairment, urinary retention, dry mouth, blurred vision (Connolly and Lang 2014). Amantadine is a weak N-methyl-D-aspartate (NMDA) glutamate receptor antagonist but also has a wider range of pharmacological actions. Amantadine can also improve tremor as well as suppressing dyskinesia (Ory-Magne et al. 2014). Zonisamide is another multifunctional drug that can be utilised for motor problems in PD (Murata 2010). While useful, these agents are of limited benefits in the treatment of PD. Many other approaches to non-dopaminergic therapy have looked effective when tested in animal models of PD but in clinical trials most have failed due to lack of efficacy or side-effects (Brotchie and Jenner 2011; Fox et al. 2008; Johnston and Brotchie 2006). These include noradrenergic antagonists, serotonin agonists, anti-epileptics, and a range of glutamate antagonists.

There remains a continuing need for improved PD therapeutics and a non-dopaminergic approach would seem a highly viable approach. The development of the A$_{2A}$ receptor antagonist istradefylline provides one such opportunity and this chapter highlights the more than two-decade challenge of adequately testing and bringing this drug to the market. Its current indication in Japan is for "Improvement of *wearing off* phenomena in patients with PD on concomitant treatment with L-DOPA containing products" (product name: NOURIAST® Tablet 20 mg) (see http://www.e-search.ne.jp/~jpr/PDF/KYOWA13.PDF). In the future, new motor and non-motor uses for istradefylline and other A$_{2A}$ receptor antagonists might be expected and these are highlighted towards the end of this Chapter.

The Identification of Istradefylline As an A$_{2A}$ Receptor Antagonist for PD

Starting in the 1980s, Kyowa Hakko Kirin Co., Ltd. (KHK) (formerly Kyowa Hakko Kogyo) was exploring the potential therapeutic use of adenosine antagonists. Initially, the focus was on discovering selective and potent xanthine-derivative adenosine A$_1$ receptor antagonists structurally related to caffeine and theophylline. One successful outcome was the development of rolofylline (KW-3902), which had a diuretic and protective action in acute renal failure (Shimada et al. 1992a). Through chemical modification of derivatives of rolofylline, KHK discovered a series of compounds that differed in their selectivity and affinity for A$_1$ and A$_2$ receptors (although at this time, the A$_{2A}$ receptor had not been cloned). KHK continued these investigations to discover highly selective and potent xanthine-derivative adenosine A$_2$ receptor antagonists (Shimada et al. 1992b). These endeavours initially were conducted for purely scientific reasons because at the time there was no concept of how A$_2$ antagonists might be used in human disease. The affinity for A$_2$ receptors was assessed using receptor biochemistry studies with brain/peripheral tissue/cell culture membrane preparations and functional assays measuring cyclic AMP production. These methods were employed to determine antagonistic activity of promising compounds. At this point, scientists at KHK determined an unexpected profile of A$_2$ antagonists in rodent behavioural pharmacology experiments (see below) that led to consideration of a use in treating movement disorders as a potential indication. Meanwhile, a successful chemistry programme developed a lead compound in the form of the selective A_{2A} antagonist KF17837 early in the 1990s (Nonaka et al. 1994; Shimada et al. 1992b). Afterwards, large numbers of exploratory pharmacological and mode of action studies, as described below, were undertaken using this molecule.

The early 1990s saw several important contributions to the field: (1) Schiffmann et al. cloned A_{2A} receptors, and it was demonstrated that the receptor was highly expressed on striatopallidal MSNs but not on striatonigral MSNs (Schiffmann and Vanderhaeghen 1993; Schiffmann et al. 1991a, b). (2) The control of normal motor function and its disruption in hypo- and hyper-kinetic motor disorders was proposed to involve dysfunction of the basal ganglia thalamo-cortical circuit (Alexander and Crutcher 1990; DeLong 1990). At this point, KHK in-house data showed that A_{2A} antagonists improved motor function in some rodent PD models that were subsequently used as a primary screening tool. At this point, there was a lot of internal discussion in KHK over the feasibility of exploiting the A_{2A} receptor as a target for PD as there were no reports in any literature that associated adenosine with PD. Dopamine replacement was the central dogma and concepts that were centred on a non-dopaminergic approach were thought to be unrealistic. However, finally KHK decided to develop A_{2A} receptor antagonists as a targeted therapeutic approach for PD because new therapy was obviously required and the team believed the 'adenosine A_{2A} concept' had future potential. At this point, a research team was assembled to characterise the mode of action of A_{2A} receptor antagonism, and to develop the preclinical package necessary to take an A_{2A} antagonist into clinical development for PD. This was accomplished with only a few people at that time having the vision to anticipate that, 20 years later, an A_{2A} antagonist could evolve into a therapy for PD. As the optimisation process for the identification of a lead compound continued, KHK synthesised KW-6002 (istradefylline), which had almost the same *in vitro* affinity profile as KF17837. However, in mice behavioural studies, KW-6002 was approximately 90 times more potent (Saki et al. 2013). The company selected KW-6002 as a lead molecule, which was the first selective, potent A_{2A} antagonist subsequently taken forward as a clinical development candidate for PD.

The Actions of Istradefylline (KW-6002) in Functional Models of Motor Impairment in PD

The localisation of the A_{2A} adenosine receptor to the basal ganglia stimulated the idea that this could be a target for PD. This was supported by studies on caffeine in rats and mice where pharmacological dissection showed that the motor-enhancing effects of caffeine were associated with its A_{2A} antagonist properties. However, the strongest impetus to investigate selective A_{2A} antagonists in relation to PD probably came from the work of Ferré and colleagues (Ferré et al. 1991), which showed that the A_{2A} adenosine agonist, CGS 21680, like haloperidol, induced catalepsy in rats. As a result, rodent models of CGS 21680- and haloperidol-induced catalepsy were subsequently used to explore the novel A_{2A} antagonist drugs being synthesised at KHK. These studies showed that KF17837, but not A$_1$ adenosine antagonists, reversed catalepsy in a dose-dependent manner as did administration of L-dopa (Kanda et al. 1994). When KF17837 was replaced by KW-6002, it too was shown to reverse haloperidol-induced catalepsy. Subsequently, both KF17837 and KW-6002 demonstrated effectiveness in several rodent models of motor impairment in PD (Shiozaki et al. 1999). Both drugs reversed reserpine- and MPTP-induced hypokinesia in mice as did L-DOPA and dopaminergic agonist drugs. The two adenosine A$_{2A}$ antagonists also potentiated the effects of dopaminergic drugs on rotation in unilateral 6-OHDA-lesioned rats (Koga et al. 2000).

As these rodent studies were in progress, the search for efficacy in PD was augmented by a move to testing in the MPTP-treated primate. The model was considered highly predictive of drug action in man, but was not available at KHK. As a consequence, KHK initiated collaboration with the Jenner laboratory at King's College in London. This was a pivotal move in the development of KW-6002 (istradefylline). Studies in this primate model of PD showed that oral administration of KW-6002 alone could produce a partial reversal of impaired locomotor activity and motor disability (Kanda et al. 1998). The benefits were dose-dependent, although no further improvement was seen at the highest doses with 5–10 mg/kg producing a maximal motor improvement. Even more exciting was the discovery that KW-6002 (administered at 5–10 mg/kg) markedly improved the effects of L-DOPA administration on motor function. Most importantly, the administration of KW-6002 by itself did not evoke involuntary movement in MPTP-treated common marmosets that had been exposed to L-DOPA to initiate dyskinesia. Administration of KW-6002 with L-dopa produced no greater involuntary movements than occurred with L-DOPA alone, despite the increased improvement in motor disability. Subsequently, it was shown that on repeated administration of istradefylline, dyskinesia was not enhanced and existing involuntary movements tended to subside (Uchida et al. 2014). To validate the Jenner laboratory findings, additional experiments were undertaken using MPTP-treated cynomolgus monkeys in the laboratories of Paul Bédard in the University of Laval, Quebec City. The cynomolgus monkey studies demonstrated, as in the common marmoset, that the improvement in motor function occurred without exacerbation of dyskinesia (Grondin et al. 1999).

The effects of KW-6002 also were evident when the drug was used in combination with a dopamine agonist, quinpirole. An improvement in motor disability was not seen with quinpirole alone whereas administration of the combination improved motor function but with no increase in dyskinesia (Kanda et al. 2000). Recently, the combined administration of KW-6002 with two clinically used dopamine agonists, ropinirole and pergolide, was shown to further improve motor function in MPTP treated primates (Uchida et al. 2015). Testing of the effects of combinations of KW-6002 with either L-DOPA or dopamine agonist monotherapy, has not been examined in man, but based on these findings, it clearly needs to be explored.

While the single administration of istradefylline with L-DOPA or quinpirole resulted in an expected interaction of improved motor function, it was an unexpected finding that when an additional dose of L-DOPA or quinpirole was administered 24 h later (with no further KW-6002), the same enhanced response occurred. Similarly, if L-DOPA or quinpirole was again administered 48 h after the administration of KW-6002, the improvement in motor function effect was still seen (Kanda et al. 2000). These findings implied that there might be some adaptive change occurring as a result of A_{2A} receptor blockade that alters the response to dopaminergic drug action, since these effects occurred at a time far beyond the biological half-life of KW-6002.

The Search for a Mechanism of Action for Istradefylline in PD

The search for a mechanism of action for KW-6002 was based on several key pieces of evidence from in vivo research, as mentioned above. These include: (1) that A_{2A} agonists induce motor dysfunction, (2) that A_{2A} antagonists ameliorate motor dysfunction in some experimental models of PD including A_{2A} agonist-induced motor dysfunction. In addition, it had been verified that A_{2A} receptors are specifically located on striatopallidal MSNs, which constitute the striatopallidal pathway in the basal ganglia circuitry. It was proposed that excessive activation of the striatopallidal pathway was induced by the loss of dopamine D_2 receptor-mediated inhibitory modulation as a consequence of the loss of nigrostriatal dopamine neurones in PD, a key pathophysiological feature explaining some of its motor symptoms (Alexander and Crutcher 1990; DeLong 1990). This led the team at KHK to consider that, instead using dopamine replacement and/or D_2 receptor stimulation, if "something" could reduce the excitability of striatopallidal pathway, it could be effective in the symptomatic treatment of PD.

Bringing all available evidence together, it seemed reasonable to speculate that an adenosine A_{2A} antagonist might offer therapeutic benefit for PD. Specifically, since adenosine acts to increase striatopallidal pathway output via adenosine A_{2A} receptors on striatopallidal MSNs, an A_{2A} antagonist like istradefylline could block the A_{2A} receptor-induced modulation of the striatopallidal pathway. The net effect would be an increase of pallidal output to the STN, resulting in restoring the balance of the basal ganglia thalamocortical circuit. To prove whether this hypothesis was correct, it was necessary to determine whether, under physiological conditions, adenosine A_{2A} receptors regulate the activity of striatopallidal MSNs. After many preliminary studies conducted at KHK (investigating the range of transmitters and channels that regulate MSN activity), the decision was made to concentrate on GABAergic modulation of MSNs. Deciding factors included: (1) GABAergic input onto MSNs is thought to be a crucial system in determining membrane excitability of MSNs which receive massive excitatory glutamatergic inputs from the cortex and thalamus (Kita 1996), (2) The collaboration with Peter Richardson's laboratory at Cambridge University had provided evidence that A_{2A} receptors regulated GABA release in *in vitro* striatal synaptosomal preparations (Kirk and Richardson 1994; Kurokawa et al. 1994). Also, KHK had already established an electrophysiological method to evaluate synaptic transmission onto single MSN in brain slice preparations in collaboration with Tokyo University (Mori et al. 1994) in the early 1990s. Finally, KHK discovered that A$_{2A}$ receptors modulate intrastriatal GABAergic synaptic transmission onto MSNs (Mori et al. 1996) in the mid-1990s. Triggered by these findings and other data, A$_{2A}$ receptor antagonists then became an important therapeutic target for PD with a physiological rationale (Richardson et al. 1997). Furthermore, and separate from this striatal modulation, it was found that adenosine A$_{2A}$ receptor activation also enhanced GABAergic transmission in the global pallidus (GP) (Shindou et al. 2001, 2002, 2003). Both physiological findings allowed KHK to hypothesise the presence of an A$_{2A}$ receptor-mediated dual modulation of the activity of striatopallidal pathway, regulating both striatal GABA input onto MSNs and GABAergic output from pallidal terminal of MSNs that occurred via A$_{2A}$ receptors located in both the striatum and external GP, respectively. This hypothesis, which developed from *in vitro* investigations, was further strengthened by *in vivo* microdialysis investigations, showing that both intrastriatal and intrapallidal application of an A$_{2A}$ agonist led to excessive GABA release from the GP of rats (Ochi et al. 2000). These *in vivo* studies also demonstrated that GABA release from the GP in 6-OHDA rats was significantly higher than that of normal rats, indicating excessive excitation of the striatopallidal pathway that might occur in PD. The excessive GABA output was significantly suppressed by oral administration of istradefylline to 6-OHDA lesioned rats. At this point, the physiological hypothesis as to how A$_{2A}$ receptors function was translated into a therapeutic mechanism of action of A$_{2A}$ antagonist therapy for improving motor function in PD (Kase et al. 2003; Mori and Shindou 2003). This mechanism of action was strongly supported by independent *in vitro* research carried out by others at the same time (Chergui et al. 2000; Mayfield et al. 1993).

However, other research groups believed that a different mode of action by A$_{2A}$ antagonists was responsible for their pharmacological effects. An alternative hypothesis was based on an A$_{2A}$-D$_2$ receptor interaction driven by the evidence for a co-localization of A$_{2A}$ and D$_2$ receptors on same striatopallidal MSNs. In support, some *in vitro* data was available, including a reciprocal interaction of both receptors at the second messenger level, and *in vivo* synergy between the effects of A$_{2A}$ antagonist and D$_2$ dopamine receptor agonists. This alternative hypothesis proposed that A$_{2A}$ receptor antagonists work through D$_2$ receptors to produce an anti-parkinsonian effect (Fuxe et al. 2001; Svenningsson et al. 1999). However, KHK considered that the apparent A$_{2A}$-D$_2$ interaction could not explain many of the *in vivo* behavioural effects produced by an A$_{2A}$ antagonist such as istradefylline, one of which was the failure to evoke dyskinesia. In addition, much of the *in vitro* data was collected in experiments involving dopamine-free conditions. At this point, KHK initiated a collaboration with Emiliana Borrelli, Institut de Génétique et de Biologie Moléculaire et Cellulaire in Strasbourg. This work demonstrated that istradefylline alone could ameliorate motor dysfunction exhibited by D$_2$ dopamine receptor knock-out mice and also reversed changes in striatopallidal marker protein mRNA (Aoyama et al. 2000). Therefore, KHK concluded that a mechanism independent of a D$_2$ interaction was the most likely explanation for the potential adenosine A$_{2A}$ receptor-related anti-parkinsonian action of istradefylline and this conclusion was subsequently supported by other research groups (Chen et al. 2001).

The Development of Istradefylline as a Drug for Use in PD

Based on the predictive nature of the MPTP treated primate model of PD, the question that needed to be answered in a clinical trial was whether istradefylline worked in conjunction with L-DOPA to extend the duration of motor improvement in those patients who showed '*wearing off*' and did not have an adequate response to further dopaminergic medication.

Thus, in 1996, KHK started a clinical development programme for istradefylline, conducting Phase I studies in Japan and outside of Japan. The profile of istradefylline evaluated through these Phase I studies, such as its pharmacokinetic characteristics, can be seen in the Japanese package insert (product name: NOURIAST® Tablet 20 mg; the English translated version is available in http://www.e-search. ne.jp/~jpr/PDF/KYOWA13.PDF). After repeated oral dosing of istradefylline once daily for 14 days in healthy subjects, the pharmacokinetic analysis indicated a dose-proportional increase in C_{max} and AUC_{0-24} in plasma, approximately 260 ng/mL and 460 ng/mL at 20 and 40 mg/day, respectively. The $T_{1/2}$ after single dosing was approximately 60–75 h and the C_{min} reached an approximate steady state after repeated dosing for 14 days.

At this point, KHK conducted three proof-of-concept (POC) studies in the US and UK. One study (6002-US-001) was a double-blind, randomised, placebo-controlled investigation over 12-weeks with dose-titration of istradefylline up to 20 or 40 mg/day that enrolled 83 L-DOPA treated patients with PD showing motor response complications. This study was used to determine a suitable clinical endpoint for the drug in PD. Among 18 efficacy endpoints, 'OFF' time reduction evaluated by patient diaries was found to be the most appropriate clinical endpoint for the effectiveness of istradefylline, providing a statistically significant change from baseline compared to placebo (Hauser et al. 2003). Another study conducted at the National Institute of Health (NIH) approached the same question from a different direction. Study 6002-US-004 enrolled 15 subjects and demonstrated that istradefylline potentiated the reversal of motor symptoms induced by a low dose of intravenous L-DOPA, and as evaluated by UPDRS part III scores (Bara-Jimenez et al. 2003). A further study (6002-EU-06) was undertaken in healthy volunteers to determine the dose of istradefylline needed to occupy A_{2A} receptors in the brain using positron emission tomography (PET) scan imaging of [^{11}C]-labelled istradefylline in the human brain, with/without oral administration of istradefylline. The results indicated a selective distribution of istradefylline to the caudate-putamen, and also showed that istradefylline 20 and 40 mg/day oral treatment seemed sufficient to occupy A_{2A} receptors in healthy subjects (Brooks et al. 2008). In parallel, the pharmaceutical development programme of KHK accumulated large amounts of non-clinical and clinical pharmacology, safety/toxicology, drug metabolism, and pharmacokinetic data indicating that istradefylline possessed an appropriate profile for full clinical Phase II/III development in PD. This included the biochemical profile of istradefylline that showed it to be a selective A_{2A} receptor antagonist, without effect/affinity for any other major transmitter receptors/transporters, (monoamine oxidase-B [MAOB], and catechol-O-methyl transferase [COMT]) relevant to PD (Saki et al. 2013) .

The Clinical Development Programme

The main drive of istradefylline clinical development has been aimed at L-dopa adjunctive therapy in PD with seven pivotal studies conducted in the North America(NA)/European Union (EU) and Japan, triggered by the successful results from 6002-US-001. For some patients with advanced PD, up to one-half of a typical day can be impaired by episodic 'OFF' states (periods when motor symptomatology re-emerges). Such 'OFF' states, which may be prolonged and unpredictable in their occurrence, also can be associated with peak-effect involuntary movements (dyskinesia) that may evolve as a consequence of chronic L-DOPA therapy. The problems of long-term L-DOPA therapy are only partially helped by conventional pharmacological strategies to gain more consistent motor control, such as using long-acting dopamine agonists, sustained-release L-DOPA formulations, or inhibitors acting on MAOB or COMT (LeWitt 2008). Therefore, all of the studies conducted to date have had almost the same design with 'OFF' time reduction as the primary objective as assessed using 24-h patient diaries. For the NA and EU pivotal development programme conducted by KHK, two Phase IIB studies (6002-US-005 (LeWitt et al. 2008) and 6002-US-006 (Stacy et al. 2008) were undertaken, followed by three Phase III studies (6002-US-013 (Hauser et al. 2008), 6002-US-018 (Pourcher et al. 2012) and 6002-EU-007) that were completed by early 2006. During these investigations, all subjects received L-DOPA at a stable dosage regimen. Other PD drugs were permitted, except entacapone in 6002-EU-007, (with the majority of study participants receiving adjunctive medication, such as dopamine agonists and a COMT inhibitor). These studies also collected data and analysed results with respect to adverse events, tolerability, and safety. The change from baseline in percentage of daily awake time spent in the 'OFF' state (% OFF time), evaluated by 24-h patient diary, was the primary endpoint; secondary endpoints included Unified Parkinson's Disease Rating Scale (UPDRS) motor examination scores as well as global assessments. Although several of the studies were quite similar in study format and drug dosage utilized, their outcomes differed, as described below and Table 13.1.

Clinical development was paused prior to starting Phase III due to preclinical toxicological findings in the rat. However after 6 months intensive and extensive examination by external experts, as well as consultation with the FDA, the clinical development program was re-started in 2004 (see press release on April, 23, 2004: http://www.kyowa-kirin.com/news_releases/kyowa/2004/er040423.html). Subsequently, Japanese Phase IIB (6002–0608) and Phase III (6002-009) studies were conducted from 2007 to 2011 (Mizuno et al. 2010, 2013). Both the NA/EU and Japanese Phase III studies included long-term follow-up safety studies that showed tolerability of istradefylline, with suggesting a sustained reduction of OFF time in

Table 13.1 Overall summary of NA/EU/Japanese randomized controlled study efficacy outcomes in pivotal development programme: The outcome regarding 'OFF' time is shown positive (+) if statistical significant separation of active arms from placebo treatment was observed in either daily OFF time hours or percentage of awake time per day spent in the 'OFF' state (daily % OFF time). (see text for references for each study)

Phase	Region	Study	Treatment duration (weeks)	Outcomes/dose (mg/day)	
				OFF time	UPDRSIII
PIIB	North America	US-005	12	40: +	40: −
		US-006	12	20: + 60: +	20: − 60: −
PIII	North America	US-013	12	20: +	20: −
		US-018	12	10: − 20: − 40: −	10: − 20: − 40: +
	EU	EU-007	16	40: − Entacapone: −	40: − Entacapone: +
PIIB	Japan	0608	12	20: + 40: +	20: + 40: +
PIII		009	12	20: + 40: +	20: − 40: +

L-DOPA treated PD patients, on continued treatment (Factor et al. 2010; Kondo et al. 2015).

The five NA/EU double blind, placebo-controlled studies of istradefylline as adjunctive therapy in PD patients treated with L-DOPA were conducted using the same protocol (except the 6002-EU-007 study), with 'OFF' time reduction in '*wearing-off*' as the primary outcome. Three studies (i.e., 6002-US-005, −006 and −013) demonstrated statistical significant separation from placebo arm in 'OFF' time reduction, but studies US-018 and EU-007 (see http://www.info.pmda.go.jp/shinyaku/P201300035/index.html) did not. Both Japanese Phase IIB and Phase III trials resulted in a positive outcome.

One of the positive NA/EU studies, 6002-US-005 (large North American multicenter Phase IIB clinical trial) investigated the adjunctive role of istradefylline 40 mg/day for the improvement of "OFF" time. This 12-week study involved 1:2 randomization of 195 subjects to regimens of either placebo or istradefylline 40 mg/day. The primary efficacy outcome was the change from baseline to the 12-week assessment in daily 'OFF' time. The highly-significant ($p=0.006$) treatment effect in daily OFF time hours was a mean of −1.8 h (95 %-CI of −1.28 to −0.08 for istradefylline) and −0.6 h (95 %-CI of −2.26 to −1.26) for placebo (LeWitt et al 2008). This corresponded to a reduction in daily % OFF time of 28 % for istradefylline and 10 % for placebo. These benefits were observed by the second week of treatment. Throughout the 12-week study, there was at least a 1-h difference between placebo and the istradefylline effect. The patient diary findings for "ON" time without dyskinesia was slightly increased over the effects of placebo, as was the findings for "ON" time with dyskinesia (which were rated predominantly as "non-troublesome" by subjective rating). Overall, drug-related treatment-emergent adverse events were greater for the istradefylline-treated subjects (66.7%) than for placebo (57.6%), with dyskinesia, dizziness, insomnia, nausea, and accidents involving falling being the most common.

In one of the negative NA/EU studies (large North American multicenter Phase III study, 6002-US-018), the goal was to determine the optimal dosing and the minimally-effective dose of istradefylline producing an anti-parkinsonian effect. Using a 1:1:1:1 randomization scheme, 610 PD subjects with motor fluctuations were assigned to placebo, 10, 20, or 40 mg/day treatments (total ITT population: 584 subjects) (Pourcher et al. 2012). As with the studies discussed above, the primary efficacy end-point of the study was to determine change in daily % OFF time. Expressed as percent change, the placebo treatment resulted in a 7.6% reduction. The three doses of istradefylline led to no statistically different changes from placebo or each other in the extent of reduction in percent "OFF" time: 5.7% for 10 mg, 6.1% for 20 mg, and 9.1% for 40 mg. Although the effects of istradefylline were numerically ordered by dose, the actual time changes from baseline in awake "OFF" time at 12 weeks were minor. No obvious explanation was discovered as to why these results differed from the other Phase IIB and Phase III studies, since similar PD subjects were enrolled and the rating methods were similar. One interpretation proposed by the authors was that the study design utilised had enhanced the magnitude of the placebo effect (which was larger than expected and more prolonged). An increase in dyskinesias (mild to moderate and not troublesome) was also observed in the istradefylline treated groups.

After completion of five NA/EU studies, KHK filed a new drug application (NDA) in 2007 with the Food and Drug Administration (FDA) with data from the pivotal clinical trials (see http://www.kyowa-kirin.com/news_releases/kyowa/2007/er070427_02.html), but received a not-approvable letter in 2008 (see http://www.kyowa-kirin.com/news_releases/kyowa/2008/er080228_01.html). KHK decided to discontinue the development program outside of Japan. However, later the same year, an independent development program in Japan reached a successful outcome from a Phase IIB study (6002-0608). The daily OFF time hours changes from baseline at endpoint were -0.66 h for placebo, -1.31 h for 20 mg/day istradefylline, and -1.58 h for 40 mg/day istardefylline. The differences from placebo were 0.65 h (p=0.013) with istradefylline at 20 mg/day and 0.92 h ($p<0.001$) with the 40 mg/day as an improvement of daily 'OFF' time (Mizuno et al. 2010). A secondary end-point, the UPDRS motor examination conducted while subjects were in an "ON" state, showed an improvement of 5.7 points compared to an improvement of 3.7 points in placebo-treated patients ($p=0.006$). KHK proceeded with the program in Japan and initiated the Phase III program. Study 6002-009 was carried out at a number of clinical sites and enrolled 373 subjects. This randomized, placebo-controlled 12-week trial investigated a primary efficacy outcome as those studied in North America and Europe (change in 'OFF' time). The results showed that both 20 and 40 mg/day of istradefylline led to reductions of a similar extent and the differences from reduction extent of placebo arm were 0.76 h with 20 mg/day ($p=0.003$) and 0.74 h with 40 mg/day ($p=0.003$) (Mizuno et al. 2013). However, only the 40 mg/day istradefylline dose resulted in an improvement in the UPDRS motor examination score ($p=0.001$) as a secondary efficacy endpoint. The istradefylline treatment regimen was well tolerated and the most frequent common adverse event showing a greater incidence than placebo was dyskinesia. The successful outcome led to KHK filing an NDA in Japan in 2012 (see http://www.kyowa-kirin.com/news_releases/2012/e20120402_01.html). In 2013 (see http://www.kyowa-kirin.com/news_releases/2013/e20130325_04.html) KHK obtained approval for marketing of istradefylline for the indication "improvement of *wearing-off* phenomenon in patients with PD on concomitant treatment with L-DOPA containing preparations". The approved dosage and regimen is 20 mg once a day oral administration and according to symptoms, 40 mg once a day can be orally administrated.

The reasons why 6002-US-018 and 6002-EU-007 failed to demonstrate statistical separation of the active treatment arms from placebo treatment in 'OFF' time reduction are unknown. With respect to the 6002-US-018 study, the placebo response around the primary endpoint was very large and was almost double in comparison with other pivotal studies. The driving factor contributing to such a high placebo change has not been identified. Several possible factors were discussed, including a high probability (75%) of being assigned to active drug (Pourcher et al. 2012). Study 6002-EU-007 study was carried out in 14 countries including Europe, South America, India, and Russia, which may suggest that variability between centres arising from differences in the medical environment and language in different territories was not adequately controlled.

What were the reasons that made a difference between the NA/EU and Japan studies? To date, there are no answers. However, the Pharmaceuticals and Medical Devices Agency (PMDA), Japan has issued their view as part of review of the istradefylline submission data in Japan:

> As the applicant discussed, despite the fact that there were no major differences in the intrinsic and extrinsic ethnic factors, patient background, or study design between Japan and overseas, foreign clinical studies failed to clearly demonstrate the efficacy of istradefylline with no consistent results with respect to the effect of istradefylline in reducing OFF time across different studies. However, in the Japanese clinical studies compared with the foreign clinical studies, the number of daily diaries assessed was increased with an aim of increasing the precision of data and furthermore, the Japanese studies were conducted under the system where the results were less likely to be affected by centre differences, compared with Study 6002-EU-007 that failed to demonstrate efficacy. Thus, the effect of istradefylline in reducing OFF time may have been assessed more accurately in the Japanese late phase II and phase III studies compared with the foreign clinical studies. In addition, in the Japanese development program, two placebo-controlled, parallel-group, comparative studies of similar design were conducted and the reproducibility of efficacy can also be assessed. Therefore, the efficacy of istradefylline should be evaluated based primarily on the data from the Japanese late phase II and phase III studies. (Review Results February 22, 2013 by PMDA, Japan: http://www.pmda.go.jp/english/service/drugs.html)

The outcome of change in UPDRS part III also showed inconsistency across all seven studies undertaken. However, there were slight differences in the timing of evaluation of UPDRS part III scores between two Phase IIB studies in NA/EU and all other studies, including the two Japanese studies. It should be noted that the former studies determine UPDRS part III score in a morning OFF state after overnight omission of PD medication at baseline, week 4 and week 12 as an endpoint, occurring after the first dose of L-DOPA that day in patients who were asked to come to the clinic in the 'OFF' state. But, the latter investigations evaluated UPDRS part III in the 'ON' state in patients who came to the clinic already 'ON'. This difference may have had some impact on baseline UPDRS part III scores between studies and may have influenced the changes from baseline.

The entire development strategy for istradefylline has been driven by the outcome of L-DOPA adjunctive studies in PD but monotherapy has also been evaluated. In a proof-of-concept monotherapy study (6002-US-051), 176 PD subjects were randomized 1:1 to either istradefylline or placebo for 12 weeks (Fernandez et al. 2010). Their motor symptoms were mild and past exposure to dopaminergic therapy was minimal. In this multi-centre investigation, the daily dose of istradefylline was 40 mg. At 12 weeks, the UPDRS motor examination did not support a lessening of motor disability compared to baseline ($p = 0.228$), although there was a numerical trend for an improvement as compared with placebo, including statistical separation observed at week 2. The study also tried to evaluate cognitive function by using some sub-scores of the California Verbal Leaning Test, Second Edition although the study was not powered for these variables, showing that istradefylline does not worsen cognitive status (Fernandez et al. 2010) (Table 13.1).

The Future Development of Istradefylline and A$_{2A}$ antagonists—Motor and Non-Motor Symptoms

After the success of the clinical trials programme in Japan and the marketing of istradefylline for PD in Japan, attention has returned to obtaining approval of the drug in the NA and in Europe. A new clinical trial has been initiated in both territories using the same basic design as in previous investigations, under special protocol assessment (SPA) agreement with the US/FDA (see http://www.kyowa-kirin.com/news_releases/2013/e20131121_01.html). The study end point is a reduction in 'OFF' time in L-DOPA-treated patients with a stable regimen of any other anti-PD therapy (MAO-B inhibitors, COMT inhibitors, dopaminergic agonists) compared to placebo (clinicaltrials.gov identifier: NCT01968031).

This study should establish the role of istradefylline as an adjunct to L-DOPA therapy to treat '*wearing-off*' in PD. However, its usefulness and efficacy in earlier treatment strategies has not been fully explored in pivotal development trials. There are only a few investigations of the clinical effect of istradefylline as monotherapy in early PD, like 6002-US-051 study (Fernandez et al. 2010), and positive results from non-human primate studies as described above. Perhaps importantly, the synergy seen between istradefylline and low doses of both L-DOPA and dopamine agonist drugs in experimental models of PD suggests two further possibilities for future A$_{2A}$ antagonist strategies. First, A$_{2A}$ antagonists might be used in an L-DOPA sparing strategy to reduce L-DOPA dosage in patients where unacceptable side-effects, such as dyskinesia, are occurring but a further improvement in motor function is required. Second, the class might be used in patients who show insufficient improvement from low dose L-DOPA or dopamine agonist monotherapy to avoid any further increase in dopaminergic medication so that the onset of motor fluctuations and motor complications can be avoided. Both potential possibilities should be explored in future clinical investigations for A_{2A} antagonist development.

In addition to the motor symptoms of PD, A_{2A} antagonists may have a potential use in the treatment of non-motor symptoms which largely respond poorly to dopaminergic medication and represent a major clinical unmet need (Chaudhuri and Schapira 2009). Recently, experimental studies undertaken using istradefylline have demonstrated a potential role in the control of neuropsychiatric components of PD. Istradefylline was shown to be active in the tail suspension test and forced swim test and on learned helplessness, suggesting antidepressant potential effects in all three paradigms that suggest a high predictive value and the possibility that not only would istradefylline be effective against depression in PD but also against depression syndromes affecting the general population (Yamada et al. 2013, 2014). In relation to cognitive impairment in PD, istradefylline improved cognitive performance in rats with a 6-OHDA lesion in prefrontal cortex (Kadowaki Horita et al. 2013).

Conclusions

The discovery of the ability of istradefylline to improve motor function in PD without worsening dyskinesia through its A_{2A} receptor antagonist activity, has provided a novel non-dopaminergic approach to the treatment of the illness. The launch of the drug for the treatment of PD in Japan is the culmination of more than 20 years of endeavour. It is a great illustration of the need to persevere in drug development and the need to overcome setbacks that might otherwise have led to the termination of the programme. However, the initial introduction of istradefylline as a therapy for PD is, in reality, only the beginning of the story. The full potential of istradefylline in the treatment of PD has yet to be explored. In particular, its use in treating early PD either as monotherapy or as an alternative to the introduction or to an increase in dopaminergic therapy, need investigation. The potential for the use of A_{2A} antagonists in treating the neuropsychiatric components of PD is only now starting to emerge, and its wider potential for the treatment of anxiety, depression, and cognitive disorders in the non-PD patient population is a real possibility for the future. The therapy for PD has very much remained dominated by approaches linked to dopamine replacement therapy, but the development of istradefylline shows that it is possible to alter basal ganglia function in areas beyond the damaged dopaminergic system by manipulating non-dopaminergic targets controlling striatal output. This raises the clear possibility the A_{2A} receptor-based approaches and other non-dopaminergic receptor populations may be the way forward in the symptomatic treatment of PD for the future.

References

Alexander GE, Crutcher MD (1990) Functional architecture of basal ganglia circuits: neural substrates of parallel processing. Trends Neurosci 13:266–271

Aoyama S, Kase H, Borrelli E (2000) Rescue of locomotor impairment in dopamine D2 receptor-deficient mice by an adenosine A$_{2A}$ receptor antagonist. J Neurosci 20:5848–5852

Bara-Jimenez W, Sherzai A, Dimitrova T et al (2003) Adenosine A(2A) receptor antagonist treatment of Parkinson's disease. Neurology 61:293–296

Braak H, Ghebremedhin E, Rub U et al (2004) Stages in the development of Parkinson's disease-related pathology. Cell Tissue Res 318:121–134

Brooks DJ, Doder M, Osman S et al (2008) Positron emission tomography analysis of [11C] KW-6002 binding to human and rat adenosine A$_{2A}$ receptors in the brain. Synapse 62:671–681

Brotchie J, Jenner P (2011) New approaches to therapy. Int Rev Neurobiol 98:123–150

Chaudhuri KR, Schapira AH (2009) Non-motor symptoms of Parkinson's disease: dopaminergic pathophysiology and treatment. Lancet Neurol 8:464–474

Chaudhuri KR, Odin P, Antonini A et al (2011) Parkinson's disease: the non-motor issues. Parkinsonism Relat Disord 17:717–723

Chen JF, Moratalla R, Impagnatiello F et al (2001) The role of the D(2) dopamine receptor (D(2)R) in A(2A) adenosine receptor (A(2A)R)-mediated behavioral and cellular responses as revealed by A(2A) and D(2) receptor knockout mice. Proc Natl Acad Sci U S A 98:1970–1975

Chergui K, Bouron A, Normand E et al (2000) Functional GluR6 kainate receptors in the striatum: indirect downregulation of synaptic transmission. J Neurosci 20:2175–2182

Connolly BS, Lang AE (2014) Pharmacological treatment of Parkinson disease: a review. JAMA 311:1670–1683

DeLong MR (1990) Primate models of movement disorders of basal ganglia origin. Trends Neurosci 13:281–285

Factor S, Mark MH, Watts R et al (2010) A long-term study of istradefylline in subjects with fluctuating Parkinson's disease. Parkinsonism Relat Disord 16:423–426

Fahn S (2008) How do you treat motor complications in Parkinson's disease: medicine, surgery, or both? Ann Neurol 64:S56–S64

Fernandez HH, Greeley DR, Zweig RM et al (2010) Istradefylline as monotherapy for Parkinson disease: results of the 6002-US-051 trial. Parkinsonism Relat Disord 16:16–20

Ferré S, Rubio A, Fuxe K (1991) Stimulation of adenosine A$_2$ receptors induces catalepsy. Neurosci Lett 130:162–164

Fox SH, Brotchie JM, Lang AE (2008) Non-dopaminergic treatments in development for Parkinson's disease. Lancet Neurol 7:927–938

Fuxe K, Stromberg I, Popoli P et al (2001) Adenosine receptors and Parkinson's disease. Relevance of antagonistic adenosine and dopamine receptor interactions in the striatum. Adv Neurol 86:345–353

Grondin R, Bedard PJ, Hadj Tahar A et al (1999) Antiparkinsonian effect of a new selective adenosine A$_{2A}$ receptor antagonist in MPTP-treated monkeys. Neurology 52:1673–1677

Hauser RA, Hubble JP, Truong DD et al (2003) Randomized trial of the adenosine A(2A) receptor antagonist istradefylline in advanced PD. Neurology 61:297–303

Hauser RA, Shulman LM, Trugman JM et al (2008) Study of istradefylline in patients with Parkinson's disease on levodopa with motor fluctuations. Mov Disord 23:2177–2185

Javoy-Agid F, Ruberg M, Taquet H et al (1984) Biochemical neuropathology of Parkinson's disease. Adv Neurol 40:189–198

Johnston TH, Brotchie JM (2006) Drugs in development for Parkinson's disease: an update. Curr Opin Invest Drugs 7:25–32

Kadowaki Horita T, Kobayashi M, Mori A et al (2013) Effects of the adenosine A$_{2A}$ antagonist istradefylline on cognitive performance in rats with a 6-OHDA lesion in prefrontal cortex. Psychopharmacology 230:345–352

Kanda T, Shiozaki S, Shimada J et al (1994) KF17837: a novel selective adenosine A_{2A} receptor antagonist with anticataleptic activity. Eur J Pharmacol 256:263–268

Kanda T, Jackson MJ, Smith LA et al (1998) Adenosine A_{2A} antagonist: a novel antiparkinsonian agent that does not provoke dyskinesia in parkinsonian monkeys. Ann Neurol 43:507–513

Kanda T, Jackson MJ, Smith LA et al (2000) Combined use of the adenosine A(2A) antagonist KW-6002 with L-DOPA or with selective D1 or D2 dopamine agonists increases antiparkinsonian activity but not dyskinesia in MPTP-treated monkeys. Exp Neurol 162:321–327

Kase H, Aoyama S, Ichimura M et al (2003) Progress in pursuit of therapeutic A2A antagonists: the adenosine A_{2A} receptor antagonist KW6002: Research and development toward a novel nondopaminergic therapy for Parkinson's disease. Neurology 61:S97–S100

Kirk IP, Richardson PJ (1994) Adenosine A_{2a} receptor-mediated modulation of striatal [3H]GABA and [3H]acetylcholine release. J Neurochem 62:960–966

Kita H (1996) Glutamatergic and GABAergic postsynaptic responses of striatal spiny neurons to intrastriatal and cortical stimulation recorded in slice preparations. Neuroscience 70:925–940

Koga K, Kurokawa M, Ochi M et al (2000) Adenosine A(2A) receptor antagonists KF17837 and KW-6002 potentiate rotation induced by dopaminergic drugs in hemi-Parkinsonian rats. Eur J Pharmacol 408:249–255

Kondo T, Mizuno Y, Japanese Istradefylline Study Group (2015) A long-term study of istradefylline safety and efficacy in patients with Parkinson disease. Clin Neuropharmacol 38:41–46

Kurokawa M, Kirk IP, Kirkpatrick KA et al (1994) Inhibition by KF17837 of adenosine A_{2A} receptor-mediated modulation of striatal GABA and ACh release. Br J Pharmacol 113:43–48

LeWitt PA (2008) Levodopa for the treatment of Parkinson's disease. N Engl J Med 359:2468–2476

LeWitt PA, Guttman M, Tetrud JW et al (2008) Adenosine A_{2A} receptor antagonist istradefylline (KW-6002) reduces "off" time in Parkinson's disease: a double-blind, randomized, multicenter clinical trial (6002-US-005). Ann Neurol 63:295–302

Martinez-Martin P, Rodriguez-Blazquez C, Kurtis MM et al (2011) The impact of non-motor symptoms on health-related quality of life of patients with Parkinson's disease. Mov Disord 26:399–406

Mayfield RD, Suzuki F, Zahniser NR (1993) Adenosine A2a receptor modulation of electrically evoked endogenous GABA release from slices of rat globus pallidus. J Neurochem 60:2334–2337

Mizuno Y, Kondo T, Japanese Istradefylline Study Group (2013) Adenosine A_{2A} receptor antagonist istradefylline reduces daily OFF time in Parkinson's disease. Mov Disord 28:1138–1141

Mizuno Y, Hasegawa K, Kondo T et al (2010) Clinical efficacy of istradefylline (KW-6002) in Parkinson's disease: a randomized, controlled study. Mov Disord 25:1437–1443

Mori A, Shindou T (2003) Modulation of GABAergic transmission in the striatopallidal system by adenosine A_{2A} receptors: a potential mechanism for the antiparkinsonian effects of A_{2A} antagonists. Neurology 61:S44–S48

Mori A, Takahashi T, Miyashita Y et al (1994) Two distinct glutamatergic synaptic inputs to striatal medium spiny neurones of neonatal rats and paired-pulse depression. J Physiol 476:217–228

Mori A, Shindou T, Ichimura M et al (1996) The role of adenosine A2a receptors in regulating GABAergic synaptic transmission in striatal medium spiny neurons. J Neurosci 16:605–611

Murata M (2010) Zonisamide: a new drug for Parkinson's disease. Drugs Today 46:251–258

Nonaka H, Ichimura M, Takeda M et al (1994) KF17837 ((E)-8-(3,4-dimethoxystyryl)-1,3-dipropyl-7-methylxanthine), a potent and selective adenosine A_2 receptor antagonist. Eur J Pharmacol 267:335–341

Obeso JA, Rodriguez-Oroz MC, Benitez-Temino B et al (2008) Functional organization of the basal ganglia: therapeutic implications for Parkinson's disease. Mov Disord 23:S548–S559

Ochi M, Koga K, Kurokawa M et al (2000) Systemic administration of adenosine A(2A) receptor antagonist reverses increased GABA release in the globus pallidus of unilateral 6-hydroxydopamine-lesioned rats: a microdialysis study. Neuroscience 100:53–62

Ory-Magne F, Corvol JC, Azulay JP et al (2014) Withdrawing amantadine in dyskinetic patients with Parkinson disease: the AMANDYSK trial. Neurology 82:300–307

Pourcher E, Fernandez HH, Stacy M et al (2012) Istradefylline for Parkinson's disease patients experiencing motor fluctuations: results of the KW-6002-US-018 study. Parkinsonism Relat Disord 18:178–184

Rektorova I, Aarsland D, Chaudhuri KR et al (2011) Nonmotor symptoms of Parkinson's disease. Parkinsons Dis 2011:351–461

Richardson PJ, Kase H, Jenner PG (1997) Adenosine A$_{2A}$ receptor antagonists as new agents for the treatment of Parkinson's disease. Trends Pharmacol Sci 18:338–344

Saki M, Yamada K, Koshimura E et al (2013) In vitro pharmacological profile of the A$_{2A}$ receptor antagonist istradefylline. Naunyn Schmiedebergs Arch Pharmacol 386:963–972

Schiffmann SN, Vanderhaeghen JJ (1993) Adenosine A$_2$ receptors regulate the gene expression of striatopallidal and striatonigral neurons. J Neurosci 13:1080–1087

Schiffmann SN, Jacobs O, Vanderhaeghen JJ (1991a) Striatal restricted adenosine A$_2$ receptor (RDC8) is expressed by enkephalin but not by substance P neurons: an in situ hybridization histochemistry study. J Neurochem 57:1062–1067

Schiffmann SN, Libert F, Vassart G et al (1991b) Distribution of adenosine A$_2$ receptor mRNA in the human brain. Neurosci Lett 130:177–181

Shimada J, Suzuki F, Nonaka H et al (1992a) 8-Polycycloalkyl-1,3-dipropylxanthines as potent and selective antagonists for A1-adenosine receptors. J Med Chem 35:924–930

Shimada J, Suzuki F, Nonaka H et al (1992b) (E)-1,3-dialkyl-7-methyl-8-(3,4,5-trimethoxystyryl) xanthines: potent and selective adenosine A$_2$ antagonists. J Med Chem 35:2342–2345

Shindou T, Mori A, Kase H et al (2001) Adenosine A(2A) receptor enhances GABA(A)-mediated IPSCs in the rat globus pallidus. J Physiol 532:423–434

Shindou T, Nonaka H, Richardson PJ et al (2002) Presynaptic adenosine A$_{2A}$ receptors enhance GABAergic synaptic transmission via a cyclic AMP dependent mechanism in the rat globus pallidus. Br J Pharmacol 136:296–302

Shindou T, Richardson PJ, Mori A et al (2003) Adenosine modulates the striatal GABAergic inputs to the globus pallidus via adenosine A$_{2A}$ receptors in rats. Neurosci Lett 352:167–170

Shiozaki S, Ichikawa S, Nakamura J et al (1999) Actions of adenosine A$_{2A}$ receptor antagonist KW-6002 on drug-induced catalepsy and hypokinesia caused by reserpine or MPTP. Psychopharmacology 147:90–95

Stacy M, Silver D, Mendis T et al (2008) A 12-week, placebo-controlled study (6002-US-006) of istradefylline in Parkinson disease. Neurology 70:2233–2240

Stocchi F, Tagliati M, Olanow CW (2008) Treatment of levodopa-induced motor complications. Mov Disord 23:S599–S612

Svenningsson P, Le Moine C, Fisone G et al (1999) Distribution, biochemistry and function of striatal adenosine A$_{2A}$ receptors. Prog Neurobiol 59:355–396

Uchida S, Soshiroda K, Okita E et al (2015) The adenosine A2A receptor antagonist, istradefylline enhances the anti-parkinsonian activity of low doses of dopamine agonists in MPTP-treated common marmosets. Eur J Pharmacol 747:160–165

Uchida S, Tashiro T, Kawai-Uchida M et al (2014) The Adenosine A$_{2A}$-Receptor Antagonist Istradefylline Enhances the Motor Response of L-DOPA Without Worsening Dyskinesia in MPTP-Treated Common Marmosets. J Pharmacol Sci 124:480–485

Yamada K, Kobayashi M, Mori A et al (2013) Antidepressant-like activity of the adenosine A(2A) receptor antagonist, istradefylline (KW-6002), in the forced swim test and the tail suspension test in rodents. Pharmacol Biochem Behav 114–115:23–30

Yamada K, Kobayashi M, Shiozaki S et al (2014) Antidepressant activity of the adenosine A$_{2A}$ receptor antagonist, istradefylline (KW-6002) on learned helplessness in rats. Psychopharmacology 231:2839–2849

Chapter 14
Adenosinergic Receptor Antagonists: Clinical Experience in Parkinson's Disease

Emmanuelle Pourcher and Robert A. Hauser

Abstract In the management of Parkinson's disease, chronic L-DOPA therapy is associated with the development of motor fluctuations and dyskinesias. Since L-DOPA is the most effective antiparkinsonian medication currently available, adjunctive medications that reduce or prevent fluctuations and dyskinesias could be of great value.

Adenosine 2A (A_{2A}) antagonists provide antiparkinsonian benefit by reducing overfiring of striatopallidal neurons. In animal models, A_{2A} antagonists have been demonstrated to provide antiparkinson benefit as monotherapy and as adjuncts to L-DOPA. In L-DOPA-primed primates, addition of an A_{2A} antagonist to a lower dose of L-DOPA allows maintenance of the antiparkinsonian response with less dyskinesia.

Three A_{2A} antagonists (istradefylline, preladenant, and tozadenant) have demonstrated efficacy in Phase II clinical trials of PD patients, reducing OFF time in patients with motor fluctuations on L-DOPA. However, preladenant failed in Phase III, and istradefylline yielded mixed results. None of the A_{2A} antagonist were to shown to be effective as monotherapy in early PD.

It seems likely that A_{2A} antagonists are efficacious as adjuncts to L-DOPA in fluctuating patients as evidenced by results of Phase II trials, but clinical trial methodologic difficulties have made it a challenge to demonstrate their efficacy in Phase III trials when larger populations, more sites, more investigators/raters are involved.

Keywords A_{2A} antagonists · Istradefylline · Preladenant · Tozadenant · Parkinson's disease · Treatment · Clinical trials

R. A. Hauser (✉)
USF Byrd Parkinson's Disease and Movement Disorders Center, Byrd Institute, University of South Florida, Tampa, USA
e-mail: rhauser@health.usf.edu

E. Pourcher
Faculty of Medicine, Clinique Sainte Anne Mémoire et Mouvement, Laval University, Quebec, QC, Canada

© Springer International Publishing Switzerland 2015

M. Morelli et al. (eds.), *The Adenosinergic System,* Current Topics in Neurotoxicity 10, DOI 10.1007/978-3-319-20273-0_14

Introduction

Pharmacological strategies targeting dopamine (DA) for the control of motor symptoms in Parkinson's disease (PD) include L-DOPA combined with DOPA-decarboxylase inhibitors, dopaminergic agonists, catechol-0-methyl transferase (COMT) inhibitors, and monoamine oxidase (MAO) B inhibitors. Pivotal studies and clinical experience have established their benefits and limitations, with the latter consisting essentially of the difficulties with any combination therapy in avoiding progressive pharmacodynamic changes in nigral and extra-nigral dopaminergic pathways. This can lead to fluctuations in motor performance occurring several times throughout the day, involuntary abnormal movements, which may be observed at rest or in overflow, and development of fluctuations of non-motor symptoms such as anxiety pain, or attention deficit.

With better knowledge of the anatomo-chemical complexity of the striatal targets of dopaminergic afferents, it seemed logical to explore L-DOPA sparing strategies involving non-dopaminergic drugs capable of downstream modulation of consequences resulting from non-physiological dopaminergic input.

Adenosine antagonists have been suggested as potential agents to treat PD since the 1990s (Kostic et al. 1999; Schwarzchild et al. 2002). However, clinical trials of non-selective antagonists, such as caffeine and theophylline, were not judged positive enough to be pursued.

The A_{2A} receptor, one of the four G-protein coupled adenosine receptor subtypes (A_1, A_{2A}, A_{2B} and A_3), has emerged as an attractive target for PD therapy for several reasons: its selective localization in the basal ganglia, its co-localization with DA D_2 receptors in heterodimers with opposing actions, and its highly selective expression on dendritic spines of GABAergic medium, spiny neurons of the indirect pathway. (Latini et al. 1996; Martinez-Mir et al. 1991; Mori et al. 1996)

In theory, such properties could give A_{2A} antagonists the ability to correct the effects of D_2 blockade on locomotion without systemic side effects (such as cardiovascular stimulation). It could also counter the imbalance between the direct (mainly D_1) and the indirect (mainly D_2) basal ganglia output pathways observed in DA-depleted Parkinsonian animals exposed to chronic L-DOPA. This imbalance is thought to be partly responsible for the motor complications of PD. These premises were favourably verified in the 6-hydroxydopamine (6-OHDA) rodent model and in the 1-methyl-4-phenyl-1,2,3,6-tetrahydropyridine (MPTP) monkey model (Grondin et al. 1999; Kanda et al. 2000; Morelli and Pinna 2001).

In support of this rationale, genetic inactivation of A_{2A} receptors in 6-OHDA hemi-parkinsonian mice, attenuates the repeated L-DOPA induced behavioral sensitization (i.e. increased contra-lateral rotation, increased grooming activity) and the observed reduction in dynorphin mRNA expression observed in the direct pathway (Chen et al. 2003). It also diminishes the increased locomotor response that is progressively observed as a result of repeated administration of amphetamine (Chen et al. 2003).

In experiments conducted by Bibbiani et al. (2003), co-administration of the A_{2A} antagonist KW-6002 (Istradefyllline) with L-DOPA blocked the progressive shortening of motor response duration in the hemi-parkinsonian rat model. However, in a study reported by another group using the same model, with the same paradigm, the dyskinetic response was not prevented (Lundblad et al. 2003).

Finally, there is potentially important evidence that A_{2A} receptors are upregulated in the brains of PD patients with dyskinesia compared to those without, in both post-mortem studies (Calon et al. 2004) and positron emission tomography (PET) imaging studies (Mishina et al. 2011; Ramlackhansingh et al. 2011).

Pre-clinical evidence to suggest that A_{2A} antagonists prevent the development of dyskinesia is scarce. In Bibbiani et al.'s experiments with Cynomolgus monkeys, a preventive effect on the development of dyskinesia was observed with the repeated administration of KW-6002 with apomorphine; L-DOPA was not studied. In critique of a paper by Wills et al. (2013) on the potential preventive effects of caffeine on the risk of dyskinesia in PD, which was based on a post-hoc analysis of the CALM–PD population, Jenner (2013) noted an unpublished study in which his group was unable to demonstrate a preventative effect of coadministration of KW-6002 with high doses of L-DOPA in the marmoset monkey.

During the last two decades, the pharmacological properties of several A_{2A} antagonists have been delineated (Yang et al. 2007). Some followed preclinical and clinical steps to Phase IIB and Phase III trials. Istradefylline (KW-6002) is the most advanced in development; it was recently approved by the Japanese Health Authorities under the brand name NOURIAST® for the treatment of PD. It is now being studied in a large global Phase III trial. Preladenant (SCH 420814) demonstrated efficacy in Phase II (Hauser et al. 2011a) but failed in Phase III and is no longer being developed as a treatment for PD. Tozadenant (SYN 115) was recently reported efficacious as an adjunct to L-DOPA in PD patients with fluctuations in a Phase II trial and Phase III trials are planned. Vipadenant (B11B014) has pursued development to Phase II trials in early PD, and as an adjunct to L-DOPA in moderate to late stages, but results of these studies have not been published and further development of Vipadenant for PD is currently not planned.

The following summary of clinical studies published to date reflects the earlier development of Istradefylline.

Istradefylline (IST, KW-6002)

Pharmacology

IST is a purine compound selectively antagonizing the A_{2A} receptor with a binding affinity (Ki value) of 12 nmol/L for the human A_{2A} receptor; it is a hundred times more selective for A_{2A} receptors than for other adenosinergic receptors. It has low or no affinity for other major neurotransmitters including DA, serotonin (5HT) and norepinephrine (NE) receptors. It has no inhibitory activity against MAO–A, MAO–

B or COMT. IST exhibits a moderate rate of absorption in healthy volunteers (Tmax: 2–5 h) and a slow elimination half-life (average half-life across several Phase I studies was 70–118 h) allowing for once daily administration (Yang et al. 2007).

Increased systemic exposure may occur following co-administration with potent CYP 3 A inhibitors such as ketoconazole. IST is devoid of pharmacokinetic interactions with L-DOPA and carbidopa. IST has no effect on the QT_c interval at dosing regimens up to 240 mg/day. Selectivity of IST binding for striatal A_{2A} receptors has been confirmed by PET-analysis (Brooks et al. 2008).

Clinical Efficacy and Safety

IST as Monotherapy

In disagreement with most pre-clinical studies, Istradefylline, as an antiparkinsonian agent on its own, did not demonstrate antiparkinsonian efficacy. In a 12-week, double-blind, placebo-controlled study, Fernandez et al. (2010) investigated the safety and efficacy of IST 40 mg/day as monotherapy in 176 PD patients (average age 63 years, average disease duration 15 months, never treated with L-DOPA or treated for no more than 4 weeks at any time). While safe and well tolerated, IST failed to separate from placebo for the primary endpoint of change in the Unified Parkinson's Disease Rating Scale (UPDRS) motor score after 12 weeks.

IST as an Adjunct to L-DOPA

IST has mainly been studied as an adjunct to L-DOPA and other established dopaminergic medications in moderate to advanced PD patients, notably those with motor fluctuations and established dyskinesia.

In a small proof-of-concept, double-blind, placebo-controlled, study of 15 patients at the NINDS, Bara-Jimenez et al. (2003) explored in a 6-week dose–escalating design (40 and 80 mg/day) the motor effects of IST in combination with a steady-state L-DOPA infusion, dose-optimized to each individual patient.

IST had no effect as monotherapy and did not add significant benefit to the motor effects of the highest optimal dose of L-DOPA infusion. Following withdrawal of the L-DOPA infusion, 80 mg/day KW-6002 prolonged the L-DOPA effective half-time (i.e. time for UPDRS motor score to decline by 50 %) by an average of 47 min (76 %; $p < 0.05$) in the 10 patients for whom data are available. In addition at a low-dose steady state L-DOPA infusion, the motor effect of 80 mg/day potentiated the antiparkinsonian response by 36 % ($p < 0.02$), but with 45 % less dyskinesia compared to that induced by the optimal dose of L-DOPA infusion alone ($p < 0.05$). Resting tremor was abated by 72 % ($p < 0.02$), rigidity by 43 % ($p < 0.01$) and bradykinesia by 38 % ($p < 0.05$). The addition of IST allowed similar antiparkinsonian response at a lower L-DOPA dose with less dyskinesia, consistent with results obtained in the MPTP primate.

Further Phase II clinical studies have analysed the efficacy and safety of IST in randomized controlled trials (RCT) of 12 weeks duration in similar populations (i.e. fluctuations and peak-dose dyskinesia).

In these studies, the primary outcome was usually assessed using a patient completed PD diary (Hauser et al. 2004) to evaluate change from baseline to endpoint at 12 weeks. In this diary, the patient must perform a self-assessment, every half-hour during waking hours, and categorize themselves into one of 4 categories: OFF—time when medication has worn off and is no longer providing benefit with regard to mobility, slowness, and stiffness. ON—time when medication is providing benefit with regard to mobility, slowness, and stiffness; ON with non-troublesome dyskinesia (involuntary twisting, turning movements are present but do not interfere with function or cause meaningful discomfort), or ON with troublesome dyskinesia (involuntary twisting, turning movements are present and interfere with function or cause meaningful discomfort).

It is important to note the subjective quality of this classification, especially for dyskinesia; peak-dose mild choreiform dyskinesias are frequently visible to the observer or caregiver, when the patient himself is unaware of them. In addition, peak-dose dyskinesia, may be seen as moderate to severe by an observer but could accurately be rated as non-troublesome by the patient if these movements are not interfering with function or causing discomfort. Trials that employ this methodology usually include training for patients to be able to recognize these states and accurately complete the diary. In addition, "concordance" testing is usually undertaken to be sure the patient has observable motor fluctuations and truly understands the PD states.

The KW-6002-US-001 study reported by Hauser et al. (2003) compared two dose-escalating groups, 5–10–20 mg/day and 10–20–40 mg/day, through weeks 1–4, 5–8, 9–12, respectively, with placebo. Overall, approximately 28 patients were randomized to each group with 22 completing the study.

Subjects assigned to IST exhibited a significant reduction in OFF time compared to placebo as observed by home diaries ($-7.1 \pm 2\%$ versus $+2.2 \pm 2.7\%$ $p=0.008$). In both treatment groups, this reduction was statistically significant. Expressed in terms of hours, the reduction of OFF time was 1.2 ± 0.3 h for the combined IST groups versus an increase of 0.5 ± 0.5 h in the placebo group ($p=0.0004$).

Assessment of the percent and hours of reduction in OFF time during an 8-hour in-office evaluation by investigators confirmed a trend for greater reduction in the combined IST groups versus placebo ($10.0 \pm 2\%$ versus $3.3 \pm 2.8\%$ ($p=0.05$), 0.8 ± 0.2 h versus 0.3 ± 0.2 h ($p=0.06$) for duration.

Overall IST was well tolerated, with the following adverse events noted in more than 5% of patients: nausea, mainly observed during upward titration and resolving in most cases after 10 days, worsening dyskinesia, dizziness, vomiting and insomnia. Only one serious adverse event, a fatal myocardial infarction, in the IST group was interpreted as potentially related to the experimental drug. Vital signs, electrocardiogram and laboratory values, except for an elevated lipase in seven patients on IST compared to zero patients on placebo, were unchanged.

As for dyskinesia, both the diaries and 8-hour in-office evaluations showed an increase in ON time with dyskinesia, although the severity, as observed by different scales such as the UPDRS part IV and a modified Abnormal Involuntary Movement Scale, was not significantly different.

Two larger North American, multicenter, Phase IIB BRCTs with very similar designs have also explored reductions in OFF time.

The 6002-US-005 study reported by Le Witt et al. (2008) compared 40 mg/day IST to placebo. The population included 129 patients on IST and 66 patients on placebo (safety data set). A total of 114 patients on IST and 58 on placebo completed the study. Patients had to have at least 2 h OFF time at baseline to be included.

Change in the percentage of OFF time from baseline to study endpoint, which was the predefined primary endpoint, as well as change in the number of waking hours spent in the OFF state significantly favoured IST: $-10.8 \pm 16.6\%$ versus $-4.0 \pm 15.7\%$ ($p=0.007$) and -1.7 ± 2.7 h versus -0.6 ± 2.7 h ($p=0.006$, respectively. This corresponds to a reduction in daily OFF time of 28% for IST and 10% for placebo.

ON time without dyskinesia showed a small increase of 15 min over placebo in the IST group, which was not significant. ON time without troublesome dyskinesia (i.e. the combination of ON time without dyskinesia and ON time with non troublesome dyskinesia) showed an increase of 0.96 h over placebo which was significant ($p=0,026$). Improvement in Clinical Global Impression (CGI) scores achieved significance for IST at 4 and 8 weeks, but not at 12 weeks (53.5% for IST versus 40.9% for placebo).

Regarding adverse effects, the most frequently reported event was mild to moderate dyskinesia, occurring in 30.2% of IST patients compared to 15.2% for placebo. The percentage of subjects discontinuing the study due to adverse effects was similar in both groups at around 7%.

The 6002US-006 study reported by Stacy et al. (2008), compared the efficacy and safety of 20 mg/day and 60 mg/day IST to placebo in a similar population, the baseline demographics and disease characteristics of which were, on average, the following: 64 years of age, 60% male, 9 years disease duration, and a 3 year history of motor complications. On average, 90% of patients were on combinations of L-DOPA and DA agonists, 40% used entacapone, 30% used amantadine, and 15% were users of selegiline. Patients averaged 6 h OFF at baseline, 3 h ON with dyskinesia, 0.5 to 1 h ON with troublesome dyskinesia and scored an average of 17 on the UPDRS III in the ON state. This population was not different from the US-005 study in which 40 mg/day IST was compared to placebo.

This study included 163 patients assigned to 20 mg/day, 155 to 60 mg/r day and 77 to placebo, of which 152, 126 and 69, respectively, completed the study.

Change in the percentage of OFF time from baseline to study end was as follows: -7.83% for IST 20 mg/day (95% confidence interval (CI): -10.0 to -5.6), -7.96% for IST 60 mg/day (95% CI: -102 to -5.6), -3.47% for placebo (95% CI: -6.68 to -0.27). Compared to placebo, the difference was significant for both 20 mg/day ($p=0.026$) and for 60 mg/day ($p=0.024$).

Thus, the primary efficacy variable, the change from baseline to study endpoint in the percentage of waking hours per day spent in the OFF state, was significantly improved for both dosages. This represented a 22 % reduction in total waking hours spent in the OFF state for 20 mg/day and 24 % for 60 mg/day, compared to only a 10 % reduction for placebo. For the change in the absolute time spent in OFF, the reductions, in hours, were as follows: -1.24 h for 20 mg (95 % CI: -1.62 to -0.86), -1.37 h for 60 mg (95 % CI: -1.77 to -0.97), -0.60 h for placebo (95 % CI: -1.15 to -0.005), $p=0.065$ for overall treatment effect.

For the change in the absolute time spent in ON time without dyskinesia, the increases (in hours) were: $+0.25$ h for 20 mg/day compared to placebo, $+0.46$ h for 60 mg/day compared to placebo, but the overall treatment effect was not significant.

As for the change in the absolute time spent in ON time without troublesome dyskinesia, the increases, in hours, were: $+0.71$ h for 20 mg/day compared to placebo, $+0.60$ h for 60 mg/day compared to placebo. Again, the overall treatment effect was not significant. With respect to the CGI and UPDRS motor scores, no significant changes were observed from baseline to study endpoint.

In terms of safety and adverse events, the most frequently reported treatment-related event was dyskinesia (23.9 % in patients on 20 mg/day IST, 22.6 % in patients on 60 mg/day IST versus 14.3 % in patients on placebo), followed by nausea (20.0 %, 10.4 % and 6.5 %) and dizziness (11.0 %, 11.0 % and 6.5 %). The percentage of patients discontinuing the study drug due to adverse effects was lower in the 20 mg/day group (3.7 %) than in the 60 mg/day group (10.3 %) and the placebo group (6.5 %).

Two additional Phase III RCTs were conducted in North American centers in populations who were very similar to the US-005 and US-006 studies.

A positive trial, *the 6002 -US-013 study,* was reported by Hauser et al. (2008) and compared the efficacy of IST 20 mg per day versus placebo for 12 weeks. The ITT population included 112 patients with 104 completing the study in the IST 20 mg group and 113 with 103 completing the study in the placebo group. The mean absolute percent reduction in OFF time was 9.3 % in IST 20 mg group versus 5 % in the placebo group (between group least mean square 4.6 % (95 % CI: 0.6–8.6)), in favor of IST ($p=0.03$). This corresponds to a reduction in daily OFF time of 24 % for IST and 14 % for placebo.

Changes in UPDRS part III motor scores were significantly different in favor of IST at the 4-week evaluation, but this difference did not persist at study end. Changes in ON time with dyskinesia and troublesome dyskinesia were not different between groups. Dyskinesia as a treatment-emergent adverse event was reported in 22.6 % of patients on IST versus 12.2 % on placebo. The incidence of nausea was not different from placebo (7.8 vs 7 %).

Contrary to these findings, *the 6002-US-018 study* reported by Pourcher et al. (2012), was negative. In this very large Phase III study, 610 patients from 73 centers were randomized to receive either placebo ($n=154$; ITT: 151; 140 completed), IST 10 mg per day ($n=155$; ITT: 153; 136 completed), IST 20 mg per day ($n=149$; ITT: 148; 131 completed), or IST 40 mg per day ($n=152$; ITT: 152; 135 completed). Overall reductions in the percentage of OFF time were 7.6 % for placebo, 5.7 % for IST 10 mg, 6.1 % for IST 20 mg and 9.1 % for IST 40 mg. Reductions in the total number of hours spent in OFF were 1.4 h for placebo, 1.1 h for IST at both 10 and 20 mg, and 1.5 h for IST 40 mg.

A large placebo response occurred in this trial and could account for the negative result. There was a numerical dose ordered response in the istradefylline groups at most visits for the percent reduction in waking hours spent in the OFF state. Furthermore, there was a statistically significant overall treatment effect for the UPDRS motor subscore in the ON state for IST compared to placebo. This observation held true at each post-baseline visit and at study endpoint ($p=0.043$). The differences in least mean square for the UPDRS motor score were statistically significant at study endpoint and at weeks 2, 4, 8 and 12 for the 40 mg per day dose ($p<0.05$).

In this study, the actual change from baseline for IST 40 mg (1.5 h) was comparable to similarly designed studies (US-005, US-006, US-013 studies) of 20 mg (1.6 h), 40 mg (1.7 h) and 60 mg (1.4 h). A reduction in OFF time of 1.4 h in the placebo group is notable compared to 0.6 h in US-005, 0.6 h in US-006 and 0.9 h in US-013.

Several factors may contribute to a placebo effect, which in Parkinson's disease has been shown to correlate with a heightened release of DA, as shown by displacement of radiolabelled raclopride binding in the striatum (De La Fuente-Fernandez 2009). Goetz et al. (2008) have also reported that an increased probability of receiving active therapy, higher baseline UPDRS motor scores and more advanced disease at baseline was associated with a larger placebo response. Another hypothetical factor for discussion was the slightly higher prevalence of selegiline users in the placebo group (17.9%) compared to the IST group (11.4%). It may be possible that the amphetaminergic properties of selegiline may contribute to heightened expectancies. Finally, as a large number of participating sites were involved in the study, suboptimal data may result from the inability to optimally select and train patients entered into the trial.

Japanese Experience with IST

A further Phase IIB study, 6002-0608 (Mizuno et al. 2010), was conducted by the Japanese Istradefylline Study Group with 363 patients randomized to receive either placebo ($n=119$; 109 completed), IST 20 mg ($n=119$; 106 completed) or IST 40 mg ($n=119$; 112 completed). Study design and analysis were identical to the previously detailed North American studies, however baseline demographics differed slightly by gender (60% female) and use of MAO-B and COMT inhibitors as concurrent medications in fluctuators. To compare, selegiline was used by 50% of patients in the Japanese study versus 15% on average in the United States and Canada. Entacapone was used in 15% of patients compared to 40%, on average, in the United States and Canada.

In this 12-week study, daily OFF time was reduced from baseline by -1.3 h for 20 mg ($p=0.013$) -1.6 h for 40 mg ($p<0.001$) and by 0.66 h for placebo. UPDRS III scores were reduced by 5.7 points in both IST groups and by 3.7 points in the placebo group ($p=0.006$). The incidence of treatment-emergent dyskinesias was rather low: 8.5% of patients receiving 20 mg IST, 6.4% of patients receiving 40 mg IST and 2.5% of patients receiving placebo.

There was no difference between groups in change in time spent in ON with dyskinesia: −0.09 h for placebo, +0.14 h for 20 mg and +0.32 h for 40 mg. As for time spent in ON with troublesome dyskinesia, it was slightly, albeit significantly, increased for 40 mg versus placebo: −0.10 h for placebo, +0.007 h for 20 mg and +0.25 h for 40 mg ($p=0.011$).

Another Phase III Japanese study (KW-6002-009) with the same design was reported by the same author (Mizuno et al. 2013). This study yielded similar results, as change in OFF time was −0.23 h for placebo, −0.99 h ($p=0.003$) for IST 20 mg, and −0.96 h ($p=0.003$) for IST 40 mg. Changes were significant compared to placebo, however there were not significant between the two doses of IST.

Dyskinesia was the most frequently reported treatment-related adverse effect in both IST groups (proportions not provided). Daily ON time without troublesome dyskinesia (ON + ON with non-troublesome dyskinesia) increased significantly compared to placebo: +0.26 h for placebo, +1.09 h for IST 20 mg ($p=0.003$) and +1.08 h for IST 40 mg ($p=0.004$).

European Experience with IST

In the 6002-EU-007 study, neither IST 40 mg nor entacapone as an active comparator separated from placebo (results not reported).

Regulatory History

Current Regulatory Status

With this "portfolio" IST was not judged approvable by the FDA in 2008, for insufficient evidence of efficacy. It has been however approved by Japanese regulatory authorities in 2013. A further large international Phase III study is presently ongoing with the same design, in the same type of population, comparing 20 mg, 40 mg and PBO, to revalidate previous positive results (Kyowa Hakko Kirin 2013).

Critical Summary of Istradefylline Studies

The development of IST for PD has been targeted mainly at reducing OFF time as an adjunct to L-DOPA and other dopaminergic medications in patients with established motor fluctuations (many of whom also had dyskinesia). More than 2000 patients were exposed to Istradefylline in doses of 20–80 mg/day. The average population had around a 9-year duration of illness and about 6 h of OFF time at baseline. In such a population, the clinician often faces the dilemma of treating OFF periods with more frequent dosing or fractionation of L-DOPA, or adding DA agonists, COMT or MAO-B inhibitors, often at the expense of increased dyskinesia, psychomotor stimulation and sleep fragmentation. IST appears to be safe and well tolerated, with some mild dose-related adverse effects, such as nausea, to which a tolerance develops, and dyskinesia and dizziness, the incidence of which plateaus at 40 mg per day. However, the probable efficacy of IST may be to reduce OFF time about 1 h compared to placebo, and this is very similar to the results obtained with available drugs assessed under the same conditions. Expectations were high based on preclinical studies, touting A_{2A} antagonists as potentially non-dyskinesigenic. However, it does appear that A_{2A} inhibitors do not reduce dyskinesia when added to an ongoing regimen and can increase it (Morelli et al. 2012). Nonetheless, since existing therapies are only partially successful in reducing ON time, additional therapies would be welcome.

Preladenant

Preladenant was the second adenosine A_{2A} antagonist to be developed for the indication of Parkinson's disease. Phase I and II studies were conducted by Schering-Plough prior to the company merging with Merck.

Pharmacology

Preladenant is a non-methylxanthine, selective, competitive antagonist of the human A_{2A} receptor with an affinity constant k_i of 0.9 nmol/L. It is more than 1000-fold more selective for A_{2A} than for the three other subtypes of adenosine receptors (A_1, A_{2B} and A_3) and various other aminergic receptors and ion channels (Neustadt et al. 2007). It has no inhibitory effect on MAO or COMT. Time to maximum concentration is about 1 h and the effective half-life is about 8 h, resulting in a twice-daily regimen.

Convincing preclinical evidence of antiparkinsonian activity comes from several models (haloperidol induced catalepsy, 6-OHDA rodent model) in different species. In the Cynomolgus monkey, rendered parkinsonian by the toxin MPTP and primed to develop dyskinesia by repetitive dosing of L-DOPA, preladenant monotherapy provided significant anti-parkinsonian activity at 3 mg/kg, equivalent to the effect of L-DOPA at 6 mg/kg, without inducing dyskinesia (Hodgson et al. 2009). When given as an adjunct to a subthreshold dose of 3 mg/kg of L-DOPA, the effect observed on motility was equivalent to the anti-parkinsonian effect of 6 mg/kg of L-DOPA without the typical level of dyskinesia (Hodgson et al. 2010). Interestingly, the same author reported that preladenant at 1 mg/kg, given repeatedly with L-DOPA, potently attenuated behavioral sensitization in the 6-OHDA rodent model, suggesting a reduced risk of progressive amplification of the motor response with time.

In Phase I studies, the most common adverse events in healthy volunteers were insomnia (26%), headache (14%), dizziness (10%), and nausea (2%), with some cases of elevated transaminases at doses higher than 25 mg/day. This last observation, coupled with cases of hepatic toxicity in dogs at doses of 100 mg/kg, heightened the level of monitoring in the of early Phase II studies, however this initial concern was eventually disproven as a signal did not arise from the cumulative safety data in humans (Cutler et al. 2012).

Cardiovascular safety was also carefully evaluated in two randomized placebo-controlled studies in healthy volunteers, as a transient increase in blood pressure was observed within a few hours of preladenant intake. In a single dose escalating study from 5 to 200 mg, as well as with repeated doses escalating from 10 to 200 mg once daily over 10 days, transient increases in blood pressure were neither cumulative nor dose related and were not maintained over time (Cutler et al. 2012).

Clinical Efficacy

A single Phase II, double-blind, randomized dose-finding trial (Hauser et al. 2011b) and its extension are the only preladenant studies to be published to date, although four Phase III RCTs have been registered.

Patients were eligible for the Phase II trial if they had 2 or more hours of OFF time per day according to three consecutive daily diaries. Key exclusion criteria included a history of hepatic dysfunction or increased liver enzymes or elevated blood pressure at baseline (systolic ≥ 180 mmHg or diastolic ≥ 105 mmHg). To start, patients were randomized in a 1:1:1:1 ratio to 1, 2, or 5 mg of preladenant or matching placebo twice daily. An independent data safety monitoring board conducted a preplanned interim analysis once 40 patients had completed the initial randomization. Authorization was then given to randomize at a 1:1:1:2:1 ratio to 1, 2, 5 or 10 mg of preladenant twice daily or matching placebo in order to create a 10 mg twice daily arm. The primary efficacy outcome was the change in mean daily OFF time from baseline to 12 weeks. A total of 253 patients were randomized and 245 received at least one dose of study medication and comprised the safety data set; the last observation carried forward (LOCF) was used for 54 (23 %) of 234 patients after exclusion of 12 for missing data.

Compared to placebo, mean daily OFF time was significantly reduced at 12 weeks in the Preladenant 5 and 10 mg twice-daily groups with a mean difference of -1.4 h ($p=0.011$) and -1.2 h ($p=0.040$), respectively. Overall, although LOCF was used in nearly 25 % of patients, week 12 results for completers were similar to LOCF results for all treatment groups except for the 1 mg group, whose average decrease in OFF time at 12 weeks was greater in those completing the study.

Post-hoc analyses were conducted to assess change in ON time, dyskinesia and UPDRS scores. No statistical adjustments were made for multiplicity in the analyses of these secondary outcome variables, but several results of interest were revealed. A significant increase in mean daily ON time between baseline and week 12 was observed in the 5 and 10 mg groups compared to placebo. No significant differences were revealed in ON time without dyskinesia, ON time with troublesome dyskinesia or the distribution of patients across dyskinesia severity categories.

As per the UPDRS part I, preladenant 5 and 10 mg twice daily were found to be significantly superior to placebo on the motivation/initiative item ($p = 0.01$); there was only a trend towards superiority in improving depression and thought disorders (both $p = 0.08$) compared to placebo.

As for UPDRS part III motor scores, which were obtained without regard to timing of L-DOPA intake, no significant improvement over baseline was observed. Epworth sleepiness scores were also decreased, however this was not significant in any group.

Preladenant was safe and well tolerated as adverse events, including worsening of dyskinesia, nausea and insomnia were no more frequent than in the placebo group, with the exception of dizziness. Adverse events leading to discontinuation were similar across treatment groups. Mild transient increases in systolic and diastolic blood pressure were considered clinically insignificant; increased liver enzymes greater than three times the upper limit of normal were not observed.

Long-term safety data are available for 140 patients who elected to enroll in a 36-week open-label extension (OLE); 106 patients (76%) completed the 36 weeks.

Comparing week 36 of the OLE phase to week 12 of the RCT, there was a lower incidence of nausea (6 vs 13%), a higher incidence of constipation (19 vs 13%) and a higher incidence of dyskinesia (33 vs 9%). Increased incidence of dyskinesia as an adverse event could be related to the longer observation period, adjustments in dopaminergic treatment or advancing disease. Laboratory blood tests and blood pressures did not reveal new long-term risks.

Critical Summary of Preladenant Studies

Preladenant, as a more selective and with a higher affinity binding constant than istradefylline and a favourable preclinical profile, was explored as L-DOPA potentiating agent with great hopes. A rigorously designed Phase II RCT confirmed that it is a safe, well-tolerated drug, significantly reducing OFF time at both 5 and 10 mg twice daily, without increasing ON time spent with troublesome dyskinesia and without increased dyskinesia as a treatment-emergent adverse event over 12 weeks. At first glance, the lack of effect on the incidence of dyskinesia seems to be different compared to istradefylline but further positive trials would be required to determine if this would be a consistent finding. A similar Phase II trial has been conducted in Japan (NCT 01294800) and two Phase III trials have been registered (NCT 01155466, NCT 01227265) with the same design, one with rasagiline as an active comparator. However, Merck is no longer actively pursuing development of preladenant as a treatment for PD since the positive Phase II results were not replicated in 2 Phase III trials (Merck—update—Phase III Clinical program—Preladenant—2013)

Tozadenant

Tozadenant (SYN 115) was developed first by Hoffmann-La Roche and then by Biotie Therapies Inc.

It is the third A_{2A} antagonist to provide Phase IIB results in a large number of PD patients.

Pharmacology

SYN 115 is a non-xanthine competitive antagonist of A_{2A} adenosinergic receptors with a binding affinity (K_1 value) of 11.5 nmol/L for the human A_{2A} receptor. Tozadenant (TOZ) is rapidly absorbed, reaching C_{max} between 1 and 5 h after oral administration with a terminal half-life of approximately 15 h from plasma, theoretically allowing once daily dosing. In healthy, older volunteers, steady state plasma concentrations were reached within 7 days with an approximative 2-fold accumulation over time between 7 and 14 days. Exposure is proportional to the increase in dose using a twice-daily regimen, while it is less than proportional using once daily dosing, suggesting a possible saturation of TOZ absorption (beyond approximately 240 mg per dose).

Clinical Summary of Studies

Cardiovascular safety of a 100 mg dose was assessed in a Phase I study with the primary aim of evaluating the psycho-stimulant properties of TOZ in 22 cocaine-dependent subjects. There was no difference in blood pressure compared to placebo (Lane et al. 2012). However, similar to preladenant, a potential for increasing systolic and diastolic blood pressure on initiation of treatment was observed, but this effect is subject to tachyphylaxis with continued dosing beyond 14 days.

The potential to regulate the excessive indirect striato-pallidal pathway activity in PD was explored with a perfusion magnetic resonance imaging study in 21 PD patients on a stable infusion of L-DOPA, but no agonists, in a double-blind placebo-controlled crossover trial (Black et al. 2010). TOZ produced a highly significant dose-dependent decrease in thalamic cerebral blood flow, consistent with reduced thalamic inhibition and correction of the "Parkinsonian network" (Eidelberg et al. 1997). Tapping speed was faster on 60 mg twice daily of TOZ than on placebo before (5%; $p=0.03$) and during (6%; $p=0.02$) an intravenous L-DOPA infusion, revealing a potential to further augment the anti-parkinsonian effect of L-DOPA.

An international, multicenter Phase IIB randomized, placebo-controlled, parallel-group, dose-finding, clinical trial of tozadenant was conducted in L-DOPA treated patients experiencing *wearing off* phenomenon and more than 2.5 h of OFF time as assessed for at least 2 days prior to study entry (Hauser et al. 2014).

The primary endpoint was change from baseline to 12 weeks in the time spent in the OFF state. Population, patient selection, tools and statistical analysis on primary and secondary endpoints were very similar to other Phase IIB studies of patients with motor fluctuations discussed above for other A_{2A} antagonists. Baseline demographics and disease characteristics were as follow: average age of 68 years, disease duration of 8.7 years, duration of L-DOPA exposure of 6.7 years, time to onset of motor complications averaging 2.4 years, UPDRS III motor score of 21.7 in ON state. A total of 420 patients from 76 centers in 6 countries (United States, Canada, Argentina, Chile, Romania and Ukraine) were randomized to five groups: 84 patients to placebo (74 completed), 85 patients to tozadenant 60 mg twice daily (74 completed; 7 early withdrawals for adverse events), 82 patients to TOZ 120 mg twice daily (65 completed; 10 early withdrawals fore adverse events), 85 patients to TOZ 180 mg twice daily (65 completed; 10 early withdrawals for adverse events), 84 patients to TOZ 240 mg twice daily(58 completed; 17 early withdrawals for adverse events).

The primary endpoint of reduction from baseline to week 12 in OFF time was significantly reduced with TOZ 120 mg twice daily, 180 mg twice daily, and the combined group of 120 and 180 mg twice daily compared to placebo: (120 mg: -1.1 h (95% CI: -1.8 to 0.4; $p=0.0039$), 180 mg: -1.2 h (95% CI: -1.9 to 0.4; $p=0.0039$), 120 mg and 180 mg combined: -1.1 h (95% CI: -1.8 to 0.5; $p=0.0006$).

Several secondary endpoints also significantly favoured tozadenant. Mean daily ON time was significantly increased in the combined 120–180 mg group and in the 120 mg group. In addition, ON time without troublesome dyskinesia at 12 weeks was significantly increased compared to placebo for the 120 mg group, and a trend towards superiority was observed in the combined 120–180 mg group. Finally, UPRDRS part III motor score in the ON state was significantly reduced in the 120 mg group: -3.2 versus -0.9 ($p=0,0325$), the 180 mg group: -3.5 versus -0.9 ($p=0.0325$) and the combined group: -3.3 versus -0.9 ($p=0.0081$). CGI-severity and CGI-improvement scores were significantly improved compared to placebo in all tozadenant groups. Patient global impression (improvement) scores were significantly improved compared to placebo in the combined 120–180 mg and 120 mg groups.

Treatment emergent adverse effects were very similar to those observed with istradefylline and preladenant and appeared to be dose related. The respective incidence of the most common adverse events observed on placebo, TOZ 120 mg b.i.d. and TOZ 180 mg b.i.d. were dyskinesia (8.3, 15.9, 20.0%), nausea (3.6, 11.0, 11.8%), dizziness (1.2, 4.9, 12.9%).

In summary, a daily dose of tozadenant 120 or 180 mg twice daily demonstrated efficacy and tolerability. 60 mg twice daily did not provide a significant abatement of OFF time and 240 mg twice daily was associated with an excessive discontinuation rate due to adverse events (20.2%).

Summary of Clinical Trial Experience with A_{2A} Antagonists in Moderately Advanced PD Patients with Motor Fluctuations

There has been remarkable consistency in results of Phase II clinical trials of A_{2A} antagonists as adjuncts to L-DOPA in PD patients with motor fluctuations. Phase II trials for istradefylline, preladenant, and tozadenant all showed about a 1 h reduction in OFF time. This reduction is consistent with what is considered clinically relevant (Hauser et al. 2011b; Schrag et al. 2006) and similar to what is achieved with MAO-B inhibitors and entacapone. However, istradefylline yielded mixed results in Phase III and preladenant failed. It seems that the most likely explanation for difficulty in these Phase III studies relates to methodologic and operational issues when going from smaller studies performed at expert centers to large studies conducted at many sites with varying degrees of expertise in PD and clinical trials.

The abundant preclinical literature promoting A_{2A} antagonists as potential non-dyskinesigenic anti-parkinsonian medications probably aroused unfounded expectations as a mild increase in dyskinesia has been seen in most fluctuator studies that also yielded a significant reduction in OFF time. In part, some of the variability in results pertaining to dyskinesia may stem from the fact that fluctuator trials include variable proportions of patients with dyskinesia or with troublesome dyskinesia at baseline.

To date, no well controlled large clinical trial has evaluated the possibility that the addition of an A_{2A} antagonist coupled with lowering of the L-DOPA dose might maintain antiparkinsonian activity while reducing dyskinesia as was suggested by the NIH study that employed L-DOPA intravenous infusions at variable doses (Bibbiani et al. 2003).

Two clinical trials of A_{2A} antagonists as monotherapy in early PD have been negative so it seems unlikely that they have efficacy in that clinical situation. Although preclinical data is mixed, the possibility that they might reduce the development of dyskinesia if introduced at the same time L-DOPA is initiated remains intriguing, if untested. Further exploration of A_{2A} antagonists for non-motor symptoms such as mood and motivation, which are partly DA dependent, appears justified considering experimental evidence (Cunha et al. 2008; Kadowaki Horita et al. 2013; Yamada et al. 2013) and the necessity for a clearer differentiation from available "anti-OFF" drugs.

References

Bara-Jimenez W, Sherzai A, Dimitrova T et al (2003) Adenosine A(2 A) receptor antagonist treatment of Parkinson's disease. Neurology 61:293–296

Bibbiani F, Oh JD, Petzer JP et al (2003) A2A antagonist prevents dopamine agonist-induced motor complications in animal models of Parkinson's disease. Exp Neurol 184:285–294

Black KJ, Koller JM, Campbell MC et al (2010) Quantification of indirect pathway inhibition by the adenosine A2a antagonist SYN115 in Parkinson's disease. J Neurosci 30:16284–16292

Brooks DJ, Doder M, Osman S et al (2008) Positron emission tomography analysis of [11C]KW-6002 binding to human and rat adenosine A2A receptors in the brain. Synapse 62:671–681

Calon F, Dridi M, Hornykiewicz O et al (2004) Increased adenosine A2A receptors in the brain of Parkinson's disease patients with dyskinesias. Brain 127:1075–1084

Chen JF, Freduzzi S, Bastia E et al (2003) Adenosine A2 receptors in neuroadaptation to repeated dopaminergic stimulation. Neurology 61:574–581

Chen Y, Choi JK, Jenkins BG (2005) Mapping interactions between dopamine and adenosine A2a receptors using pharmacologic MRI. Synapse 55:80–88

Cunha RA, Ferré S, Vaugeois JM et al (2008) Potential therapeutic interest of adenosine A2A receptors in psychiatric disorders. Curr Pharm Des 14:1512–1524

Cutler DL, Tendolkar A, Grachev ID (2012) Safety, tolerability and pharmacokinetics after single and multiple doses of preladenant (SCH420814) administered in healthy subjects. J Clin Pharm Ther 37:578–587

Eidelberg D, Moeller JR, Kazumata K et al (1997) Metabolic correlates of pallidal neuronal activity in Parkinson's disease. Brain 120:1315–1324

Fernandez HH, Greeley DR, Zweig RM et al (2010) Istradefylline as monotherapy for Parkinson's disease: results of the 6002-US-051 trial. Parkinsonism Relat Disord 16:16–20

De la Fuente-Fernandez R (2009) The placebo-reward hypothesis: dopamine and the placebo effect. Parkinsonism Relat Disord 15:S72–S74

Goetz CG, Wuu J, McDermott MP et al (2008) Placebo response in Parkinson's disease: comparisons among 11 trials covering medical and surgical interventions. Mov Disord 23: 690–699

Grondin R, Bédard PJ, HadjTahar A et al (1999) Antiparkinsonian effect of a new selective adenosine A2A receptor antagonist in MPTP-treated monkeys. Neurology 52:1673–1677

Hakko KK (2013) Pharma initiates a global phase III trial of Istradefylline (KW-6002) for Parkinson's disease. 2013 Nov 21. http://www.Kyowa-Kirin.com/news_releases/2013/e2013ll21_01.html. Accesses 19 March 2014

Hauser RA, Hubble JP, Truong DD et al (2003) Randomized trial of the adenosine A(2 A) receptor antagonist istradefylline in advanced PD. Neurology 61:297–303

Hauser RA, Deckers F, Lehert P (2004) Parkinson's disease home diary: further validation and implications for clinical trials. Mov Disord 19:1409–1413

Hauser RA, Shulman LM, Trugman JM et al (2008) Study of istradefylline in patients with Parkinson's disease on levodopa with motor fluctuations. Mov Disord 23:2177–2185

Hauser RA, Cantillon M, Pourcher E et al (2011a) Preladenant in patients with Parkinson's disease and motor fluctuations: a phase 2, double-blind, randomised trial. Lancet Neurol 10:221–229

Hauser RA, Avinger P, Parkinson Study group (2011b) Determination of minimal clinically important change in early and advanced Parkinson's disease. Mov Disord 2011:813–818

Hauser RA, Olanow CW, Keiburtz KD et al (2014) Tozadenant (SYN115) in patients with Parkinson's disease who have motor fluctuations on levodopa: a phase 2 b, double blind randomized trial. Lancet Neurol 13:767–776

Hodgson RA, Bertorelli R, Varty GB et al (2009) Characterization of the potent and highly selective A2A receptor antagonists preladenant and SCH 412348 [7-[2-[4–2,4-difluorophenyl]–1-piperazinyl]ethyl]– 2-(2-funaryl)-7 H-pyrazolo[4,3-e][1,2,4]triazolo[1,5-c] pyrmidin-5 amine] in rodent models of movement disorders and depression. J Pharmacol Exp Ther 330:294–303

Hodgson RA, Bedard PJ, Varty GB et al (2010) Preladenant, a selective A(2 A) receptor antagonist, is active in primate models of movement disorders. Exp Neurol 225:384–390

Jenner P (2013) A cup of coffee a day keeps dyskinesia away? Mov Disord 28:26–27

Kadowaki Horita T, Kobayashi M, Mori A et al (2013) Effects of the A2A antagonist istradefylline on cognitive performance in rats with 6-OHDA lesion in prefrontal cortex. Psychopharmacology 230:345–352

Kanda T, Jackson MJ, Smith LA et al (2000) Combined use of the adenosine A(2 A) antagonist KW-6002 with L-Dopa or with selective D1 or D2 dopamine agonists increases antiparkinsonian activity but not dyskinesia in MPTP-treated monkeys. Exp Neurol 162:321–327

Kostic VS, Svetel M, Sternic N et al (1999) Theophylline increases "on" time in advanced parkinsonian patients. Neurology 52:1916

Lane S, Green C, Steinberg J et al (2012) Cardiovascular and subjective effects of the novel adenosine A(2 A) receptor antagonist SYN115 in cocaine dependent individuals. J Addict Res Ther 28; S1. pii 009

Latini S, Pazzagli M, Pepeu G et al (1996) A2 adenosine receptors: their presence and neuromodulatory role in the central nervous system. Gen Pharmacol 27:925–933

LeWitt PA, Guttman M, Tetrud JW et al (2008) Adenosine A2A receptor antagonist istradefylline (KW-6002) reduces "off" time in Parkinson's disease: a double-blind, randomized, multicenter clinical trial (6002-US-005). Ann Neurol 63:295–302

Lundblad M, Vaudano E, Cenci MA (2003) Cellular and behavioural effects of the adenosine A2a receptor antagonist KW-6002 in a rat model of I-DOPA-induced dyskinesia. J Neurochem 84:1398–1410

Martinez-Mir MI, Probst A, Palacios JM (1991) Adenosine A2 receptors: selective localization in the human basal ganglia and alterations with disease. Neuroscience 42:697–706

Merck (2013) Merck provides update on phase III clinical program for preladenant the company's investigational Parkinson's disease medicine. http://www.business.com/news/home/20130523006358/en/Merck-update-PhaseIII-clinical-program-Preladenant. Accessed 21 May 2013)

Mishina M, Ishiwata K, Naganawa M et al (2011) Adenosine A(2 A) receptors measured with [C] TMSX PET in the striata of Parkinson's disease patients. PLoS One 28:e17338

Mizuno Y, Hasegawa K, Kondo T et al (2010) Clinical efficacy of istradefylline (KW-6002) in Parkinson's disease: a randomized, controlled study. Mov Disord 25:1437–1443

Mizuno Y, Kondo T, Japanese Istradefylline Study Group (2013) Adenosine A2A receptor antagonist istradefylline reduces daily OFF time in Parkinson's disease. Mov Disord 28: 1138–1141

Morelli M, Pinna A (2001) Interaction between dopamine and adenosine A2A receptors as a basis for the treatment of Parkinson's disease. Neurol Sci 22:71–72

Morelli M, Blandini F, Simola N et al (2012) A(2A) Receptor antagonism and dyskinesia in Parkinson's disease. Parkinson Dis:489853

Mori A, Shindou T, Ichimura M et al (1996) The role of adenosine A2a receptors in regulating GABAergic synaptic transmission in striatal medium spiny neurons. J Neurosci 16:605–611

Neustadt BR, Hao J, Lindo N et al (2007) Potent selective and orally active adenosine A2A receptor antagonists: arylpiperazine derivatives of pyrazolo[4,3-e]- 1, 2, 4-triazolo[1,5-c] pyrimidines. Bioorg Med Chem Lett 17:1376–1380

Pourcher E, Fernandez HH, Stacy M et al (2012) Istradefylline for Parkinson's disease patients experiencing motor fluctuations: results of the KW-6002-US-018 study. Parkinsonism Relat Disord 18:178–184

Ramlackhansingh AF, Bose SK, Ahmed I et al (2011) Adenosine 2 A receptor availability in dyskinetic and nondyskinetic patients with Parkinson's disease. Neurology 76:1811–1816

Schrag A, Sampaio C, Counsell N et al (2006) Minimal clinically important change on the unified Parkinson's disease rating scale. Mov Disord 21:1200–1207

Schwarzschild MA, Chen JF, Ascherio A (2002) Caffeinated clues and the promise of adenosine A(2 A) antagonists in PD. Neurology 58:1154–1160

Stacy M, Silver D, Mendis T et al (2008) A 12-week, placebo-controlled study (6002-US-006) of istradefylline in Parkinson disease. Neurology 70:2233–2240

Wills AM, Eberly S, Tennis M et al (2013) Caffeine consumption and risk of dyskinesia in CALM-PD. Mov Disord 28:380–383

Yamada K, Kobayashi M, Mori A et al (2013) Antidepressant-like activity of the adenosine A2A receptor antagonist, istradefylline (KW-6002), in forced swim test and the tail suspension test in rodents. Pharmacol Biochem Behav 114:23–30

Yang M, Soohoo D, Soelaiman S et al (2007) Characterization of the potency, selectivity, and pharmacokinetic profile for six adenosine A2A receptor antagonists. Naunyn Schmiedebergs Arch Pharmacol 375:133–144

Chapter 15
Adenosinergic Regulation of Sleep–Wake Behavior in the Basal Ganglia

Michael Lazarus and Yoshihiro Urade

Abstract Sleep is the most mysterious brain function and seems to exist in all organisms that have a central nervous system. Human sleep habits are unique, because we often defy sleep and stay awake for occupational and recreational reasons or other life-style choices, despite experiencing fatigue during that time. The motivation to stay awake and active in modern societies is increasing and often accompanied by the use of psychoactive substances, most prominently caffeine. The basal ganglia (BG) consists of subcortical nuclei involved in motor function, habit formation, and reward/addictive behaviors and play a key role in mediating the arousal effect of caffeine. The identification of neural circuits through which the BG integrates fundamental striatal functions with arousal is a rapidly growing field in neurobiology. In this chapter, we discuss mechanistic models of sleep–wake regulation for the BG and propose that the nucleus accumbens is a key node between emotive behaviors and the circuitry for sleep and wakefulness.

Keywords Direct/indirect pathway · Locomotion · Movement disorders · Sleep · Caffeine · A_1 receptor (A_1R) · A_{2A} receptor (A_{2A}R) · D_2 receptor · Modafinil · Nucleus accumbens (NAc)

Introduction

Sleep is the most mysterious brain function (Urade and Lazarus 2013). Sleep or sleep-like states seem to exist in all complex organisms that have a central nervous system. The sleeping habits of humans are unique in the sense that we often defy sleep and stay awake for occupational and recreational reasons or other life-style choices, despite experiencing fatigue during that time. The motivation to stay awake and active in our modern society is increasing and often accompanied by the use of psychoactive substances, most prominently caffeine. The basal ganglia (BG) is the key area governing these social behaviors and contain subcortical nuclei involved in

M. Lazarus (✉) · Y. Urade
International Institute for Integrative Sleep Medicine (WPI-IIIS), University of Tsukuba,
1-1-1 Tennodai, Tsukuba, Ibaraki 305-8575, Japan
e-mail: lazarus.michael.ka@u.tsukuba.ac.jp

© Springer International Publishing Switzerland 2015
M. Morelli et al. (eds.), *The Adenosinergic System,* Current Topics in Neurotoxicity 10,
DOI 10.1007/978-3-319-20273-0_15

motor function, habit formation, and reward/addictive behaviors. The investigation of the neural network through which the BG integrates these fundamental striatal functions with arousal is a rapidly growing field in neurobiology.

BG consist of four major nuclei namely the striatum, globus pallidus (GP), subthalamic nucleus (STN), and substantia nigra (SN) (Crittenden and Graybiel 2011) and are strongly connected with the cortex, thalamus, amygdala, and midbrain dopaminergic neurons. BG act as a cohesive functional unit in the process of optimizing behavior and regulating the vigilance state of wakefulness. However, little attention has been received to the specific role of the two efferent, direct striatonigral *versus* indirect striatopallidal, pathways of the BG in integrating wakefulness, motor, and behavior.

Adenosine promotes sleep through the activation of adenosine A_1 and A_{2A} receptors (A_1R and $A_{2A}R$) (Basheer et al. 2004; Huang et al. 2007). $A_{2A}R$ are densely co-expressed with dopamine D_2 receptors (D_2R) on the striatopallidal neurons, whereas A_1R are colocalized with dopamine D_1 receptors (D_1R) on the striatonigral neurons. For decades many laboratories have studied the mesolimbic dopamine system from the midbrain to the striatum for motor control and motivational behavior. However, experimental evidence has only recently started to emerge for the intrinsic roles of adenosine and dopamine in the BG for sleep–wake regulation. Our recent studies clearly demonstrated that caffeine induces strong wakefulness by blocking the action of adenosine at $A_{2A}R$ in the shell of the nucleus accumbens (NAc) (Huang et al. 2005; Lazarus et al. 2011). In this chapter, we discuss anatomical and molecular mechanistic models of sleep–wake regulation for the BG and propose that the NAc is a key node between motivational behaviors (Ikemoto 2007) and the circuitry for sleep and arousal (Fort et al. 2009; Saper et al. 2010). The dorsal striatum, NAc, and GP are key structural elements for the regulation of wakefulness. These findings have a major impact on our understanding of where and how $A_{2A}R$ antagonists or D_2R agonists, most commonly used for the treatment of Parkinson's disease (PD), affect sleep and wakefulness.

The Role of BG in Control of the Sleep–Wake Cycle: Evidence from Clinical Findings of Sleep Abnormality in PD and Huntington's Disease (HD)

The importance of the BG in sleep control is suggested by clinical findings that BG-related neurodegenerative and neuroinflammatory disorders, such as PD, HD, and Encephalitis Lethargica, are associated with abnormalities of sleep and waking (Adler and Thorpy 2005; Dale et al. 2004; Goodman and Barker 2010).

Disturbance of sleep is very common in patients with PD: up to 88 % of PD patients present primary sleep abnormalities, such as REM sleep behavior disorder, periodic limb movements during sleep, and restless legs syndrome (Wetter et al. 2000). For example, PD patients present disturbed sleep that is the result of underlying symptoms of PD, such as depression, anitparkinsonian medication, akinesia, pain, and dystonia. Furthermore, they reportedly experience severe excessive daytime sleepiness and sleep attacks for which nocturnal sleep disturbances, motor and cognitive impairment, or antiparkinsonian treatment are not responsible, but may be directly related to the pathology of the disease, likely the loss of dopaminergic neurons in the midbrain (Arnulf et al. 2002). In fact, modafinil, a wakefulness-promoting compound, is commonly used for treatment of excessive sleepiness in PD and other sleep disorders, such as narcolepsy, shift-work sleep disorder, and obstructive sleep apnea/hyponea syndrome, that do not result from low dopamine conditions (Hogl et al. 2002; Minzenberg and Carter 2007; Zeitzer et al. 2006). Modafinil enhances extracellular levels of dopamine in the NAc and medial prefrontal cortex (mPFC), although this compound affects multiple neurotransmitter systems such as catecholamines, serotonin, glutamate, γ-aminobutyric acid (GABA), hypocretin/orexin, and histamine (Murillo-Rodriguez et al. 2007). Furthermore, the arousal effect of modafinil is abolished in knockout mice for the dopamine transporter through which dopamine is primarily cleared from the synapses (Wisor et al. 2001). Our recent finding by using a D_1R antagonist and D_2R knockout mice suggests that the arousal effect of modafinil is exclusively mediated by the D_1R and D_2R, with D_2R being the receptor of primary importance (Qu et al. 2008).

Patients with HD also commonly show sleep disturbances at night and daytime somnolence (Videnovic et al. 2009). HD is a progressive neurodegenerative disorder caused by a genetic defect on chromosome 4 that affects muscle coordination and leads to cognitive decline and dementia (Walker 2007). Although the dorsal striatum and GP are preferentially affected in this disease, neurodegeneration is more widespread, including also areas associated with sleep and circadian rhythm regulation in the hypothalamus and brainstem (Goodman and Barker 2010; Kassubek et al. 2004). Similar to PD, however, there is a lack of association between poor overnight sleep and excessive daytime sleepiness in HD patients. The disturbed daytime alertness may be caused by parkinsonian mechanisms based on dysfunctional dopamine signaling in the striatum, but more detailed studies in animal models of HD would be required to support this conclusion. In fact, increased nocturnal and decreased diurnal activity in HD patients may also be attributed to neurodegeneration of neurons in the suprachiasmatic nucleus (SCN) of the hypothalamus, leading to alteration of the circadian rhythm (Morton et al. 2005).

Evidence from Anatomical Lesioning Studies for Distinct Roles of the Four Major BG Nuclei in the Sleep–Wake Cycle

Studies based on neurotoxic lesioning of the BG indicate a significant role of the BG in regulating the sleep–wake cycle. Bilateral lesions made in the striatum result in a significant reduction of wakefulness and fragmentation of both sleep and wakefulness. However, when these lesions include the NAc, the effect of striatal lesions on wakefulness is attenuated (Qiu et al. 2010). Consistent with this observation, lesions restricted to the NAc produce an increase in wakefulness and reduced durations of non-rapid eye movement (non-REM, NREM) sleep bouts.

These findings suggest that the dorsal and ventral striatum play opposing roles in sleep–wake regulation (i.e., the caudate-putamen (CPu) enhances wakefulness while the NAc promotes sleep).

Cell body-specific lesioning of the external GP (GPe) in rats leads to insomnia, specifically, a dramatic increase (~45 %) in total wakefulness and pronounced fragmentation of NREM sleep and wakefulness, including more sleep transitions and shortened sleep bouts, (Qiu et al. 2010). Moreover, the loss of neurons in the SN, but not the internal GP or the STN, affected sleep–wake behavior towards increased wakefulness (Gerashchenko et al. 2006; Qiu et al. 2010). Such findings suggest that the loss of dopaminergic input from the SN to the dorsal striatum in PD may contribute, at least in part, to insomnia observed in these patients (Adler and Thorpy 2005). Interestingly, lesions in the CPu, NAc, and GP produce a generalized slowing of the cortical electroencephalogram (EEG), with less theta and more delta power during wakefulness, REM and NREM sleep (Qiu et al. 2010), a phenomenon that is also observed in PD patients (Morita et al. 2009).

The GPe contains direct cortical-projecting neurons, therefore, a dorsostriato-pallido-cortical loop is hypothesized as a likely mechanism by which the dorsal striatum and the GPe contribute to sleep–wake behavior (Fig. 15.1; Vetrivelan et al. 2010). The model predicts that GABAergic GPe neurons suppress cortical activity—regardless of the sleep–wake state—by modulating activity of layer five pyramidal neurons and interneurons in the cerebral cortex. Disinhibition by loss of GPe neurons (e.g., via neurotoxic lesioning) or dopamine input (e.g., in PD patients) may therefore lead to similar cortical activity that occurs during wakefulness.

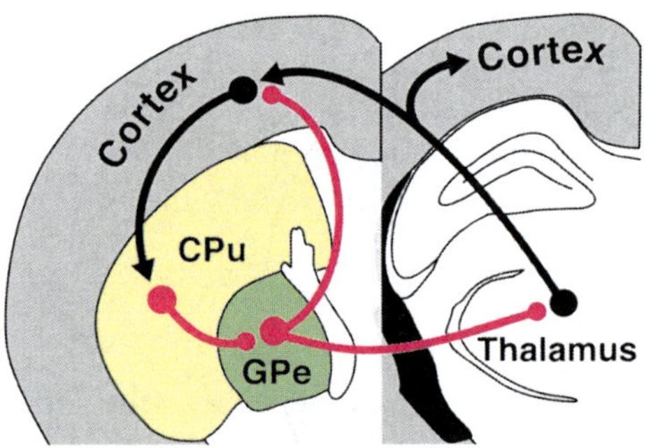

Fig. 15.1 A model in which a dorsostriato-pallido-cortical loop regulates sleep–wake behavior and cortical activation. The caudate-putamen (*CPu*) projects to the external globus pallidus (*GPe*), which in turn projects directly or via the thalamus (mainly the mediodorsal thalamic nucleus) to the cerebral cortex. Therefore, activity of layer V pyramidal neurons and interneurons in the cerebral cortex is modulated through inhibition by GABAergic *GPe* neurons (Qiu et al. 2010; Vetrivelan et al. 2010). Disinhibition of the cerebral cortex by inhibition (loss of dopaminergic input to the CPu) or neurotoxic lesioning of GPe neurons may therefore lead to cortical activity as occurs in wakefulness. *Black arrows* represent excitatory glutamatergic synapses; *magenta round-headed lines* represent inhibitory GABAergic synapses. (Adapted from Lazarus et al. 2012)

The Role of the Dopaminergic System for Regulating the Sleep–Wake Cycle

Researchers have long attempted to elucidate the role of dopamine in regulating sleep and wakefulness. Seminal findings based on electrolytic lesioning of neurons in the midbrain of cats showed that dopamine-containing neurons of the SN and ventral tegmental area (VTA) are involved in the maintenance of behavioral arousal and reactivity but not in electrocortical awakening (Jones et al. 1973). These findings may account, to some extent, for the slow progress observed thus far in this area of research. Lesioning approaches can, however, produce collateral damage to adjacent brain structures such as medial VTA glutamatergic neurons with afferents to the mPFC (Hur and Zaborszky 2005). Such damage may, in turn, produce effects on sleep behavior and cortical EEG beyond those resulting from lesions to dopaminergic midbrain neurons.

The results of *in vivo* microdialysis experiments in combination with polysomnographic recording indicate that extracellular dopamine levels in the mPFC and NAc are high during wakefulness and REM sleep, but significantly lower during NREM sleep (Léna et al. 2005). This observation of high dopamine levels during REM sleep in the NAc may indicate that dopamine can cause arousal that is independent of movement. By contrast, there is evidence that movement is inhibited during REM sleep by brainstem mechanisms that produce spinal atonia as animals with pontine lesions show active behavior during REM sleep (Jouvet and Delorme 1965; Vetrivelan et al. 2009). Thus, it is possible that NAc neurons are active during REM sleep, but their impact on movement is blunted by the actions of pontine atonia mechanisms.

In fact, the deletion of D_2R from the entire animal leads to a significant decrease in wakefulness with a concomitant increase in NREM and REM sleep in addition to drastically lower NREM sleep delta power (Qu et al. 2010). Such D_2R knockout mice frequently enter sleep after short periods of wakefulness during the nocturnal phase. These studies clearly show that the D_2R has a crucial role in maintaining wakefulness during the normal wake phase, however it is impossible to identify the neural substrates involved in dopaminergic modulation of behavioral states. A previous study found a similar range of decreased wakefulness after neurotoxic lesions were made in ventral periaqueductal gray (vPAG) dopaminergic neurons, but this effect was observed throughout the sleep–wake cycle (Lu et al. 2006a). Therefore, the effect observed during only the nocturnal phase in global D_2R knockout mice may not be exclusively regulated by the vPAG, but may also result from activation of D_2R in additional areas such as the striatum. This assumption is supported by the fact that the D_2R agonist quinelorane when directly infused into the NAc increases wakefulness in rats, whereas a D_2R antagonist induces sleep when injected into this same region (Barik and de Beaurepaire 2005).

The Role of the Adenosinergic System for the Sleep–Wake Cycle

Adenosine promotes sleep by acting through A_1R and $A_{2A}R$, however the relative contribution of these receptors to sleep induction remains controversial (Basheer et al. 2004; Huang et al. 2007). The brain substrates through which adenosine acts on inhibitory A_1R and excitatory $A_{2A}R$ to produce sleep are not well understood. Adenosine acting via A_1R was shown to induce sleep by inhibiting arousal-related cell groups surrounding the striatum such as the horizontal limb of the diagonal band of Broca, the substantia inominata (Alam et al. 1999; Strecker et al. 2000), and hypocretin/orexin neurons in the lateral hypothalamus (LHA) (Thakkar et al. 2008). A previous study suggested that activating A_1R in the tuberomammilary nucleus (TMN) also promotes NREM sleep by inhibiting the histaminergic arousal system (Oishi et al. 2008). By contrast, stimulating A_1R in the lateral preoptic area of the hypothalamus promotes wakefulness (Methippara et al. 2005), supporting the idea that A_1R-mediated effects on sleep and wakefulness are region-specific. For example, lateral ventricle infusions of the A_1R agonist N6-cyclopentyladenosine (CPA) in mice does not change the amounts of observed NREM and REM sleep (Urade et al. 2003), which may indicate opposing effects on sleep and wakefulness in different areas of the brain. CPA can, however, produce dose-dependent increases in EEG slow-wave activity during NREM sleep when administered systemically or intracerebroventricularly in rats (Benington et al. 1995).

Adenosine deaminase, an enzyme that catabolizes adenosine to inosine, is predominantly localized in the TMN of the brain. Moreover, the TMN is enriched in histaminergic neurons containing A_1R thereby suggesting that the histaminergic arousal system is actively regulated by adenosine in the TMN. In fact, bilateral injections of the A_1R agonist CPA into the rat TMN significantly increases the amount of NREM sleep (Oishi et al. 2008). Bilateral injections of adenosine or coformycin, an inhibitor of adenosine deaminase, into the rat TMN also increases NREM sleep, an increase that is completely abolished by co-administration of the selective A_1R antagonist 1,3-dimethyl-8-cyclopenthylxanthine. These results indicate that endogenous adenosine in the TMN suppresses the histaminergic system via A_1R to promote NREM sleep. Interestingly, single-nucleotide polymorphism analyses have identified a human genetic variant of adenosine deaminase with low enzymatic activity that is linked to the enhancement of deep sleep and slow-wave activity during sleep (Rétey et al. 2005).

To date, the neural and cellular basis of "sleep drive" remains unresolved although it is often conceptualized as a homeostatic pressure that builds during the waking period and is dissipated by sleep. One or more endogenous somnogenic factors are thought to comprise the cellular basis of this homeostatic process or so-called "sleep homeostat". Ishimori (1909), Kubota (1989) and Pieron (Legendre and Pieron 1913) had proposed the existence of sleep-promoting chemicals more than 100 years ago. These hypnogenic substances or so-called 'hypnotoxins' were hypothesized to accumulate as a result of prolonged periods of wakefulness (Inoué et al. 1995). Since then prostaglandin (PG) D_2 has been implicated in the sleep homeostatic process [(Ueno et al. 1982), reviewed in detail: (Urade and Hayaishi 2011)], together with substances such as cytokines [reviewed in detail: (Krueger et al. 2011)], anandamide (Garcia-Garcia et al. 2009), and peptides including urotensin II (Huitron-Resendiz et al. 2005). However, our current understanding of the mechanisms and brain substrates through which PGD_2 acts to produce sleep in healthy humans is rudimentary. When PGD_2 is infused into the subarachnoid space of the basal forebrain (BF) [the region in which PGD_2 type DP_1 receptors (DP_1R) are most abundant] in wild-type mice, extracellular adenosine concentration increases in a dose-dependent manner (Mizoguchi et al. 2001). This increase is absent in DP_1R knockout mice, indicating that the increase in adenosine in the subarachnoid space depends on DP_1R.

CGS 21680, a highly selective $A_{2A}R$ agonist, produces profound increases in NREM and REM sleep after infusions into the subarachnoid space underlying the ventral surface region of the rostral BF in rats or into the lateral ventricle of mice (Satoh et al. 1996; Urade et al. 2003). *In vivo* microdialysis experiments demonstrated that infusions of the CGS 21680 into the BF inhibits the release of histamine in both the frontal cortex and medial preoptic area in a dose-dependent manner, and increases the release of GABA in the TMN of the hypothalamus but not in the frontal cortex (Hong et al. 2005). CGS 21680-induced blocking of histamine release is antagonized when the TMN is perfused with the GABA antagonist picrotoxin, suggesting that the $A_{2A}R$ agonist induces sleep by inhibiting the histaminergic system through an increase in GABA release in the TMN. It had been previously proposed that sleep is promoted by activating sleep neurons in the ventrolateral preoptic area (VLPO) and reciprocal suppression of histaminergic wake neurons in the TMN through GABAergic and galaninergic inhibitory projections. The existence of two distinct types of VLPO neurons in terms of their responses to serotonin and adenosine was demonstrated by the intracellular recording of VLPO neurons in rat brain slices. VLPO neurons are inhibited uniformly by the arousing neurotransmitters noradrenaline and acetylcholine, and primarily inhibited by an A_1R agonist. Serotonin inhibits type-1 neurons but excites type-2 neurons, whereas an $A_{2A}R$ agonist postsynaptically excites type-2, but not type-1 neurons. These results implicate type-2 neurons in the initiation of sleep; whereas type-1 neurons contribute to sleep consolidation as they are only activated in the absence of inhibitory effects from the arousal systems (Gallopin et al. 2005).

However, the administration of CGS 21680 into the rostral BF produces c-fos expression not only in the VLPO, but also within the shell of the NAc and the medial portion of the olfactory tubercle (Satoh et al. 1999; Scammell et al. 2001). Interestingly, the direct perfusion of the $A_{2A}R$ agonist into the NAc shell induces NREM and REM sleep that corresponds to about three-quarters of the amount of sleep measured when the $A_{2A}R$ agonist is infused into the suparachnoid space (Satoh et al. 1999). These results can be interpreted to indicate that $A_{2A}R$ within or close to the NAc shell predominantly promotes sleep.

Acting opposite to adenosine, caffeine enhances wakefulness because it acts as an antagonist for both A_1R and $A_{2A}R$ subtypes (Fredholm et al. 1999). Experiments using global genetic knockouts of A_1R and $A_{2A}R$ revealed that the $A_{2A}R$, but not the

A_1R, mediates the arousal-inducing effect of caffeine (Fig. 15.2a, b; Huang et al. 2005). The specific role of $A_{2A}R$ in the BG was investigated by using powerful tools for site-specific gene manipulations, such as conditional knockout mice for the $A_{2A}R$ based on the Cre/lox technology or local infection with adeno-associated virus carrying short-hairpin RNA of $A_{2A}R$ to silence expression of the receptor subtype (Lazarus et al. 2011). Deletion of $A_{2A}R$ selectively in the NAc shell eliminated caffeine-induced wakefulness (Fig. 15.2c). Excitatory $A_{2A}R$ within the NAc shell must be tonically activated by adenosine for caffeine to be effective as an $A_{2A}R$ antagonist. This tonic activation probably occurs in the NAc shell because sufficient levels of adenosine are available under basal conditions and $A_{2A}R$ are abundantly expressed throughout the striatum, including the NAc shell (Rosin et al. 1998; Svenningsson et al. 1999). Thus, activation of $A_{2A}R$ in the NAc shell contributes to the restraint of the arousal system, whereby caffeine overrides the "adenosine brake" to promote wakefulness. Interestingly, the deletion of the dopamine transporter reduces NREM sleep, increases wakefulness, and unmasks hypersensitivity to the wake-promoting effects of caffeine (Wisor et al. 2001). The last observation may indicate that the expression of accumbal D_2R working opposite to $A_{2A}R$ are involved in the arousal effect of modafinil. Despite the fact that stimulating $A_{2A}R$ leads to decreased affinity for dopamine at D_2R via intramembrane interactions and to a reduction in G_i-protein coupling of the D_2R for the inhibition of cAMP production (Fuxe et al. 2003), adenosine and its antagonists, such as caffeine, can modulate the activity of medium spiny projection neurons in the striatum via $A_{2A}R$ independently of D_2R (Aoyama et al. 2000; Chen et al. 2001).

Integrating the NAc into the Sleep–Wake Regulatory Network

Technical advances have often precipitated quantum leaps in our understanding of neurobiological processes. For example, Hans Berger's discovery in 1929 that electrical potentials recorded from the human scalp took the form of sinusoidal waves, the frequency of which was directly related to the level of wakefulness of the person, which then led to rapid advances in our understanding of sleep–wake regulation in both animals and humans. To this day the EEG, in conjunction with the electromyogram (EMG), represents the data "backbone" of nearly every experimental and clinical assessment that seeks to correlate behavior and physiology with the activity of cortical neurons in freely behaving animals, including humans. In most basic sleep research laboratories these EEG/EMG recordings are performed using a cable-based system wherein acquired data is subjected off-line to pattern and spectrum analysis [e.g., Fast Fourier transform (FFT)] to determine the vigilance state of the subject under recording (Kohtoh et al. 2008). Over the years, and on the basis of EEG interpretation, several models of sleep–wake regulation, both circuit- and humoral-based, were proposed (Fig. 15.3).

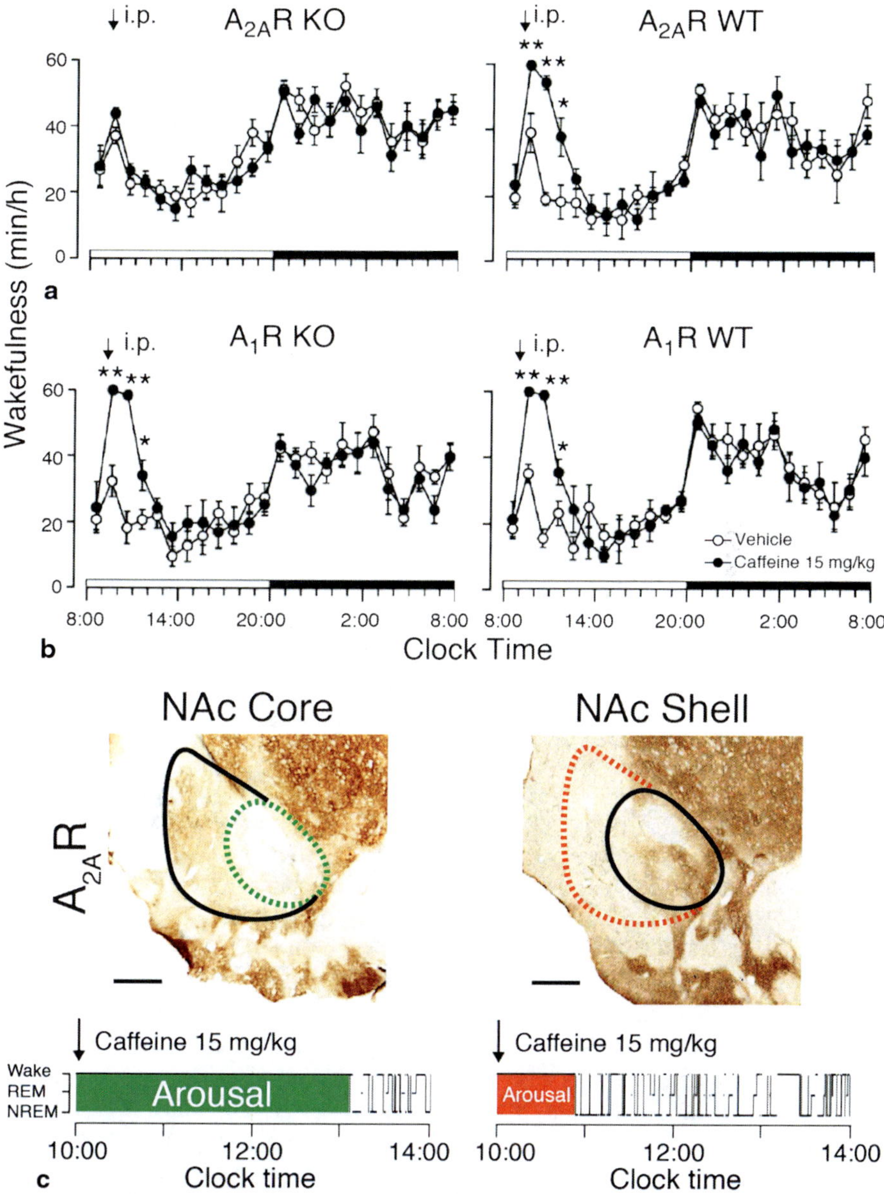

Fig. 15.2 Caffeine-induced arousal in adenosine receptor gene-manipulated mice. Caffeine (15 mg/kg, i.p.) induced arousal in wild-type (*WT*) and A$_1$ receptor knockout (*A$_1$R KO*) mice (**b**), but not in A$_{2A}$ receptor Regulkout (*A$_{2A}$R KO*) mice (**a**). **c** To identify the neurons in which caffeine acts to produce arousal, A$_{2A}$R were focally depleted by bilateral injections of adeno-associated virus carrying short-hairpin RNA for the A$_{2A}$R into the NAc core (*dashed green line* in the *left* panel) or shell (*dashed red line* in the *right* panel) of rats. Typical hypnograms that show the time course of changes in wakefulness and in rapid eye movement (*REM*) and non-REM (*NREM*) sleep after administration of caffeine at a dose of 15 mg/kg indicate that rats with a shell, but not a core, knockdown of the A$_{2A}$R showed strongly attenuated caffeine mediated arousal. *Green and red* areas in the hypnograms represent wakefulness after caffeine administration that correspond to the depletion of A$_{2A}$R in the respective core and shell of the NAc

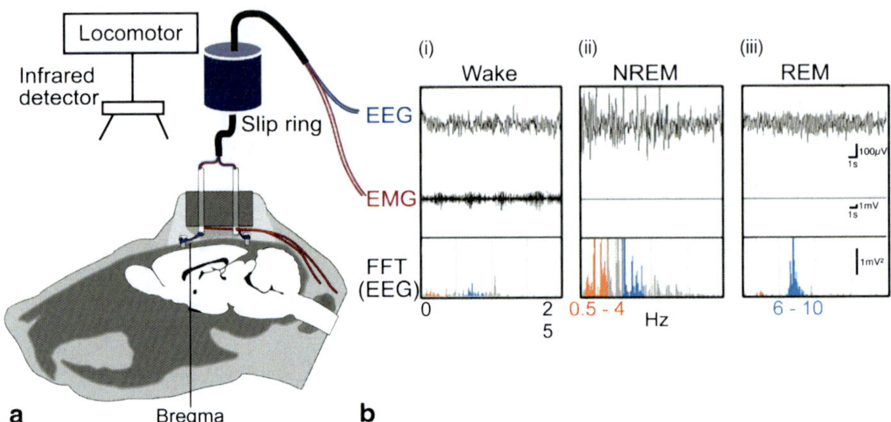

Fig. 15.3 Sleep bioassay system for rodents. **a** To monitor electroencephalogram (*EEG*) signals, stainless steel screws are implanted epidurally over the frontal cortical area and the parietal area of one hemisphere. In addition, electromyogram (*EMG*) activity is monitored by stainless steel, teflon-coated wires bilaterally placed into both trapezius muscles. **b** In contrast to sleep stages, wakefulness (*i*) is characterized by low to moderate voltage *EEG* and the occurrence of *EMG* activity. *NREM* sleep (*ii*) can be identified by the appearance of large, slow brain waves with a delta rhythm below 4 Hz (*orange* frequencies in the fast Fourier transform, *FFT*, of the *EEG*). At the transition from *NREM* to *REM* sleep (*iii*), there is a shift from low-frequency delta activity to a rapid low-voltage *EEG* in the theta range between 6 and 10 Hz (*blue* frequencies in *FFT* of the *EEG*)

Several interactions between sleep- and wake-active neurons are proposed at the systems level in models of sleep–wake regulation (Fig. 15.4). For instance, sleep is promoted by inhibiting cholinergic neurons in the BF whereby slow wave sleep is caused through inhibiting acetylcholine release by adenosine (Jones 2004). Another contemporary systems-level model of NREM-sleep/wake regulation describes a "flip-flop" switching mechanism involving mutually inhibitory interactions between sleep-promoting neurons in the VLPO and wake-promoting neurons in the brainstem and hypothalamus. The latter model includes the histaminergic TMN, noradrenergic LC, serotonergic dorsal raphe nucleus (DR), and cholinergic pontine (pedunculopontine and laterodorsal tegmental, PPT/LDT) nuclei (Fort et al. 2009; Saper et al. 2005, 2010). Aminergic neurons in the TMN, LC, and DR promote wakefulness by direct excitatory effects on arousal systems in the thalamus, hypothalamus, BF and cerebral cortex, in addition to the inhibition of sleep-promoting neurons in the VLPO. During sleep, the VLPO inhibits these arousal-promoting regions through GABAergic and galaninergic projections. The flip-flop model also predicts that hypocretin/orexin neurons of the LHA prevent unwanted transitions into sleep and thus stabilize wakefulness. However, even lesions to the entire VLPO only leads to a reduction in the amount of sleep by about 50 % for a minimum duration of 3 weeks in rats (Lu et al. 2000), suggesting that other areas of the brain can also restrain the arousal system and promote sleep.

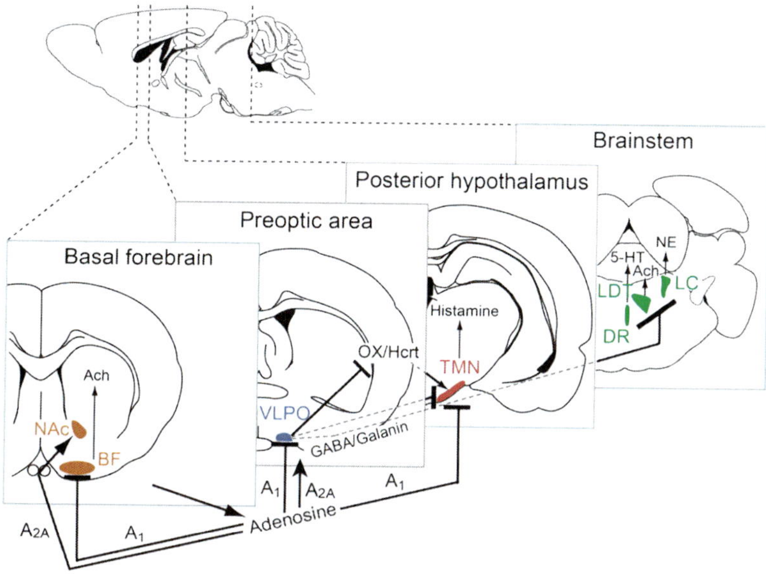

Fig. 15.4 Molecular mechanisms of sleep–wake regulation. Sleep is promoted by the inhibition of cholinergic neurons in the basal forebrain, whereas slow-wave sleep is caused by inhibition of acetylcholine release by adenosine. Another contemporary systems-level model of *NREM* sleep–wake regulation describes a flip–flop switching mechanism involving mutually inhibitory interactions between sleep-promoting neurons in the ventrolateral preoptic area (*VLPO*) and wake-promoting neurons in the hypothalamus, including the histaminergic tuberomammilary nucleus (*TMN*), and the brainstem, including the noradrenergic locus coeruleus (*LC*), serotonergic dorsal raphe nucleus (*DR*), and cholinergic pontine (pedunculopontine and laterodorsal tegmental, *PPT/LDT*) nuclei. The flip-flop switch of sleep-wakefulness regulation between the *VLPO*, hypothalamus and brainstem is stabilized by orexin/hypocretin (OX/Hcrt)-mediated activation. Adenosine is known to act as an endogenous somnogen that promotes sleep via inhibitory A_1 receptors (A_1) in the basal forebrain, VLPO,TMN and via excitatory A_{2A} receptors (A_{2A}) in the nucleus accumbens (NAc) and VLPO (Huang et al. 2007, 2011; Lazarus et al. 2012, 2013). *Ach* acetylcholine, *5-HT* serotonin, *NE* norepinephrine

A mutually inhibitory interaction between the vPAG, lateral pontine tegmentum, and sublaterodorsal nucleus (SLD) in the brainstem is currently proposed to act as an interacting system responsible for switching organisms in and out of REM sleep (Lu et al. 2006b). The REM sleep-promoting area in the SLD also contains two distinct populations of glutamatergic neurons of which one set projects to the BF and regulates EEG components of REM sleep, whereas the other set projects to the medulla and spinal cord and regulates muscle atonia during REM sleep. REM atonia-inducing SLD cells are hypothesized to activate GABA- and glycine-containing neurons in the ventral and alpha gigantocellular reticular nucleus, which then induce atonia by inhibiting spinal motor neurons (Vetrivelan et al. 2009). Moreover, it is thought that neurons in the SLD may produce atonia by activating GABA- and glycinergic spinal cord interneurons that may inhibit skeletal motor neurons

(Lu et al. 2006b). Modulation of REM *versus* NREM sleep is provided by cholinergic and monoaminergic systems in the PPT/LDT, LC, DR, as well as by the VLPO and the hypocretin/orexinergic LHA. While flip-flop models have proven to be a valuable heuristic and have provided an important interpretative framework for studies in sleep research, a fuller understanding of the sleep-switch system will require a more complete knowledge of its components.

The NAc has the unique capability to integrate locomotion with motivational behavior through dopaminergic inputs, contextual information from the hippocampus, emotional content from the amygdala and executive/cognitive information from the prefrontal cortex. These multiple and integrated functions may be dissociable at neurotransmitter and neuromodulator levels, since dopamine, adenosine and glutamate are clearly associated with controlling motor function and modulating learning by feedback reinforcement. Thus far identified, efferents provide evidence that the NAc is capable of regulating sleep and wakefulness through inhibition of neuronal populations in the VP, the LHA), the parabrachial nucleus (PB), and the VTA (Fig. 15.5). The circuit originating from the VP includes the thalamus and mPFC,

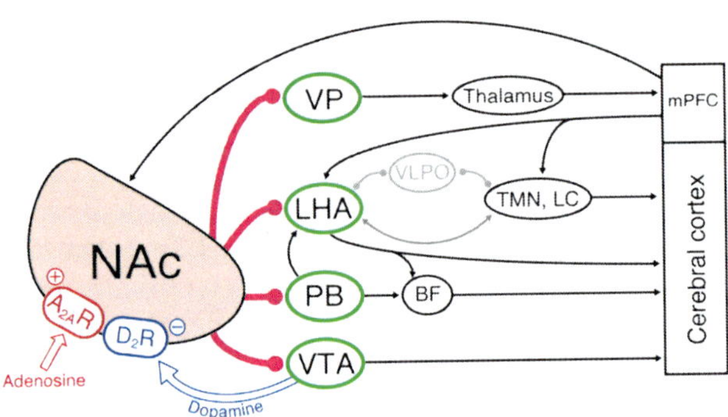

Fig. 15.5 A model in which the nucleus accumbens (NAc) plays an intrinsic role in the sleep/wake regulatory network. Inhibitory output projections of the NAc modulate activity of neuronal populations in the ventral pallidum (*VP*), the lateral hypothalamus (*LHA*), the parabrachial nucleus (*PB*), and the ventral tegmental area (*VTA*), which may be a major source of arousal. The NAc can modulate the medial prefrontal cortex (*mPFC*) via a pathway through the VP and thalamus and, in turn, the mPFC projects to arousal-promoting neurons in the hypothalamic tuberomammillary nucleus (*TMN*), the LHA, and the locus coeruleus (*LC*). Subserving the NAc, the orexinergic and glutamatergic neurons in the LHA send major projections to the basal forebrain (*BF*) and cerebral cortex. The LHA is also reciprocally connected to the flip–flop switch between non-rapid eye movement sleep and wakefulness (shown in *gray*), including the ventrolateral preoptic area (*VLPO*), the TMN, and the LC. The PB, an important component of the ascending arousal system, is known to be strongly connected to the BF and LHA. Glutamatergic neurons at the border between the VTA and supramammillary nucleus (SUM) may also relay the waking stimulus from the NAc to the cerebral cortex. Adenosine acting on excitatory $A_{2A}R$, opposite to the inhibitory dopamine/D_2R system, can modulate the activity of GABAergic output neurons in the NAc to inhibit arousal and promote sleep. *Black arrows*, excitation; *red round-headed lines*, inhibition; lines with *two round-headed ends* represent reciprocal inhibitory connections; *circled areas* with *green background*, neuronal populations with cortical projections (*light green arrows*). (Adapted from Lazarus et al. 2013)

a key executive interface between cognition and emotion and uniquely sensitive to sleep and sleep need (Chee and Choo 2004, Chuah et al. 2006; Koenigs et al. 2010; Muzur et al. 2002); and it can provide top-down modulation through its descending projections to arousal-promoting neurons in the TMN, LHA, and LC. The hypocretin/orexinergic and glutamatergic neurons in the LHA send major projections to the BF and cerebral cortex, but are also reciprocally connected to the NREM/wake flip-flop switch, including the VLPO, TMN, and LC (Hur and Zaborszky 2005; Sano and Yokoi 2007; Saper et al. 2010; Yoshida et al. 2006). The NAc shell, but not the NAc core, sends projections to the PB (Li et al. 2012; Usuda et al. 1998), which is an important component of the ascending arousal system and is known to be strongly connected to the BF and LHA. The NAc projects to the medial part of the VTA with a field of cortically projecting glutamatergic neurons (Heimer et al. 1991; Hur and Zaborszky 2005), which are likely the tail end of a larger group of neurons of the supramammillary nucleus (SUM). Interestingly, caffeine induces c-fos expression in non-dopaminergic neurons of the medial VTA (Deurveilher et al. 2006), however it remains to be clarified if it is possible that the VTA/SUM cell group relays the waking stimulus from the NAc to the cerebral cortex.

Concluding Remarks

It is now widely accepted that sleep is regulated by homeostatic (i.e., sleep pressure), circadian (i.e., daily rhythms), and allostatic (i.e., food availability or stress) factors. In the first case, the homeostatic process is controlled by sleep propensity, which increases during the course of wakefulness and dissipates during sleep (Borbely 1982). Endogenous somnogenic substances, such as adenosine, PGD_2 or cytokines, are thought to comprise the molecular basis of this so-called "sleep homeostat" that interacts with the sleep regulatory network (Huang et al. 2011; Krueger et al. 2011; Urade and Hayaishi 2011). In contrast, the circadian process is controlled by an internal pacemaker and is independent of prior sleep and waking. In mammals, this pacemaker is the SCN in the hypothalamus; it influences not only the timing of sleep and wakefulness but also a wide range of other behaviors and physiological functions (Achermann and Borbely 2003; Saper et al. 2010). Stressful situations, such as a lack of food, predator confrontation, mating pressure, and seasonal migration require a rapid adjustment of the wake-sleep state towards high arousal for which specific networks in the mPFC, amygdala, hypothalamus, and brainstem are known to exist (Cano et al. 2008; Yamanaka et al. 2003). The ventral striatum has the unique capability to integrate behavioral functions and emotional events and has plausible efferents that may contribute to the regulation of sleep and waking. It may be an ideal site where wakefulness is promoted by behavioral processes that require consciousness, whereas locomotor and arousal systems are inhibited during sleep. We propose that motivation is a fourth fundamental principle in the sleep–wake regulation in addition to homeostatic, circadian, and allostatic factors.

Despite the fact that many questions remain, the most pressing issue is to know which of the output projections of the striatum and the NAc relay the waking stimulus from the BG to the sleep–wake regulatory network and lead to cortical awakening. This task will be solved by using recently developed molecular biological technologies for systems-level sleep research in freely behaving animals. These technical advances include a wide range of approaches, from conditional gene deletion based on the Cre/loxP technology to RNA interference (Lazarus et al. 2007, 2011) to modulating neuronal activity using genetically engineered optical switches (e.g. channel rhodopsin) (Adamantidis et al. 2007; Deisseroth et al. 2006) to *in vivo* reversible silencing (e.g., nonmammalian Cl channels) (Lerchner et al. 2007) and activation (e.g., stimulatory GPCRs) (Alexander et al. 2009; Nawaratne et al. 2008) of neurons.

Many sleep abnormalities are reported in patients with PD and HD due to the dysfunction of the BG (Adler and Thorpy 2005; Goodman and Barker 2010). However, in almost all instances, their etiological causes are unclear, because that the neuronal mechanisms subserving these pathogenesis remain unresolved. The BG are a prime example for a sleep-disease connection that creates a vicious circle. Initially, movement disorders, psychiatric disorders, and substance abuse disorders disturb sleep, and the resulting sleep abnormalities further exacerbate the original BG dysfunction. A new level of anatomic and molecular analysis of the BG circuitry regulating sleep–wake behavior may shed new light into the underlying mechanisms in addition to potential treatment strategies for sleep disturbances associated with BG disorders.

Acknowledgements We thank Kristopher McEowen (University of Tsukuba) for editorial help with this manuscript. Our research was supported by a Japan Society for the Promotion of Science Grant 24300129 (to M.L.) and 22300133 (to Y.U.); World Premier International Research Center Initiative (WPI) from the Ministry of Education, Culture, Sports, Science, and Technology (to M.L. and Y.U.); a grant from the Nestlé Nutrition Council, Japan (to M.L.); grants from the Ministry of Health, Labor and Welfare (to Y.U.) and the National Agriculture and Food Research Organization; and grants from Ono Pharmaceutical Co. (to Y.U.) and Takeda Pharmaceutical Co. (to Y.U.).

References

Achermann P, Borbely AA (2003) Mathematical models of sleep regulation. Front Biosci 8:
 s683–693
Adamantidis AR, Zhang F, Aravanis AM et al (2007) Neural substrates of awakening probed with optogenetic control of hypocretin neurons. Nature 450:420–424
Adler CH, Thorpy MJ (2005) Sleep issues in Parkinson's disease. Neurology 64:S12–S20
Alam MN, Szymusiak R, Gong H et al (1999) Adenosinergic modulation of rat basal forebrain neurons during sleep and waking: neuronal recording with microdialysis. J Physiol 521:679–
 690
Alexander GM, Rogan SC, Abbas AI et al (2009) Remote control of neuronal activity in transgenic mice expressing evolved G protein-coupled receptors. Neuron 63:27–39
Aoyama S, Kase H, Borrelli E (2000) Rescue of locomotor impairment in dopamine D2 receptor-deficient mice by an adenosine A2A receptor antagonist. J Neurosci 20:5848–5852

Arnulf I, Konofal E, Merino-Andreu M et al (2002) Parkinson's disease and sleepiness: an integral part of PD. Neurology 58:1019–1024

Barik S, de Beaurepaire R (2005) Dopamine D3 modulation of locomotor activity and sleep in the nucleus accumbens and in lobules 9 and 10 of the cerebellum in the rat. Prog Neuropsychopharmacol Biol Psychiatry 29:718–726

Basheer R, Strecker RE, Thakkar MM et al (2004) Adenosine and sleep–wake regulation. Prog Neurobiol 73:379–396

Benington JH, Kodali SK, Heller HC (1995) Stimulation of A1 adenosine receptors mimics the electroencephalographic effects of sleep deprivation. Brain Res 692:79–85

Borbely AA (1982) A two process model of sleep regulation. Hum Neurobiol 1:195–204

Cano G, Mochizuki T, Saper CB (2008) Neural circuitry of stress-induced insomnia in rats. J Neurosci 28:10167–10184

Chee MWL, Choo WC (2004) Functional imaging of working memory after 24 hr of total sleep deprivation. J Neurosci 24:4560–4567

Chen J-F, Moratalla R, Impagnatiello F et al (2001) The role of the D2 dopamine receptor (D2R) in A2A adenosine receptor (A2AR)-mediated behavioral and cellular responses as revealed by A2A and D2 receptor knockout mice. Proc Natl Acad Sci U S A 98:1970–1975

Chuah YML, Venkatraman V, Dinges DF et al (2006) The neural basis of interindividual variability in inhibitory efficiency after sleep deprivation. J Neurosci 26:7156–7162

Crittenden JR, Graybiel AM (2011) Basal ganglia disorders associated with imbalances in the striatal striosome and matrix compartments. Front Neuroanat 5:59

Dale RC, Church AJ, Surtees RAH et al (2004) Encephalitis lethargica syndrome: 20 new cases and evidence of basal ganglia autoimmunity. Brain 127:21–33

Deisseroth K, Feng G, Majewska AK et al (2006) Next-generation optical technologies for illuminating genetically targeted brain circuits. J Neurosci 26:10380–10386

Deurveilher S, Lo H, Murphy JA et al (2006) Differential c-Fos immunoreactivity in arousal-promoting cell groups following systemic administration of caffeine in rats. J Comp Neurol 498:667–689

Fort P, Bassetti CL, Luppi PH (2009) Alternating vigilance states: new insights regarding neuronal networks and mechanisms. Eur J Neurosci 29:1741–1753

Fredholm BB, Bättig K, Holmén J et al (1999) Actions of caffeine in the brain with special reference to factors that contribute to its widespread use. Pharmacol Rev 51:83–133

Fuxe K, Agnati LF, Jacobsen K et al (2003) Receptor heteromerization in adenosine A2A receptor signaling: relevance for striatal function and Parkinson's disease. Neurology 61:S19–S23

Gallopin T, Luppi PH, Cauli B et al (2005) The endogenous somnogen adenosine excites a subset of sleep-promoting neurons via A2A receptors in the ventrolateral preoptic nucleus. Neuroscience 134:1377–1390

Garcia-Garcia F, Acosta-Pena E, Venebra-Munoz A et al (2009) Sleep-inducing factors. CNS Neurol Disord-Drug Targets 8:235–244

Gerashchenko D, Blanco-Centurion CA, Miller JD et al (2006) Insomnia following hypocretin2-saporin lesions of the substantia nigra. Neuroscience 137:29–36

Goodman A, Barker R (2010) How vital is sleep in Huntington's disease? J Neurol 257:882–897

Heimer L, Zahm DS, Churchill L et al (1991) Specificity in the projection patterns of accumbal core and shell in the rat. Neuroscience 41:89–125

Hogl B, Saletu M, Brandauer E et al (2002) Modafinil for the treatment of daytime sleepiness in Parkinson's disease: a double-blind, randomized, crossover, placebo-controlled polygraphic trial. Sleep 25:905–909

Hong Z-Y, Huang Z-L, Qu W-M et al (2005) An adenosine A2A receptor agonist induces sleep by increasing GABA release in the tuberomammillary nucleus to inhibit histaminergic systems in rats. J Neurochem 92:1542–1549

Huang Z-L, Qu W-M, Eguchi N et al (2005) Adenosine A2A, but not A1, receptors mediate the arousal effect of caffeine. Nat Neurosci 8:858–859

Huang Z-L, Urade Y, Hayaishi O (2007) Prostaglandins and adenosine in the regulation of sleep and wakefulness. Curr Opin Pharmacol 7:33–38

Huang ZL, Urade Y, Hayaishi O (2011) The role of adenosine in the regulation of sleep. Curr Top Med Chem 11:1047–1057

Huitron-Resendiz S, Kristensen MP, Sánchez-Alavez M et al (2005) Urotensin II modulates rapid eye movement sleep through activation of brainstem cholinergic neurons. J Neurosci 25:5465–5474

Hur EE, Zaborszky L (2005) Vglut2 afferents to the medial prefrontal and primary somatosensory cortices: A combined retrograde tracing in situ hybridization. J Comp Neurol 483:351–373

Ikemoto S (2007) Dopamine reward circuitry: two projection systems from the ventral midbrain to the nucleus accumbens-olfactory tubercle complex. Brain Res Rev 56:27–78

Inoué S, Honda K, Komoda Y (1995) Sleep as neuronal detoxification and restitution. Behav Brain Res 69:91–96

Ishimori K (1909) True cause of sleep: a hypnogenic substance as evidenced in the brain of sleep-deprived animals. Tokyo Igakkai Zasshi 23:429–457

Jones BE (2004) Activity, modulation and role of basal forebrain cholinergic neurons innervating the cerebral cortex. Prog Brain Res 145:157–169

Jones BE, Bobillier P, Pin C et al (1973) The effect of lesions of catecholamine-containing neurons upon monoamine content of the brain and EEG and behavioral waking in the cat. Brain Res 58:157–177

Jouvet M, Delorme F (1965) Locus coeruleus et sommeil paradoxal. C R Soc Biol (Paris) 159:895–899

Kassubek J, Juengling FD, Kioschies T et al (2004) Topography of cerebral atrophy in early Huntington's disease: a voxel based morphometric MRI study. J Neurol Neurosurg Psychiatry 75:213–220

Koenigs M, Holliday J, Solomon J et al (2010) Left dorsomedial frontal brain damage is associated with insomnia. J Neurosci 30:16041–16043

Kohtoh S, Taguchi Y, Matsumoto N et al (2008) Algorithm for sleep scoring in experimental animals based on fast Fourier transform power spectrum analysis of the electroencephalogram. Sleep Biol Rhythms 6:163–171

Krueger JM, Clinton JM, Winters BD et al (2011) Involvement of cytokines in slow wave sleep. Prog Brain Res 193:39–47

Kubota K (1989) Kuniomi Ishimori and the first discovery of sleep-inducing substances in the brain. Neurosci Res 6:497–518

Lazarus M, Yoshida K, Coppari R et al (2007) EP3 prostaglandin receptors in the median preoptic nucleus are critical for fever responses. Nat Neurosci 10:1131–1133

Lazarus M, Shen H-Y, Cherasse Y et al (2011) Arousal effect of caffeine depends on adenosine A2A receptors in the shell of the nucleus accumbens. J Neurosci 31:10067–10075

Lazarus M, Huang Z-L, Lu J et al (2012) How do the basal ganglia regulate sleep–wake behavior? Trends Neurosci 35:723–732

Lazarus M, Chen J-F, Urade Y et al (2013) Role of the basal ganglia in the control of sleep and wakefulness. Curr Opin Neurobiol 23:780–785

Legendre R, Pieron H (1913) Recherches sur le besoin de sommeil consécutif Ã une veille prolongée. Z Allegem Physiol 14:235–262

Léna I, Parrot S, Deschaux O et al (2005) Variations in extracellular levels of dopamine, noradrenaline, glutamate, and aspartate across the sleep–wake cycle in the medial prefrontal cortex and nucleus accumbens of freely moving rats. J Neurosci Res 81:891–899

Lerchner W, Xiao C, Nashmi R et al (2007) Reversible silencing of neuronal excitability in behaving mice by a genetically targeted, ivermectin-gated Cl− channel. Neuron 54:35–49

Li C-S, Chung S, Lu D-P et al (2012) Descending projections from the nucleus accumbens shell suppress activity of taste-responsive neurons in the hamster parabrachial nuclei. J Neurophysiol 108:1288–1298

Lu J, Greco MA, Shiromani P et al (2000) Effect of lesions of the ventrolateral preoptic nucleus on NREM and REM sleep. J Neurosci 20:3830–3842

Lu J, Jhou TC, Saper CB (2006a) Identification of wake-active dopaminergic neurons in the ventral periaqueductal gray matter. J Neurosci 26:193–202

Lu J, Sherman D, Devor M et al (2006b) A putative flip–flop switch for control of REM sleep. Nature 441:589–594

Methippara MM, Kumar S, Alam MN et al (2005) Effects on sleep of microdialysis of adenosine A1 and A2a receptor analogs into the lateral preoptic area of rats. Am J Physiol-Regul Integr Comp Physiol 289:R1715–R1723

Minzenberg MJ, Carter CS (2007) Modafinil: a review of neurochemical actions and effects on cognition. Neuropsychopharmacology 33:1477–1502

Mizoguchi A, Eguchi N, Kimura K et al (2001) Dominant localization of prostaglandin D receptors on arachnoid trabecular cells in mouse basal forebrain and their involvement in the regulation of non-rapid eye movement sleep. Proc Natl Acad Sci U S A 98:11674–11679

Morita A, Kamei S, Serizawa K et al (2009) The relationship between slowing EEGs and the progression of Parkinson's disease. J Clin Neurophysiol 26:426–429

Morton AJ, Wood NI, Hastings MH et al (2005) Disintegration of the sleep–wake cycle and circadian timing in Huntington's disease. J Neurosci 25:157–163

Murillo-Rodríguez E, Haro R, Palomero-Rivero M et al (2007) Modafinil enhances extracellular levels of dopamine in the nucleus accumbens and increases wakefulness in rats. Behav Brain Res 176:353–357

Muzur A, Pace-Schott EF, Hobson JA (2002) The prefrontal cortex in sleep. Trends Cog Sci 6:475–481

Nawaratne V, Leach K, Suratman N et al (2008) New insights into the function of M4 muscarinic acetylcholine receptors gained using a novel allosteric modulator and a DREADD (Designer Receptor Exclusively Activated by a Designer Drug). Mol Pharmacol 74:1119–1131

Oishi Y, Huang Z-L, Fredholm BB et al (2008) Adenosine in the tuberomammillary nucleus inhibits the histaminergic system via A1 receptors and promotes non-rapid eye movement sleep. Proc Natl Acad Sci U S A 105:19992–19997

Qiu M-H, Vetrivelan R, Fuller PM et al (2010) Basal ganglia control of sleep–wake behavior and cortical activation. Eur J Neurosci 31:499–507

Qu W-M, Huang Z-L, Xu X-H et al (2008) Dopaminergic D1 and D2 receptors are essential for the arousal effect of modafinil. J Neurosci 28:8462–8469

Qu W-M, Xu X-H, Yan M-M et al (2010) Essential role of dopamine D2 receptor in the maintenance of wakefulness, but not in homeostatic regulation of sleep, in mice. J Neurosci 30:4382–4389

Rétey JV, Adam M, Honegger E et al (2005) A functional genetic variation of adenosine deaminase affects the duration and intensity of deep sleep in humans. Proc Natl Acad Sci U S A 102:15676–15681

Rosin DL, Robeva A, Woodard RL et al (1998) Immunohistochemical localization of adenosine A2A receptors in the rat central nervous system. J Comp Neurol 401:163–186

Sano H, Yokoi M (2007) Striatal medium spiny neurons terminate in a distinct region in the lateral hypothalamic area and do not directly innervate orexin/hypocretin- or melanin-concentrating hormone-containing neurons. J Neurosci 27:6948–6955

Saper CB, Scammell TE, Lu J (2005) Hypothalamic regulation of sleep and circadian rhythms. Nature 437:1257–1263

Saper CB, Fuller PM, Pedersen NP et al (2010) Sleep state switching. Neuron 68:1023–1042

Satoh S, Matsumura H, Suzuki F et al (1996) Promotion of sleep mediated by the A2a-adenosine receptor and possible involvement of this receptor in the sleep induced by prostaglandin D2 in rats. Proc Natl Acad Sci U S A 93:5980–5984

Satoh S, Matsumura H, Koike N et al (1999) Region-dependent difference in the sleep-promoting potency of an adenosine A2A receptor agonist. Eur J Neurosci 11:1587–1597

Scammell TE, Gerashchenko DY, Mochizuki T et al (2001) An adenosine A2a agonist increases sleep and induces Fos in ventrolateral preoptic neurons. Neuroscience 107:653–663

Strecker RE, Morairty S, Thakkar MM et al (2000) Adenosinergic modulation of basal forebrain and preoptic/anterior hypothalamic neuronal activity in the control of behavioral state. Behav Brain Res 115:183–204

Svenningsson P, Fourreau L, Bloch B et al (1999) Opposite tonic modulation of dopamine and adenosine on c-fos gene expression in striatopallidal neurons. Neuroscience 89:827–837

Thakkar MM, Engemann SC, Walsh KM et al (2008) Adenosine and the homeostatic control of sleep: effects of A1 receptor blockade in the perifornical lateral hypothalamus on sleep–wakefulness. Neuroscience 153:875–880

Ueno R, Ishikawa Y, Nakayama T et al (1982) Prostaglandin D2 induces sleep when microinjected into the preoptic area of conscious rats. Biochem Biophys Res Commun 109:576–582

Urade Y, Hayaishi O (2011) Prostaglandin D2 and sleep/wake regulation. Sleep Med Rev 15:411–418

Urade Y, Lazarus M (2013) Prostaglandin D2 in the regulation of sleep. In: Shaw PJ, Tafti M, Thorpy MJ (eds) The genetic basis of sleep and sleep disorders. Cambridge University, New York, pp 73–83

Urade Y, Eguchi N, Qu WM et al (2003) Sleep regulation in adenosine A2A receptor-deficient mice. Neurology 61:S94–S96

Usuda I, Tanaka K, Chiba T (1998) Efferent projections of the nucleus accumbens in the rat with special reference to subdivision of the nucleus: biotinylated dextran amine study. Brain Res 797:73–93

Vetrivelan R, Fuller PM, Tong Q et al (2009) Medullary circuitry regulating rapid eye movement sleep and motor atonia. J Neurosci 29:9361–9369

Vetrivelan R, Qiu M-H, Chang C et al (2010) Role of basal ganglia in sleep–wake regulation: neural circuitry and clinical significance. Front Neuroanat 4:145

Videnovic A, Leurgans S, Fan W et al (2009) Daytime somnolence and nocturnal sleep disturbances in Huntington disease. Parkinsonism Relat Disord 15:471–474

Walker FO (2007) Huntington's disease. Lancet 369:218–228

Wetter TC, Collado-Seidel V, Pollmacher T et al (2000) Sleep and periodic leg movement patterns in drug-free patients with Parkinson's disease and multiple system atrophy. Sleep 23:361–367

Wisor JP, Nishino S, Sora I et al (2001) Dopaminergic role in stimulant-induced wakefulness. J Neurosci 21:1787–1794

Yamanaka A, Beuckmann CT, Willie JT et al (2003) Hypothalamic orexin neurons regulate arousal according to energy balance in mice. Neuron 38:701–713

Yoshida K, McCormack S, España RA et al (2006) Afferents to the orexin neurons of the rat brain. J Comp Neurol 494:845–861

Zeitzer JM, Nishino S, Mignot E (2006) The neurobiology of hypocretins (orexins), narcolepsy and related therapeutic interventions. Trends Pharmacol Sci 27:368–374